FERROMAGNETIC RESONANCE IN ORIENTATIONAL TRANSITION CONDITIONS

FERROMAGNETIC RESONANCE IN ORIENTATIONAL TRANSITION CONDITIONS

V.G. Shavrov

V.I. Shcheglov

CRC Press
Taylor & Francis Group
Boca Raton London New York

CISP

CRC Press is an imprint of the
Taylor & Francis Group, an **informa** business

Translated from Russian by V.E. Riecansky

First Edition published 2022
by CRC Press
6000 Broken Sound Parkway NW, Suite 300, Boca Raton, FL 33487-2742

and by CRC Press
2 Park Square, Milton Park, Abingdon, Oxon, OX14 4RN

© 2022 by CISP

CRC Press is an imprint of Taylor & Francis Group, LLC

ISBN: 978-0-367-49056-0 (hbk)
ISBN: 978-0-367-49604-3 (pbk)
ISBN: 978-1-003-04683-7 (ebk)

Contents

Introduction

Among numerous and various dynamic phenomena in magnetic media, a special place is occupied by ferromagnetic resonance (FMR), which is manifested in the resonance dependence of the magnetic susceptibility of the medium on the frequency of the applied alternating magnetic field.

The unique properties of FMR in magnetodielectric solids are widely used to create highly efficient analog information processing devices in the microwave range. Such devices include filters, delay lines, phase shifters, non-reciprocal and non-linear devices, and others. Films and ferrite plates are used as the magnetodielectric medium, first of all, an yttrium–iron garnet (YIG), which has record-breaking magnetic losses.

The various properties of FMR are significantly complicated by the structuring and non-uniform nature of the distribution of the magnetization of the medium, the appearance of new spectrum and anomalous dependences on the magnetic field.

Of particular difficulty are the above effects in an anisotropic medium, including during orientational transitions, when the vector magnetization under the action of an applied field reoriented from one equilibrium position to another.

These circumstances were the impetus for the study of some properties of FMR under conditions of transitions, the description of the results of which is devoted to this monograph. The information given in the monograph, in no way claiming to be exhaustive, no more than reflects the range of scientific interests of the authors.

The main volume of the monograph is a summary of the main scientific and applied results obtained by the second group from 1990 to 2015. The monograph is the third of a cycle devoted to magnetostatic waves, ferromagnetic resonance and the phenomena accompanying them.

The first monograph [1] is called 'Magnetostatic waves in inhomogeneous fields' and covers a range of issues related to the propagation of magnetostatic waves (MSW) in ferrite films with free surfaces (FFS), as well as in some structures that are homogeneous in the film plane, first of all – ferrite–dielectric–metal (FDM) and ferrite–metal (FM), magnetized by a non-uniform field.

The second monograph [2] is called 'Magnetostatic and Electromagnetic Waves in Complex Structures' and covers a range of issues related to the propagation of MSW and partially electromagnetic waves (EMW) in complex structures, which are based on magnetic media with low attenuation, first of all – iron–yttria films garnet. The propagation of magnetostatic waves in periodic structures with a different ratio between the structure period and the wavelength is considered. Special attention is paid to the effect of the transformation of magnetostatic waves into electromagnetic waves on the field heterogeneity, which manifests itself as the emission of electromagnetic waves from ferrite. Some possibilities of practical application and further study of the described phenomena are discussed.

This monograph, the third in the aforementioned series, is devoted to considering the properties of ferromagnetic resonance in specific words of orientational transitions that take place in anisotropic transitions. This monograph is mainly theoretical. Experimental results are given to the extent that they concern sets out theoretical theoretical issues, as well as are the source material for the formulation of the theoretical problems under consideration. Some experiments are presented with the aim of expanding the picture of observable phenomena, which at the present time have not yet found, that is, expected future researchers, an exhaustive theoretical interpretation.

The specific experience of the authors with students and beginning researchers shows that upon initial familiarization the subject of noticeable difficulties is the development of the mathematical apparatus, much of which is scattered in articles that are not always easy to match with each other. In order to correct this situation, the initial chapters of the monograph are devoted to a detailed description of the mathematical apparatus used in the calculation of low-level magnetic field properties. Other, more specific methods of calculation, are also considered in the corresponding chapters with sufficient detail.

The structure of the monograph is as follows.

The first chapter is a brief review of the literature, designed to introduce the reader to the range of issues addressed in subsequent chapters. The review does not in any way cover the whole subject of

FMR and orientational transitions, but only refers to some works that are important for the subsequent presentation. Any overlap with the reviews given in previous monographs of this the series [1, 2], is also absent, only references to the works really needed here are given.

The second chapter is also largely of a reviewing nature and is intended to familiarize the reader with the basic principles and principles of the science of crystallographic symmetry and anisotropic solid-state properties. There is some information about the magnetic anisotropy of real materials, important for further presentation.

The third chapter is devoted to the description of the basic mathematical apparatus used in the calculation of anisotropic solids of a solid. The method of working with transition matrices between different coordinate systems is described in detail, as well as methods for obtaining such matrices.

The fourth chapter considers specific examples of transformation matrices for various orientations of uniaxial anisotropy, and also for the most important axial orientations of the cubic anisotropy axes. Analytical expressions for the energy density of uniaxial and cubic magnetic anisotropy are obtained for various orientations as a cubic cell.

The fifth chapter refers to the description of orientational transitions in magnetic fields with uniaxial anisotropy. The results of studying transitions in garnet ferrite films with a deviation of the axis of uniaxial anisotropy from the normal to the film plane are presented.

The sixth chapter is devoted to the study of the properties of ferromagnetic resonance in an anisotropic medium. The free and forced oscillations of the magnetization during the orientational transition are considered. In the linear approximation, the dynamic susceptibility tensor is found for the magnetic environment, combining both symmetric and antisymmetric properties.

The seventh chapter refers to an important case for practice – ferromagnetic resonance in films with strong uniaxial anisotropy, deviated from the normal to the film plane. Analytical expressions for resonant frequencies and fields are obtained, an overview is given for some experimental results.

The eighth chapter is devoted to the description of the resonance properties of the composite medium, consisting of a set of ferrite spheres, the anisotropy axes of which are oriented arbitrarily relative to each other. Using the averaging model, the components of the dynamic magnetic susceptibility tensor are obtained under conditions of partial and complete axial disorder.

The ninth chapter examines the precession of magnetization in the field preceding the complete termination of the orientational transition, representing second-order two-periodic precession. Five possible modes of such precession are revealed, the conditions of excitation in the absence of symmetry of the system are considered. This chapter partially includes the material published earlier in the monograph [3], written by one of the authors together with V.S. Vlasov and L.N. Kotov.

The tenth chapter is devoted to the description of the precession of the equilibrium position of the magnetization in an anisotropic medium. The cases of uniaxial and cubic anisotropy are considered for various orientation orientations of the easy magnetization axes. Transition kinetics between some precession modes are investigated. Recommendations for setting up experiments are given and possible areas of practical application are noted.

As in the first monographies of this series, numerous co-authors participated in the works described here. So in theoretical works concerning resonance in anisotropic and compositional media, an important role belongs to V.I. Zubkov, experimental works on films of mixed ferrites–garnets were made jointly with F.V. Lisovsky, in the FMR experiments in such films V.A. Osika took part, the films themselves were grown by I.G. Avaeva and V.B. Kravchenko. The works devoted to the study of the properties of the precession of the equilibrium state of magnetization (second-order precession) were carried out jointly with V.S. Vlasov, M.S. Kirushev and L.N. Kotov. In more detail the participation of the co-authors of the work the basis of the monograph is reflected in the list of references.

Stimulated attention and numerous useful remarks by S.V. Yakovlev contributed significantly to the work. Numerous discussions with fruitful discussions were held with the participation of P.E. Zil'berman. In numerous open discussions concerning the physics of phase transitions A.F. Kabychenkov and V.V. Koledov took part. A number of useful notes, including those related to the methodological nature of the presentation of the material, was made by S.V. Tarasenko. V.A. Kotov took part in the discussion of a number of questions on the technology of films of mixed ferrites–garnet films. Questions concerning the technique of measuring weak magnetic fields, the effect of dissipation, and the possibility of experimentally observing the precession of the equilibrium state of magnetization were discussed with P.M. Vetoshko.

The fp;;pwing scientists took part in numerous aspects of the original works which are the basis of this monograph: A.V. Vashkovsky, E.G. Locke, S.V. Gerus, Yu.I. Bespyatykh, I.E. Dickstein, V.V. Tarasenko, V.D. Kharitonov, E.I. Nefedov, G.S.Makeeva, F.V. Lisovskiy, E.G. Mansvetova, G.V. Arzamastseva, N.N. Kiryukhin, A.V. Voronenko, D.G. Shakhnazaryan, V.V. Kildishev, L.A. Krasnozhen.

The most important role in creating favourable conditions for work, constant attention to it and the repeated provision of administrative and scientific assistance belong to the Yu.V. Gulyaev, Academician of the Russian Academy of Sciences.

The direct execution of the work, including the writing of this monograph, was made possible thanks to the help of the S.A. Nikitov, the Corresponding Member of the Russian Academy of Sciences. Almost all the work was carried out with the active participation of numerous technical staff, without whose help the execution of the task would be absolutely impossible.

The authors express their deep gratitude to all the participants and co-authors of the works, give a deep bow and bring the deepest gratitude. Thank you very much, dear colleagues,friends and helpers!

1

General properties of ferromagnetic resonance and orientational transitions (literature review)

The subject of consideration of this monograph are some features of ferromagnetic resonance and orientational transitions in an anisotropic medium. In this case, the main emphasis is placed on the properties of the ferromagnetic resonance under the conditions of the orientational transition. In this regard, this chapter is a brief overview of the basic principles of ferromagnetic resonance and orientational transitions needed to get the reader up to date and to provide an understanding of the material of subsequent years.

The given data, without in any way pretending to be complete, rely on the main literary sources known to the authors and with rare exceptions do not include the results of the work, Completed with amimi authors, the subsequent chapters of the monograph will be devoted to the presentation

There is no overlap with the reviews cited in previous monographs of this series [1, 2] as far as possible, only references to the works necessary for further consideration.

The review in the course of the presentation contains links to the original sections of this monograph, developing the considered provisions.

This chapter consists of two parts: the first is devoted to ferromagnetic resonance, primarily in an anisotropic medium, the second is an orientational transition as such. Consider these parts in sequence.

1.1. Ferromagnetic resonance

Consideration of ferromagnetic resonance (FMR) will begin with the general equation of the dynamics of magnetization, including in the historical aspect.

1.1.1. General equation of magnetization dynamics

Ferromagnetic resonance as a physical phenomenon has a considerable history. The first attempts to study the dynamic properties of magnetic materials, mainly the frequency dependence of magnetic permeability, refer to the beginning of the 20th century. A brief overview of the first papers can be found, for example, in [4, pp. 772–773]. Among the fundamental theoretical works, we should first of all mention the article [5], written before the discovery of the FMR in the experiment and intended to rather interpret the structure and dynamic properties of domain structures. However, it was in this work that the equation of motion of magnetization was first proposed, which underlies the entire phenomenological theory of dynamic phenomena in magnets, including ferromagnetic resonance and magnetostatic waves [1, 2], referred to as the Landau–Lifshitz equation.

A scheme was also given to obtain the effective fields that enter into this equation, which determine dynamic phenomena, namely, it is shown that such fields can be calculated as derivatives of the magnetic energy density of a magnetic material by the components of magnetization. In view of its immense importance, we present here the Landau–Lifshitz equation with a dissipative term in the Hilbert form [6–9]

$$\frac{\partial \mathbf{M}}{\partial t} = -\gamma [\mathbf{M} \times \mathbf{H}_e] + \frac{\alpha}{M_0} \left[\mathbf{M} \times \frac{\partial \mathbf{M}}{\partial t} \right], \qquad (1.1)$$

Here $\gamma > 0$, while the free precession of the magnetization is right; $\mathbf{H}_e$ is an effective field equal to

$$\mathbf{H}_e = -\frac{\partial U}{\partial \mathbf{M}}, \qquad (1.2)$$

where U is the energy density.

The FMR issues are discussed in sufficient detail in a number of review monographs, of which, first of all, it should be noted [6–8]. A detailed analysis of equation (1.1) and its solutions is also contained in [1, 2].

1.1.2. Ferromagnetic resonance in an anisotropid medium

The study of ferromagnetic resonance in an anisotropic medium began quite soon after the discovery of the phenomenon of ferromagnetic resonance as such. The anisotropic properties of ferromagnetic resonance were studied in particular detail in single crystals of an iron–yttrium garnet (IYG) with a cubic symmetry, which was associated with high prospects for the practical application of this material with record low magnetic losses [6–8].

Separately, one should note the study of resonance properties of magnetic materials at high power levels, that is, under conditions of non-linear ferromagnetic resonance (NFMR) [10–12]. Here, the more or less stationary behaviour of the precession of magnetization at low ignals is complicated by the parametric excitation of the small waves, which leads to a significant increase in losses and expansion of the resonance line. Especially sharply magnetic anisotropy manifests itself in the behaviour of the magnetization beyond the threshold parametric excitation, where phenomena of self-modulation nature are observed.

Thus, according to [10, p. 213], with transverse pumping (that is, with a variable microwave field perpendicular to a constant), the maximum self-modulation amplitude occurs when the crystal is magnetized along the [001] crystallographic axis, and when magnetized along the [011] axes and [111] falls by about 10 and 100 times, respectively. As far as the authors of this monograph are aware, the theoretical interpretation of such strong anisotropy of the threshold to self-modulation is still missing.

We briefly list some works on FMR in an anisotropic medium that are important for further consideration. Linear FMR has been widely studied in anisotropic plates with cubic symmetry, primarily garnet ferrites ($R_3Fe_5O_{12}$, where R is yttrium or rare-earth metal) [8, 13–26] and ferrite spinels (RFe_2O_4, where R is a non-magnetic metal) [27–29]. The studies [22–25] present the results of the investigation of FMR in films of mixed ferrites–garnets with simultaneously cubic and sufficiently strong uniaxial anisotropy, including under the conditions of significant deviation of the anisotropy axis from the normal to the film plane [30].

The article by Smit and Beljers [31], who experimentally studied FMR under conditions of an orientational transition in a highly anisotropic barium hexaferrite, should be considered as one of the most important research papers ($BaFe_{12}O_{19}$). The measurements were performed at a frequency of 23.9 GHz in a field of up to 25 kOe. The field of uniaxial anisotropy in the material under study was 17 kOe, which corresponds to a natural resonance frequency of about 50 GHz. Magnetization was carried out along the normal to the anisotropy axis, and a significant decrease in the resonance frequency was observed when the orientation of the magnetization vector to the direction of the field was approached. The resonant frequency was calculated using the Landau–Lifshitz equation, written in a spherical coordinate system. A brief overview of the applied mathematical apparatus (Smit –Beljers–Sul equation) can be found, for example, in [7, 8]. It should be noted, however, that, despite the 'heroic' nature of the work [31], the registration of the domain structure was carried out very approximately, as a demagnetizing factor of the sample as a whole. The Smit–Sul method is discussed in greater detail in Section 2.3 of this monograph.

The fundamental properties of FMR in anisotropic magnetics and antiferromagnetics were analyzed in numerous reviews and monographs, of which [7,8,32–34] should be mentioned. The last two reviews largely affect not only homogeneous FMR, but also wave processes of both dipole and exchange nature.

In connection with the works on cylindrical magnetic domains (CMD) for memory devices [35–43], the resonance properties of materials with strong uniaxial anisotropy, including orthoferrites and mixed garnets, were studied [22, 23, 25, 44–46]. The monograph [40] presents three main tasks that were solved in relation to such devices through the FMR: detection of CMD, nucleation of CMD, pressure on the wall of CMD.

1.1.3. Frequency gap of the ferromagnetic resonance spectrum

Separately, one should note the study of the resonant properties of antiferromagnets with weak ferromagnetism, whose anisotropy is so strong that in the FMR spectrum is a significant energy gap due to spontaneous deformations due to the magnetoelastic interaction [47, 48].

The size of the gap, for example, in hematite (α-Fe_2O_3), is about 10 GHz, so to observe FMR, we need quite significant magnetic fields.

On the other hand, the existence of such a gap leads to a very strong dependence of the elastic wave velocity on the magnetic field, even in fields from 0.5 to 2000 Oe, 20% of the initial value [25.49–52].

According to the historically first theoretical model [25,49,50] (built independently a little later in [51]), in the absence of a gap, the velocity of an elastic wave increases smoothly as the field grows. However, in real solids in a field up to tens of kiloersted, this increase is insignificant and does not exceed fractions of a percent. In this case, the maximum increase in velocity occurs already in insignificant fields (Oe units), and then the steepness of dependence drops sharply, and it goes very hollow. Such a decrease in the dependence steepness occurs long before the domain structure disappears; therefore, no multidomain samples are observed in experiments.

In the weak ferromagnets, in the presence of an energy gap of a magnetoelastic nature, the external field is multiplied by a coefficient equal to the ratio of the Dzyaloshinsky field to the exchange field, which is about 10^{-3}, so that the initial part of the sharp growth of the elastic velocity across the field is strongly stretched. Such a stretching occurs just at the field actually observed in the experiment of the order of several thousand Oersted, which looks like a strong dependence of the elastic wave velocity on the field.

In the subsequent more detailed theoretical model [53], it was shown that in a small field the magnetization of a weak ferromagnet undergoes a second-order phase transition, accompanied by a 'softening' of the system, as a result of which the elastic wave velocity tends to zero. We note, however, that in the experiment, a similar 'vanishing' of the elastic wave velocity was not observed, apparently due to the interfering action of the domain structure.

These investigations were further developed in numerous studies, including those devoted to orthoferrites [54], Rayleigh surface waves [55], parametric excitation of elastic waves in the vicinity of phase transition points [56], some of which were summarized in review [33].

1.1.4. Measurement of parameters of magnetic materials

It should be noted that along with the study of resonant properties proper, the use of FMR as a tool for measuring the parameters of materials to which anisotropy primarily relates, is of great importance.

The tendency of such measurements, apparently, originates from the metal films [57], with the magnetization being often measured simultaneously with anisotropy [13, 20.22–26, 58–65].

When measuring the parameters of films with uniaxial anisotropy perpendicular to the film plane, the problem of separate measurement of anisotropy and magnetization arises. If the anisotropy axis is in the film plane, as is often the case in metal films [57], the problem of separate measurement is solved by measuring the FMR frequency when the film is magnetized parallel and perpendicular to the anisotropy axis. In films of ferrite–garnets, especially used for CMD devices, the anisotropy axis is usually perpendicular to the film plane, so that the demagnetization and anisotropy energy densities are described by the same expressions, which makes separate measurement of constants difficult. The position is simplified in the case of at least a small deviation of the anisotropy axis from the normal to the film plane [25, 30]. In this case, the direction of magnetization of the film in the plane along and across the projection of the axis anisotropy on this plane becomes non-equivalent, and separate measurement of the anisotropy constant and magnetization becomes possible [22–25]. In [66] for the same separation it is proposed to use surface magnetostatic waves, whose demagnetizing fields, having a frequency of that order as the wavelength, they allow to some extent to neutralize the effect of the demagnetizing factor of the film, which has a significant value due to the extremely small ratio of the film thickness to its longitudinal dimensions. Measurements of the constants of the magnetoelastic interaction in magnetic strips are described in [25, 67–73]. In [74–77], FMR methods are used to study the domain structure and distribution magnetization in the medium for magnetic recording.

1.1.5. Magnetic composite media

Recently, work has been actively developing on the creation and study of composite materials with promising electrodynamic properties for practice. Such environments are in most cases performed on the basis of discrete current elements that represent both segments of the conductor, annular, straight, omega-shaped, and other shapes [78–83]. These environments allow for a very wide variation of parameters by choosing the shape and the relative position of the discrete elements. An important example are chiral media [84, 85], consisting of spiral elements, as well as various bi-isotropic and bi-anisotropic media [86]. A set of current spontaneous elements allows to create environments in which effective the dielectric and magnetic permeabilities of which are simultaneously negative [80, 81]. In such media, it becomes possible to propagate inverse electromagnetic waves [87–89], whose properties are very different from the properties of direct waves, which gives hope on the creation of information processing devices on new principles [90–97].

At the same time, there is another possibility to create conditions for the circulation of the reverse waves. These are magnetic structures based on ferrites [6–8], where both bulk and surface reverse waves can propagate [98–106]. Moreover, ferrites possess gyrotropic properties, which makes it possible to create information transmission devices with nonreciprocity in the microwave range [107]. The most important feature of ferrites is the strong dependence of resonance properties on the magnetic field, which allows to create devices that are controlled by changing the external field [6–8, 107–115]. Special advantages can be given by the use of anisotropic ferrites in which the implementation of an orientational transition greatly expands the possibilities of magnetic control.

Thus, the use of ferrites as a basis for compositional editing would allow the creation of nonreciprocal information processing devices, including on the basis of reverse waves, with a wide range of parameters depending on the control field. However, despite the significant number of works cited, the possibility of creating such parameters and the variety of their parameters investigated insufficiently.

In this regard, in Chapter 8 of this monograph, one such composition medium is considered, which is a lattice of anisotropic ferrite spheres, first proposed in [116, 117]. Such a medium should have the properties of gyrotropy, that is, be non-reciprocal with respect to propagating waves, and the parameters of the medium can be either initially set by selecting the ferrite material or changed during the operation of the device by changing the external magnetic field. Based on it, it is possible to create structures capable of supporting reverse electromagnetic waves. Chapter 8 is devoted to the calculation of one of the parameters of such a medium, primarily determining the possibility of practical applications, namely, its dynamic magnetic susceptibility.

1.1.6. Some features of non-linear ferromagnetic resonance

The above-considered properties of ferromagnetic resonance relate primarily to linear oscillatory phenomena that occur with a small amplitude of oscillations. However, with a large amplitude, that is, in a non-linear mode, the oscillations acquire a number of features that are significantly different from linear ones. This monograph does not set as its goal the study of a more or less complete circle of non-linear phenomena accompanying ferromagnetic resonance, because it is incredibly extensive and requires separate attention. However, for some orientation of the reader, we still give here a brief overview, which, according to the authors, is useful for further go review.

The study of microwave non-linear phenomena in ferrites is the subject of a significant number of works, partially generalized in [6–8, 10–12, 118]. Studied in detail as stable non-linear processes – multiplication and frequency conversion, detection, generation and amplification of electromagnetic signals [6, 7], as well as various instabilities, associated primarily with parametric excitation of exchangeable spin waves, which have found their use in power limiters, noise suppressors, generators and other devices [8, 118]. With the introduction to the practice of materials with record low losses, primarily iron–yttrium garnet (IYG) [6–8], it was found that unstable nonlinear phenomena associated with the parametric excitation of exchange pins in waves, arise already at extremely small magnitudes of magnetization precession, when the angle of deviation of the magnetization vector from the equilibrium position does not exceed one to two degrees [10]. In this case, the development of spin-wave stability makes it difficult to manifest other non-linear processes that require a deviation of the magnetization at much larger angles, including up to several tens of degrees. Apparently, the only possibility for the realization of such precession angles is the use of a ferrite film (plate), magnetized perpendicular to its plane, when the frequency of the uniform ferromagnetic resonance (FMR) mode falls on the lower boundary (bottom) of the spectrum of exchangeable pin waves, as a result of which parametric excitation becomes impossible. In the experiment, such conditions were realized in [119–122], from which indirectly one can judge about the possibility of reaching precession angles up to 20–25 degrees and more without any excitation of exchange spin waves.

Such a circumstance gave rise to a significant amount of theoretical work [123–131], which considers various non-linear modes of magnetization precession at high magnitudes. Both stochastic and complex regular ones were identified, including self-oscillation precession modes, static and dynamic bistability, which lead to dynamic film reversal, are considered, bifurcation diagrams are constructed, demonstrating the wide possibilities of controlling the nonlinear dynamics of magnetization in thin-film lines by changing external magnetic fields.

In most of the works listed, the magnetization vector in the steady state was assumed to be oriented along the constant magnetic field applied to the film. At the same time, the fact that in anisotropic bands the vector of magnetization in the process of increasing the field rotates from the direction of the anisotropy axis to the direction of the field, making an orientational transition [132–134], was ignored. It should be expected that such a change in the orientation of the equilibrium magnetization leads to a change in the nature of the nonlinear precession. Chapters 9 and 10 of this monograph are devoted to clarifying this issue.

1.2. Orientational and phase transitions

The main subject of consideration of this monograph is the ferromagnetic resonance under the conditions of an orientational transition; therefore, we now briefly discuss some transitions that are important for the further properties of transitions as such.

1.2.1. Concepts of orientation and phase transitions

Immediately, we note that in the definition of an orientational transition, today's terminology is somewhat unsteady. The terms 'orientational transition' and 'orientational phase transition' are often used as synonymous [132], although a gradual change in the orientation of the magnetization vector may occur without changing the phase, for example, when magnetizing a uniaxial magnet in the direction perpendicular to the anisotropy axis.

The term 'spin-reorientation transition' [133] is often used, in essence, also as a synonym for the above expressions.

The monograph [134] directly indicates different interpretations in the concept of 'angular phase' [134, p. 10, footnote]. There is also a definition, which, apparently, should be brought here literally [134, p. 14]: 'The orientational phase transition is a transition between the magnetic phases possible in a given crystal, which are determined by the conditions of the minimum of its thermodynamic potential.' Such a definition, apparently, should be considered a classic. From this it follows that this transition is precisely between different phases. But the phase is considered to be the state of a magnetic, characterized by a certain symmetry of the distribution of magnetization [135]. In the monograph [134, p. 15] the phase is called 'symmetric', when the magnetization vectors are oriented along certain crystallographic directions, and 'angular', when the orientation of the vector can change with changing external parameters, such as the field. In this case, the symmetry properties of the 'symmetric' and 'angular' phases differ significantly. Thus, the transition between the 'symmetric' and 'angular''phases is a phase transition in full accordance with the meaning of this term. However, in the 'angular' phase with a change in the field, the orientation of the magnetization vector can change quite smoothly, while no change of symmetry occurs. That is, in this case, despite the change in the orientation of the magnetization vector, the transition from one phase to another does not occur. Thus, such a change in orientation called the 'phase transition' is not quite correct. It can be assumed that the definition of an 'orientational transition' without inserting a 'phase' is more appropriate here.

It is this terminology that the authors intend to adhere to in this monograph. That is, a transition with a change in orientation, but without a change in symmetry, will be referred to as 'orientational transition', and the transition with phase change 'phase transition'.

1.2.2. Phase transitions of the first and second kind

In view of the importance of distinguishing the concepts of 'orientational transition' and 'phase transition', and the main subject of consideration will be the orientational transition first, we first briefly discuss the second one, namely the phase transition.

So, the phase change, that is, the degree of ordering of magnetic dielectrics, occurs through phase transitions of the first or second kind. Since orientational transitions, as a rule, are accompanied by phase transitions of the second kind, Landau's theory [135–138] is used for the theoretical interpretation of such phase transitions, consisting in the decomposition of the potential in a small parameter (as a rule, one of the components or the magnetization module) in the vicinity of the transition point. Temperature or external field relative to th potential is considered as a perturbation leading to the transformation of its extremes (minima), the number of which (about one or two) determines the transition from a homogeneous phase to a non-uniform one. According to Landau's theory, the thermodynamic potential, which describes the state of the body, during phase transitions is continuous, and its derivatives with respect to temperature and magnetic field undergo a discontinuity. The first-order phase transition is characterized by discontinuities of the first derivative. The second-order phase transition is the discontinuity of the second derivative, while maintaining the continuity of the first. During a phase transition of the first kind, both the state of the body and its symmetry change by a leap. During a phase transition of the second kind, the state of the body changes smoothly, and its symmetry undergoes a jump.

Magnetic phase transitions of the first kind are usually accompanied by a jump in magnetization (in magnitude or direction). Typical examples of such transitions are the transfer of the magnetization vector from one anisotropy axis to another with a change in temperature or external field, as well as the magnetization reversal of a uniaxial ferromagnet by the field antiparallel to the initial equilibrium position of the magnetization. With such transitions in a certain interval of a field change, two (and more) different phases can coexist simultaneously. Spatially bounded regions with different phase states (for example, with different directions of magnetization) are domains separated by phase separation boundaries – domain boundaries. In the case of a uniform initial state of magnetization, the formation of the second phase occurs by the appearance of nuclei followed by their subsequent growth. The difficulty in the formation or the initial absence of the nuclei of a new phase in the remagnetization causes hysteretic phenomena.

During transitions of the second kind, the magnetization of the body changes smoothly, and the symmetry changes abruptly. The state of the body is characterized by the 'order parameter' η, where $0 \leq \eta \leq 1$, with $\eta = 0$ corresponds to a more symmetrical state, and $\eta \neq 0$ – to a less symmetrical one. At the transition point $\eta = 0$. In magnetics, the order parameter is usually the normalized magnetization:

$$\eta = M/M_0 \qquad (1.3)$$

In the vicinity of the phase transition point, the thermodynamic potential decomposes in a series in even powers of η:

$$\Phi = \Phi_0 + a\eta^2 + b\eta^4 \qquad (1.4)$$

where the coefficients $a, b,...$ determine the nature of the transition. Typical examples of second-order phase transitions are the transition from a ferromagnetic to a paramagnetic state at the Curie point and an orientational transition in a uniaxial magnet with a field perpendicular to the axis of anisotropy. In the vicinity of the second-order phase transition, there can be a domain structure originating from fluctuations of the order parameter without hysteresis. The Landau theory allows one to accurately determine the existence of a transition point, and also predicts an anomalous increase in the heat capacity, magnetic susceptibility, and some of the parameters of the medium as it approaches this point. However, a more detailed study shows that in the closest neighbourhood of the transition point, the order parameter undergoes significant fluctuations the geometric size of which tends to infinity. The theory of critical indexes is used to describe the special features, which give a power dependence of the fluctuation correlation radius from the reciprocal of the order parameter. A more detail with this theory can be found, for example, in the monograph [139].

1.2.3. Properties of orientational transitions

We now return to the main subject of this monograph, the orientational transition, and briefly describe some of the papers that are important for further consideration. A very detailed description of the properties of orientational transitions in magnetic materials can be found in the review [133] and monograph [134]. A significant number of experimental results, interpreted mainly in the framework of Landau's theory, is given. Quite remarkable is the conclusion that in a multi-sublattice ferromagnet (for example, gadolinium garnet ferrite), the domain structure can exist in several slots of a change in the external field. So, after saturation in a relatively small field (less than 1 kOe), the configuration of the angular phase is transformed, so that the resulting magnetization vector regains its ability to orient in two opposite directions, with the result that the domains appear again in a fairly significant field (about 15 kOe).

Experimental observation of a similar effect was described in Ref. [140], as well as in a number of other articles cited in Ref. [134]. According to [134], the interpretation of the appearance of domains in a strong field may be similar to that for antiferromagnets [141]. That is, the criterion for the existence of a domain structure is not the smallness of the field, but the presence of simultaneously existing several energy-favourable states in a certain range of temperatures and fields [134, p. 243].

A detailed illustration of another variant of dissimilar homogeneous structures, replacing each other as the successive orientational transitions take place, is given in [142].

From the point of view of the problems of FMR and MSW considered in this monograph, the pioneering work of Smit–Beljers [31], as well as the subsequent work on FMR under conditions of orientational transitions, partially generalized in [6–8], are of considerable interest. Chapters 6 and 7 of this monograph will be devoted to a detailed study of such processes.

The consideration of orientational transitions in discrete compositional structures [143], including representing a set of freely oriented dipoles [144, 145]. The behaviour of magnetic susceptibility in composite structures consisting of arbitrarily misoriented anisotropic ferrite spheres will be discussed in Chapter 8 of this monograph.

1.3. Domain structure

The orientational transition, which is one of the main subjects of consideration in this monograph, often occurs in conditions where the magnetization has two or more equilibrium orientations, having equal or close to each other levels of energy. For example, in the case of uniaxial anisotropy, the magnetization with the same success can orient in both the positive and negative directions of the axis. In the case of a limited sample, the tendency to a decrease in the demagnetization energy leads to its splitting into domains, within which the magnetization is oriented in one or another of the possible directions. In this case, the main condition for the formation of a domain structure is the equilibrium between the energy of the demagnetization field and the energy of the domain cells of the sample generally. The domain structure as such is not a subject of consideration in this monograph, however, in view of its importance for the physics of the orientational transition, we present a brief summary of the main information on domains that are useful for further considerations.

The beginning of the study of the domain structure in magnetic materials can apparently be considered the discovery of 'Barkhausen jumps', a detailed description of which can be found, for example, in [4, pp. 210–214]. Following numerous studies of the domain structure by the method of powder figures [4, pp. 214–225], the discovery of the possibility of observing domains in transparent dielectrics, including ferrites, on the Faraday effect [146], was made. A detailed review of the results of such studies can be found, for example, in [147]. A special surge in work on the domain structure occurred after the discovery of cylindrical ferrous domains (CFD) [35, 37] in connection with the possibility of using them to create memory systems for computers [148–151]. The rapid development of domain physics in those years is reflected in numerous reviews and monographs, among which, without pretending to be complete, one can mention [38–41]. We present a brief review of the main results that are important for further consideration.

1.3.1. Physics of domain structures

Let us briefly list the main physical properties of domain structures. The main reason for splitting a limited sample into domains is the desire to minimize demagnetization fields, that is, such which are formed at the exit of a uniformly oriented magnetization on some obstacles that violate such a uniform orientation. The boundaries of the sample or boundaries of other domains with a different magnetization orientation. An equilibrium domain structure is formed as a result of a balance between the energy of the demagnetization fields formed by uniformly magnetized areas of the sample and the energy of the domain boundaries, where the magnetization sharply changes the orientation, triggering the strong energy of the non-uniform exchange interaction.

In the general case, the configuration of the domain structure can be determined by minimizing the expression for the total energy of the magnet W. For a uniaxial magnet with an anisotropy axis that coincides with the Oz axis, W has the form [4]

$$W = \int_V \left\{ \frac{A_0}{2aM_0^2} \sum_{i,k} \left(\frac{\partial M_i}{\partial x_k} \right)^2 + \frac{K}{M_0^2} \left(M_x^2 + M_y^2 \right) - \frac{1}{2} \mathrm{grad} \int \frac{\mathrm{div}\mathbf{M}}{r} dx \right\} dV, \quad (1.5)$$

where V is the sample volume, A_0 is the inhomogeneous exchange interaction constant, a is the crystal lattice constant, M_0 is the saturation magnetization, K is the uniaxial anisotropy constant. Here, the first term corresponds to the energy of non-uniform exchange, the second to the anisotropy energy and the third to the energy of the demagnetizing field.

The application of the variational principle (Euler equation [152]) to the problem of the domain boundary lying in the Oyz plane (i.e. to the first two terms of expression (1.5)) leads to the dependence of the angle θ between M and Oz of the coordinate z, having the form [4]

$$\theta = \arccos\left(-\mathrm{th}\frac{x}{\delta_0} \right). \quad (1.6)$$

Here δ_0 is the thickness of the domain boundary, determined by the expression

$$\delta_0 = \sqrt{\frac{A_0}{2aK}}. \quad (1.7)$$

The magnetization vector in such a boundary, called the 'Bloch' one, as the x coordinate changes, unfolds in the Oyz plane, which is perpendicular to this coordinate. For very thin films, the 'Néel' domain wall may be energetically more favourable, in which the magnetization vector unfolds in the Oxy plane, that is, in the plane of films [39, 40, 57].

In weakly anisotropic uniaxial films, the anisotropy axis of which is oriented perpendicular to the film plane, the domain structure has the form of bands with closing domains near the surface [4].

A convenient characteristic of a uniaxial magnetic material is the 'quality factor' Q, defined by the expression

$$Q = \frac{K}{2\pi M_0^2},$$ (1.8)

The criterion for the existence of a through structure (without closing domains) is the fulfillment of the inequality: $Q > 1$.

In the absence of anisotropy in the film plane, the domain structure has a disordered (labyrinth) character. The application of a constant magnetic field perpendicular to the plane of the film causes a banding of the banded domains in cylindrical magnetic domains (CMD), whose properties, first studied in [35, 37, 153, 154], are considered in detail in numerous reviews and monographs [38, 155–182].

The cylindrical domain exists in the interval between the elliptic instability field H_e, below which it stretches to banded, and the field of collapse H_c, above which it disappears. These fields correspond to the diameter of the elliptical instability d_e and the diameter of collapse d_c (obviously, $H_e < H_c$ and $d_e > d_c$).

Along with the isolated through cylindrical domains, depending on the nature of the anisotropy, the layered structure, the coercivity and the prehistory of the sample, there may be more complex types of domains (oblique, non-through, ring) and domain structures (linear band, hexagonal or amorphous CMD lattice and others [35, 38,155–171].

In films of mixed garnet ferrites (MGF), along with 'normal' CMD, which have a purely Bloch domain boundary, there are so-called 'hard' CMDs, whose circular boundary consists of alternating segments with Bloch and Néel distribution of magnetization [39, 172–179]. Short Néel segments are commonly referred to as 'vertical Bloch lines' (VBL). With a large number of VBLs, the stability interval of the CMD over the magnetic field increases significantly. VBL have a significant impact on the dynamic properties of CMD. The experimental detection of VBL in MFG films by optical methods is difficult due to diffraction boundaries [180]. VBL was experimentally observed by electron microscopy in films of cobalt and some orphic materials [181–183], as well as by the method of powder coating on thin permalloy films covering MGF films [184]. Along with vertical Bloch lines, horizontal Bloch lines (HBL) can also exist in the domain wall, significantly affecting its dynamic properties. In [185–192], it was shown that the Bloch domain boundary in a thin film is not realized, and the distribution of magnetization in the centre of the film has a quasi-Bloch character, and near the surfaces is a quasi-Néel one (twisted boundary). A solution of this type, however, does not have a limit to the case of thick samples, and in [193] it is called into question. There is also an opinion on the feasibility of revising contemporary ideas about the dynamic properties of fixed boundaries based on the 'twisting' model. However, at present, the concept of a 'twisted' domain wall with HBL is generally accepted when interpreting dynamic experiments with CMD.

The dynamic properties of the complex structure of the CMD are clearly manifested in the very beautiful effect of the 'ballistic aftereffect'. The effect manifests itself in the motion of the CMD 'by inertia' that is occurring after the end of the action of the field propulsing pulse. During the forward movement of a domain under the action of such a pulse, HBL is generated in the frontal part of the CMD, as a result of which two systems appear on the lateral sides where there are no HBL tightly twisted in VBL, curled in a spiral like a rubber motor of a flying model [194, 195]. After the end of the propulsive pulse, both spirals unwind, which ensures the generation of HBL in the opposite direction, which causes the forward movement of the CMD 'by inertia' forward a further considerable distance. A more detailed description of the effect of ballistic aftereffect with numerous references to original works can be found, for example, in [40, 41].

In addition to isolated CMD, labyrinth and strip domain structures in films of mixed garnet ferrites with normal uniaxial anisotropy, there are numerous types of periodic lattices consisting of isolated, limited in length domains.

In addition to a simple hexagonal lattice from round centres of CMD [6–8], there are structures of dumbbell-shaped, three-lobed and other symmetric and isolated domains that organize themselves into regular lattices [199–202]. The dynamics of such patterns is characterized by a great variety, including the so-called 'anger state', which is the formation of complex structures that remain only in the dynamics [197, 198,203–206].

Recently, studies have appeared on the dynamics of domain structures excited by ultrashort light pulses from a femtosecond laser [207]. It can be assumed that such domain motion occurs as a result of the implementation of a three-step conversion of a light pulse into heat, which causes thermal expansion of the medium, giving rise to elastic displacements, which in turn act through magnetization on magnetostriction ('three-temperature model'). More details on the mechanisms of action of light on magnetization can be found, for example, according to [208–228].

1.3.2. Domain structure at orientational transitions

As an initial stage of studying the domain structure at the orientational transitions should probably be indicated in Ref. [36], where a change in the period of linear polar domains was noted when the magnetic plate with a normal magnetic field was tangentially magnetized by uniaxial anisotropy. The number of pioneering works should also include [31], where a change in the frequency of ferromagnetic resonance can be traced when the plate is tangently magnetized to the phase transition point. However, the role of the domain structure in this work, although it is mentioned, has not been sufficiently identified.

Extensive study of domains with orientational approaches developed during the CMD epoch when studying orthoferrites, and then mixed ferrite garnet. Observation of the domain structure in transparent films with films and similar materials, carried out using the Faraday effect was a powerful tool for studying various ideas of magnetic anisotropy. The starting point here was the fact that the domain structure disappeared upon tangential magnetization of films with normal anisotropy, which occurs during the phase transition of the second kind in a field close to the field of uniaxial anisotropy.

To a large extent, such studies are summarized in [25,38–44], but here we mention only some works that are important for further consideration.

The magnetization of a single sublattice uniaxial magnetic dielectric by a field perpendicular to the EMA (easy magnetization action), occurs by turning of the magnetization vector **M** to the direction of the field and ends when the magnetization is completely aligned along the field. The moment of alignment is a second-order phase transition. For a field below the transition, there are two energetically equivalent directions of magnetization; for a field above, there is only one. The existence of two possible directions in a limited sample leads to its splitting into domains. Thus, in a sample having a plate shape perpendicular to the EMA, a domain structure appears, as a rule, of a strip nature. As the field increases, the vectors the magnetization in the domains of both signs tend to the direction of the field, so that the demagnetization energy decreases. At the same time, the rotation angle of the magnetization vector inside the domain wall, equal to the difference of angles between the directions of the magnetization vectors of neighbouring domains also decreases. That is, the energy of the domain wall, also decreases. The balance between the demagnetization energy density and the domain energy density leads to the change in the period of the strip domain structure.

The ratio of the rates of change of both energies with a change in the field leads to the fact that the period of the domain structure decreases with increasing field. The experimentally observed decrease in the period at the phase transition point, relative to the period in the absence of a field, is up to two to three times. In this case, the fact of complete disappearance of the structure is interpreted as the moment of equality or sufficient proximity of the external field to the value of the anisotropy field. [25,160,161,229–236]. The features of the disappearance of the structure in films with a deviation of the magnetic anisotropy axis from the normal to the film plane were considered in [25, 30]. This issue is discussed in more detail in Chapter 5 of this monograph.

An interesting tool for the study of anisotropy is the observation of the orientational transition along elastic oscillations excited by magnetostriction during the motion of the domain structure [237,238].

A theoretical analysis of the dependence of the period of the domain structure on the field in [229, 232–234] was performed under the assumption that the magnetization distribution is uniform over the sample thickness.

However, it was shown in [230] that such a consideration is valid only far enough from the transition point. Near the transition point, due to the increase in magnetic susceptibility and action demagnetizing field, the characteristic size of the inhomogeneity of the magnetization distribution approaches the sample thickness, therefore allowance for thickness heterogeneity becomes necessary. In this case, according to [229], the period of the domain structure tends to zero as it approaches the phase transition point, whereas in [231] it is argued that the period at this point tends to a finite limit, significantly from zero to different.

A definite resolution of such a contradiction is contained in [25], where, by detailed experimental research analyzing the derivative of the period dependence on the field near the point transition, preference was given to the model considered in [230], which takes into account the non-uniform distribution of the magnetization over the sample thickness.

Orientational transitions with a field perpendicular to the EMA are also observed in multi-sublattice-ground-axis magnets. In this case, due to the existence of several lattices, instead of a single phase transition in a single-sublattice magnet, there are several in multi-sublattice magnetic transitions.

Transitions in magnets with a point of temperature compensation are of particular variety. Near such a temperature, the transition fields sharply increase, which allows one to observe the domain structure in strong (with equal exchange) magnetic fields.

The phase diagram of a two-sublattice uniaxial ferrimagnet upon its magnetization perpendicular to the EMA is given in [239]. The existence of two-linear phases – low-temperature with ordinary domains and high-temperature with high-field domains – is noted. Between these phases is a non-collinear phase, where domains are missing. In this case, there may be two intervals across the field, where the sample is divided into domains separated by intervals, where domains are missing. The boundaries of the transitions between phases with a change in temperature and field were obtained in [231, 240–245]. The most consistent consideration of the domain structure of a uniaxial two sublattice ferrimagnet near a second-order phase transition was carried out in Ref. [246]. In [247–251], the existence of a stable domain structure in strong fields – up to the 'collapse' of the magnetization vectors of both sublattices ($\sim 10^6$ Oe). The monodomain interval near the compensation temperature was studied in [252– 254]. An experimental study of the described range of the phenomena is associated with the difficulty of obtaining strong magnetic fields (10^5–10^6 Oe); A review of methods for producing such fields can be found, for example, in [255–257].

Conclusions on Chapter 1

In this chapter, a brief review of the literature concerning ferromagnetic resonance under conditions of the polarization transition is made.

The main issues covered in this chapter are as follows.

1. The dominant role of the Landau–Lifshitz equation as the main equation of motion of magnetization in describing phenomena of ferromagnetic resonance was investigated. The importance of the study of ferromagnetic resonance in materials possessing cubic magnetic anisotropy, to which the iron–yttrium garnet with record low magnetic losses belongs, is noted. The original and survey works devoted to the study of ferromagnetic resonance in anisotropic materials, including mixed garnet ferrites, spinels, orthoferrites, as in bulk samples, and in thin stripes are noted. Data are given concerning the especially strong manifestation of the anisotropic magnetic resonance properties under conditions of the parametric excitation

of exchange spin waves. The main tasks are solved by ferromagnetic resonance in devices located on cylindrical magnetic domains, such as the detection and nucleation of domains, as well as the promotion of domain boundaries.

2. The presence of an energy gap in the magnetic resonance spectrum of antiferromagnets with weak ferromagnetism due to uniaxial anisotropy in combination with magnetostriction is noted. The effect of the gap on the effect of a strong dependence of the elastic wave velocity on an external magnetic field is discussed, and one of the interpretation options is the excitation of a soft mode at the second-order phase transition point. Ferromagnetic resonance plays an important role as a very effective tool for measuring the parameters of magnetic materials, primarily anisotropy, magnetization and magnetoelastic properties, as well as the behaviour and character of the domain structure.

3. The basic electrodynamic properties of the known composite media consisting of discrete current elements are briefly considered. The possibility of creating media in which the effective dielectric and magnetic permeabilities at the same time are negative, is noted, so that in such environments it becomes possible to propagate backward waves that are useful for creating microwave information processing devices. The possibility of creating composite compositions based on ferrites is noted, which makes it possible to control the parameters of the media by changing the magnetic field. Special advantages may come from the use of ferrites possessing significant magnetic anisotropy, the implementation of the orientational transition in which greatly expands the possibilities of magnetic control.

4. The main phenomena accompanying ferromagnetic resonance in a non-linear mode are briefly mentioned. The difference between stable and unstable non-linear processes is noted. Among the unstable processes, parametric excitation of exchange spin waves is highlighted, hindering the development of stable processes of considerable amplitude. As a means of preventing such excitation, the magnetization of a thin ferrite plate by a field perpendicular to its plane has been proposed. The possibility of excitation of magnetization in such a geometry with significant angles of deviation from the field direction, up to several tens of degrees, is reflected. Some limitations of the known works are noted, consisting in insufficient consideration of high-amplitude precession in terms of the orientational transition.

5. The relationship between the concepts of orientational and phase transitions is considered, characteristic distinctions of both are noted. The terminology used is used later in this monograph. The difference is noted and the main features of phase transitions of the first and

second kind are given. A brief overview is presented of the main provisions of Landau's theory, based on the decomposition into a series of thermodynamic potentials in powers of the order parameter. The theory of critical indexes is mentioned, supplementing Landau's theory with regard to fluctuations of the order parameter. Examples of phase transitions of the first are given and the second kind in magnetics.

6. The main works concerning orientational transitions in magnetic materials interpreted on the basis of Landau's theory are considered. The role of domains is noted and the possibility of their existence in several instances of a change in the external field, including due to the formation of the angular phase, is indicated. Mentioned is the consideration of orientational transitions in discrete compositional structures, which are a set of freely oriented dipoles. The pioneering role of Smit–Beljers work in the experimental study of ferromagnetic resonance during an orientational transition in a uniaxial magnet is noted. Other works that follow are mentioned on the same topic, concerning both theory and experiment.

7. Some features of the domain structure that are important for orientational transitions are considered. It is noted that the main condition for the formation of a domain structure is the equilibrium between the energy demagnetization fields and domain boundary energy. The general assumptions underlying the physics of domain structures are presented. Briefly listed are the methods for observing domains, such as historically the first – by Barkhausen jumps, the powder-figure method following it, and the main one now – by the Faraday magneto-optical effect.

8. The significant development of the physics of micromagnetism was noted in connection with the work on the use of cylindrical magnetic domains (CMD) for memory systems of computational buses. The main types of domain structures in films of mixed ferrite granites with normal anisotropy – linear band, labyrinth, and CMD lattices – are listed. The 'hard' cylindrical magnetic domains, the vertical and horizontal Bloch lines are mentioned, some properties of the dynamics of the domains, including the effect of a ballistic aftereffect, are noted. Some works are presented concerning the study of the motion of domains and, in general, the local magnetization reversal of magnetic films under the action of ultrashort powerful light pulses from a femtosecond laser.

9. The behaviour of the domain structure of magnetic materials under the conditions of the orientational transition is considered. The high efficiency of observation of domains by the Faraday effect near the second-order phase transition, which is an important tool for

measuring the magnetic anisotropy parameters of iron garnet films, is noted. The reduction of the period of the strip domain structure with increasing tangential field is considered. The period tendency is noted at the phase point transition to the final value, as a rule, two to three times less than the period in the demagnetized state. The possibility of studying the anisotropy of films by excitation of other vibrations in them is noted. under the action of an alternating field. The theoretical model, which takes into account the non-uniform distribution of the magnetization over the film thickness, is briefly mentioned. Its advantage in interpreting experimentative results on the phase transition in comparison with the model in which this distribution is not taken into account is noted. A specific feature of the implementation of the orientational transition in multi-sublattice magnets is noted, which consists in the possibility of the existence of several hase transitions in temperature and field with corresponding intervals of the existence of domain structures, including super-strong magnetic fields (up to $\sim 10^6$ Oe).

2

Mathematical apparatus used in calculating ferromagnetic resonance

Brief information concerning the mathematical apparatus used in the following exposition when calculating the parameters of ferromagnetic resonance is described in this chapter. Since in previous monographs of this series [1, 2] the mathematical apparatus received considerable attention, here we confine ourselves to a brief illustration of some of the mathematical techniques needed later. The material of the chapter is predominantly based on literary sources, among which the main ones may be noted [4, 6–8, 258]. The remaining references, due to their diversity and multiplicity, are given in the text.

2.1. Energy density and effective fields

Since the most important tool for the theoretical analysis of ferromagnetic resonance and orientational transitions is the minimization of the energy density of the structure under consideration, we briefly list the main types of such density important for further discussion. The following general expressions are present in a variety of review papers and training monographs, to list at least a small part of which is not possible, therefore we note only [4.6–8, 40, 134, 135, 258, 259] as the most closely corresponding to the nature of the subsequent presentation.

2.1.1. The total energy density of a magnetic crystal

We assume that the total energy density of a magnetic crystal is:

$$U = U_m + U_e + U_{me}, \qquad (2.1)$$

where U_m is the density of magnetic energy, U_e is the density of elastic energy, U_{me} is the density of energy of the magnetoelastic interaction. The density of magnetic energy is determined by the expression

$$U_m = U_{ex} + U_H + U_M + U_a \qquad (2.2)$$

where U_{ex} is the energy density of the inhomogeneous exchange interaction, U_H is the interaction energy density of magnetization with an external field, U_M is the density of the energy of interaction of magnetization with the demagnetization field, U_a is the density of the energy of magnetic anisotropy.

Consider these types of energy in more detail.

Non-uniform exchange interaction. U_{ex} is the energy density of a non-uniform exchange interaction, which has several types of records [4, 7, 8, 40]:

$$U_{ex} = \frac{A_0}{2aM_0^2}\sum_{i,k}\left(\frac{\partial M_i}{\partial x_k}\right)^2 = \frac{1}{2}q_0\sum_{i,k}\left(\frac{\partial M_i}{\partial x_k}\right)^2 = \frac{A}{M_0^2}\sum_{i,k}\left(\frac{\partial M_i}{\partial x_k}\right)^2, \qquad (2.3)$$

where A_0, q_0, A are various types of recordings of the exchange interaction constant, a is the lattice constant of the magnetic crystal;

Interaction of magnetization with the external field. U_H is the energy density of the interaction of magnetization with the external field **H**:

$$U_H = -\mathbf{M\cdot H} = -\left(M_x H_x + M_y H_y + M_z H_z\right). \qquad (2.4)$$

Interaction of magnetization with the demagnetization field. U_M is the density of internal magnetostatic energy:

$$U_M = -\frac{1}{2}\mathbf{M\cdot H}_p, \qquad (2.5)$$

where $\mathbf{H}_p$ is the demagnetization field, defined as the gradient of the potential of the bulk and surface 'magnetic charges' [4, p. 291]:

$$\mathbf{H}_p = -\mathrm{grad}\left(\int_V \frac{\mathrm{div}\,\mathbf{M}}{R}dV - \oint_S \frac{\mathrm{Div}\,\mathbf{M}}{R}dS\right), \qquad (2.6)$$

In the case of a uniform internal field (sample in the form of an ellipsoid), the demagnetization field is conveniently expressed in terms of the demagnetizing factor tensor:

$$\mathbf{H}_p = -\ddot{N}\mathbf{M}, \qquad (2.7)$$

whose form for an infinite plate in the Cartesian coordinate system *Oxyz*, whose plane coincides with the plane *Oxy*, is

$$\ddot{N} = \begin{pmatrix} 0 & 0 & 0 \\ 0 & 0 & 0 \\ 0 & 0 & 4\pi \end{pmatrix} \qquad (2.8)$$

At the same time, U_M for a thin plate in the coordinate system with the axis *Oz*, perpendicular to the plane of the plate, has the form:

$$U_M = 2\pi M_z^2 \qquad (2.9)$$

Magnetic anisotropy. U_a is the magnetic anisotropy energy density, which in the general case it is convenient to present as series expansion in powers of the normalized components of the magnetization ($m_i = M_i/M_0$):

$$U_a = \sum_{i=1}^{3} a_i m_i + \sum_{i,k,l=1}^{3} a_{ikl} m_i m_k m_l + ..., \qquad (2.10)$$

where a_i, a_{ik}, a_{ikl} are the anisotropy constants of the corresponding order. In this form, the first term describes uniaxial anisotropy, and the second and third include the cubic one.

The energy density of the uniaxial anisotropy with the axis along *Oz* is determined by the expression:

$$U_{a0} = K(m_x^2 + m_y^2) \qquad (2.11)$$

where K is the uniaxial anisotropy constant.

The energy density of cubic anisotropy with the coordinate axes along the edges of the cube has the form:

$$U_{ak} = K_1 \left(m_x^2 m_y^2 + m_y^2 m_z^2 + m_z^2 m_x^2 \right) + K_2 m_x^2 m_y^2 m_z^2, \qquad (2.12)$$

where K_1 and K_2 are the 'first' and 'second' constants of cubic anisotropy, respectively.

Elasticity and magnetoelasticity. To complete the picture, we give the expressions for the densities of elastic U_e and magnetoelastic U_{me} energies also for the case of cubic symmetry with the same orientation of the cell:

$$U_e = \frac{1}{2} c_{11} \left(u_{xx}^2 + u_{yy}^2 + u_{zz}^2 \right) + c_{12} \left(u_{xx} u_{yy} + u_{yy} u_{zz} + u_{zz} u_{xx} \right) +$$
$$+ 2 c_{11} \left(u_{xy}^2 + u_{yz}^2 + u_{zx}^2 \right), \quad (2,13)$$

$$U_{me} = B_1 \left(m_x^2 u_{xx} + m_y^2 u_{yy} + m_z^2 u_{zz} \right) +$$
$$2 B_2 \left(m_x m_y u_{xy} + m_y m_z u_{yz} + m_z m_x u_{zx} \right), \qquad (2.14)$$

where u_{ik} is the strain tensor:

$$u_{ik} = \frac{1}{2} \left(\frac{\partial u_i}{\partial x_k} + \frac{\partial u_k}{\partial x_i} \right), \qquad (2.15)$$

c_{11}, c_{12}, c_{44} are the constants (moduli) of elasticity, B_1, B_2 are the 'first' and 'second' constants of the magnetoelastic interaction.

In more detail, various types of energy density are discussed as necessary in the relevant sections of this monographs.

2.1.2. Effective fields

When calculating various effects involving magnetization, the above expressions for the energy density are used to find 'effective' fields. Without going into details, which can be found, for example, from [7, p. 68], we give only the standard formula for such a calculation:

$$H_e = -\frac{\partial U}{\partial \mathbf{M}} + \sum_{p=1}^{3} \frac{\partial}{\partial x_y} \left[\frac{\partial U}{\partial \left(\frac{\partial \mathbf{M}}{\partial x_y} \right)} \right]. \qquad (2.16)$$

In this expression, the first term corresponds to the effective fields that vary in space in a rather smooth manner when the exchange interaction does not yet manifest itself. The second term relates precisely to the effective field, which varies in space so dramatically that the exchange interaction manifests itself significantly. In the case of wave processes, the first term allows one to work with waves whose order of length is 10^{-4} cm, that is, with waves longer than one micron. For shorter voltages, the exchange interaction reflected by the second term should be taken into account. (2.16). It is necessary to take into account the exchange interaction when calculating the structure of the domain boundaries, since their content in real materials, as a rule, is tenths and hundredths of microns [4,40,258].

2.2. Dynamic magnetic susceptibility

The most important stage in the analysis of dynamic phenomena in magnetic materials is the calculation of dynamic magnetic susceptibility. According to the classical definition [4, 258, 260], magnetic susceptibility is a coefficient of proportionality between the volume-averaged total magnetization and the external magnetization acting on this magnetization magnetic field:

$$\chi = \frac{M}{H}. \tag{2.17}$$

If a constant magnetic field is applied to an isotropic medium, then the magnetization vector in the static state assumes a certain equilibrium position determined by this field in combination with other constant fields (for example, demagnetization or magnetostriction) acting inside the medium.

If now in addition to the constant a small alternating field is applied, the magnetization vector precesses around its equilibrium position in accordance with the law determined by the Landau–Lifshitz equation.

In the general case, such a precession is non-linear, so that the magnetization is proportional to the field, not in the first, but to a higher degree. However, at a small amplitude of the variable field, the precession can be considered linear, for the description of which linearization must be carrie out in the Landau–Lifshitz equation.

Following mainly [6–9], and also [1, 2], we show how this can be done in the case of linear oscillations.

2.2.1. Landau – Lifshitz equation and linearization procedure

Let us proceed from the Landau–Lifshitz equation with the dissipative term in the Hilbert form [6–8]:

$$\frac{\partial \mathbf{M}}{\partial t} = -[\mathbf{M} \times \mathbf{H}_e] + \frac{\alpha}{M_0}\left[\mathbf{M} \times \frac{\partial \mathbf{M}}{\partial t}\right]. \tag{2.18}$$

Here, $\gamma > 0$, while the free precession is right; $\mathbf{H}_e$ is an effective field equal to

$$\mathbf{H}_e = -\frac{\partial U}{\partial \mathbf{M}}, \tag{2.19}$$

where U is the energy density.

Opening the vector products in equation (2.18), we obtain:

$$\frac{\partial M_x}{\partial t} = -\gamma\left(M_y H_z - M_z H_y\right) + \frac{\alpha}{M_0}\left(M_y \frac{\partial M_z}{\partial t} - M_z \frac{\partial M_y}{\partial t}\right);$$

$$\frac{\partial M_y}{\partial t} = -\gamma\left(M_z H_x - M_z H_z\right) + \frac{\alpha}{M_0}\left(M_z \frac{\partial M_x}{\partial t} - M_x \frac{\partial M_z}{\partial t}\right); \tag{2.20}$$

$$\frac{\partial M_z}{\partial t} = -\gamma\left(M_x H_y - M_y H_x\right) + \frac{\alpha}{M_0}\left(M_x \frac{\partial M_y}{\partial t} - M_y \frac{\partial M_x}{\partial t}\right). \tag{2.21}$$

Now suppose

$$\mathbf{H} = \left\{h_z, h_y, H_0 + h_z\right\}; \tag{2.23}$$

$$\mathbf{M} = \left\{m_x, m_y, H_0 + m_z\right\}, \tag{2.24}$$

where H_0 and M_0 are constant.

Substituting (2.23)–(2.24) into (2.20)–(2.22) gives:

$$\frac{\partial m_x}{\partial t} = -\gamma\left[m_y\left(H_0 + h_z\right) - \left(M_0 + m_y\right)h_y\right] +$$

$$+ \frac{\alpha}{M_0}\left[m_y \frac{\partial m_z}{\partial t} - \left(M_0 + m_z\right)\frac{\partial m_y}{\partial t}\right]; \tag{2.25}$$

$$\frac{\partial m_x}{\partial t} = -\gamma \Big[m_y \big(H_0 + h_z \big) - \big(M_0 + m_z \big) h_y \Big] +$$

$$+ \frac{\alpha}{M_0} \left[m_y \frac{\partial m_z}{\partial t} - \big(M_0 + m_z \big) \frac{\partial m_y}{\partial t} \right]; \qquad (2.26)$$

$$\frac{\partial m_y}{\partial t} = -\gamma \Big[m_y \big(M_0 + m_z \big) h_x - m_x \big(H_0 + h_z \big) \Big] +$$

$$+ \frac{\alpha}{M_0} \left[\big(M_0 + m_z \big) \frac{\partial m_z}{\partial t} - m_x \frac{\partial m_z}{\partial t} \right]; \qquad (2.27)$$

$$\frac{\partial m_z}{\partial t} = -\gamma \big(m_x h_y = m_y h_x \big) + \frac{\alpha}{M_0} \left(m_x \frac{\partial m_y}{\partial t} - m_y \frac{\partial m_x}{\partial t} \right).$$

These are the complete equations of motion for the magnetization vector with attenuation taken into account.

Now let's do the linearization of these equations, for which we set

$$h_z \sim h_y \sim h_z \sim m_x \sim m_y \sim m_z \sim \frac{\partial m_x}{\partial t} \sim \frac{\partial m_y}{\partial t} \sim \frac{\partial m_z}{\partial t} \ll H_0 \sim M_0. \quad (2.28)$$

Linearization consists in neglecting in expressions (2.25)–(2.27) all terms of the order $m_i m_k$, $m_i \dot{m}_k$, $m_i h_k$, as a result of which, instead of the full equalities (2.25)–(2.27), linearized are obtained:

$$\frac{\partial m_x}{\partial t} = -\gamma \big(m_y H_0 - M_0 h_y \big) - \alpha \frac{\partial m_y}{\partial t}; \qquad (2.29)$$

$$\frac{\partial m_t}{\partial t} = -\gamma \big(M_0 h_x - H_0 m_x \big) + \alpha \frac{\partial m_x}{\partial n}; \qquad (2.30)$$

$$\frac{\partial m_y}{\partial t} = 0. \qquad (2.31)$$

The third equation here speaks of the constancy of the z-component of the magnetization vector, that is, it can be disregarded in dynamics. Now we set the harmonic time dependence for the field and magnetization in the form

$$\mathbf{h} = \mathbf{h}_0 e^{i\omega t}; \qquad (2.32)$$

$$\mathbf{m} = \mathbf{m}_0 e^{i\omega t}. \qquad (2.33)$$

To simplify recording, the subscripts '0' at amplitudes of variable values are omitted below.

Substituting (2.32)–(2.33) into (2.29)–(2.30) and performing simplifying algebraic transformations, we get

$$i\omega m_x + (\gamma H_0 + i\omega\alpha) m_y = \gamma M_0 h_y; \qquad (2.34)$$

$$(\gamma H_0 + i\omega\alpha) m_x - i\omega m_y = \gamma M_0 h_x. \qquad (2.35)$$

This is a system of two-linear equations with two unknowns. Solving it, for example, using the extended matrix method (Cramer's rule [261, p. 45, formula (1.9-3)], and also [262, p. 55, formula (12), remark on p. 67]), we get

$$m_x = -\frac{\gamma M_0 (\gamma H_0 + i\omega\alpha)}{\omega^2 - (\gamma H_0 + i\omega\alpha)^2} h_x - \frac{i\omega\gamma M_0}{\omega^2 - (\gamma H_0 + i\omega\alpha)^2} h_y; \qquad (2.36)$$

$$m_y = -\frac{i\omega\gamma M_0}{\omega^2 - (\gamma H_0 + i\omega\alpha)^2} h_x - \frac{\gamma M_0 (\gamma H_0 + i\omega\alpha)}{\omega^2 - (\gamma H_0 + i\omega\alpha)^2} h_y; \qquad (2.37)$$

2.2.2. Magnetic susceptibility tensor

We introduce the magnetic susceptibility tensor, which in the classical representation has the form [6–8, 263]

$$\ddot{X} = \begin{pmatrix} X & iX_a & 0 \\ -iX_a & X & 0 \\ 0 & 0 & 0 \end{pmatrix} \qquad (2.38)$$

where

$$\chi = -\frac{\gamma M_0 (\gamma H_0 + i\omega\alpha)}{\omega^2 - (\gamma H_0 + i\omega\alpha)^2}; \qquad (2.39)$$

$$\chi_a = -\frac{\omega\gamma M_0}{\omega^2 - (\gamma H_0 + i\omega\alpha)^2}. \qquad (2.40)$$

In this case, (2.36)–(2.37) is written in the form

$$\begin{pmatrix} m_x \\ m_y \\ m_z \end{pmatrix} = \ddot{X} \begin{pmatrix} h_x \\ h_y \\ h_z \end{pmatrix} \tag{2.41}$$

or, taking into account the zero (as a result of linearization) values of the components of the field vectors and magnetization, in a shortened form:

$$\begin{pmatrix} m_x \\ m_y \end{pmatrix} = \begin{pmatrix} X & iX_a \\ -iX_a & X \end{pmatrix} \begin{pmatrix} h_x \\ h_y \end{pmatrix}, \tag{2.42}$$

that is:

$$m_x = \chi h_x + i\chi_a h_y; \tag{2.43}$$

$$m_y = -i\chi_a h_a + \chi h_y. \tag{2.44}$$

Comment. Hereinafter, when recording tensors and vectors in the form of tables, the dotted lines have a demarcation character and are introduced for convenience.

2.2.3. Magnetic permeability tensor

From the above expressions it can be seen that the susceptibility due to the gyromagnetic properties of the medium has a tensor character. At the same time, permeability, defined as

$$\ddot{\mu} = 1 + 4\pi\ddot{\chi}, \tag{2.45}$$

is also a tensor:

$$\ddot{\mu} = \begin{pmatrix} 1+4\pi X & i4\pi X_a & 0 \\ -i4\pi X_a & 1+4\pi X & 0 \\ 0 & 0 & 1 \end{pmatrix} = \begin{pmatrix} \mu & i\mu_a & 0 \\ -i\mu_a & \mu & 0 \\ 0 & 0 & 1 \end{pmatrix}. \tag{2.46}$$

where:

$$\mu = 1 + 4\pi\chi \tag{2.47}$$

$$\mu_a = 4\pi\chi_a \tag{2.48}$$

Comment. From the expressions (2.38) and (2.46), it can be seen that the diagonal susceptibility component χ_{33} is zero and the corresponding permeability component μ_{33} is equal to one. However, this circumstance certainly occurs only when choosing a dissipative term of the Landau–Lifshitz equation (2.18) in the Hilbert form [9]. When using other dissipative member terms, these components assume slightly different zero and unit values. In this case, the susceptibility (2.38) and permeability (2.46) are written in the form

$$
\ddot{X} = \begin{pmatrix} X & iX_a & 0 \\ -iX_a & X & 0 \\ 0 & X_{\parallel} & \end{pmatrix} \tag{2.49}
$$

$$
\ddot{\mu} = \begin{pmatrix} \mu & i\mu_a & 0 \\ -i\mu_a & \mu & 0 \\ 0 & 0 & \mu_{\parallel} \end{pmatrix} \tag{2.50}
$$

With a small dissipation the value $\chi_{\parallel}$ and $\mu_{\parallel}$ differ little from zero and one respectively. The explicit form of the expressions for $\chi_{\parallel}$ and $\mu_{\parallel}$ can be found, for example, in [6–8]. It also states that the choice of a dissipative term in a form different from the Hilbert form is required if the length of the magnetization vector is changed during the precession. In this case, there are so-called 'longitudinal' and 'transverse' relaxation. Distinguishing the two types of relaxation is important when considering paramagnetic resonance, as well as in the interpretation of some self-modulation phenomena at a non-linear ferromagnetic resonance [10]. Within the framework of this monograph, the length of the magnetization vector is always considered constant, so that it can be assumed that the use of a dissipative term in the Hilbert form is justified.

2.2.4. *Different types of component records of susceptibility tensor*

Note some shorter types of recording of received expressions. Thus, the following notation is often used [6–8]

$$
\omega_H = \gamma H_0 \tag{2.51}
$$

At the same time, expressions (2.36) and (2.37), after disclosing brackets in the denominator, take the form:

$$m_x = \frac{\gamma_0\left(\omega_H + i\omega\alpha\right)}{\omega_H^2 - \left(1+\alpha^2\right)\omega^2 + 2i\alpha\omega\omega_H} h_x + \frac{i\omega\gamma M_0}{\omega_H^2 - \left(1+\alpha^2\right)\omega^2 + 2i\alpha\omega\omega_H} h_y; \quad (2.52)$$

$$m_y = \frac{i\omega\gamma M_0}{\omega_H^2 - \left(1+\alpha^2\right)\omega^2 + 2i\alpha\omega\omega_H} h_x + \frac{\gamma M_0\left(\omega_H + i\omega\alpha\right)}{\omega_H^2 - \left(1+\alpha^2\right)\omega^2 + 2i\alpha\omega\omega_H} h_y; \quad (2.53)$$

so that the components of the susceptibility tensor are written as

$$\chi = \frac{\gamma M_0\left(\omega_H + i\omega_\alpha\right)}{\omega_H^2 - \left(1+\alpha^2\right)\omega^2 + 2i\alpha\omega\omega_H}; \quad (2.54)$$

$$\chi_a = \frac{\omega\gamma M_0}{\omega_H^2 - \left(1+\alpha^2\right)\omega^2 + 2i\alpha\omega\omega_H}. \quad (2.55)$$

These are classical expressions that coincide with those given in [6–8].

When considering magnetostatic waves [1, 2], a definite abbreviation of the record gives the introduction of normalized bases:

$$\Omega = \frac{\omega}{4\pi\gamma M_0}; \quad (2.56)$$

$$\Omega_H = \frac{H_0}{4\pi M_0}. \quad (2.57)$$

The components of the susceptibility tensor are

$$\chi = \frac{1}{4\pi} \frac{\left(\Omega_H + i\alpha\Omega\right)}{\Omega_H^2 - \left(1+\alpha^2\right)\Omega^2 + 2i\alpha\Omega\Omega_H}; \quad (2.58)$$

$$\chi_a = \frac{1}{4\pi} \frac{\Omega}{\Omega_H^2 - \left(1+\alpha^2\right)\Omega^2 + 2i\alpha\Omega\Omega_H}. \quad (2.59)$$

In the absence of attenuation, that is, when $\alpha = 0$, these expressions become the following:

$$\chi = \frac{1}{4\pi} \frac{\Omega_H}{\Omega_H^2 - \Omega^2}, \quad (2.60)$$

$$\chi_a = \frac{1}{4\pi} \frac{\Omega}{\Omega_H^2 - \Omega^2}, \quad (2.61)$$

It is these expressions for susceptibility that are used when considering the dispersion of magnetostatic waves in ferrite films and structures based on them [1,2].

2.2.5. Frequency dependence and resonance frequency

From the structure of expressions (2.39) and (2.40), as well as equivalent to them (2.54)–(2.55) and (2.58)–(2.59), one can see that they have a resonant character with respect to the frequency ω or the field H_0.

In the absence of attenuation (for example, for (2.60) –(2.61)), there is a divergence at the resonant frequency, if there is attenuation, all listed expressions remain finite. In this case, the resonant frequency is exactly Ω_H or, taking into account with (2.56) and (2.57):

$$\omega_{res} = \gamma H_0. \qquad (2.62)$$

However, in the presence of attenuation, the resonant frequency varies somewhat. Consider such a change on the example of (2.54)–(2.55).

From these expressions, getting rid of the imaginary in the denominator and writing down the real and imaginary parts separately, we get:

$$\chi' = \frac{\gamma M_0 \left\{ \omega_H \left[\omega_H^2 - \left(1+\alpha^2\right)\omega^2 \right] + 2\alpha^2\omega^2\omega_H \right\}}{\left[\omega_H^2 - \left(1+\alpha^2\right)\omega^2 \right]^2 + \left(2\alpha\omega\omega_H\right)^2}; \qquad (2.63)$$

$$\chi'' = \frac{\alpha\omega\gamma M_0 \left\{ \left[\omega_H^2 - \left(1+\alpha^2\right)\omega^2 \right] + 2\omega_H^2 \right\}}{\left[\omega_H^2 - \left(1+\alpha^2\right)\omega^2 \right]^2 + \left(2\alpha\omega\omega_H\right)^2}; \qquad (2.64)$$

$$\chi_a' = \frac{\omega\gamma M_0 \left[\omega_H^2 - \left(1+\alpha^2\right)\omega^2 \right]}{\left[\omega_H^2 - \left(1+\alpha^2\right)\omega^2 \right]^2 + \left(2\alpha\omega\omega_H\right)^2}; \qquad (2.65)$$

$$\chi_a'' = \frac{2\alpha\omega^2\omega_H\gamma M_0}{\left[\omega_H^2 - \left(1+\alpha^2\right)\omega^2 \right]^2 + \left(2\alpha\omega\omega_H\right)^2}. \qquad (2.66)$$

As an example Fig. 2.1 shows the dependences of the real $\chi'(a)$ and imaginary $\chi''(b)$ parts of the susceptibility tensor component χ of frequency. Curves are constructed according to the formulas (2.63), (2.64) with several values of the attenuation parameter.

From Fig. 2.1 it is clear that the real part χ' of the components of susceptibility χ, starting at a level near zero, as the frequency increases, first increases, after which, after passing through a maximum, decreases, then passes through a minimum and increases again, asymptotically tending to zero. The passage of the curve through the zero on the site a sharp decline corresponds to resonance. In the absence of attenuation at this point there is a divergence, so that on both sides of it the curve tends to this or that infinity, in accordance with formula (2.60). As the attenuation increases (from curve 1 to curve 3) the extremes of the curve are respectively smoothed, so that the dependence gradually acquires an aperiodic character, at which extremes are completely absent.

Figure 2.1 b shows that the imaginary part χ'' of the components of susceptibility χ, starting at a level close to zero, as the frequency increases, first decreases, after which, after passing through a minimum, it tends to zero again. The minimum just corresponds to the resonance, so it has The position is approximately at the same frequency as the curve passing through zero in Fig. 2.1 a. Note that the frequency correspondence is, although close, but not quite accurate, as will be discussed later. With increasing attenuation, the depth of the minimum falls, so that the entire curve as a whole tends to zero.

The dependence shown in Fig. 2.1 reflects the behaviour of component χ only. The χ_a component behaves in a similar way. The nature of her behavior can be judged from the similarity of formulas (2.60) and (2.61). In more detail, the character of the χ and χ_a components can be found, for example, according to [6–8].

2.2.6. Two definitions of the resonant frequency

Let us now consider the question of the correspondence of the resonant values of frequencies according to (2.63) and (2.64) in somewhat more detail. From the structure of these expressions, it can be seen that for all values of ω_H and ω in the case $\alpha \neq 0$ the denominator is always non-zero, so there is no divergence. With a small attenuation it can be approximately put that the resonance corresponds to the zero value of the first term of the denominator, that is, the condition:

36

a

b

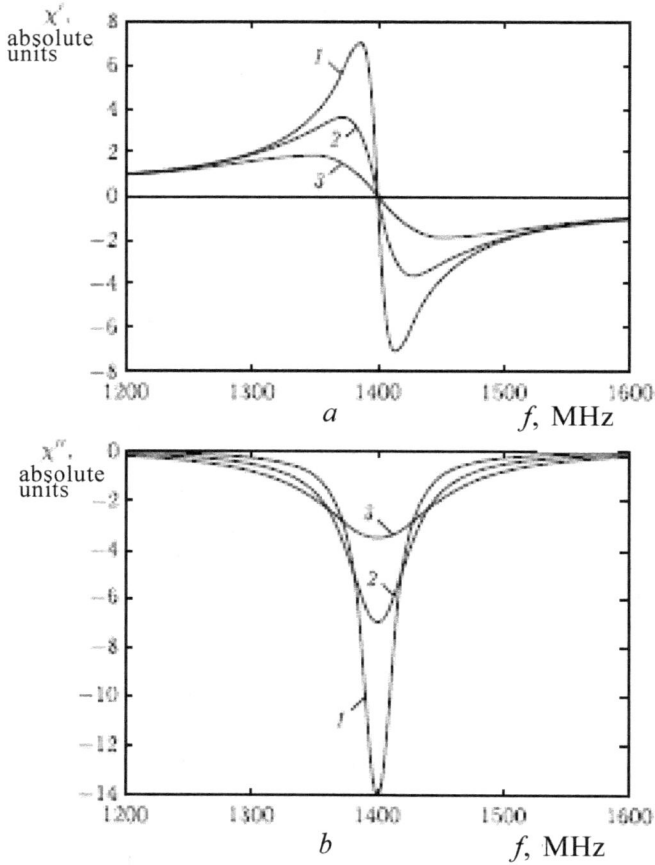

Fig. 2.1. Dependences of the real (*a*) and imaginary (*b*) parts of the susceptibility tensor component χ on the frequency for different values of the parameter damping α: 1 – 0.01; 2 – 0.02; 3 – 0.04. Parameters: $4\pi M_0$ = 1750 G; H = 500 Oe

$$\omega_H^2 - (1+\alpha^2)\omega^2 = 0, \tag{2.67}$$

so for a given field H_0, the resonant frequency is:

$$\omega_{\text{res}} = \frac{1}{\sqrt{1+\alpha^2}}\gamma H_0, \tag{2.68}$$

and at a given frequency ω, the resonant field is:

$$H_{\text{res}} = \sqrt{1+\alpha^2}\,\frac{\omega}{\gamma}. \tag{2.69}$$

From (2.68) and (2.69) it can be seen that with increasing attenuation α, the resonance frequency decreases, and the resonance field becomes stronger. Such a conclusion, especially regarding frequency (2.68), appears naturally, if we take into account that with increasing damping, the resistance to movement of the system increases (the friction force increases), so that the free oscillations of the system, which determine the resonance frequency, become slower.

Not quite so, however, is the case in a similar consideration of the frequency dependence of the real and imaginary parts.of the components of the susceptibility tensor. So, from consideration of Fig. 2.1 it can be seen that at resonance the real part of the susceptibility component χ' passes through zero and changes the sign, and the imaginary part χ'' passes through the minimum, the sign without changing. As a definition of resonance two options can be considered:

1) the maximum amplitude of oscillations, that is, the midpoint between the extremes of the curve χ' or the point of passage of this curve through zero;

2) the maximum absorption of vibration energy, that is, the minimum of the curve χ''.

In Fig. 2.1 the difference between these options is not visible due to the relative roughness of the drawing (with the accuracy up to the width of the lines, both points for all the attenuation parameters taken there are almost the same), so we turn to the numerical data. A detailed calculation performed with the technical parameters, which is shown in Fig. 2.1, with an accuracy of seven significant digits, shows that the passage of the curve χ' through zero, with the attenuation parameter $\alpha = 0.01$, occurs at a frequency of 1400.070 MHz, with the attenuation parameter $\alpha = 0.02$ at a frequency of 1400.280 MHz, and with the attenuation parameter $\alpha = 0.04$, at a frequency of 1401.122 MHz. That is, as the attenuation increases, the frequency of passing the curve χ' through zero increases.

At the same time, the minimum of the curve χ'' at $\alpha = 0.01$ takes place at a frequency of 1399.929 MHz, at $\alpha = 0.02$ at a frequency of 1399.720 MHz and at $\alpha = 0.04$ at a frequency of 1398.825 MHz. That is, as increasing the damping the frequency of the minimum of the curve χ'' goes down.

Thus, if the outcome is from the qualitative physical meaning of the problem, that is, from the fact that as the attenuation increases, the system becomes more difficult to move, and the natural oscillation frequency decreases, then we can conclude that more accurate is the determination of the resonance frequency according to the second of the noted distributions, that is, the minimum χ'', which coincides with the conclusion made above based on consideration (2.67) and (2.68).

Comment. The main role of χ'' and χ_a'' in the absorption of energy, generally speaking, from the physical side is intuitive. In fact, as can be seen from (2.64) and (2.66), both of these components are proportional to the attenuation parameter α, that is, in the absence of attenuation, they are zero. At the same time, χ' and χ_a'', that is, (2.63) and (2.65), in the absence of damping from zero, are different. Since the damping of oscillations occurs due to the absorption of their energy by the medium, it can be assumed that the components χ'' and χ_a'' are responsble A more detailed consideration of this issue, based on the concept of the Poynting vector, can be found by the inquisitive reader, for example, in monographs [7, pp. 257–259, 8, pp. 119–122].

2.2.7. Two types of susceptibility tensor recording

The most traditional form of recording the dynamic susceptibility tensor for a gyrotropic medium is (2.38) or (2.49):

$$\ddot{X} = \begin{pmatrix} X & iX_a & 0 \\ -iX_a & X & 0 \\ 0 & 0 & 0 \end{pmatrix} \tag{2.70}$$

in which some components are obviously set to zero. However, as will be shown later (Chapters 6–8), under the conditions of a transitional transition, the susceptibility tensor may have non-zero all nine components, and the symmetry conditions between them can be quite complex. Therefore, in some cases it makes sense to record the susceptibility tensor in a more general form with all nine components:

$$\ddot{X} = \begin{pmatrix} X_{xx} & X_{xy} & X_{xz} \\ X_{yz} & X_{yy} & X_{yz} \\ X_{zx} & X_{zy} & X_{zz} \end{pmatrix} \tag{2.71}$$

The calculation of the explicit form of these component will be left to the next sections, and here we will show the connection between both notation systems.

We assume that all components have both real and imaginary parts and consider as an example the first two components of the first row of each tensor. Thus, we will assume that:

$$\chi = \chi' + i\chi'' \; ; \tag{2.72}$$

$$\chi_a = \chi_a' + i\chi_a'' \; ; \tag{2.73}$$

and:

$$\chi_{xx} = \chi_{xx}' + i\chi_{xx}'' \tag{2.74}$$

$$\chi_{xy} = \chi_{xy}' + i\chi_{xy}'' \tag{2.75}$$

From the condition of equality of components by their position in the tensor matrix, that is, $\chi \to \chi_{xx}$ and $i\chi_a \to \chi_{xy}$, we get:

$$\chi_{xx}' = \chi' \tag{2.76}$$

$$\chi_{xx}'' = \chi'' \tag{2.77}$$

and:

$$\chi_{xy}' = - \chi_a'' \tag{2.78}$$

$$\chi_{xy}'' = \chi_a' \tag{2.79}$$

The relationship between χ and χ_{xx} components is evident from relations (2.76) and (2.77), and the relationship between the χ_a and χ_{xy} components is illustrated in Fig. 2.2.

From this figure one can see that the real part χ_a' is similar to imaginary part χ_{xy}'', and the imaginary part χ_a'' is similar to real part χ_{xy}' with the opposite sign. This means that the imaginary parts of χ_a and χ_{xy} change roles, and the imaginary part still changes its sign. Such a change is due to the fact that when writing a tensor in the form (2.70), the components symmetric with respect to the main diagonal contain the imaginary unit i as a multiplier, then as with the recording of the tensor in the form (2.71), this factor is absent for the elements located at the same location.

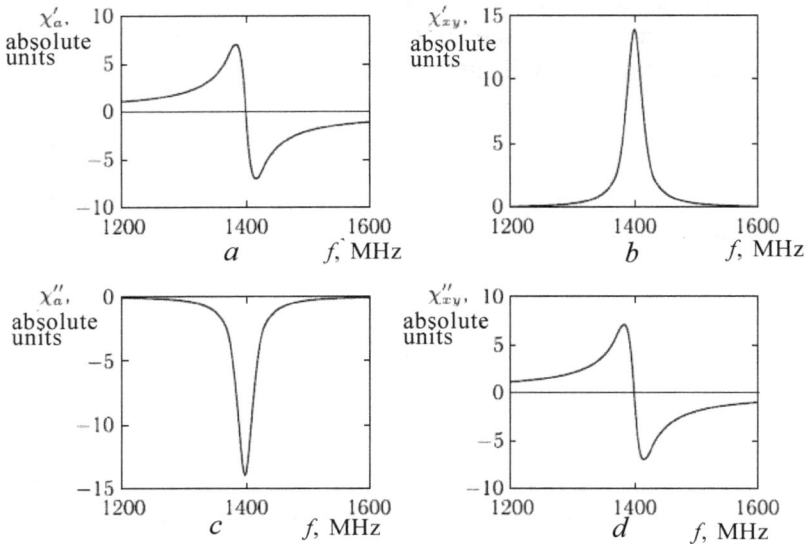

Fig. 2.2. Illustration of the relationship between real and imaginary parts of the components χ_a and χ_{xy}. Scale on vertical axes - conditional. Parameters: $4\pi M_0 = 1750$ G; $H = 500$ Oe, $\alpha = 0.01$.

2.2.8. A note on the dependence of tensor components on the field

The dependences given here consider the variation of the susceptibility tensor on frequency. However, for example, from formulas (2.60)–(2.61), one can see that such a change can be achieved by leaving the frequency constant, but changing the field. In these rmulaht aka the possibility is traced in the most obvious way, since attenuation does not prevent its visual detection, although in all other expressions (2.54), (2.55), (2.58), (2.59), and also (2.63)–(2.66) also holds to the same extent. The dependences of the components of the tensor χ and χa on the field have approximately the same form as the dependences on frequency, up to the opposite sign. That is, for example, the dependence χ? from H at a fixed frequency as the field increases, it first falls, then passes a section of a sharp rise, then falls again, so that has the same appearance as the curves in Fig. 2.1, but with the opposite sign. Dependence χ'' also has a single extremum, but being in the positive range of values, instead of the minimum – the maximum. The same goes for the dependences of the real and imaginary parts of the component χ_a on the field, they have a form similar to those of frequency, but the opposite sign. Some differences relate to the size of the curves on the vertical, but their cut-away nature is fully preserved. The relationship between the frequency and field dependencies of the components of the susceptibility tensor can be found, for example, in [8, p. 21].

It should be noted that both types of dependence are fully equivalent, however, in practical use, one or another type of dependency, as a rule, is given a certain preference.

Thus, the majority of experiments on ferromagnetic resonance are carried out using spectrometers operating at a fixed frequency, when the field changes. Such spectrometers include, for example, the experiments used in the section described below to observe FMR in films of ferrites–garnets with an inclination of the anisotropy axis (section 7.6) of the EPA-2M spectrometer operating at a fixed frequency of 9500 MHz in the range of changes in the field from 0 to 5000 Oe. On the other hand, most experiments on the propagation of electromagnetic waves, especially magnetostatic [1, 2], are performed using spectrometers operating in the frequency sweep mode with a fixed field created by an external permanent or electromagnet. Such spectrometers are used in most studies on the propagation of magnetostatic waves in inhomogeneous fields and complex structures [1, 2], a measurer of complex coefficients of transmissions of type P4-23 or similar panoramic meter of type P2-52–P2-58 [1, pp. 125–132].

In this section, preference is given to the dependence of the components of susceptibility not on the field, but on frequency, because later the chapter 8 describes the composition medium consisting of anisotropic ferrite particles, the main purpose of which is to use it not so much purely resonant as waveguide properties for which descriptions of the frequency dependences are more convenient.

2.2.9. Complete equations of motion for magnetization

In the previous sections, equations of motion for the magnetization in the form of (2.25)–(2.27) are obtained, of which linearized equations (2.29)–(2.31) are derived. It is this linearized form that allowed us to find the dynamic susceptibility tensor (2.38) with its components (2.39), (2.40) or in other records (2.54), (2.55), (2.58), (2.59), including in the complex form (2.63)–(2.66). All these expressions are obtained in a linear approximation, that is, for small values of lithodes of recession magnetization. However, there are a number of problems where a linear approximation is insufficient, so it is required to consider the precession of magnetization under conditions of large amplitude. That is, in the system of equations of motion one cannot limit oneself to only two equations for transverse magnitudes, but one must also take into account the longitudinal one oriented along the constant field. We present a brief derivation of the full equations for this case.

Let us proceed from the equation of motion of the Landau–Lifshitz magnetization vector with the dissipative term in the Hilbert form (2.18):

$$\frac{\partial \mathbf{M}}{\partial t} = -\gamma \left[\mathbf{M} \times \mathbf{H}_e \right] + \frac{\alpha}{M_0} \left[\mathbf{M} \times \frac{\partial \mathbf{M}}{\partial t} \right]. \tag{2.80}$$

Here, $\gamma > 0$, while the free precession is right; $\mathbf{H}_e$ is an effective field equal to:

$$\mathbf{H}_e = -\frac{\partial U}{\partial \mathbf{M}}, \tag{2.81}$$

where U is the energy density.

We introduce the normalized magnetization in the form

$$\mathbf{m} = \frac{\mathbf{M}}{M_0}. \tag{2.82}$$

At the same time (2.80) takes the form

$$\frac{\partial \mathbf{m}}{\partial t} = -\gamma \left[\mathbf{m} \times \mathbf{H}_e \right] + \alpha \left[\mathbf{m} \times \frac{\partial \mathbf{m}}{\partial t} \right], \tag{2.83}$$

and:

$$\mathbf{H}_e = -\frac{1}{M_0} \frac{\partial U}{\partial \mathbf{m}}. \tag{2.84}$$

Here it is assumed that the full length of the magnetization vector is preserved, i.e.

$$M_x^2 + M_y^2 + M_z^2 = M_0^2, \tag{2.85}$$

$$m_x^2 + m_y^2 + m_z^2 = 1. \tag{2.86}$$

In the Cartesian coordinate system, the vector products in (2.83) have the form

$$\left[\mathbf{m} \times \mathbf{H}_e\right] = \mathbf{i}\left(m_y H_{ez} - m_z H_{ey}\right) + \mathbf{j}(m_x H_{ez}) + \\ \mathbf{k}\left(m_x H_{ey} - m_y H_{ex}\right); \tag{2.87}$$

$$\left[\mathbf{m} \times \frac{\partial \mathbf{m}}{\partial t}\right] = \mathbf{i}\left(m_y \frac{\partial m_z}{\partial t} - m_z \frac{\partial m_y}{\partial t}\right) + \mathbf{j}\left(m_z \frac{\partial m_x}{\partial t} - m_x \frac{\partial m_z}{\partial t}\right) + \\ \mathbf{k}\left(m_x \frac{\partial m_y}{\partial t} - m_y \frac{\partial m_x}{\partial t}\right). \tag{2.88}$$

Substituting (2.87) and (2.88) into (2.83), we obtain the equations of motion for $m_{x,\,y,\,z}$:

$$\frac{\partial m_x}{\partial t} = -\gamma\left(m_y H_{ez} - m_z H_{ey}\right) + \alpha\left(m_y \frac{\partial m_z}{\partial t} - m_z \frac{\partial m_y}{\partial t}\right); \tag{2.89}$$

$$\frac{\partial m_y}{\partial t} = -\gamma\left(m_z H_{ex} - m_x H_{ez}\right) + \alpha\left(m_z \frac{\partial m_x}{\partial t} - m_x \frac{\partial m_z}{\partial t}\right); \tag{2.90}$$

$$\frac{\partial m_z}{\partial t} = -\gamma\left(m_x H_{ey} - m_y H_{ex}\right) + \alpha\left(m_x \frac{\partial m_z}{\partial t} - m_y \frac{\partial m_x}{\partial t}\right). \tag{2.91}$$

It can be seen that these equations, up to a normalization on M_0, coincide with equations (2.20)–(2.22) or (2.25)–(2.27).

Comment. To avoid confusion, one should take into account that there the small letter $m_{x,y}$ denoted a small addition to the value of the vector magnetization, where the small size of the letter underlined the smallness of the additive itself, that is, it was assumed $m_{x,y} \ll M_0$, and also $m_z = 0$. Such assumption was necessary for the successful implementation of the subsequent linearization. Here, the small letter $m_{x,y,z}$ means the normalized component of magnetization along the corresponding coordinate, with the value of this component not related in any way, that is, any of the components m_x, m_y, and also m_z can have the same order of magnitude as M_0, it is only necessary that relation (2.86) is fulfilled.

Select from the equations (2.89)–(2.91) time derivatives:

44

$$\frac{\partial m_x}{\partial t} + \alpha m_z \frac{\partial m_y}{\partial t} - \alpha m_y \frac{\partial m_x}{\partial t} = -\gamma\left(m_y H_{ez} - m_z H_{ey}\right); \quad (2.92)$$

$$-\alpha m_x \frac{\partial m_x}{\partial t} + \frac{\partial m_y}{\partial t} + \alpha m_x \frac{\partial m_z}{\partial t} = -\gamma\left(m_z H_{ex} - m_x H_{ex}\right); \quad (2.93)$$

$$-\alpha m_x \frac{\partial m_x}{\partial t} + \frac{\partial m_y}{\partial t} + \alpha m_y \frac{\partial m_x}{\partial t} - \alpha m_x \frac{\partial m_y}{\partial t} + \frac{\partial m_z}{\partial t} = -\gamma\left(m_x H_{ey} - m_y H_{ex}\right).$$

$$(2.94)$$

We solve the resulting system of equations (2.92)–(2.94) with respect to time derivatives, for example, using the Cramer rule [261, 262]. Its extended matrix is:

$$\begin{vmatrix} 1 & \alpha m_z & -\alpha m_y \\ -\alpha m_z & 1 & \alpha m_x \\ \alpha m_y & -\alpha m_x & 1 \end{vmatrix} \begin{vmatrix} -\gamma\left(m_y H_{ez} - m_z H_{ey}\right) \\ -\gamma\left(m_z H_{ex} - m_x H_{ez}\right) \\ -\gamma\left(m_x H_{ey} - m_y H_{ex}\right) \end{vmatrix} \quad (2.95)$$

The determinant given (2.86) is equal to:

$$D_0 = \begin{vmatrix} 1 & \alpha m_z & -\alpha m_y \\ -\alpha m_z & 1 & \alpha m_x \\ \alpha m_y & -\alpha m_x & 1 \end{vmatrix} = 1 + \alpha^2. \quad (2.96)$$

The solution is

$$\frac{\partial m_x}{\partial t} = -\frac{\gamma}{1+\alpha^2}\left[\left(m_y + \alpha m_x m_z\right)H_{ez} - \left(m_z - \alpha m_y m_x\right)H_{ey} - \right.$$
$$\left. -\alpha\left(m_y^2 + m_z^2 H_{ex}\right)\right]; \quad (2.97)$$

$$\frac{\partial m_y}{\partial t} = -\frac{\gamma}{1+\alpha^2}\left[\left(m_z + \alpha m_y m_z\right)H_{ex} - \left(m_x - \alpha m_z m_y\right)H_{ez} - \right.$$
$$\left. -\alpha\left(m_z^2 + m_x^2\right)H_{ey}\right]; \quad (2.98)$$

$$\frac{\partial m_z}{\partial t} = -\frac{\gamma}{1+\alpha^2}\Big[\big(m_x+\alpha m_z m_y\big)H_{ey}-\big(m_y-\alpha m_x m_z\big)H_{ex}-$$
$$-\alpha\big(m_x^2+m_y^2\big)H_{ex}\Big];$$

(2.99)

This is the basic system of equations for the components of magnetization in the case of considering nonlinear oscillations [3,264–266]. Further it will be used in chapters 7–10.

Comment. From the structure of the equations (2.97)–(2.99) one can see that even in the absence of attenuation, including taking into account (2.32), (2.33), they can not be reduced to a form similar to (2.36) and (2.37), that is,

$$m_{x,y,z} = F_{x.y.z}\big(h_x,h_y,h_z\big),$$

(2.100)

where the right side depends only on the field components with constant coefficients. This difficulty is due to their linear form, where, even in the absence of attenuation, products like $m_i h_k$, and in the case of damping also $m_i m_k h_l$. Therefore, obtaining susceptibility (2.17) in the classical form, similar to (2.38), (2.49), (2.70), is hardly possible. Nevertheless, one can hope that the solution of equations of the type (2.97)–(2.99) is nevertheless feasible, first of all, in numerical form (which is undoubted), and in addition, in some specific ways, for example, with a proper choice of the coordinate system or a certain ratio of the quantities involved in these equations (as it turns out, for example, with linearization).

2.3. Smit–Sul method

The Smit–Sul method [7, 8, 31, 267] is a solution of the linearized Landau–Lifshitz equation, performed in a spherical coordinate system. In this case, the effective fields are not calculated separately, and the formula for the frequency of ferromagnetic resonance is expressed through the derivatives of the energy density in spherical coordinates.

The method is convenient for calculating the dependences of the resonant frequency for the value of the constant field in the case when the orientation of such the field changes at large angles, up to a complete rotation around some given axis, which is convenient to combine with the polar axis of the spherical coordinate system. We give a simplified derivation of the basic Smit–Sul formula, in general following [7].

2.3.1. Equations of motion of magnetization in the Cartesian coordinate system

Let us proceed from the equation of motion for the Landau –Lifshitz magnetization without decay [6–8]

$$\frac{\partial \mathbf{M}}{\partial t} = -\gamma [\mathbf{M} \times \mathbf{H}_e], \tag{2.101}$$

where γ is the gyromagnetic constant ($\gamma > 0$), $\mathbf{M}$ is the magnetization vector, in the Cartesian coordinate system having the form

$$\mathbf{M} = \mathbf{i} M_x + \mathbf{j} M_y + \mathbf{k} M_z, \tag{2.102}$$

$\mathbf{H}_e$ is an effective field, in accordance with (2.16), equal to

$$\mathbf{H}_e = -\frac{\partial t}{\partial \mathbf{M}} = -\mathbf{i}\frac{\partial U}{\partial M_x} - \mathbf{j}\frac{\partial U}{\partial M_y} - \mathbf{k}\frac{\partial U}{\partial M}, \tag{2.103}$$

where $\mathbf{i}, \mathbf{j}, \mathbf{k}$ are unit vectors (orts) of the Cartesian coordinate system $Oxyz$, U is the magnetic energy density, which is a function of the components of the magnetization vector.

We write the vector product by components (for simplicity, we omit the 'e' index from the field):

$$[\mathbf{M} \times \mathbf{H}] = \mathbf{i}(M_y H_z - M_z H_y) +$$
$$+ \mathbf{j}(M_z H_x - M_x H_z) + \mathbf{k}(M_x H_y - M_y H_z). \tag{2.104}$$

The equation of motion of the magnetization vector (2.101), written by component, takes the form:

$$\frac{\partial M_x}{\partial t} = -\gamma(M_y H_z - M_z H_y); \tag{2.105}$$

$$\frac{\partial M_y}{\partial t} = -\gamma(M_z H_x - M_x H_z); \tag{2.106}$$

$$\frac{\partial M_z}{\partial t} = -\gamma(M_x H_y - M_y H_x). \tag{2.107}$$

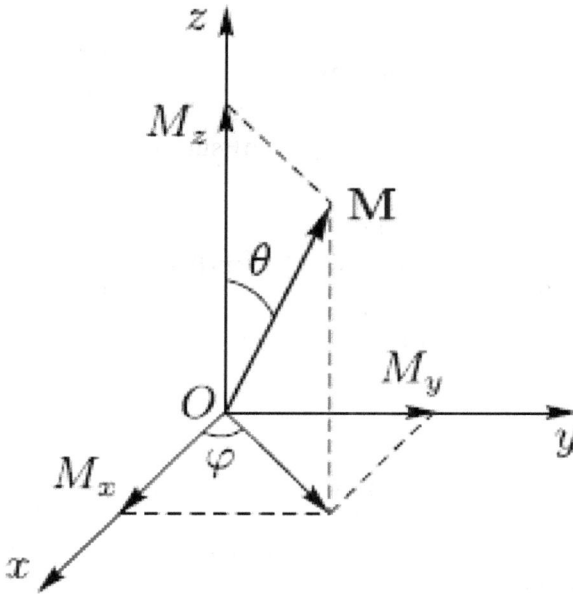

Fig. 2.3. Magnetization vector in the spherical coordinate system

2.3.2. Transition to a spherical coordinate system

For further consideration, we turn to the spherical coordinate system shown in Fig. 2.3.

The components of the magnetization vector in such a system are:

$$M_x = M \sin \theta \cos \theta \qquad (2.108)$$

$$M_y = M \sin \theta \cos \varphi \qquad (2.109)$$

$$M_z = M \cos \theta \qquad (2.110)$$

where M is the total length of the magnetization vector. The system of equations of magnetization motion (2.105)–(2.107) contains three variables M_x, M_y, M_z, and there are no restrictions on the length of the vector **M**. The spherical system contains two angular variables – the polar and azimuth angles θ and φ, and the third variable here is the length of the magnetization vector M, all of which may depend on time. The equations of motion (2.105)–(2.107) in the left parts contain time derivatives of the Cartesian magnetization components. We write these derivatives in a spherical system, taking into account

48

the time dependence of all the relative variables M, θ and φ:

$$\frac{\partial M_x}{\partial t} = \frac{\partial}{\partial t}(M\sin\theta\cos\varphi) = (\sin\theta\cos\varphi)\frac{\partial M}{\partial t} +$$
$$+(M\cos\theta\cos\varphi)\frac{\partial\theta}{\partial t} - (M\sin\theta\sin\varphi)\frac{\partial\varphi}{\partial t}; \qquad (2.111)$$

$$\frac{\partial M_y}{\partial t} = \frac{\partial}{\partial t}(M\sin\theta\sin\varphi) = (\sin\theta\sin\varphi)\frac{\partial M}{\partial t} +$$
$$+(M\cos\theta\sin\varphi)\frac{\partial\theta}{\partial t} - (M\sin\theta\cos\varphi)\frac{\partial\varphi}{\partial t}; \qquad (2.112)$$

$$\frac{\partial M_z}{\partial t} = \frac{\partial}{\partial t}(M\cos\theta) = (\cos\theta)\frac{\partial M}{\partial t} - (M\sin\theta)\frac{\partial\theta}{\partial t}. \qquad (2.113)$$

For the right parts of the equations of the system (2.105)–(2.107), we introduce the auxiliary notation:

$$P_x = -\gamma\left(M_y H_z - M_z H_y\right); \qquad (2.114)$$

$$P_y = -\gamma\left(M_z H_x - M_x H_z\right); \qquad (2.115)$$

$$P_x = -\gamma\left(M_x H_y - M_y H_x\right); \qquad (2.116)$$

Substituting (2.111)–(2.113) and (2.114)–(2.116) into (2.105)–(2.107), we get:

$$(\sin\theta\cos\varphi)\frac{\partial M}{\partial t} + (M\cos\theta\cos\varphi)\frac{\partial\theta}{\partial t} - (M\sin\theta\sin\varphi) = P_x; \quad (2.117)$$

$$(\sin\theta\cos\varphi)\frac{\partial M}{\partial t} + (M\cos\theta\sin\varphi)\frac{\partial\theta}{\partial t} + (M\sin\theta\cos\varphi)\frac{\partial\varphi}{\partial t} = P_y; \quad (2.118)$$

$$(\cos\theta)\frac{\partial M}{\partial t} - (M\sin\varphi)\frac{\partial\theta}{\partial t} = P_z. \qquad (2.119)$$

It is a system of three equations for three known $\partial M/\partial t$, $\partial\theta/\partial t$, $\partial\varphi/\partial t$. Solving it, we find:

$$\frac{\partial M}{\partial t} = \left(P_x \sin\theta\cos\varphi + P_y \sin\theta\sin\varphi + P_z \cos\theta \right);$$

(2.120)

$$\frac{\partial\theta}{\partial t} = \frac{1}{M} \left(P_x \cos\theta\cos\varphi + P_y \cos\theta\sin\varphi + P_z \sin\theta \right);$$

(2.121)

$$\frac{\partial\varphi}{\partial t} = \frac{1}{M\sin\theta} \left(-P_x \sin\varphi + P_y \cos\varphi \right).$$

(2.122)

2.3.3. Preservation of the length of the magnetization vector

Up to this point, no boundaries on the length of the magnetization vector M were imposed. That is, the resulting system is suitable for solving problems in which the length is not preserved. The number of such problems include, for example, taking into account the attenuation in the Bloch form [6–8] or simply the oscillations of the magnetization of sufficiently large amplitude, including at nonlinear ferromagnetic resonance [10]. The magnetization can also change as a result of some highly unsteady processes, for example, when the electronic system of a conducting magnet is excited by a strong pulse light from a femtosecond laser [208–228]. However, in most of the classical problems related to ferromagnetic resonance, the length of the magnetization vector remains constant, so for the time being we will not dwell on this exotic, but assume that the length of the vector **M** is preserved, and we will denote it as M_0. In this case, from equation (2.120), under the condition $\partial M/\partial t = 0$, we obtain

$$P_z = -P_x \frac{\sin\theta\cos\varphi}{\cos\theta} - P_y \frac{\sin\theta\sin\varphi}{\cos\theta}.$$

(2.123)

Substituting this expression into (2.121) and (2.122), we obtain a system of two equations for two remaining variables:

$$\frac{\partial\theta}{\partial t} = \frac{1}{M_0\cos\theta} \left(P_x \cos\varphi + P_y \sin\varphi \right);$$

(2.124)

$$\frac{\partial\varphi}{\partial t} = \frac{1}{M_0\sin\theta} \left(-P_x \sin\varphi + P_y \cos\varphi \right).$$

(2.125)

This is a system of equations characterizing the motion of the magnetization vector, provided that its length is preserved, written in a spherical coordinate system.

Here, they are not disclosed with respect to the spherical alternating parameters P_x and P_y. We now turn to their calculation in explicit form. According to (2.114) and (2.115), these parameters contain effective fields determined by the energy density derivatives of the magnetization components (2.103). In accordance with the general rule for finding the partial derivatives [268, 269], we replace the differentiation of the energy density with respect to the components of the magnetization by differentiation with respect to spherical variables:

$$H_x = -\frac{\partial U}{\partial M_x} = -\frac{\partial U}{\partial \theta}\frac{\partial \theta}{\partial M_x} - \frac{\partial U}{\partial \varphi}\frac{\partial \varphi}{\partial M_x};$$
(2.126)

$$H_y = -\frac{\partial U}{\partial M_y} = -\frac{\partial U}{\partial \theta}\frac{\partial \theta}{\partial M_y} - \frac{\partial U}{\partial \varphi}\frac{\partial \varphi}{\partial M_y};$$
(2.127)

$$H_z = -\frac{\partial U}{\partial M_z} = -\frac{\partial U}{\partial \theta}\frac{\partial \theta}{\partial M_z} - \frac{\partial U}{\partial \varphi}\frac{\partial \varphi}{\partial M_z};$$
(2.128)

From (2.108)–(2.110) we express the spherical variables through the components of magnetization:

$$\theta = \arccos\left(\frac{M_z}{M_0}\right);$$
(2.129)

$$\varphi = \operatorname{arc} tg\left(\frac{M_y}{M_x}\right);$$
(2.130)

For differentiation, we use the table formulas [268–270]:

$$\frac{d}{dx}(ar\cos x) = -\frac{1}{\sqrt{1-x^2}},$$
(2.131)

$$\frac{d}{dx}(\operatorname{arctg} x) = \frac{1}{1+x^2}.$$
(2.132)

Finding by means of (2.129)–(2.132) the partial derivatives $\partial\theta/\partial M_{x,y,z}$ and $\partial\varphi/M_{x,y,z}$, substituting them in (2.126)–(2.128) and performing the reduction of similar terms, taking into account (2.108)–(2.110), we get

$$H_x = \frac{\sin\varphi}{M_0 \sin\theta} \frac{\partial U}{\partial\varphi};$$

(2.133)

$$H_y = \frac{\cos\varphi}{M_0 \sin\theta} \frac{\partial U}{\partial\varphi};$$

(2.134)

$$H_z = \frac{1}{M_0 \sin\theta} \frac{\partial U}{\partial\theta};$$

(2.135)

Substituting the obtained field values into (2.114) and (2.115) and writing the components of magnetization through spherical variables (2.108)–(2.110), we get

$$P_x = -\gamma\left(\sin\varphi\frac{\partial U}{\partial\theta} + \frac{\cos\theta\cos\varphi}{\sin\theta}\frac{\partial U}{\partial\varphi}\right)$$

(2.136)

$$P_y = -\gamma\left(\sin\varphi\frac{\partial U}{\partial\theta} + \frac{\cos\theta\sin\varphi}{\sin\theta}\frac{\partial U}{\partial\varphi}\right).$$

(2.137)

Substituting these expressions into (2.124) and (2.125) and performing the reduction of similar lines, we get:

$$\frac{\partial\theta}{\partial t} = -\frac{\gamma}{M_0 \sin\theta}\frac{\partial U}{\partial\varphi},$$

(2.138)

$$\frac{\partial\varphi}{\partial t} = -\frac{\gamma}{M_0 \sin\theta}\frac{\partial U}{\partial\theta}.$$

(2.139)

These are the desired equations of motion of the magnetization vector in a spherical coordinate system, obtained under the condition of preserving its length. In these equations, there are some restrictions on the values angles θ and φ are absent, that is, it can be assumed that they can be used at any amplitude of magnetization oscillations, including at any level of nonlinearity.

Nevertheless, it should be noted that the equations obtained both contain in the denominators of fractions $\sin\theta$, that is, at $\theta \to 0$, we can expect some divergences. However, this will only happen if the derivatives $\partial U/\partial\varphi$ and $\partial U/\partial\theta$ are non-zero. In the real case, it is necessary to look at a specific task, that is, a specific form of the function $U(\theta, \varphi)$, which at $\theta \to 0$, can also tend to zero. So it will be, for example, in the case of uniaxial anisotropy, the axis of which coincides with the polar axis of the coordinate system. In this case, instead of divergence there will be an uncertainty of the type of division of zero by zero. In solving the problem, such uncertainty must be disclosed. In more detail, the issue of divergence requires a separate study, which was not carried out in this work. In practice, one can follow the recommendation given in the monograph [7, p. 78]: "If we want to use the Smit–Sul method without hesitation", then the polar axis should be selected so that $\theta_0 \neq 0$. This is always possible, 'because in the Smit–Sul method, there are no bounds on the choice of axis'.

2.3.4. Linearization of the equations of motion

So, let us return to the consideration of nonlinear equations (2.138) - (2.139) and perform their linearization. A necessary condition for linearization is the smallness of the amplitude of oscillations in the vicinity of the equilibrium point. That is, we first need to find the equilibrium position of the magnetization vector, and then consider small deviations of the magnetization vector from this position. The traditional way to find equilibrium is minimization of the energy density, which consists in equating to zero the first derivatives with respect to the coordinates of the energy density [7, 8, 134].

Thus, the first step in using the Smit–Sul method is to find the equilibrium position of the magnetization vector, that is, to solve the system of equations

$$\frac{\partial U}{\partial \theta} = 0; \qquad (2.140)$$

$$\frac{\partial U}{\partial \varphi} = 0; \qquad (2.141)$$

If these equations are solved, this means that the equilibrium values of the polar and azimuthal angles θ_s and φ_s are found. We further represent the derivatives of the energy density with respect to the angular coordinates through small deviations $\Delta\theta$ and $\Delta\varphi$ from their equilibrium positions as the first term of the expansion in a series of powers:

$$\frac{\partial u}{\partial \theta} = \left(\frac{\partial U}{\partial \theta} \right)_s + \frac{\partial}{\partial \theta} \left(\frac{\partial U}{\partial \theta} \right) \bigg|_S \Delta\theta + \frac{\partial}{\partial \varphi} \left(\frac{\partial U}{\partial \theta} \right) \bigg|_S \Delta\varphi =$$

$$= \left(\frac{\partial^2 U}{\partial \theta^2} \right) \bigg|_S \Delta\theta + \left(\frac{\partial^2 U}{\partial \theta \partial \varphi} \right) \bigg|_S \Delta\varphi; \quad (2.142)$$

$$\frac{\partial u}{\partial \varphi} = \left(\frac{\partial U}{\partial \varphi} \right)_s + \frac{\partial}{\partial \theta} \left(\frac{\partial U}{\partial \varphi} \right) \bigg|_S \Delta\theta + \frac{\partial}{\partial \varphi} \left(\frac{\partial U}{\partial \varphi} \right) \bigg|_S \Delta\varphi =$$

$$= \left(\frac{\partial^2 U}{\partial \theta \partial \varphi} \right) \bigg|_S \Delta\theta + \left(\frac{\partial^2 U}{\partial \varphi^2} \right) \bigg|_S \Delta\varphi, \quad (2.143)$$

where the equalities of the zero of the first derivatives at the equilibrium position (2.140) and (2.141) are taken into account, and the subscript 'S' means that the derivatives are taken at the point of the equilibrium position. Suppose now that the small oscillations of the magnetization in the vicinity of the equilibrium position are harmonic:

$$\Delta\theta = \theta_0 e^{i\omega t}; \quad (2.144)$$

$$\Delta\varphi = \varphi_0 e^{i\omega t}, \quad (2.145)$$

where θ_0 and φ_0 are the corresponding sufficiently small amplitudes. Substituting now (2.14)–(2.143) and (2.144)–(2.145) into equations (2.138)–(2.139), reducing the time factor and citing similar terms, we get:

$$\left[i\omega + \frac{\gamma}{M_0 \sin \theta_S} \left(\frac{\partial^2 U}{\partial \theta \partial \varphi} \right)_S \right] \theta_0 + \frac{\gamma}{M_0 \sin \theta_S} \left(\frac{\partial^2 U}{\partial \varphi^2} \right)_S \varphi_0 = 0; \quad (2.146)$$

$$-\frac{\gamma}{M_0 \sin \theta_S} \left(\frac{\partial^2 U}{\partial \theta^2} \right)_S \theta_0 + \left[i\omega - \frac{\gamma}{M_0 \sin \theta_S} \left(\frac{\partial^2 U}{\partial \theta \partial \varphi} \right)_S \right] \varphi_0 + \frac{\gamma}{M_0 \sin \theta_S} \left(\frac{\partial^2 U}{\partial \varphi \partial \varphi} \right)_S \varphi_0 = 0.$$

$$(2.147)$$

2.3.5. Solving equations of motion. Smit – Sul formula

Formulas (2.146) and (2.147) are a system of homogeneous equations with respect to amplitudes θ_0 and φ_0. The condition for a nonzero solution is that its determinant is zero, and by opening and extracting the square root from the resulting expression, we obtain the frequency of own magnetization oscillations:

$$\omega = \frac{\gamma}{M_0 \sin\theta_S} \sqrt{\left(\frac{\partial^2 U}{\partial\theta^2}\right)_S \left(\frac{\partial U}{\partial\varphi^2}\right)_S - \left(\frac{\partial^2 U}{\partial\theta\partial\varphi}\right)_S^2}, \qquad (2.148)$$

where the index S means that the derivatives and the value of the angle θ are taken in the equilibrium position. This is the famous Smit–Sul formula, which makes it possible to find the frequency of ferromagnetic resonance depending on the orientation of the magnetic field. As noted above, the orientation of the polar axis of the spherical coordinate system should be chosen so that the angle θ_S differs as much as possible from zero. If the orientation of the field changes in some plane, then the optimal choice is the polar axis perpendicular to this plane, that is, so that the angle θ_S is equal to 90°. We also note that often for convenience of writing, the index S is omitted, and derivatives are written using lower indexes. This was done, for example, in [7,8, 31, 267]. In this case (2.148) takes the form:

$$\omega = \frac{\gamma}{M_0 \sin\theta_0} \sqrt{U_{\theta\theta} U_{\varphi\varphi} - U_{\theta\varphi}^2}. \qquad (2.149)$$

In this case, the index '0' with θ plays the same role as S in (2.148), that is, corresponds to the equilibrium position.

2.4. Methods for solving high degree equations

When calculating the characteristics of ferromagnetic resonance in an anisotropic medium, including under the conditions of orientational transitions, it often becomes necessary to find the equilibrium position of the magnetization vector. Such a situation, as a rule, is determined by the action of several factors: an anisotropy field, a demagnetization field, and an external field. In a magnetoelastic medium in the formation of the equilibrium state of magnetization internal, external, or spontaneous deformations can also be involved. The determination of the equilibrium position of the magnetization is necessary both for determining the characteristics of the orientational transition itself, and to calculate the conditions of ferromagnetic resonance, which is the phenomenon of precession of the magnetization around this equilibrium position.

The procedure for determining the equilibrium position of the magnetization is to minimize the density of the total energy of the magnet, which in most cases is the sum of the energy densities of anisotropy, the external field and the demagnetization field. The densities of magnetoelastic and elastic energies are included sometimes also included here. This minimization involves equating to zero the first derivative of the total energy density with respect to magnetization, which usually leads to the appearance of the equations of high temperatures, first of all the third and fourth, which require a specific solution. In the overwhelming majority of cases, such a solution is carried out by numerical methods: by searching for zero or by the method of determination discussed in Chapter 5 of this monograph. However, both of these methods, consisting in a large number (tens, hundreds and more) of uniform operations, require a certain machine time, significantly increasing in the case when the equilibrium position has to be find many times.

Such problems, for example, include the features of the orientational transition in films of mixed ferrite garnet (Chapter 5), the orientational dependences of the frequency of ferromagnetic resonance in films with uniaxial and cubic anisotropy (Chapter 6), in films with an inclination of the axis of easy magnetization (Chapter 7), the frequency properties of the composition medium consisting of individual particles with different orientations of the anisotropy axes (Chapter 8) and many others that go beyond the scope of this monograph.

In a number of similar cases, repeated use of numerical methods leads to an irrational increase in computer time. due to the significant number of steps required to obtain a given accuracy. However, to solve the equations of the third and fourth degrees, except numerical, there are certain analytical methods that allow one to immediately obtain the required solution in a single pass without resorting to repeated repetitions of the same approximate algorithm. It can be assumed that the use of analytical methods for solving such a system would significantly reduce machine time, that is, make it easier to solve very cumbersome problems.

Unfortunately, the use of analytical methods for solving equations of the third and fourth degrees in the practice of calculations of ferromagnetic resonance known to the authors of this monograph and orientational transitions are extremely rare. On the other hand, even in specific mathematical literature, it is not always possible to find a sufficiently detailed description of such methods. Therefore, the authors found it very useful to include in this monograph this section devoted specifically to analytical methods for solving similar equations. Consideration is limited to only two methods: for equations of the third degree – the Cardano method, for equations of the fourth – the Ferrari method. The exposition will mainly follow the methodology described in the monograph [262], as well as the reference data given in [261].

2.4.1. Third degree equations

Consider a complete third degree equation:

$$a_0x^3 + a_1x^2 + a_2x + a^3 = 0. \tag{2.150}$$

According to common practice [261, p. 44, 262, p. 198], we assume that all coefficients a_0, a_1, a_2 and a_3 are real (the method of solving equation (2.150) with complex coefficients is not know by the authors of this monograph is not known).

In the equation written in the form of (2.150) there are all degrees of the unknown x. Let us show that in any case (for $a_0 \neq 0$) the second degree can be excluded from this equation. To do this, we replace the variable, for which we represent x as

$$x = y + \varepsilon \tag{2.151}$$

where y is a new variable, and ε is a constant value that needs to be defined. In this case, we obtain

$$x_2 = y^2 + 2\varepsilon y = \varepsilon^2, \tag{2.152}$$
$$x^3 = y^3 + 3y^2\varepsilon + 3y\varepsilon^2 + \varepsilon^3 \tag{2.153}$$

Substituting (2.151)–(2.153) into (2.150) and dividing by a_0, we get

$$y^3 + \frac{3a_0\varepsilon + a_1}{a_0}y^2 + \frac{3a_0\varepsilon + 2a_0\varepsilon + a_2}{a_0}y + \frac{a_0\varepsilon + a_1\varepsilon^2 + a_2\varepsilon + a_3}{a_0} = 0. \tag{2.154}$$

We demand equality

$$\frac{3a_0\varepsilon + a_1}{a_0} = 0, \tag{2.155}$$

where do we get:

$$\varepsilon = -\frac{a_1}{3a_0}.$$ (2.156)

That is, (2.151) takes the form

$$x = y - \frac{a_1}{3a_0}.$$ (2.157)

Substituting (2.157) into (2.150), after reducing such terms, allows equation (2.150) to be reduced to the form:

$$y^3 + ay + b = 0,$$ (2.158)

where the notation is entered:

$$a = \frac{a_2}{a_0} - \frac{a_1^2}{3a_0^2};$$ (2.159)

$$b = \frac{2a_1^3}{27a_0^3} - \frac{a_1 a_2}{2a_0^2} + \frac{a_3}{a_0}.$$ (2.160)

Note that since all the components of expressions a and b are real, then these expressions themselves are also valid. Thus, the resulting equation (2.158), being cubic, does not contain a second degree unknown. Next, we will deal with the solution of this particular equation.

Comment. Besides the direct use of formulas (2.159) and (2.160), direct substitution (2.157) into (2.150) is possible, which is often technically simpler. However, if the second term in (2.157) is fractional, then when trying to bring such the terms in the equation (2.150) there will appear a common factor that can be put out of the bracket. However, a reduction by this factor may result in a coefficient other than unity at y^3 in equation (2.158), which is unacceptable, since it leads to a distortion of the values a and b. Therefore, in the event of the appearance of such a coefficient, it is necessary to divide all the terms of the resulting equation into it so that the coefficient at y^3 is necessarily equal to one.

We now represent one unknown y as a sum of two unknowns u and v, that is, we put

$$y = u + v.$$ (2.161)

Substituting (2.161) into (2.158), we get

$$u^3 + v_3 + (u + v)(3uv + a) + b = 0.$$ (2.162)

Since instead of one unknown y, two are introduced: u and v, then between them, in addition to (2.161), one can introduce some connection. That is, there will be a system of two equations, in which u and v will be unknown, and y will act as a parameter. Pick these unknown so that the relation is satisfied

$$3uv + a = 0, \tag{2.163}$$

$$uv = -\frac{a}{3}. \tag{2.164}$$

In this case, equation (2.162) takes the form

$$u^3 + v^3 = -b \tag{2.165}$$

At the same time, from (2.164) we get

$$u^3v^3 = -\frac{a^3}{27}, \tag{2.166}$$

Thus, we have the sum of the values of u^3 and v^3 (2.165), as well as their production (2.166). That is, these unknowns can be considered as the roots of a quadratic equation

$$z^2 + bz - \frac{a^3}{27} = 0. \tag{2.167}$$

whose solution has the form

$$z = -\frac{b}{2} \pm \sqrt{\frac{b^2}{4} + \frac{a^3}{27}}, \tag{2.168}$$

so that

$$u^3 = -\frac{b}{2} + \sqrt{\frac{b^2}{4} + \frac{a^3}{27}}; \tag{2.169}$$

$$v^3 = -\frac{b}{2} - \sqrt{\frac{b^2}{4} + \frac{a^3}{27}}, \tag{2.170}$$

from where

$$u^3 = \sqrt[3]{-\frac{b}{2} + \sqrt{\frac{b^2}{4} + \frac{a^3}{27}}} ;$$

(2.171)

$$v = \sqrt[3]{-\frac{b}{2} - \sqrt{\frac{b^2}{4} + \frac{a^3}{27}}} .$$

(2.172)

Substituting the obtained expressions into (2.161), we obtain the solution of equation (2.158) in the form of the well-known Cardano formula:

$$y = \sqrt[3]{-\frac{b}{2} + \sqrt{\frac{b^2}{4} + \frac{a^3}{27}}} + \sqrt[3]{-\frac{b}{2} - \sqrt{\frac{b^2}{4} + \frac{a^3}{27}}} .$$

(2.173)

This formula is fundamental when solving equations of the third degree in radicals. Its validity can be verified by direct substitution into equation (2.158).

In the formula (2.173) under the sign of the radical of the third degree is the extraction of the square root of the expression:

$$D = \frac{b^2}{4} + \frac{a^3}{27},$$

(2.174)

which is called the 'discriminant' of equation (2.158).

Comment. The monograph [262, p. 197] considers the expression (2.174) with a minus sign, and it is called the 'discriminant'. However, the same expression with a plus sign is considered in [261, pp. 45–50]. However, in both cases, an essentially identical analysis of the sign of this expression is carried out. The authors of this monograph find it difficult to give preference to one or another interpretation, but then, proceeding from a certain convenience and taking into account the identity of the final sections of the ultats, it is the expression (2.174) that is used as a 'discriminant'.

So, depending on the sign of the expression (2.174), there may be various solutions, that is, it may be necessary to extract the cube root from both the real and the complex number.

Let us consider the possible solutions, determined by the sign of the discriminant of equation (2.158), separately.

Option number 1. Let the values of a and b be such that $D > 0$. The result of the extraction of the square root will be valid. If, when extracting cubic boxes in formula (2.173), we confine ourselves to their real values, then the resulting solution will also be valid. At the same time, if the conditions of the problem require obtaining a valid solution, then this can be limited to.

If it is also required to obtain complex solutions, then it should be noted that the cube root of any number has three values – one real and two complex conjugates, that is, with the actual value of the cube root $\xi = \sqrt[3]{\alpha}$ there are two more complex conjugates $\omega\xi$ and $\omega^2\xi$ where ω and $\omega^2\xi$ where ω and ω^2 are imaginary cubic roots of one (the notation is borrowed from [262, p. 196]):

$$\omega = \frac{-1 + i\sqrt{3}}{2};$$

(2.175)

$$\omega^2 = \frac{-1 - i\sqrt{3}}{2}.$$

(2.176)

The formulas (2.175) and (2.176) are obtained under the assumption that the equation

$$x^3 = 1 = 0$$

(2.177)

has one real root $x = 1$, so after dividing (2.177) by $(x - 1)$ we get a quadratic equation

$$x^2 + x + 1 = 0,$$

(2.178)

whose solutions are the formulas (2.175) and (2.176). By direct verification, we can verify that $\omega^3 = 1$ and $(\omega^2)^3 = 1$, that is, equation (2.177) is satisfied.

Thus, subject to the assumption of imaginary, each of the terms of formula (2.173) has three meanings, the possible combinations of which have nine solutions. However, when choosing the proper solutions, we must take into account the condition (2.164), which admits the presence of only three possible ways of combining the values of u and v, that is, the cubic roots in the formula (2.173). This can be confirmeed by direct substituting nine possible combinations into condition (2.164). Thus, for y, there are only three possibilities:

$$y_1 = u + v; \tag{2.179}$$

$$y2 = \omega u + \omega^2 v; \tag{2.180}$$

$$y3 = \omega^2 u + \omega v. \tag{2.181}$$

By direct verification, we can verify that all three expressions (2.179)–(2.181) are the roots of equation (2.158).

Accordingly, there are three solutions to the original equation (2.150):

$$x_1 = -\frac{a_1}{3a_0} + u + v; \tag{2.182}$$

$$x_2 = -\frac{a_1}{3a_0} + \omega u + \omega^2 v; \tag{2.183}$$

$$x_3 = -\frac{a_1}{3a_0} + \omega^2 u + \omega v; \tag{2.184}$$

Substituting the formulas (2.171) and (2.172) into these expressions, taking into account (2.159) and (2.160), we can obtain a solution written through the parameters of the original equation (2.150). Here, the authors of this monograph do not give these expressions because of their cumbersomeness combined with obvious triviality.

Option number 2. Now suppose that the values of a and b are such that $D = 0$. In this case, the result of extracting the square root will be equal to zero, so from formula (2.173) we get

$$y = 2\sqrt[3]{-\frac{b}{2}}. \tag{2.185}$$

We further take into account that when $D = 0$, the equality

$$\frac{b^2}{4} = -\frac{a^3}{27}, \tag{2.186}$$

from which we get

$$\left(-\frac{b}{2}\right)^2 = -\frac{a^3}{27}. \tag{2.187}$$

We next produce the following chain of consecutive conversions performed according to (2.187):

$$\sqrt[3]{-\frac{b}{2}} = \sqrt[3]{\frac{\left(-\dfrac{b}{2}\right)^3}{\left(-\dfrac{b}{2}\right)^2}} = -\frac{b}{2}\frac{1}{\sqrt[3]{\left(-\dfrac{b}{2}\right)^2}} = -\frac{b}{2}\frac{1}{\sqrt[3]{-\dfrac{a^3}{25}}} = \frac{b}{2}\frac{1}{\dfrac{a}{3}} = \frac{3b}{2a}. \qquad (2.188)$$

Substituting the resulting expression into (2.185), we get:

$$y_1 = \frac{3b}{a}, \qquad (2.189)$$

The other two roots are obtained taking into account the complex roots from unity, as in (2.180) and (2.181):

$$y_2 = \frac{3b}{a}\left(\omega + \omega^2\right) = -\frac{3b}{a}, \qquad (2.190)$$

$$y_3 = \frac{3b}{a}\left(\omega^2 + \omega\right) = -\frac{3b}{a}, \qquad (2.191)$$

In this case, in accordance with (2.157), we obtain:

$$x_1 = -\frac{a_1}{3a_0} + \frac{3b}{a}; \qquad (2.192)$$

$$x_2 = -\frac{a_1}{3a_0} - \frac{3b}{a}; \qquad (2.193)$$

$$x_3 = -\frac{a_1}{3a_0} - \frac{3b}{a}. \qquad (2.194)$$

Substituting these expressions (2.159) and (2.160) also allows us to obtain a solution written through the parameters of the original equation (2.150).

Option number 3. Now suppose that the values of *a* and *b* are such that $D < 0$. In this case, the result of extracting the square root will be imaginary, that is:

$$\sqrt{\frac{b^2}{4} + \frac{a^3}{27}} = i\sqrt{-\frac{b^2}{4} - \frac{a^3}{27}}. \qquad (2.195)$$

In this case, the use of formula (2.173) requires the calculation of the cubic root of a complex number, which, as shown in [262, p. 199], is equivalent to solving a cubic equation of the same form as (2.158). That is, it turns out to be a 'vicious circle', from which direct application of the algebraic formula (2.173) does not give amything [262, p. 198].

However, in this case, one can use the convenient reception (in the days of Cardano that did not yet exist), allowing to replace the exponentiation of the complex number in the trigonometric form multiplying its argument by a number that is an indication of the degree So, the record of the complex number z in the trigonometric form (Euler formula) has the form [261, p. 32]:

$$z = x + iy = r (\cos \varphi + i\sin \varphi), \tag{2.196}$$

where

$$r = \sqrt{x^2 + y^2}; \tag{2.197}$$

$$\varphi = \operatorname{arctg}\left(\frac{y}{x}\right). \tag{2.198}$$

Raising the power of such a number is given by the Moivre formula [261, p. 33]

$$z^n = (x+iy)^n = r^n\left[\cos(n\varphi) + i\sin(n\varphi)\right], \tag{2.199}$$

where n is an integer.

Now let's return to the solution of equation (2.158) in the form (2.173). In this case, from (2.171) and (2.172) in accordance with (2.195) we get

$$u = \sqrt[3]{-\frac{b}{2} + i\sqrt{\frac{b^2}{4} - \frac{a^3}{27}}}; \tag{2.200}$$

$$v = \sqrt[3]{-\frac{b}{2} - i\sqrt{\frac{b^2}{4} - \frac{a^3}{27}}}. \tag{2.201}$$

We introduce the notation:

$$\alpha = -\frac{b}{2}; \tag{2.202}$$

$$\beta = \sqrt{-\frac{b^2}{4} - \frac{a^3}{27}},$$

(2.203)

With these designations (2.200) and (2.201) take the form

$$u = \sqrt[3]{\alpha + i\beta};$$

(2.204)

$$v = \sqrt[3]{\alpha - i\beta};$$

(2.205)

By virtue of the equality of the real parts of these expressions, and also taking into account relation (2.164), the right side of which is real, it follows that u and v are complex conjugate. Erecting (2.204) into a cube and presenting the resulting complex the number in accordance with the Euler formula (2.196), we get

$$u^3 = \alpha + i\beta = r(\cos\gamma + i\sin\gamma),$$

(2.206)

where

$$r = \sqrt{\alpha^2 + \beta^2};$$

(2.207)

$$\gamma = \text{arctg}\left(\frac{\beta}{\alpha}\right),$$

(2.208)

where the values of r and γ are known, since they can obviously be determined on the basis of relations (2.202) and (2.203). We also set

$$u = \rho(\cos\varphi + i\sin\varphi),$$

(2.209)

where ρ and ϕ are unknown quantities to be determined. Raising (2.207) in a cube in accordance with the formula of Moivre (2.199), we obtain

$$u^3 = \rho^3(\cos 3\varphi + i\sin 3\varphi)$$

(2.210)

Putting the modules equal to the complex numbers of villages (2.206) and (2.210) also equal, we get

$$\rho^3 = \sqrt{\alpha^2 + \beta^2}.$$

(2.211)

Substituting α and β in accordance with (2.202) and (2.203), we obtain

$$\rho^3 = \sqrt{-\frac{a^3}{27}}.$$

(2.212)

Taking the cubic root of both sides of this expression, we get

$$\rho = \sqrt{-\frac{a}{3}}.$$

(2.213)

that is, the quantity ρ appearing in (2.209) is already defined. We now turn to the definition of φ. Comparing the real parts (2.206) and (2.210), we get

$$\alpha = \rho^3 \cos 3\varphi,$$

(2.214)

where do we get

$$\cos 3\varphi = \frac{\alpha}{\rho^3}.$$

(2.215)

Substituting into this formula (2.202) and (2.212), we get

$$\cos 3\varphi = \frac{-\dfrac{b}{2}}{\sqrt{-\dfrac{a^3}{27}}},$$

(2.216)

i.e.

$$\varphi = \frac{1}{3}\arccos\left(\frac{-\dfrac{b}{2}}{\sqrt{-\dfrac{a^3}{27}}}\right),$$

(2.217)

Now both parameters ρ and φ included in the formula (2.209) are defined, so that it can be written as

$$u = \sqrt{-\frac{a}{3}}\left(\cos \varphi + i\sin \varphi\right),$$

(2.218)

where the value of φ, defined by the formula (2.217), is not disclosed here, so as not to create excessive bulkiness.

Considering the complex conjugacy of u and v noted at consideration of the formulas (2.204) and (2.205), we can write the value v in the form

$$v = \sqrt{-\frac{a}{3}}\left(\cos\varphi = i\sin\varphi\right). \qquad (2.219)$$

Thus, taking into account (2.179), that is, folding (2.218) and (2.219), it is possible to obtain the first solution of equation (2.158) in the form

$$y_1 = 2\sqrt{-\frac{a}{3}}\cos\varphi, \qquad (2.220)$$

where φ is defined by the formula (2.217).

Recording two other solutions in the form of (2.180) and (2.181), using (2.175) and (2.176), we obtain:

$$y_2 = 2\sqrt{-\frac{a}{3}}\cos\left(\varphi+\frac{\pi}{3}\right); \qquad (2.221)$$

$$y_3 = 2\sqrt{-\frac{a}{3}}\cos\left(\varphi-\frac{\pi}{3}\right), \qquad (2.222)$$

where φ is still defined by the formula (2.217).

In this case, in accordance with (2.157), we obtain:

$$x_2 = -\frac{a_1}{3a_0} + 2\sqrt{-\frac{a}{3}}\cos\left(\varphi+\frac{\pi}{3}\right); \qquad (2.223)$$

$$x_2 = -\frac{a_1}{3a_0} + 2\sqrt{-\frac{a}{3}}\cos\left(\varphi+\frac{\pi}{3}\right); \qquad (2.224)$$

$$x_3 = -\frac{a_1}{3a_0} + 2\sqrt{-\frac{a}{3}}\cos\left(\varphi-\frac{\pi}{3}\right); \qquad (2.225)$$

Substituting into these expressions, as well as into their expression (2.217), the formulas (2.171) and (2.172) with regard to (2.159) and (2.160), also allows one to obtain a solution written through the parameters of the original equation (2.150). We note that this way of solving the equation (2.158) in the case of the negativeness of its discriminant (2.174) is usually called the 'trigonometric solution'

[261,262]. However, from (2.200)–(2.203) one can see that it is not independent, but relies on the same basic Cardano formula (2.173), 'refined' with the help of the Euler formulas (2.196) and Moivre that did not exist during Cardano's time (2.199).

2.4.2. Fourth degree equations

We now turn to an analytical solution of a fourth degree equation. There are several ways to solve such equations [261, 262]. Historically, the first and ideologically closest to the Cardano solution is the third degree equation, for the fourth degree equation, apparently, it is the Ferrari method (student of Cardano), to consider which we proceed further. Consider a complete fourth degree equation

$$a_0x^4 + a_1x^3 + a_2x^2 + a_3x + a_4 = 0. \qquad (2.226)$$

Similar to the equations of the third degree, we will assume that all coefficients a_0, a_1, a_2, a_3 and a_4 are real (the method of solving equation (2.226) with complex coefficients is unknown to the authors of this monograph).

The equation written in this way contains all the degrees of the unknown x. Let us show that in any case (for $a_0 \neq 0$) the third degree can be excluded from this equation. To do this, we replace the variable, for which we represent x as

$$x = y + \varepsilon, \qquad (2.227)$$

where y is a new variable, and ε is a constant value that needs to be defined. In this case, we obtain:

$$x^2 = y^2 = 2\varepsilon y + \varepsilon^2 \qquad (2.228)$$

$$x^3 = y^3 + 3y^2\varepsilon + 3y\varepsilon^2 + \varepsilon^3. \qquad (2.229)$$

$$x^4 = y^4 + 4y^3\varepsilon + 6y^2\varepsilon^2 + 4y\varepsilon^3 + \varepsilon^4. \qquad (2.230)$$

Substituting (2.227)–(2.230) into (2.226) and dividing by a_0, we get:

$$y^4 + \frac{4a_0\varepsilon + a_1}{a_0}y^3 + \frac{6a_0\varepsilon^2 + 3a_1\varepsilon + a_2}{a_0}y^2 + \frac{4a_0\varepsilon^2 + 3a_1\varepsilon^2 + 2a_2\varepsilon + a^3}{a_0} +$$

$$+ \frac{a_0\varepsilon^4 + a_1\varepsilon^3 + a_2\varepsilon^2 + a_3\varepsilon + a_4}{a_0} = 0.$$

$$(2.231)$$

We demand equality

$$\frac{4a_0\varepsilon + a_1}{a_0} = 0, \qquad (2.232)$$

where do we get

$$\varepsilon = -\frac{a_1}{4a_0}. \qquad (2.233)$$

That is, (2.227) takes the form

$$x = y - \frac{a_1}{4a_0}. \qquad (2.234)$$

Substituting (2.234) into (2.226), after reducing such terms, allows equation (2.226) to be reduced to

$$y^4 + ay^2 + by + c = 0, \qquad (2.235)$$

where the notation is entered:

$$a = \frac{a_2}{a_0} - \frac{3a_1^2}{8a_0^2}; \qquad (2.236)$$

$$b = \frac{a_3}{a_0} - \frac{a_1 a_2}{2a_0^2} + \frac{a_1^3}{8a_0^3}; \qquad (2.237)$$

$$c = \frac{a_4}{a_0} - \frac{a_1 a_3}{4a_0^2} + \frac{a_1^2 a_2}{15a_0^3} - \frac{3a_1^4}{256a_0^4}, \qquad (2.238)$$

Note that since all the components of the expressions for a, b, and c are valid, these expressions themselves are also valid.

Thus, the resulting equation (2.235), being a fourth-degree equation, does not contain a third degree unknown. Next, we will deal with the solution of this particular equation. The main idea of the Ferrari method is to represent the fourth-degree equation (2.235) in the form of the difference of the squares of two polynomials of the second and first degree with respect to y, then decompose the difference of squares into the sum and difference of such polynomials, as a result, equation (2.235) will be divided into two square equations for y, which can be solved means. Select from the equation (2.235) the first complete square, for which we represent it in the form

$$\left(y^2 + \delta + r\right)^2 + g = 0.$$

(2.239)

where the value δ is assumed constant, and r we will assume to be an auxiliary variable, which is required to be defined further. For the time being, we also consider the value g as ancillary, and we assume that it has absorbed everything that remains from equation (2.235) after writing it in the form (2.239). In expression (2.239), we open the parentheses and write it down in powers of y, and what remains is in powers of r:

$$y^2 + 2\delta y^2 + 2y^2 x + f^2 + 2\delta r + \delta^2 + g = 0.$$

(2.240)

Comparing the first two terms of this equation with the first two terms of equation (2.235), we find that they are identical provided

$$\delta = \frac{a}{2}.$$

(2.241)

Substituting (2.241) into (2.240), we separate the first two terms (2.240) into the left part, and the rest into the right part:

$$y^4 + ay^2 = -2y^2 r - r^2 - ar - \frac{a^2}{4} - g.$$

(2.242)

Perform a similar separation in equation (2.235):

$$y^4 + ay^2 = -by - c.$$

(2.243)

In these expressions, the left parts are equal, that is, the right parts are also equal. Equating the right parts, we get

$$2y^2r + r^2 + ar + \frac{a^2}{4} + g = by + \varepsilon. \tag{2.244}$$

Express from this equality the value of g:

$$g = by + c - 2y^2r - r^2 - ar - \frac{a^2 4}{}. \tag{2.245}$$

Substituting this expression into (2.239), and also, taking into account (2.241), we reduce equation (2.239) to the form:

$$\left(y^2 + \frac{a}{2} + r\right)^2 + by + c - 2y^2r - r^2 - ar - \frac{a^2}{4} = 0. \tag{2.246}$$

This equation already contains one square with respect to y^2, that is, to form the difference of squares, the entire remaining part of this equation with the opposite sign must also be represented as a square relatively to y already in the first degree. To do this, we need to define an auxiliary unknown r so that from this part we get exactly the full square with respect to y.

We assume that the required second square, in accordance with (2.239) equal to g, can be represented as

$$g = -\alpha^2(y + \beta)^2 \tag{2.247}$$

where α and β are unknown parameters so far that we define from the condition of the possibility of writing g in the form (2.247).

By opening the brackets in this expression, we get

$$g = -\alpha^2 y^2 - 2\alpha^2 y\beta - \alpha^2\beta^2. \tag{2.248}$$

Arrange the terms (2.245) containing y in powers of y, and those terms that do not contain y in powers of r:

$$g = -2y^2r + by - r^2 - ar - \frac{a^2}{4} + c, \tag{2.249}$$

In expressions (2.248) and (2.249), only the first terms depend on y^2. There is no other dependence on y^2 in either expression, so we can assume that in both expressions these dependences are identical, from which we get the condition

$$\alpha^2 = 2r, \qquad (2.250)$$

i.e
$$\alpha = \sqrt{2r}. \qquad (2.251)$$

At the same time (2.248) takes the form

$$g = 2ry^2 - 4ry\beta - 2r\beta^2 \qquad (2.252)$$

Transferring to (2.249) and (2.252) the first addend to the left, we get:

$$g + 2y^2 r = by - r^2 - ar - \frac{a^2}{4} + c, \qquad (2.253)$$

$$g + 2ry^2 = -4ry\beta - 2r\beta^2. \qquad (2.254)$$

In these cases, the cold parts are equal, and therefore the right parts are equal. Equating the right parts, we get

$$by - r^2 - ar - \frac{a^2}{4} + c = -4ry\beta - 2r\beta^2. \qquad (2.255)$$

Transferring everything to the left side, writing down the powers of β, y, r and dividing everything by $2r$, we get

$$\beta^2 + 2\beta y + \frac{b}{2r} y - \frac{r}{2} - \frac{a}{2} - \frac{a^2}{8r} + \frac{c}{2r} = 0. \qquad (2.256)$$

This equation contains two unknown quantities: y and r, as well as the free parameter β. Since there are unknowns, and the equation is only one, due to the choice of this parameter, one of them can be eliminated so that only one equation remains with one unknown. Since the main unknown is y, and r is auxiliary, we exclude the principal unknown y, in order to obtain the auxiliary equation for the auxiliary unknown r. Equation (2.256) containstwo terms – second and third – which depend on y. We select the so far free parameter β so that terms mutually copemsate each other, i.e. the conditions

$$2\beta y = -\frac{b}{2r}\,y.$$

(2.257)

is fulfilled, From this expression, we obtain the necessary value for the free parameter β, so that now it becomes quite definite:

$$\beta = -\frac{b}{4r}.$$

(2.258)

Substituting the obtained value of β in (2.256), we obtain

$$\frac{b^2}{16r^2} - \frac{r}{2} - \frac{a}{2} - \frac{a^2}{8r} + \frac{c}{2r} = 0.$$

(2.259)

Only the quantity r enters into this expression as the unknown, that is, it can be considered as an equation for determining this quantity. Multiplying (2.259) by $16r^2$ and writing down the powers of r, we get

$$8r^3 + 8ar^2 + 2\left(a^2 - 4c\right)r - b^2 = 0.$$

(2.260)

This equation contains, in addition to the unknown value r, only the coefficients of the original equation (2.235), namely: a, b and c, defined by formulas (2.236)–(2.238). The resolution of this equation, that is, the expression of r through these coefficients, can be considered as a preliminary stage for solving the basic equation (2.235). Equation (2.260) is a complete third-degree equation, so it can be solved by the Cardano method described in the previous section. In the classic [262], this equation is called the 'resolving cubic equation' or 'cubic resolvent' for a fourth-degree equation (2.235).

If equation (2.260) is resolved, then equation (2.239), taking into account (2.247), can be represented as the desired difference of two-quadrats:

$$\left(y^2 + \delta + r\right)^2 - \alpha^2(y + \beta)^2 = 0.$$

(2.261)

In this equation, δ, α and β are defined in terms of the coefficients a, *b, c* of the equation (2.235) using the relations (2.241), (2.251) and (2.258), respectively. The parameter *r*, being a solution to equation (2.260), is also determined by these factors.

So, substituting (2.241), (2.251), (2.258) into (2.261), we reduce this equation to the form

$$\left(y^2 + \frac{a}{2} + r\right)^2 - \left[\sqrt{2r}\left(y - \frac{b}{4r}\right)\right]^2 = 0.$$

(2.262)

By decomposing the obtained difference of squares into the product of the sum and difference of the expressions contained in them and equating each of the obtained multipliers to zero, we obtain two equations:

$$y^2 + \frac{a}{2} + r + \sqrt{2r}\left(y - \frac{b}{4r}\right) = 0;$$

(2.263)

$$y^2 + \frac{a}{2} + r - \sqrt{2r}\left(y - \frac{b}{4r}\right) = 0;$$

(2.264)

Positioning in powers of *y* we get:

$$x_{1,2} = -\frac{\sqrt{2r}}{2} \pm \sqrt{-\frac{r}{2} - \frac{a}{2} + \frac{b}{2\sqrt{2r}} - \frac{a_1}{4a_0}};$$

(2.265)

$$y^2 - \sqrt{2r}\,y + \frac{a}{2} + r - \frac{2}{2\sqrt{2r}} = 0.$$

(2.266)

This is a two quadratic equation with respect to *y*, provided that *r* is already determined from the solution of equation (2.260).

Each of these equations has two solutions of the following form:

$$y_{1,2} = -\frac{\sqrt{2r}}{2} \pm \sqrt{-\frac{r}{2} - \frac{a}{2} + \frac{b}{2\sqrt{2r}}};$$

(2.267)

$$y_{3,4} = \frac{\sqrt{2r}}{2} \pm \sqrt{-\frac{r}{2} - \frac{a}{2} - \frac{b}{2\sqrt{2r}}};$$

(2.268)

Thus, equation (2.235) has four solutions, as it should be for a fourth-degree equation. From solutions (2.267) and (2.268) in accordance with formula (2.234), four solutions of the original equation (2.226) are obtained, having the form:

$$x_{1,2} = -\frac{\sqrt{2r}}{2} \pm \sqrt{-\frac{r}{2} - \frac{a}{2} + \frac{b}{2\sqrt{2r}}} - \frac{a_1}{4a_0};$$

(2.269)

$$x_{3,4} = \frac{\sqrt{2r}}{2} \pm \sqrt{-\frac{r}{2} - \frac{a}{2} - \frac{b}{2\sqrt{2r}}} - \frac{a_1}{4a_0},$$

(2.270)

where r is a solution of equation (2.260), and the coefficients a and b are defined by formulas (2.236), (2.237). Thus, with a proper choice of the solution of equation (2.260), the original equation of the fourth degree (2.226) has four solutions defined by formulas (2.269), (2.270).

Comment. Generally speaking, the resolving cubic equation (2.260) for r has not one, but three solutions, so each of these, by substituting in (2.265), (2.266), gives its four solutions for y, that is, there can be twelve solutions, while equation (2.235), being of the fourth order, should have only four solutions. In the well-known authors of this monograph literature. A detailed interpretation of the choice of the required four solutions from the resulting twelve is missing. So in [262, p. 202] it is stated that it suffices to find only one of the roots of equation (2.260), and in [261, p. 44] it is stated that an arbitrary root of the equation is suitable for substitution in (2.265), (2.266) 2.260). The authors of this monograph did not trace the authenticity of these outlines thoroughly; however, in practical calculations, it is recommended to verify the correctness of the solutions obtained in the form (2.267), (2.268) directly by substitution into the original equation (2.235).

Conclusions on Chapter 2
This chapter provides a brief overview of the basic mathematical apparatus used in solving problems concerning ferromagnetic resonance. The main results obtained in this chapter are as follows.

1. The energy densities of various ions that are involved in the formation of the dynamic behaviour of magnetization under the action of a field are considered. It is noted that the total energy density of a magnetic crystal, within the framework of the questions considered in this monograph, has three components: the magnetic,

elastic, and magnetoelastic energies. The density of magnetic energy includes the following components: the energy of the inhomogeneous exchange interaction, the interaction energy of the magnetization with the external field, the internal magnetostatic energy, that is, the interaction energy of the magnetization with the demagnetizing field, and the density anisotropy energies, of which the main ones are uniaxial and cubic. For all of the listed energy density types, analytical expressions are given in the systems of the coordinates, associated with the direction of the field, the shape of the sample and the axes of the anisotropy. A general formula is given for calculating the effective magnetization of the effective particles, and the characteristics of its structure are given in connection with the degree of non-uniformity of the magnetization distribution in the amount of material.

2. The dynamic susceptibility of the magnetic medium is considered. A general definition of susceptibility as a coefficient is given. proportionality between the volume-averaged total magnetization and the magnetic field acting on it. There is a difference between the action of constant and variable fields, consisting that the constant field determines the equilibrium position of the magnetization, whereas the variable field determines its precession around this equilibrium position. The Landau–Lifshitz equation with a dissipative term in the Hilbert form is given, the procedure linearization, resulting in a linearized system of equations of motion of magnetization is given. From the solution of the obtained system of equations, the components of the dynamic susceptibility tensor are determined as the coefficients of proportionality between the components of the magnetization and the coordinate components of the variable field. In accordance with the classical definition of magnetic permeability, the dynamic tensor is obtained from the susceptibility tensor permeability. The real and imaginary parts of the components of the susceptibility tensor are obtained, and their cut-off nature is revealed. Resonant values of frequency and field are found, limb of components is marked susceptibility tensor at resonance. Two different variants of determining the resonant frequency are given and it is shown that the definition of the components of susceptibility by the extremum of imaginary parts of components is most consistent with the physical meaning. There are two types of entries. susceptibility tensor matrices: shortened and complete, the relations between them are determined. The possibility of determining the dependence of the components of the susceptibility tensor on the

field is noted; the equivalence of such a definition of the dependence on frequency, the predominant areas of applicability of both types of dependences in the experiment are indicated. The equations of motion for the magnetization in general form are obtained, which do not contain the assumption that the transverse components are small, that is, suitable for solving problems on nonlinear oscillations.

3. The solution of the linearized Landau–Lifshitz equation, performed in a spherical coordinate system, known as the Smit–Sul method, is considered. The convenience of the method for calculating the dependences of the resonant frequency on the field in the case where the orientation is constant field changes to large angles. A simplified derivation of the Smit–Sul formula is given, which consists in expressing the time derivatives in the Landau–Lifshitz equation in the spherical coordinate system with the subsequent elimination of the derivative of the magnetization under the condition that the full length of its vector is preserved. The result of this transformation is a system of two equations for the polar and azimuthal angles of the magnetization vector. The linearization of the equations obtained with respect to the equilibrium position, taking into account the harmonic dependence on time, allows us to obtain a system of two linear equations for the amplitudes of the polar and azimuthal angles of the magnetization vector. The condition that the solution to such a system is non-trivial gives the dependence of the resonance frequency on the field, known as the Smit–Sul formula. The limitation of the Smit–Sul method, consisting in the divergence of the obtained formula with a small equilibrium polar angle, is noted. the magnetization vector, which in practice requires the choice of the orientation of the polar axis of the spherical coordinate system far from the direction of the constant field.

4. As a mathematical complement, we considered an analytical solution of algebraic equations of the third and fourth degrees, which are very often encountered when finding the equilibrium position of the magnetization vector in problems of orientational transitions and ferromagnetic resonance in anisotropic mediums. It is noted that the expediency of applying analytical solutions of such equations consists in significant (several orders of magnitude) savings. computer time in the numerical solution of the considered problems. The solution of equations of the third degree is considered in accordance with the Cardano method. A method for reducing a complete third degree equation to a shortened equation, in which the second degree is absent, is given. The solution of such an equation is obtained in

the form of the classical Cardano formula. The discriminant of a shortened equation is given, and depending on its sign, three possible options for obtaining a final solution are described, including its trigonometric form. Based on the solution of the shortened equation, a solution is obtained original complete third degree equation. The solution of the fourth degree equation is considered in accordance with the Ferrari method. A method for reducing a fourth-degree full equation to a shortened equation, in which the third degree is absent, is given. By introducing an auxiliary variable, the possibility of representing a shortened equation in the form of the difference of the squares of two expressions, including the first and first degrees of the main variable, is shown. It was established that a necessary condition for such a representation is the resolution of a third-degree equation for an auxiliary variable, for which the Cardano method can be used. It is shown that in this case, the original shortened equation decays on the product of two-dimensional equations, which give four possible solutions. It is shown that, on the basis of the quadratic equations obtained, the shortened equation can be obtained by solving the original complete fourth-degree equation.

3

Mathematical apparatus used when working with crystals of various symmetry

This chapter is devoted to a brief description of the main mathematical methods used in working with anisotropic crystals. The main types of lattice symmetry are given. The features of some of the magnetic anisotropy are considered. The apparatus of transition matrices between different coordinate systems is analyzed in detail, examples of forward and reverse transition matrices are given. A cubic crystallographic cell is considered and examples of obtaining transformation matrices corresponding to the rotation of the cell around its basic symmetry are given. The material of the chapter is mainly of an overview nature, partly revised in accordance with the subject of this monograph. Among the main sources concerning the nature of the symmetry of the crystal lattice, one can mention [135, 271–274]. Magnetic description anisotropy follows [4, 6–8]. The remaining links, including those relating to the properties of real materials, are indicated in the text.

3.1. Lattice symmetry

The most important objects of consideration in this monograph are magnetically ordered media having a crystalline structure, the atoms in which form a regular periodic lattice, that is, the environment has a certain symmetry.

The first type of symmetry is the ability of the lattice to be combined with itself, when parallel perpendicular to certain distances in certain directions, which sets its periodic nature. This is the translational symmetry inherent in the lattices of all species. In addition, the lattice may have a symmetry with respect to wearing different twists and reflections. It is this symmetry that determines the anisotropic (that is, differing in different directions) crystalline properties. Symmetry means the invariance of crystal properties when reflected in certain planes (mirror symmetry) or when rotated around certain axes (axial symmetry).

Let us consider the main types of symmetry of crystal lattices.

3.1.1. Types of lattice symmetry

In the crystal lattice, there are several different types or systems of symmetry, the schemes of the main of which are shown in Fig. 3.1. The figure shows the forms of crystalline cells, the periodic repetition of which forms the internal structure crystal. Below, under each of the cells, the structure of the lattice projection, formed by a set of similar cells, is shown on a plane parallel to one of the cell planes (in the figure, the bottom one). For cells 4 and 7 on the right, a similar projection on the plane is shown, parallel to the side face of the cell. Consider the symmetry properties of reduced cells.

1 – triclinic system

The cell of such a system is a parallelepiped formed by three pairs of mutually parallel planes, and the orientation of pairs of planes relative to each other is arbitrary.

All faces of the cell are parallelepipeds, with the ABCD plane parallel to the EFGH plane, the ADHE plane parallel to the BCGF plane, and the ABFE plane parallel to the DCGH plane. The symmetry of the system is manifested in the preservation of the cell when it is reflected in any of the planes and then rotated 180° around an axis perpendicular to the plane in which the reflection was made. The projection of the lattice on the plane (in this case, EFGH or ABCD) is a set of neighbouring with each other parallelograms, with the upper and lower edges of the cell forming the same structure of the parallelograms shifted relative to each other.

2 - monoclinic system

The cell of such a system has a parallelogram at the base, of the same type as in the triclinic system, for example, EFGH, and the pairs of opposite torons of such parallelogram differ from each other.

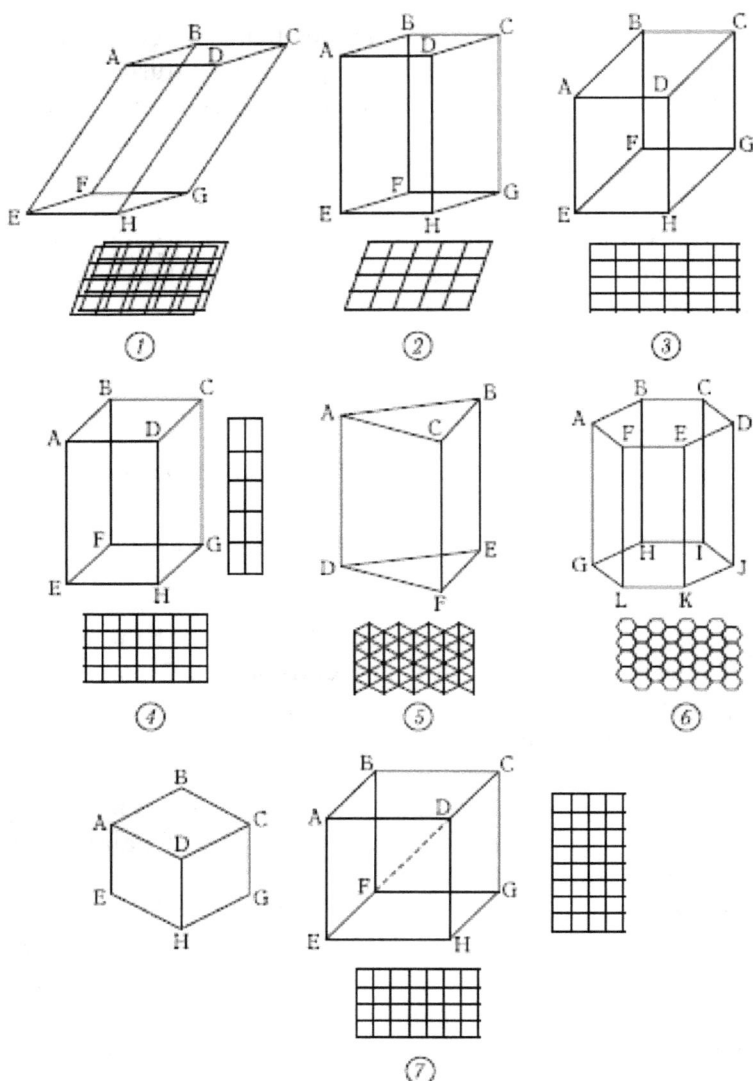

Fig. 3.1. Main types of crystallographic symmetry. 1 – triclinic; 2 – monoclinic; 3 – rhombic (orthorhombic); 4 – tetragonal (square); 5 – trigonal; 6 – hexagonal; 7 – cubic.

The difference from the triclinic system is that the edges emerging from the corners of this parallelogram (EA, FB, GC, HD) are perpendicular to its plane EFGH. Due to the parallelism of the upper and lower planes of the cell, the upper base of the ABCD is a parallelogram identical to the lower EFGH. The lateral faces of the cell are rectangles, the rectangle ABFE being equal to the rectangle

DCGH, and the plane of them is parallel, the same applies to the rectangles ADHE and BCGF.

The symmetry of the system is manifested in the possibility of a cell reflecting in one or another of the lateral planes with subsequent rotation volume of 180° around an axis perpendicular to the plane in which the reflection was made, as well as in the reflection in any of the planes of the base without subsequent rotation. It is also permissible to rotate 180° around any of the vertical ridges of the cells.

The lattice projection onto the EFGH or ABCD plane is a set of parallelograms adjacent to each other, and, unlike the triclinic parallelogram grid structure formed by the upper and lower edges of the cell, completely coincide.

3 – rhombic (orthorhombic) system. The cell of such a system has a rectangle EFGH at the base, that is, the edges EF and GH, as well as EH and FG, being pairwise equal, may differ from each other. All side edges of the cell (EA, FB, GC, HD) are perpendicular to the base plane and are equal to each other. Accordingly, the upper base of the ABCD is equal to the lower EFGH. That is, the cell is a classic rectangular parallelepiped, a typical e0th a square base.

The symmetry properties of the system are the same as in the rhombic system, however, due to the square shape of the base, the ability to rotate around any of the lateral pivots by 90° is now added. The same rotation by 90° is also permissible with respect to the axis passing through the intersection point of the diagonals of any of the bases in a direction perpendicular to the plane of the base. The cell is also symmetric with respect to the reflection in any of the planes passing through the diagonal of squares, that is, ACGE and BDHF. Symmetry is also preserved when reflected in any of the two-sided points parallel to the side faces passing through the above-mentioned central axis.

The lattice projection on the EFGH or ABCD plane is a set of adjacent squares, and the projection on the plane of any of the side walls (shown in the figure to the right of the cell) is a set of adjacent rectangles, one pair of sides which is different from the other.

Until the monoclinic, orthorhombic and tetragonal systems were considered, only the base of the cell changed its shape – from a parallelogram through a rectangle to a square. The height of the cell remained with the parameters of the base is not related. The continuous coverage of the plane with the bases of the neighboring cells corresponded to the reflection of other planes in the perpendicular planes or rotation around axes perpendicular to the baseline at 180° and 90°. Such a coating corresponds to a continuous arrangement on the plane of geometrical figures – parallelograms, rectangles and squares, of which the last is in accordance with the geometric terminology, belongs to the 'regular polygons'. However, there are two more possible options for a continuous covering of a plane with regular polygons, which are equilateral triangles and hexagons. These two types of regular polygons generate two more types of crystal cell systems – trigonal and hexagonal. The names of these systems are related tetragonal. So 'tetra' means four, that is, at the base of the cell, which is a square, there are four corners (the 'noose' is the corner). The trigonal system in its name contains 'three', so that at the base there are three corners, that is, it is a triangle. The name hexagonal contains the particle 'hex' meaning six, so the base contains six corners and is a hexagon. Consider next the last two systems.

5 – trigonal system

The cell of such a system at the base has a regular triangle DEF, the sides of which DE, EF and FD are equal to each other. Equal lateral edges DA, EB and FC are perpendicular to the base plane. Accordingly, the upper base ABC is equal to the lower DEF and is equally oriented with it. That is, the cell is a classic straight prism with a regular triangular base. The symmetry of the system, besides the possibility of reflection in planes of one or other bases, manifests itself in two ways. First, the cell goes into itself when rotated 120° around the axis, passing through the centre of a base perpendicular to the base itself (where the centre can be considered the common intersection point of bisectors, medians or heights of an equilateral triangle).

Further, the lattice projection on the DEF plane, shown below the cell, is a dense set of identical *equal* sided triangles adjacent to each other. Such a grid of triangles has additional symmetry properties That is, it allows mirror reflection in any of the planes passing through the side plane of any cell, as well as rotated 60° around the axis passing through the side edge of any cell. The lattice also allows reflection with respect to any of the planes parallel to the side edges of the prism through the two opposite vertices of any triangles with a common side of the bottom (for example, relative to a plane whose trace is horizontal in the figure, as well as other similar patches).

6 – hexagonal system

The cell of such a system at the base has a regular hexagon GHIJKL, the sides of which GH, HI, IJ, JK, KL, LG are equal to each other. Equal lateral edges GA, HB, IC, JD, KE, LF are perpendicular to the plane of the base. Accordingly, the upper base ABCDEF is equal to the lower GHIJKL and is equally oriented with it. That is, the cell is a classic straight prism with a regular hexagonal base.

The symmetry of the system, besides the possibility of reflection in the planes of one or another base, is also manifested in the fact that the cell switches itself when rotated 60° around an axis passing through the centre of one or another base perpendicular to the base itself (where the centre can be considered a common point intersections of the main diagonals of the hexagon). In addition, the cell allows reflection in any of its own planes, as well as rotation around any its rib by 120°.

The projection of the lattice on the plane of any of the bases is a set of neighboring elements with the same hexagons that completely fill the lug like 'bee from'?????.

In the systems considered above, the degree of symmetry, as the transition from system to system, as a rule, increased, so that the number of symmetry operations leaving the cell unchanged increased. A more rigorous proof of this statement, based on the mathematical theory of groups, can be found for example in [135, 271–273]. A very clear presentation with a rather popular detailed explanation of the main principles is contained in [274]. Here we confine ourselves only to the statement of the fact that in all the systems considered here only the shape of the base of the cell changed, and the height remained arbitrary. However, there is another system where height becomes critical. This is a very common cubic system among magnetic reals, to consider which further proceed.

7 – cubic system

The cell of such a system at the base has a square EFGH, that is, all its sides EF, FG, GH and HE are equal to each other and the sides having a common point, each other is perpendicular. All the edges of the side frames EA, FB, GC and HD are perpendicular to the base plane and are equal to each other. Accordingly, the upper base ABCD is equal to lower EFGH and equally oriented. In all this, the system under consideration is completely identical with the tetragonal one (No. 4). However, the difference is that the lengths of the side edges EA, FB, GC and HD are equal to the sides of the square lying at the base, that is, the edges EF, FG, GH and HE, and also AB, BC, CD and DA. That is, all the edges of such a cell are equal to each other, and the edges that have a common point, mutually perpendicular, so the cell is a regular cube. The symmetry properties of the system are the same as in the rhombic system, however, due to the completely cubic form, it is now realized the ability to rotate around any of the edges, both side and base at 90 на. The same rotation by 90° is permissible relative to the axis passing through the intersection point of the diagonals of any of the faces of the cube in the direction perpendicular to the plane of the given face. The cell is also symmetric with respect to the reflection in any of the planes passing through the diagonals of any of the faces. The projection of a set of cells on the plane of any of the faces is a grid of adjacent squares with a friend of the same squares. Examples of such a grid on the EEFGH planes, as well as DCGH shown below and to the right of the cell.

An additional property of symmetry, which is absent in all the cells above the cells, is the possibility of rotation around any of the spatial coordinates of the cube through an angle of 120°. That is, such a rotation is possible around the axes EC, AG, FD and BH.

An example of such a rotation relative to the spatial diagonal FD is shown to the left of the cell (Fig. 3.1, 7). The view along this diagonal from the side of point D, at which the edges DA, DC and DH converge is presented. To these edges are adjacent the faces of DABC, DCGH and DHEA. It can be seen that turning the pattern by 120 на around point D, for example, clockwise causes the DA edge to switch to DC, DC to DC, DH, to DH to DA. At the same time, the DABC face moves to the place of the DCGH face, the DCGH face to the DHEA place, and the DHEA face to the DABC place. Similarly, other edges and faces are transformed adjacent to the other end of the DF axis, that is, point F. Thus, it can be assumed that a cubic cell has the highest degree of symmetry of all discussed above.

3.1.2. Magnetic anisotropy

The crystallographic anisotropy considered in the previous sections relates primarily to the crystal structure of the material, that is, to the geometric arrangement of atoms in the periodic lattice. Changing the position of atoms causes forces seeking to return them to an equilibrium position. Therefore, any displacement of the atoms relative to each other obeys exactly the symmetry properties considered. Such a shift occurs during mechanical action on the lattice, for example, during any deformation or elastic wave propagation. That is, when considering problems related to elasticity, it is necessary to take into account the crystallographic anisotropy of the medium.

However, for magnetic media, the position is somewhat different. First of all, it concerns magnetically-ordered media, that is, those with ferromagnetism, antiferromagnetism, or ferrimagnetism, in which the fernite atoms are connected by exchange interaction, leading to a collinear orientation of their pins relative to each other. Other, resulting in an orderliness of orientation at significant distances ('long-range order') [4, 138]. So, in many materials, especially of complex composition, not all atoms have magnetic properties, therefore the displacement. Some non-magnetic atoms cannot change the magnetic state of a substance. However, it should be taken into account that in some cases the magnetic properties are due to the interaction of magnetic volumes through non-magnetic, for example, two iron atoms through an oxygen atom located between them. This interaction is called indirect exchange and also causes a long order of orientation magnetization. However, due to the 'dilution' of the magnetic lattice by non-magnetic atoms, in some cases the magnetic anisotropy has a higher degree of symmetry than crystallographic, that is, rotation or reflection of a cell corresponding to crystallographic symmetry does not lead to a change in magnetic symmetry. Let us consider some variants of such magnetic anisotropy.

3.1.3. Easy axis and easy plane anisotropy

The simplest example of magnetic symmetry is an easy-axis and easy-plane anisotropy. So, for example, if magnetic atoms in a rhombic cell are located along an axis parallel to one of the edges, then the rotation of the cell around this edge at an arbitrary angle of magnetic rods does not change, so that The magnetic properties of a substance relative to this direction remain the same. A similar situation may occur for other types of cells.

If the magnetization vector in an equilibrium position is oriented along the same axis, the rotation around which the magnetic state of the substance does not change, then an anisotropy of the 'easy axis' type takes place. Such anisotropy is often called 'uniaxial'. Wherein in the absence of an external field, the magnetization is oriented in the positive or negative direction of a single axis. If the magnetization vector in an equilibrium position is oriented perpendicular to the axis, the rotation around which the magnetic state of the substance does not change, then there is an anisotropy of the 'easy plane' type. Such anisotropy is often called 'easy-plane'. In this case, in the absence of an external field, the magnetization can be located in any direction in a plane perpendicular to the axis mentioned. In the general case, the degree of symmetry of the easy axis and easy plane anisotropies is the same, the difference consists only in the sign of the corresponding constant. That is, when changing its sign, for example as the temperature changes, the light axis can move to the easy plane and back. The energy density of this type of anisotropy is of the second order in magnetization.

3.1.4. Other types of magnetic anisotropy

Among other types of magnetic anisotropy one should not first note the uniaxial anisotropy but that of a higher order than that considered in the previous section. The energy density of such anisotropy has a fourth, sixth or even more magnetization high even order. At the same time, if the second order anisotropy can have only a single axis of a single direction, then there can be several fourth order anisotropy axes. That is, the crystal can have several axes of easy magnetization, differing only in direction. In general, one second order axis and one or more fourth order axes can coexist simultaneously. A typical example of multi-axis fourth order anisotropy is cubic. For example, the unit cell of iron has three equivalent axes. light magnetization parallel to the edges of the cube, and a nickel or yttrium iron garnet cell has four axes parallel to the spatial diagonals of the same cube [4, 6–8]. Other types of magnetic cubic anisotropy are possible, for example, the corresponding diagonals of a cell face. In addition to cubic properties, a rhombic or tetragonal cell, where the coexistence of several easy magnetizations can also occur, can have similar properties. the type of anisotropy of cubic passengers is uniaxial, having an arbitrary direction relative to the edges of the cubic cell. An example of such anisotropy is the so-called 'growth' anisotropy of mixed ferrite garnet crystals [40, 240, 275–277], due to elastic stresses and misorientation of the crystal growth plane relative to the (111) plane.cubic cell. Some phenomena accompanying such anisotropy are described, for example, in [22–25, 30, 278].

Another feature of cubic crystals, primarily of films of mixed ferrite garnet, is the rhombic or orthorhombic anisotropy, the result of which is a different value of a constant field applied in the film plane, corresponding to the nucleation (or disappearance) of the domain structure. A brief overview of the manifestation of such anisotropy can be found in [40, p. 25]. The difference between oblique 'growth' and 'orthorhombic' anisotropy, determined by the nucleation of a domain structure, is described in some detail in [276]. It is noted that the role of oblique uniaxial anisotropy in the saturation of films, with magnetization in their plane, significantly exceeds the role of orthorhombic anisotropy, so that the difference between them often does not go beyond the limits of experimental errors. In crystals with trigonal and hexagonal symmetries, typical is the uniaxial magnetic anisotropy, oriented along the axis of the third or sixth order. An example of a crystal with such anisotropy is hematite, a crystalline cell of which has trigonal symmetry. The magnetic anisotropy of hematite has a clearly defined uniaxial character along the trigonal axis in combination with a small hexagonal anisotropy in a plane perpendicular to this axis. Barium hexaferrite, having a hexagonal cell, has a very strong uniaxial anisotropy.

In some crystalls that have any one selected direction, the magnetization can take the form of a spiral or helical structure, the coils of which are screwed onto the axis corresponding to this direction. In this case, any predominant direction of orientation of the magnetization perpendicular to the screw axis may be absent. The period of such a structure along the helical axis can significantly exceed the period of the main crystal cell. A certain idea of such magnetic symmetry types can be obtained, for example, from [32,279,280] and the literature cited there.

3.1.5. Features of the structure of real materials

The seven types of crystal cells considered in the previous section exhaust all possibilities of the existence of crystallographic anisotropy. However, it should be noted that the cell configurations presented here do not mean that the atoms in each the cell are located at the points where the edges of the adjacent faces converge, that is, at the points at the edges of these cells. The configuration of cells means only a set of bounding planes defining the surfaces along which other similar ones adjoin this cell. Atoms belonging to a given cell can be located in any of its places, it is only important that these places from cell to cell can be repeated. For example, a cubic cell, in addition to the atoms located at the vertices of the cube, may have atoms located at the intersection of the diagonals of each face, as well as at the intersection point of all four spatial cages of the cube. That is, a cubic cell, except for the location of atoms only at the vertices of the cube, can also be 'face-centred' and 'body-centred'. The most important example of such types of the lattice is the ordinary iron.

Moreover, despite the apparent simplicity of the elementary cells (Fig. 3.1), the real arrangement of atoms in each cell can be very complicated. The main requirement here is repeatability of the internal structure of the cell with the corresponding operations of reflection and rotation. Let us consider the features of some important practices for magnetic materials, for which we begin with the same iron.

Iron. Pure iron at a temperature below 910°C (α-phase) has a body-centred cubic lattice with a lattice constant (cube edge) equal to 2.87 Å, and in the range from 910°C to 1400°C (γ-phase) – face-centered cubic lattice with a constant 3.64 Å. At temperatures above 1400°C up to the melting point of 1539°C (δ-phase), the lattice again becomes body-centered, with a constant of 2.94 Å. Above the melting point, the crystal lattice is destroyed, and the iron becomes amorphous [281]. In all these cases, except, of course, the amorphous state, the crystal lattice of iron remains cubic, that is, according to the symmetry corresponding to cell No. 7 in Fig. 3.1. However, when passing through a temperature of 910°C, the arrangement of atoms inside cell changes. Such a transition from a body-centered cubic structure to a face-centered, again cubic, is a typical example of a structural phase transition [135–139].

Hematite. Another example of the arrangement of atoms that does not correspond to the meeting points of the edges of the cell is the widespread hematite mineral (a special type of iron ore that makes up for example, the basis of the Kursk magnetic anomaly). The chemical composition of hematite is iron oxide α-Fe_2O_3. The symmetry of the hematite cell is trigonal (as opposed to γ-Fe_2O_3, where the cell is purely cubic [8]). Such a cell is a cube stretched along one of its spatial diagonals. It is this stretching that preserves the cell when rotated around this diagonal by 120°, that is, there is a symmetry of the same type as shown in Fig. 3.1 to the left of the cubic cell (No. 7). The extension axis coincides with the vertical direction of the trigonal cell 5 in Fig. 3.1. In the first approximation, it can be assumed that the iron ions are located along the centres of triangular bases of the cell, and ions oxygen – in the middle of the lateral ridges. The arrangement of atoms in the hematite lattice is schematically shown in [8, p. 71, Fig. 3.3], and the structure is studied in more detail in [282].

Orthoferrites. We begin our consideration with the orthoferrite, whose chemical formula is: $RFeO_3$, where R is yttrium or a rare earth element. From the point of view of crystallography, orthoferrite has a somewhat distorted structure of the perovskite mineral $CaTiO_3$, which is widespread in nature (ferroelectrics $BaTiO_3$ and $LiNbO_3$, which are also widely used in acoustics) have a similar structure.

Orthoferrite has a rhombic structure. However, the arrangement of atoms in the cell is very difficult. A definite idea of this arrangement can be obtained from [134, p. 38, Fig. 2.1]. Relative to the rhombic cell, the iron atoms are in the middle of the edges of the base. The atoms of yttrium or a rare earth metal – in the middle of the height of the central axis. Oxygen atoms surround each iron atom with an octahedron, the diagonals of which are oriented along the edges of the main cell. In a slightly different perspective, the same structure is presented in [138, p. 301, Fig. 6.29]. The result of this arrangement is the antiferromagnetic ordering with weak ferromagnetism combined with strong uniaxial anisotropy.

Ferrites–spinels. A typical example of cubic symmetry is the structure of a spinel. In the natural state, a spinel is a mineral having the composition AB_2O_4, where A is Mg, Zn, Mn, Fe^{2+},. . . B – Al, Fe^{3+}, Cr, Mn,. . . [283]. An idea of the arrangement of atoms in a spinel-type structure is given in [284, pp. 117–119, Fig. 3.20], as well as [138, p. 294, Fig. 6.24]. Classical spinel – $MgAl_2O_4$ – non-magnetic mineral, widely distributed in nature. When aluminium is replaced by iron, the structure is preserved; magnesium ferrite $MgO \cdot (Fe_2O_3)$ is obtained, which has magnetic properties. The unit cell of the spinel structure contains 32 oxygen ions, 16 Fe^{3+} iron ions and 8 non-magnetic metal ions, such as Mg, Mn, Zn or Fe^{2+} iron. A complete spinel cell has cubic symmetry.

Of greatest interest for this consideration are spinels containing magnetic ions, primarily iron. An example of magnetic spinel is the widespread magnetite mineral $FeO \cdot Fe_2O_3$ (or how often Fe_3O_4 is recorded) [285]. This is a special type of iron ore (Magnitogorsk deposit), the magnetic properties of which were known to the ancient Greeks. Modern microwave technology is widely used nickel–zinc $(Ni, Zn) O \cdot Fe_2O_3$ and manganese–zinc $(Mn, Zn) O \cdot Fe_2O_3$, as well as magnesium–manganese (Mg, Mn) O · Fe_2O_3 ferrites [286], which have a relatively low saturation magnetization, which is important for devices in the decimeter-wave band [284, p. 123].

Ferrites–garnets. The most important representatives of materials with a cubic structure are garnet–ferrites. In nature, garnet is a widespread mineral of composition $R_3^{2+}R_2^{3+}$ $(SiO_4)_3$, where R_3^{2+}, R_2^{3+} are various bivalent trivalent metals, giving the mineral a huge variety of species and physical substances that are of widespread use (from technical abrasives to jewelry) [287].

In radioelectronics, the garnets are widely used, in which silicon takes the place of iron and which are called ferrites–garnets. The general formula of such materials is: $M_3Fe_5O_{12}$, where M is yttrium or a rare-earth metal. The cubic unit cell of a garnet ferrite contains eight formulas of composition $M_3Fe_2(FeO_4)_3$. Sixteen Fe^{3+} ions occupy octahedral sites, twenty-four Fe^{3+} ions occupy tetrahedral sites. Twenty-four M^{3+} ions occupy nodes of an eight-vertex twelve-sided polyhedron. Each unit cell contains 96 oxygen ions occupying the nodes, which are the vertices of one tetrahedron, one octahedron, and two polyhedra [284, pp. 127–129, Fig. 3.28-3.31]. In 1970–1990 garnet ferrites containing rare-earth elements instead of yttrium, so-called 'mixed garnet ferrites', whose characteristic property is the presence of strong uniaxial anisotropy, were widely studied. Such garnets in the form of epitaxial plates grown on gadolinium–gallium substrates garnet $(Gd_3Ga_5O1_2)$ were considered as a promising material for memory devices on cylindrical magnetic domains (CMD) [40]. Some properties of such garnets are discussed further in sections 5.2, 5.3.

Iron yttrium garnet. A special place among garnet ferrites with cubic symmetry is occupied by the $Y_3Fe_5O_{12}$ iron yttrium garnet (IYG) (the formula of which is sometimes written as Y_3Fe_2 $(FeO_4)_3$ or $(Y_2O_3)_3 · (Fe_2O_3)_5$). The crystal structure of an iron–yttrium garnet is described in [288,289], as well as in [284, pp. 127–129, Fig. 3.28–3.31]. The mutual arrangement of oxygen atoms relative to iron atoms is presented, for example, in [138, p. 297, Fig. 6.26].

The special role of iron–yttrium garnet in radioelectronics is associated with its record-breaking low losses in ferromagnetic resonance (FMR). The width of the resonance line of the IYG can be small up to up to 0.5 Oe, whereas the typical FMR line width in spinel ferrites is 10–20 Oe, and in orthoferrites, 100–200 Oe [8, 284].

Barium hexaferrite. Another material important for further discussion is barium hexaferrite, which has the composition $BaFe_{12}O_{19}$

or BaO · $(Fe_2O_3)_6$. In technology, this material is widely used for the manufacture of permanent magnets called magnetoplumbite or ferroxdur.

The crystal structure of this material is pretty complex and is an ordered alternation of spinel type layers containing iron ions with cubic symmetry, separated with oxygen from barium layers with hexagonal symmetry. In this case, the magnetic properties are determined by the interaction of the iron ions of the spinel lattice through the oxygen ions in the barium layer; therefore, the hexagonal symmetry of the barium layer imposes its symmetry to the spinel layer. As a result, the material acquires strong uniaxial anisotropy (axis c) in combination with relatively weak hexagonal anisotropy in the plane perpendicular to axis c [284, pp. 139–141, Fig. 3.42–3.44]. Thus, a characteristic feature of this material is the presence of pronounced uniaxial anisotropy (anisotropy field up to 19 kOe), which allows one to create permanent magnets on its basis. The resonant properties of the barium hexaferrite are significantly worse than thaose of the garnet ferrites, but this did not prevent the implementation of a very effective study of ferromagnetic resonance under orientation transition [31].

Comment. From the above list of a very small number of real structures, one can see a large variety of arrangement of concrete volumes within a single unit cell. However, this diversity is repeated periodically, which makes it possible to judge the many properties of the crystal. To end this section, we would like to quote from [273, 290], which very colourfully describes the state of affairs in practical crystallography: "The lattice gives us the size and shape of a repeating unit of a structure, its unit cell, but does not determine what is the location of the substance inside the unit cell itself. At the first stage it and does not matter. The steel frame of the building must exist before the discussion of interior decoration or furniture begins (Lonsdale, 1952)."

3.2. The basic technique of working with anisotropy

The orientational transition consists in moving the magnetization vector from one stable position to another, also stable. In this case, the stability of both positions is determined by the minimum of the total energy of the magnet, in which the most important place is occupied by the magnetic anisotropy energy density. At the same time, the recording forms of the energy density of the anisotropy

in different systems of the couple also differ. Since for most tasks
the greatest convenience is represented by a coordinate system, one
of the axes of which is associated with the magnetization vector, in
the process of orientational transition, the orientation of this system
relative to the crystal lattice of the magnet may vary. At the same
time, the simplest form of the energy density of anisotropy is just
in the system associated with the lattice. Therefore, to successfully
solve such problems, an apparatus is needed that allows one to switch
from one system to another. The basis of such an apparatus is the
concept of a transition matrix, which allows working with anisotropy
according to the rules of matrix algebra. Therefore, we will begin
consideration with the definition of such matrices.

3.2.1. Definiton of the transition matrix

Consider the basic principles of choosing a coordinate system for
the task, which consists in determining the behaviour of the system
in accordance with the law of motion.

We will assume that we are in a laboratory coordinate system,
which is called the 'original' one. Let, in addition, there is another
coordinate system, somehow rotated relative to the initial one, which
we will call it 'turned'.

Let there be some expression that is a function of one or several
vectors, given in the original coordinate system (for example, energy
density).

Suppose there is also a certain functional rule operating with this
expression (for example, the law of motion), and the form of this
rule in a rotated system is quite simple, and in the original system
has a more complex form. For this reason, it is more convenient to
apply the required rule in a rotated system, and then transfer the
result to the original system. For this, it is necessary to translate the
functional expression, initially specified in the original coordinate
system, into a rotated system. Since a functional expression is a
collection of vectors specified in the original coordinate system, for
its translation into a rotated system, the same vectors must be written
in the rotated system. Further, since now the functional expression
will already be written in the rotated system, the required functional
rule can be applied to it with sufficient convenience. The result of
this application will be recorded still in the rotated system. That is,
now, for the final solution of the problem, it must be transferred
back to the original system, for which the vector transformation is

the opposite of the original one. Thus, two types of transformations of the components of the vectors are required:

1) write the components specified in the initial system through the components specified in the rotated system;

2) record the components specified in the rotated system through the components specified in the original one.

We will call the first transformation direct, the second reverse.

The opposite is also possible: the law of motion has a simple form in the original system, and the functional expression is rotated. Then it is necessary to transfer the components of the vectors in the rotated system to the original one, after which the law of motion should be used already in the original system. In this case, it is enough to use only the second type. transformation, that is, the reverse.

If, however, the final result is required in a rotated system, then the solution obtained in the original system must be translated. rotated, for which you should use direct conversion. Thus, to solve any problem, only two types of transformation are sufficient – direct and inverse.

3.2.2. Direct conversion example

We give a typical example of a problem that requires direct conversion. Consider the precession of magnetization in an anisotropic uniaxial medium in the presence of a constant field, the orientation of which with the direction of the anisotropy axis does not match.

The magnetic anisotropy energy density has the simplest form, being expressed through the magnetization vectors specified in the coordinate system associated with the anisotropy axis, which will be consider the original one.

The application of a constant field to such a system in statics causes the establishment of the magnetization vector in a certain equilibrium position, the orientation of which is determined by the combination of two factors – anisotropy and field, both of which are distributed in the original system.

The motion of the magnetization is conveniently regarded as a deviation from the equilibrium position in the system associated with this equilibrium position, that is, rotated. Therefore, the preliminary stage of solving the problem should be to find such an equilibrium position relative to the initial system, which will further determine the orientation of the rotated system.

So, if the orientation of the anisotropy and the direction of the constant field are specified in the original system, then in order to be able to solve the problem in a rotated system, it is necessary to express the components of the magnetization vectors specified in the original system through the same vectors specified in a rotated system, that is, to carry out a direct transformation. After substituting the components of the vectors obtained in this way into the expression for the energy density specified in the original system, we obtain the energy density expressed in terms of the components of the vectors in the rotated system, which allows us to further solve the problem in this system.

We explain the same thing in more detail.

The density of the interaction energy of magnetization with the field is given in terms of the components of the magnetization vectors and the field in the original system. The equation of motion of the magnetization vector for a given the field (Landau–Lifshitz) is a vector relation independent of the choice of the coordinate system. In general, it reduces to a system of three-linear differential equations of the first order, the solution of which is quite complicated.

However, when writing in the coordinate system, one of the axes of which is directed along the equilibrium position of the magnetization, it is easy to linearize the equations of motion for relatively small components of the magnetization vector, perpendicular to the direction of the equilibrium position. This linearization makes it possible to reduce a problem containing three nonlinear equations to a system of two linear equations, which is much simpler than the original one.

Thus, for the possibility of linearization of the problem, it is necessary to translate the expression for the energy density, specified in the original coordinate system, into a system, one of whose axes is parallel to the direction equilibrium orientation of magnetization.

To do this, it is necessary to express the components of the field vectors and magnetization, specified in the original coordinate system, through the components of technical vectors in the rotated system, and then substitute the obtained expressions into the initial expression for the energy density. Thus, the initial energy density will be expressed in terms of the components of the field vectors and the magnetization in the rotated system.

After such a transformation, the problem (that is, the Landau–Lifshitz equation which requires finding effective fields) is already solved in a rotated coordinate system, with the normalized component of magnetization parallel to the equilibrium orientation of the magnetization to be linearized, assumed to be equal to one, and both perpendicular components are so small that their squares and the product of each other can be neglected.

3.2.3. Example of the inverse transform

Suppose that in the course of the previous example, the equations of motion of the magnetization in a rotated coordinate system are obtained. If the motion of magnetization is forced, then the solution of this problem allows us to find the magnetic susceptibility tensor, the components of which will again be expressed in a rotated coordinate system. The solution of the problem of the precession of magnetization through this tensor will be expressed in a rotated system. However, if it is necessary to find susceptibility in the original system, then two things must be done.

First, to express the components of the magnetization vector in the rotated system through the components in the original system, that is, to perform the inverse transformation relative to this vector.

Second, to express the components of the susceptibility tensor in a rotated system through similar components in the original system, that is, to perform the inverse transformation relative to the components of the tensor.

Then the obtained components of the magnetization vector and the susceptibility tensor are substituted into the expression obtained in the rotated system, as a result of which everything will be obtained in the original system. An example of such a transformation will be given below in Section 6.4, which is devoted to forced oscillations of magnetization under the conditions of an orientational transition. Leave for now the transformation of the tensor before the mentioned section, and consider in more detail the technique of vector conversion.

3.2.4. Transition matrix in general

In the general case, it is necessary to express the vector components specified in one coordinate system through the components of the same vector in another system. That is, to put it in terms of mathematics, it is necessary to make the transition from one basis to another. According to the rules of such a transition [291], the transformation of the vector components when the coordinate system is rotated is performed using transition matrix in accordance with

96

the formula

$$\mathbf{a}_1 = \ddot{A}_{21}\mathbf{a}_2 \qquad (3.1)$$

where $\mathbf{a}_1$ is a vector in the first coordinate system, $\mathbf{a}_2$ is a vector in the second coordinate system, $\ddot{A}_{21}$ is the transition matrix from the second system to the first obe. The components of the transition matrix are the cosines of the angles between the axes of the first and second coordinate systems.

The reverse transition is carried out using the inverse matrix

$$\mathbf{a}_2 = \ddot{A}_{12}\mathbf{a}_1, \qquad (3.2)$$

the components of which are also cosines, but now between the axes of the second and first coordinate systems. That is, in the first case, each line of the direct matrix corresponds to the angles between one of the coordinates of the first system from each of the coordinates of the second system, and in the second case – each line of the reverse matrices correspond to angles between one of the coordinates of the second system with each of the coordinates of the first system.

Let us clarify this in more detail using the example of two Cartesian systems of coordinates.

3.2.5. Transition from one Cartesian coordinate system into another

An important particular case of the transition is the transformation of the vector $\mathbf{a}'$, given in the rotated coordinate system $Ox'y'z'$, into the vector $\mathbf{a}$, specified in the original single system $Oxyz$ [292,293]. Such a transition is carried out using direct conversion

$$\mathbf{a} = \ddot{A}\mathbf{a}'. \qquad (3.3)$$

In this case, the table of the components of the transition matrix is

	x'	y'	z'
x	$\cos(\angle xOx')$	$\cos(\angle xOy')$	$\cos(\angle xOz')$
y	$\cos(\angle yOx')$	$\cos(\angle yOy')$	$\cos(\angle yOz')$
z	$\cos(\angle zOx')$	$\cos(\angle zOy')$	$\cos(\angle zOz')$

$$(3.4)$$

that is, the direct transformation matrix itself has the form

$$\ddot{A}^{-1} = \begin{pmatrix} \cos(\angle xOx') & \cos(\angle xOy') & \cos(\angle xOz') \\ \cos(\angle yOx') & \cos(\angle yOy') & \cos(\angle zOy') \\ \cos(\angle zOx') & \cos(\angle yOz') & \cos(\angle zOz') \end{pmatrix} \qquad (3.5)$$

The inverse transition, that is, the transformation of the vector **a** given in the Cartesian coordinate system $Oxyz$ into the vector **a'**, specified in the system $Ox'y'z'$, is performed using the inverse transformation

$$\mathbf{a}' = \ddot{A}^{-1}\mathbf{a}. \qquad (3.6)$$

In this case, the table of the components of the transition matrix is

	x	y	z
x'	$\cos(\angle x'Ox)$	$\cos(\angle x'Oy)$	$\cos(\angle x'Oz)$
y'	$\cos(\angle y'Ox)$	$\cos(\angle y'Oy)$	$\cos(\angle y'Oz)$
z'	$\cos(\angle z'Ox)$	$\cos(\angle z'Oy)$	$\cos(\angle z'Ozx)$

(3.7)

The corresponding inverse transform matrix is

$$\ddot{A}^{-1} = \begin{pmatrix} \cos(\angle x'Ox) & \cos(\angle x'Oy) & \cos(\angle x'Oz) \\ \cos(\angle y'Ox) & \cos(\angle y'Oy) & \cos(\angle y'Oz) \\ \cos(\angle z'Ox) & \cos(\angle z'Oy) & \cos(\angle z'Oz) \end{pmatrix} \qquad (3.8)$$

Comparing (3.5) and (3.8), we can see that the mutual transformation of the matrices of both transitions into each other is obtained by replacing the rows of one matrix with columns of another and vice versa. Based on the fact that the sequential application of direct and inverse transformations should bring the vector to its original value, it follows that the product of the direct and inverse transformation matrices should be equal to the identity matrix leaving the vector unchanged, that is, the equality

$$\ddot{A}\cdot\ddot{A}^{-1} = \begin{pmatrix} 1 & 0 & 0 \\ 0 & 1 & 0 \\ 0 & 0 & 1 \end{pmatrix} \qquad (3.9)$$

This property, which characterizes the rotation of one Cartesian system relative to another, both of which are right and have a common the beginning is fairly universal. It is performed under the condition that the basis of each of the matrices (that is, the set of direction vectors of the coordinate axes) is orthonormal, where the normalization is equal to one unit of the sum of squares of each of these vectors [261, p. 447, formula (14.10.3)]. That is, the sum of the squares of the components of each direction vector of one system, decomposed along the axes of another system, must be equal to one. Thus, for each of the axes of each system, a unit vector is specified, and the matrix components are the cosines of the angles between unit vectors of one system and unit vectors of another. So, the main task of finding the rotation matrix is to determine the unit directions of the axes of one system relative to another. The components of the forward and reverse transition matrices are usually easier to obtain in the original $Oxyz$ coordinate system. So, if we denote the unit vectors of the axes of the coordinate system $Ox'y'z'$ in $\mathbf{l}_{x'}, \mathbf{l}_{y'}, \mathbf{l}_{z'}$, then the cosines of these vectors in the $Oxyz$ system will be the required elements of the matrices A and A^{-1}.

An important case of specifying the relative position of the original and rotated coordinate systems is one in which the position of the $Ox'y'z'$ system relative to the $Oxyz$ system, it is uniquely determined by the orientation of the Oz' axis, given by the guiding vector $\mathbf{l}z'$ in the system $Oxyz$, as well as the angle ε of the rotation of the system $Ox'y'z'$ around this axis. This angle should be counted in the plane $Ox'y'$ from the axis the direction vector of which relative to the system $Oxyz$ remains fixed and can be determined in advance. Thus, the task of constructing the transition matrix is to express all angles between (3.5) or (3.8) between the axes of the systems $Oxyz$ and $Ox'y'z'$ through the angles that determine the orientation of the axis Oz' and the axis of reference of the angle of rotation ε, as well as through the value of the angle ε itself. For this purpose it is convenient to use the Euler angle system [261, p. 450].

In this case, if the matrix, for example, of a direct transition is obtained, then the inverse transition matrix can be obtained from it by replacing the rows with the corresponding columns. The same applies to obtaining the forward transition matrix from the inverse matrix.

3.2.6. Matrix direct and reverse

Consider the transformation of the component vectors from one system to another in the most general form.

Matrix DIRECT converts a vector from the $Ox'y'z'$ system to the $Oxyz$ system:

$$\mathbf{a} = \ddot{A}\mathbf{a}'. \tag{3.10}$$

Its components are the direction cosines of the axes of the system $Ox'y'z'$ relative to the axes of the system $Oxyz$, that is, to construct a direct matrix, it is necessary to express the components of the unit axes of the axes Ox', Oy', Oz' in the $Oxyz$ system. Denote the unit direction vectors of both coordinate systems through $\mathbf{l}_i$, and the cosines of the angles between the vectors of the one and the other systems through $(\mathbf{l}_i)_k$, where $i,\ k$ are the corresponding coordinates.

The table specifying the components of the DIRECT matrix is:

	$\mathbf{l}_{x'}$	$\mathbf{l}_{y'}$	$\mathbf{l}_{z'}$
x	$(\mathbf{l}_{x'})_x$	$(\mathbf{l}_{y'})_x$	$(\mathbf{l}_{z'})_x$
y	$(\mathbf{l}_{x'})_y$	$(\mathbf{l}_{y'})_y$	$(\mathbf{l}_{z'})_y$
z	$(\mathbf{l}_{x'})_z$	$(\mathbf{l}_{y'})_z$	$(\mathbf{l}_{z'})_z$

(3.11)

i.e

$$\overleftrightarrow{A} = \begin{pmatrix} (\mathbf{l}_{x'})_x & (\mathbf{l}_{y'})_x & (\mathbf{l}_{z'})_x \\ (\mathbf{l}_{x'})_y & (\mathbf{l}_{y'})_y & (\mathbf{l}_{z'})_y \\ (\mathbf{l}_{x'})_z & (\mathbf{l}_{y'})_z & (\mathbf{l}_{z'})_z \end{pmatrix}. \tag{3.12}$$

Transformation of vector $\mathbf{a}$ from the $Ox'y'z'$ system to the $Oxyz$ system has the form:

$$\begin{pmatrix} a_x \\ a_y \\ a_z \end{pmatrix} = \begin{pmatrix} ((\mathbf{l}_{x'})_x & (\mathbf{l}_{y'})_x & (\mathbf{l}_{z'})_x \\ (\mathbf{l}_{x'})_y & (\mathbf{l}_{y'})_y & (\mathbf{l}_{z'})_y \\ (\mathbf{l}_{x'})_z & (\mathbf{l}_{y'})_z & (\mathbf{l}_{z'})_z \end{pmatrix} = \begin{pmatrix} a_{z'} \\ a_{y'} \\ a_{z'} \end{pmatrix} \tag{3.13}$$

The REVERSE matrix converts a vector from the $Oxyz$ system to the $Ox'y'z'$ system:

$$\mathbf{a}' = \ddot{A}^{-1}\mathbf{a}. \qquad (3.14)$$

Its components are the direction cosines of the axes of the $Oxyz$ system relative to the axes of the $Ox'y'z'$ system, that is, to construct a direct matrix, it is necessary to express the components of the unit vectors of the axes Ox, Oy, Oz in the $Ox'y'z'$ system.

The table specifying the components of the REVERSE matrix is:

i.e

	$\mathbf{1}_x$	$\mathbf{1}_y$	$\mathbf{1}_z$
x'	$(1_x)_{x'}$	$(1_y)_{x'}$	$(1_z)_{x'}$
y'	$(1_x)_{y'}$	$(1_y)_{y'}$	$(1_z)_{y'}$
z'	$(1_x)_{z'}$	$(1_y)_{z'}$	$(1_z)_z$

$$(3.15)$$

$$\overleftrightarrow{A}^{-1} = \begin{pmatrix} (1_x)_{x'} & (1_y)_{x'} & (1_z)_{x'} \\ (1_x)_{y'} & (1_y)_{y'} & (1_z)_{y'} \\ (1_x)_{z'} & (1_y)_{z'} & (1_z)_{z'} \end{pmatrix}. \qquad (3.16)$$

The transformation of vector $\mathbf{a}$ from the $Oxyz$ to the $Ox'y'z'$ system has the form:

$$\begin{pmatrix} a_{x'} \\ a_{y'} \\ a_{z'} \end{pmatrix} = \begin{pmatrix} (1_x)_{x'} & (1_y)_{x'} & (1_z)_{x'} \\ (1_x)_{y'} & (1_y)_{y'} & (1_z)_{y'} \\ (1_x)_{z'} & (1_y)_{z'} & (1_z)_{z'} \end{pmatrix} \begin{pmatrix} a_x \\ a_y \\ a_z \end{pmatrix}. \qquad (3.17)$$

When solving problems on the precession of magnetization, usually the $Oxyz$ system is a laboratory system associated with a constant magnetic field or the equilibrium position of the magnetization vector along which the axis Oz is directed. The anisotropy energy density is usually given in the system $Ox'y'z'$, associated with the axis of anisotropy or with the edges of a cubic unit cell, since in these cases the expression for energy density has the simplest form. In this case, the energy density is expressed through the components of the vector od magnetization $m_{x'}$, $m_{y'}$, $m_{z'}$ specified in the $Ox'y'z'$ system. To ensure the possibility of solving the problem in the laboratory system $Oxyz$, it is necessary to have an expression for the energy density in this system. For this we need the components of the magnetization vector $m_{x'}$, $m_{y'}$, $m_{z'}$ to be expressed through the components of the same vector m_x, m_y, m_z, given in the $Oxyz$ system, then substitute them into the expression for the energy density given in the $m_{x'}$, $m_{y'}$, $m_{z'}$ system. As a result, the energy density will be determined through the components m_x, m_y, m_z, which will allow work further in the $Oxyz$ system. Thus, it is necessary to carry out the transformation:

$$\mathbf{m'} = \bar{B}\mathbf{m}. \tag{3.18}$$

where $\bar{B}$ is the required transition matrix. Comparing this expression with (3.6), we see that $\bar{B}$ is the inverse matrix defined by the expression (3.16), that is

$$\mathbf{m'} = \ddot{A}^{-1}\mathbf{m}. \tag{3.19}$$

So, to obtain the energy density in the $Oxyz$ system, if it is given in the $Ox'y'z'$ system, a REVERSE matrix of the form (3.16) must be used. However, it is usually technically easier to find a direct matrix of the form (3.12), since this requires finding the direction cosines unit vectors of the $Ox'y'z'$ system in the $Oxyz$ system, that is, at the stage of finding a direct matrix, one can work in the laboratory coordinate system, which is more convenient. After finding the direct matrix find the inverse matrix of it is possible by replacing rows with columns. The validity of such a replacement is ensured by checking the product of the direct and inverse matrices, which should be equal to the identity matrix.

3.3. General rules for solving transition problems between different coordinate systems

We first note some general rules for solving problems on the transition between differently oriented coordinate systems. In the case of an anisotropic medium, two fundamentally different learning ways are possible.

. **Case number 1.** The problem is solved in the original (laboratory) coordinate system. In this case, the anisotropy energy density should be initially recorded in the system where its form is the simplest, after which this energy density should be transformed into the original system. This case is convenient for uniaxial anisotropy. The problem is solved in the laboratory system $Oxyz$. The energy density has the simplest form in the system $Ox'y'z'$ the axis Oz' of which is directed along the anisotropy axis.

The energy density of uniaxial anisotropy in the system $Ox'y'z'$ has the appearance

$$U_a = K\left(m_{x'}^2 + m_{y'}^2\right),$$

(3.20)

where in the case of a light axis, $K > 0$. To express this energy density in the $Oxyz$ system, one needs the components of the vector **m'** express through the components of the vector **m**, that is, to carry out the transformation

$$\mathbf{m'} = \overset{\leftrightarrow}{A}^{-1}\mathbf{m},$$

(3.21)

where $\overset{\leftrightarrow}{A}^{-1}$ is the transition matrix from the $Oxyz$ system to the $Ox'y'z'$ system.

This transformation is done using an inverse matrix. After this, it is necessary to substitute the components of the vector **m** into the expression (3.20), as a result of which the energy density is obtained, expressed through the components of the vector **m**.

Next, the problem is solved in the original $Oxyz$ system.

Case number 2. The problem is solved in a rotated coordinate system $Ox'y'z'$, in which the axis Oz' is oriented along some important direction for this consideration, and the anisotropy energy density has the simplest form in another system related with some characteristic direction for anisotropy. In this case, the anisotropy energy density should be initially recorded in the system where its form is the simplest, after which this energy density should be transformed into a rotated system.

This case is suitable for cubic anisotropy, when the axis Oz' of the rotated coordinate system is oriented along some characteristic directions of a cubic cell, for example, an edge, a diagonal of a face, or a spatial diagonal of a cube. The energy density has the simplest form in the $Oxyz$ system, the axes of which Ox, Oy and Oz are directed along the edges of the cube.

This energy density has the form:

$$U_c(001) = -K_1\left(m_x^2 m_y^2 - m_y^2 m_z^2 + m_z^2 m_x^2\right),$$

(3.22)

where $K_1 > 0$ corresponds to the easy axes of the type (111).

To express this energy density in the system $Ox'y'z'$, it is necessary to express the components of the vector **m** through the components of the vector **m'**, That is, to carry out the transformation:

$$\mathbf{m} = \overset{\leftrightarrow}{A}\mathbf{m'},$$

(3.23)

where $\overleftrightarrow{A}$ is the transition matrix of the system $Ox'y'z'$ to the $Oxyz$ system. This transformation is carried out using a direct matrix. After this, it is necessary to substitute the components of the vector **m** into the expression (3.22), as a result of which the energy density is obtained, expressed through the components of the vector **m'**. Further, the problem is solved in the rotated system $Ox'y'z'$, however, in order not to overload the recording with strokes, they are omitted and it is assumed that the rotated system is a laboratory one. This is equivalent to setting the cubic cell in the laboratory coordinate system in such a way that the selected cell direction (edge, face diagonal or the spatial diagonal of the cube) is oriented along the Oz axis of the laboratory $Oxyz$ coordinate system.

3.3.1. Auxiliary geometric problems used to obtain a coordinate transformation matrix

The coordinate transformation matrix is obtained by rotating the coordinate system. The components of the matrix are the cosines of the angles between the old and new axes. To find them, we need to solve a few auxiliary problems known from analytic geometry. *A list of auxiliary tasks.*

1. The known vector of arbitrary length. Find the corresponding unit vector.

2. Two vectors are known. Find the angle between them.

3. Two vectors are known. Find the components of a vector perpendicular to both source vectors.

4. The known vector. Find the equation of a line parallel to this vector passing through a given point.

5. The vector of the normal to the plane and the distance from the origin to the plane are known. Find the equation of the plane.

6. The coordinates of the three-sided plane are known. Find the equation of the plane.

7. The equation of the plane is known. Find the components of the normal vector to this plane.

8. The equations of two planes are known. Find the angle between them.

9. Equations of two planes are known. Find the equation of the beam of planes passing through the line of intersection of the original planes. We consider these problems separately, mainly following [261,292,293].

Problem No. 1. A vector of arbitrary length is known. To find a unit vector corresponding to it.

Source vector:

$$\mathbf{a} = \{X, Y, Z\}. \tag{3.24}$$

Normalization factor:

$$\mu = \frac{1}{\sqrt{X^2 + Y^2 + Z^2}}. \tag{3.25}$$

The required unit vector is

$$\mathbf{l}_a = \{\mu X, \ \mu Y, \ \mu Z\}, \tag{3.26}$$

or

$$\mathbf{l}_a = \left\{ \frac{X}{\sqrt{X^2 + Y^2 + Z^2}}, \ \frac{Y}{\sqrt{X^2 + Y^2 + Z^2}}, \ \frac{Z}{\sqrt{X^2 + Y^2 + Z^2}} \right\}. \tag{3.27}$$

Problem number 2. Two vectors are known. Find the angle between them.

Source vectors:

$$\mathbf{a} = \{X_1, \ Y_1, \ Z_1\}, \tag{3.28}$$

$$\mathbf{b} = \{X_2, \ Y_2, \ Z_2\} \tag{3.29}$$

The angle between the two vectors is from the scalar product:

$$\cos\varphi = \frac{X_1 X_2 + Y_1 Y_2 + Z_1 Z_2}{\sqrt{X_1^2 + Y_1^2 + Z_1^2} \cdot \sqrt{X_2^2 + Y_2^2 + Z_2^2}}. \tag{3.30}$$

Problem number 3. Two vectors are known. Find the components of a vector perpendicular to both source vectors.

Source vectors:

$$\mathbf{a} = \{X_1, \ Y_1, \ Z_1\}, \tag{3.31}$$

$$\mathbf{b} = \{X_2, \ Y_2, \ Z_2\} \tag{3.32}$$

The components of a vector perpendicular to the two original ones are found using the vector product:

$$[\mathbf{a} \times \mathbf{b}] = \begin{vmatrix} \mathbf{i} & \mathbf{j} & \mathbf{k} \\ X_1 & Y_1 & Z_1 \\ X_2 & Y_2 & Z_2 \end{vmatrix}. \tag{3.33}$$

The desired unit vector, perpendicular to both source vectors, has the form

$$\mathbf{m} = \frac{[\mathbf{a}\times\mathbf{b}]}{\|[\mathbf{a}\times\mathbf{b}]\|}. \tag{3.34}$$

Problem number 4. Known vector. Find the equation of a line, parallel to this vector passing through a given point.
Source vector

$$\mathbf{a} = \{l,\ m,n\}. \tag{3.35}$$

Set point
$$(x_0,\ y_0,\ z_0). \tag{3.36}$$
The sought equation of a line parallel to vector $\mathbf{a}$ passing through the point $(x_0,\ y_0,\ z_0)$ has the form

$$\frac{x-x_0}{l} = \frac{y-y_0}{m} = \frac{z-z_0}{n}, \tag{3.37}$$

These are the two equations of the planes. The desired line is the intersection of these planes.

Problem number 5. The vector of the normal to the plane and the distance from the origin to the plane are known. Find the equation of the plane.
Normal vector

$$\mathbf{n} = \{n_x,\ n_y,\ n_z\}, \tag{3.38}$$

Distance from the origin of the coordinates to the plane: p.
The desired equation of the plane:

$$xn_x + yn_y + zn_z - p = 0. \tag{3.39}$$

Problem number 6. The coordinates of three points of the plane are known. Find the equation of the plane.
Coordinates of points:

$$(x_1,\ y_1,\ z_1); \tag{3.40}$$
$$(x_2,\ y_2,\ z_2); \tag{3.41}$$

$$(x_3, y_3, z_3); \qquad (3.42)$$

The sought equation plane

$$\begin{vmatrix} x & y & z & 1 \\ x_1 & y_1 & z_1 & 1 \\ x_2 & y_2 & z_2 & 1 \\ x_3 & y_3 & z_3 & 1 \end{vmatrix} = 0. \qquad (3.43)$$

Problem No. 7. The equation of the plane is known. Find components of the normal vector to this plane.
The equation of the original plane

$$Ax + By + Cz + D = 0. \qquad (3.44)$$

The normalization factor

$$\mu = \frac{-\dfrac{D}{|D|}}{\sqrt{A^2 + B^2 + C^2}}. \qquad (3.45)$$

The required normal vector

$$\mathbf{n} = \{\mu A,\ \mu B,\ \mu C\}, \qquad (3.46)$$

Due to the two-sidedness of the plane, there is another normal vector, equal to

$$\mathbf{n}^{(-)} = -\mathbf{n}. \qquad (3.47)$$

Problem No. 8. The equations of two planes are known. To find the angle between them.
The equations of the original planes:
$$A_1 x + B_1 y + C_1 z + D_1 = 0; \qquad (3.48)$$
$$A_2 x + B_2 y + C_2 z + D_2 = 0; \qquad (3.49)$$
The angle between two planes is equal to the angle between the normal vectors to these planes.
Normalizing factors:

$$\mu_1 = \frac{-\dfrac{D_1}{|D_1|}}{\sqrt{A_1^2 + B_1^2 + C_1^2}};$$

(3.50)

$$\mu_2 = \frac{-\dfrac{D_2}{|D_2|}}{\sqrt{A_2^2 + B_2^2 + C_2^2}}.$$

(3.51)

Normal vectors:

$$\mathbf{n}_1 = \{\mu_1 A_1, \mu_1 B_1, \mu_1 C_1\};$$

(3.52)

$$\mathbf{n}_2 = \{\mu_2 A_2, \mu_2 B_2, \mu_2 C_2\};$$

(3.53)

The required angle is determined by the scalar product:

$$\cos\varphi = \frac{n_{1x} n_{2x} + n_{1y} n_{2y} + n_{1z} n_{2z}}{\sqrt{n_{1x}^2 + n_{1y}^2 + n_{1z}^2}\sqrt{n_{2x}^2 + n_{2y}^2 + n_{2z}^2}}.$$

(3.54)

After reducing the normalized multipliers we get

$$\cos\varphi = \frac{A_1 A_2 + B_1 B_2 + C_1 C_2}{\sqrt{A_1^2 + B_1^2 + C_1^2}\sqrt{A_2^2 + B_2^2 + C_2^2}}.$$

(3.55)

Problem No. 9. The equations of two planes are known. To find the equation of a beam of planes passing through the line of intersection of the original planes.

The equations of the original planes:

$$A_1 x + B_1 y + C_1 z + D_1 = 0;$$

(3.56)

$$A_2 x + B_2 y + C_2 z + D_2 = 0;$$

(3.57)

The desired equation for the beam planes:

$$(A_1 x + B_1 y + C_1 z + D_1) + \lambda(A_2 x + B_2 y + C_2 z + D_2) = 0.$$

(3.58)

where λ is a free parameter.

This equation defines all planes, except the second of the original ones.

3.4. Transformation matrix in the case of uniaxial anisotropy

We first consider the simplest type of anisotropy – uniaxial, but suppose that the anisotropy axis is oriented relative to the original (laboratory) coordinate system in an arbitrary way.

3.4.1. General geometry of the problem with arbitrary orientation of the anisotropy axis

The task is to find the transformation matrix from the coordinate system $Oxyz$ to the coordinate system $Ox'y'z'$, the Oz' aix s of which is directed arbitrarily, and the Ox' and Oy' axes are rotated around this axis also at an arbitrary angle. The overall geometry of the problem is illustrated in Fig. 3.2. Here is shown the initial coordinate system $Oxyz$, as well as the $Ox'y'z'$ rotated system, which has a common beginning with the first at the point O. The Oz' axis of the rotated system has an arbitrary direction coinciding with the segment OM. The projections of this segment on the axes Ox, Oy and Oz correspond to the segments OT, OS and ON. The projection of the segment OM on the plane Oxy is denoted by OP. The straight line AB passes through the point O and lies in the plane Oxy, perpendicular

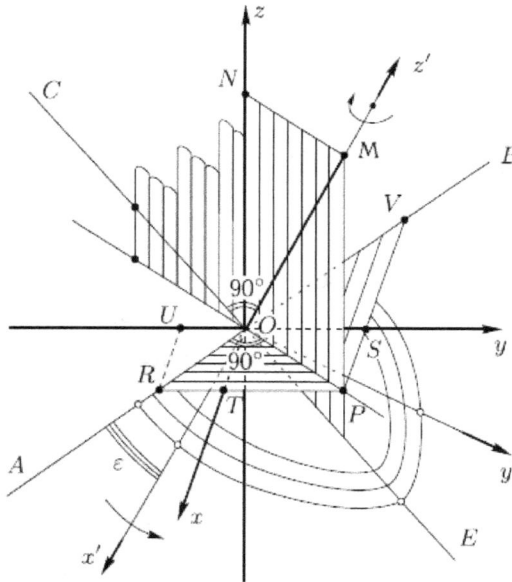

Fig.3.2. General geometry of the arbitrary orientation of axis of anisotropy

to the segment *OM*, and therefore also to the segment *OP*. Thus, ∠AOP = 90°. Point *R* corresponds to the intersection extending the line *PT* with line *AB*. The point *U* corresponds to the projection of the point *R* on the axis *Oy*. Point *V* corresponds to the intersection of the continuation of the line *PS* from the straight line *AB*. The *MOC* plane passes through the *Oz* and *Oz'* axes. The straight *COE* lies in the *MOC* plane and makes up with the axis *Oz'* the angles ∠MOC = ∠MOE = 90°. Due to the perpendicularity of the straight lines *AB* and *CE* to the axis *Oz'*, the plane *ACBEO* is perpendicular to the axis *Oz'*. So the axes of the rotated system *Ox'* and *Oy'* lie in the *ACBEO* plane. In this case, the rotation of the system *Ox'y'z'* around the *Oz'* axis is determined by the angle ε. This angle is determined by the relative position of the axis *Ox'* and straight *OA* and is counted from straight *OA* to straight *OE*. By virtue of the mutual perpendicularity of the axes *Ox'* and *Oy'*, as well as direct *OA* and *OE*, the angle between the straight line *OE* and the axis *Oy'* is also equal to ε.

Note on the direction of reference of the angle of rotation. In Fig. 3.2 the angle ε is represented as the angle between the axis *Ox'* and direct *OA*. We will consider as positive the direction of reference of the angle ε from the straight line *OA* to the straight line *OE*, as indicated by the arrow on the axis *Ox'*. This choice of reference direction is convenient if we rotate the *Ox'y'z'* coordinate system around the axis *Oz'*. At the same time, if we consider the *Ox'y'z'* system as fixed and the *Oxyz* system rotated relative to it, then with the same direction of reference of the angle ε, the cell will rotate in the opposite direction, as indicated by the arrow on the *Oz'* axis. Thus, if we look from the direction of the positive direction of the axis *Oz'*, then the angle ε is counted counterclockwise, and the system *Oxyz* rotates clockwise. If we look from the negative direction of the axis *Oz'*, then the angle ε is counted clockwise, and the system *Oxyz* rotates anti-clockwise.

3.4.2. The overall structure of the direct transition matrix

The transformation of the vector components from one system to another is carried out using the transition matrix and the formula:

$$\mathbf{a} = \ddot{A}\mathbf{a}', \tag{3.59}$$

where **a** is a vector in the original coordinate system, **a'** is the same vector in the rotated coordinate system, $\ddot{A}$ is the transition matrix from the rotated system to the initial one.

The components of the matrix $\ddot{A}$ are the reference cosines of the unit vector $\mathbf{l}_{x'}$, $\mathbf{l}_{y'}$, $\mathbf{l}_{z'}$ of the axes of the $Ox'y'z'$ coordinate system, relative to the $Oxyz$ system, that is, the first task is to find these different factors.

3.4.3. Coordinates of the main points

To find the unit vectors, we need the coordinates of some of the original points of the general geometry of the problem.

It will be assumed that the segment OM, which determines the direction of the axis Oz', is given initially. We denote the length of this segment by B and write its projections on the axes Ox, Oy and Oz:

$$OT = B_x; \qquad\qquad (3.60)$$
$$OS = B_y; \qquad\qquad (3.61)$$
$$ON = B_z; \qquad\qquad (3.62)$$

wherein:

$$B = \sqrt{B_x^2 + B_y^2 + B_z^2}. \qquad\qquad (3.63)$$

We will determine further the coordinates of the characteristic points through these projections B_x, B_y and B_z.

So, the coordinates of points O, M, N, T, S, P are equal:

$$O \rightarrow (0,0,0); \qquad\qquad (3.64)$$
$$M \rightarrow (B_z, B_y, B_z); \qquad\qquad (3.65)$$
$$N \rightarrow (0,0,B_z); \qquad\qquad (3.66)$$
$$T \rightarrow (B_x, 0, 0); \qquad\qquad (3.67)$$
$$S \rightarrow (0, B_y, 0); \qquad\qquad (3.68)$$
$$P \rightarrow (B_x, B_y, 0). \qquad\qquad (3.69)$$

To find the coordinates of the remaining points, we consider their location on the Oxy plane, shown in Fig. 3.3 (top view from the positive direction of the Oz axis).

Here: $OT = B_x$; $OS = B_y$. From $RP\|Oy$, $VP\|Ox$, $OP \perp AB$, follows the similarity of the following squares: $\triangle OUR$, $\triangle ORT$, $\triangle OTP$, $\triangle OPS$, $\triangle OSV$. In this case, we obtain:

$$R \rightarrow \left(B_x, -\frac{B_x^2}{B_y}, 0 \right),$$
$$\qquad\qquad (3.70)$$

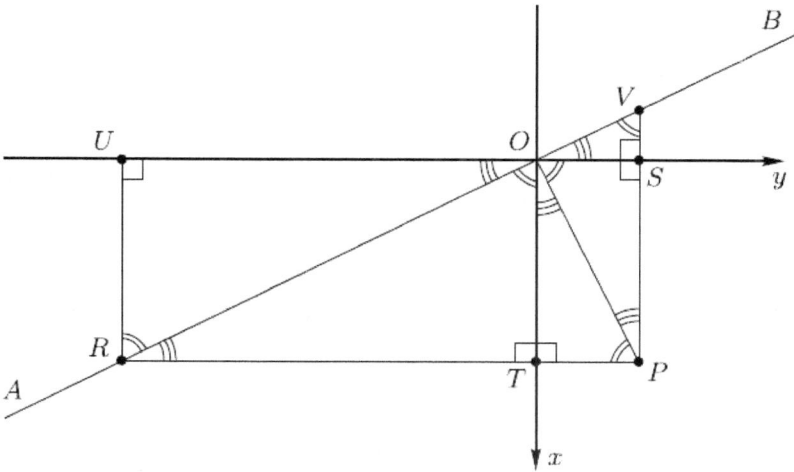

Fig. 3.3. The layout of the main points on the plane Oxy.

$$U \to \left(0, -\frac{B_x^2}{B_y}, 0\right),\qquad(3.71)$$

$$V \to \left(-\frac{B_y^2}{B_x}, B_y, 0\right).\qquad(3.72)$$

3.4.4. Determination of the rotation angle around the anisotropy axis

Components of unitary axes of rotated coordinate system $\mathbf{l}_{x'}$, $\mathbf{l}_{y'}$, $\mathbf{l}_{z'}$ are determined according to the rules of analytical geometry, with the help of auxiliary tasks Nos. 1–9.

The unit vector of the anisotropy axis $\mathbf{l}_{z'}$. We first find the unit vector of the $O_{z'}$ axis, that is, the vector $\mathbf{l}_{z'}$. In the direction this vector coincides with the segment OM. With the help of problem No. 1, one component of the vector $\mathbf{l}_{z'}$ if found as:

$$\mathbf{l}_{z'} = \left\{\frac{B_x}{B}, \frac{B_y}{B}, \frac{B_z}{B}\right\} =$$

$$= \left\{\frac{B_x}{\sqrt{B_x^2 + B_y^2 + B_z^2}}, \frac{B_y}{\sqrt{B_x^2 + B_y^2 + B_z^2}}, \frac{B_z}{\sqrt{B_x^2 + B_y^2 + B_z^2}}\right\}.\qquad(3.73)$$

The independence of this vector from the angle of rotation of the system $Ox'y'z'$ around the Oz' axis is consistent with the general formulation of the problem. It is easy to see that the sum of the squares of the components of the vector $\mathbf{l}_z$, equal to one, which corresponds to its single character.

Angle of rotation around the axis of anisotropy. Vector $\mathbf{l}_z$ defines only one of the axes of the coordinate system $Ox'y'z'$ relative to the axis $Oxyz$. The position of the double axes $O_{x'}$ and $O_{y'}$ determined by the rotation of the $Ox'y'z'$ systems around the $O_{z'}$ at an angle ε, which is also initially given. Since the main task is the expression of the unit vectors $\mathbf{l}_{y'}$ and $\mathbf{l}_{x'}$ of the $Ox'y'z'$ systems through this angle, consider its relationship with the values of B_x, B_y and B_z, which determine the direction of the axis $O_{z'}$.

Define the angle ε as the angle between the lines of intersection of the planes MOx' and $AOBM$ with the $ACBEO$ plane. The possibility of such a determination follows from the perpendicularity of both these planes to the plane $ACBEO$, which in turn follows from the passage of both these planes through the line OM, that is, the axis $O_{z'}$, which is perpendicular to the plane $ACBEO$. In this case, ε is measured in the $ACBEO$ plane from the Ox axis towards the OE line.

To find the connection of the angle ε with the parameters of the mentioned planes, we use problem No. 8, according to which the cosine of the angle between the planes is determined through the normal vectors to these planes. The angle ε consists of two planes: $AOBM$ and MOx'.

We can find the equation of the $AOBM$ plane using problem # 6 from the condition of passing the plane through three points: $O(0, 0, 0)$, $M(B_x, B_y, B_z)$ and $R(B_x^2 - B_x^2/B_y, 0)$.

So we get

$$\begin{vmatrix} x & y & z & 1 \\ 0 & 0 & 0 & 1 \\ B_x & B_y & B_z & 1 \\ B_x & -B_x^2/B_y & 0 & 1 \end{vmatrix} = 0 \qquad (3.74)$$

from where we get the $AOBM$ plane equation as

$$B_x^2 B_z x + B_x B_y B_z y - B_x \left(B_x^2 + B_y^2 \right) z = 0 / \qquad (3.75)$$

Dividing by B_x, we get

$$B_x B_z x + B_y B_z y - \left(B_x^2 + B_y^2\right)z = 0. \tag{3.76}$$

To find the equation of the plane MOx' we will construct a bundle of planes passing through the axis Oz', that is, the line OM, after which we will select the desired plane from this bundle. As the reference planes of the beam, we select the $AOBM$ and MON planes intersecting along the OM line. The equation for the $AOBM$ plane is (3.76).

The MON equation of the plane is found using problem No. 6, considering that this plane passes through three points: $O(0, 0, 0)$, $M(B_x, B_y, B_z)$ and $N(0, 0, B_z)$, that is, from the relation

$$\begin{vmatrix} x & y & z & 1 \\ 0 & 0 & 0 & 1 \\ B_x & B_y & B_z & 1 \\ 0 & 0 & B_z & 1 \end{vmatrix} = 0 \tag{3.77}$$

where do we get the equation for the MON plane as

$$B_x B_z x - B_x B_z y = 0. \tag{3.78}$$

Dividing by B_z gives

$$B_y x - B_x y = 0. \tag{3.79}$$

Thus, to build a beam of planes passing through the Oz' axis, we obtain two equations of the $AOBM$ and MON planes:

$$B_x B_z x + B_y B_z y - (B_x^2 + B_y^2)z = 0. \tag{3.80}$$

$$B_y x - B_x y = 0. \tag{3.81}$$

Using problem No. 9, we obtain the beam equation in the form

$$B_x B_z x + B_y B_z y - \left(B_x^2 + B_y^2\right)z + \lambda\left(B_y x - B_x y\right) = 0 \tag{3.82}$$

or:

$$\left(B_x B_z + \lambda B_y\right)x + \left(B_y B_z - \lambda B_x\right)y - \left(B_x^2 + B_y^2\right)z = 0, \tag{3.83}$$

where λ is a parameter that must be determined from the condition of equality of the angle ε between the *AOBM* plane and the other, not yet known *MOx'* plane contained in the beam (3.83).

So, we will consider two planes *AOBM* and *MOx'*, having the equations:

$$B_x B_z x + B_y B_z y - \left(B_x^2 + B_y^2\right) z = 0, \tag{3.84}$$

$$\left(B_x B_z + \lambda B_y\right) x + \left(B_y B_z - \lambda B_x\right) y - \left(B_x^2 + B_y^2\right) z = 0, \tag{3.85}$$

and using problem No. 8 we find the cosine of the angle ε between them:

$$\cos \varepsilon = \frac{B}{\sqrt{B^2 + \lambda^2}}, \tag{3.86}$$

where *B* is defined by the expression (3.63).

From (3.86) we get

$$\sin \varepsilon = \pm \frac{\lambda}{\sqrt{B^2 + \lambda^2}}. \tag{3.87}$$

These expressions can be considered as equations for determining λ, based on a given value of the angle ε. At the same time we get

$$\lambda_{1,2} = \pm B \ \text{tg} \ \varepsilon \tag{3.88}$$

3.4.5. Getting guided vectors of a rotated system of coordinates through the rotation angle around the anisotropy axis

Vector $\mathbf{l}_{y'}$. We now find the vector $\mathbf{l}_{y'}$, which depends on the angle ε of rotation of the *Ox'y'z'* system around the *Oz'* axis.

Using task No. 7, we will find this vector as a normal to the plane *MOx'*, whose equation has the form (3.85). The normalizing factor is

$$\mu = -\frac{1}{\sqrt{\left(B_x^2 + B_y^2\right)\left(B^2 + \lambda^2\right)}}. \tag{3.89}$$

In this case, we obtain two possible normals:

$$\mathbf{I}_{y'}^{(+)} = \{\mu\left(B_xB_z + \lambda B_y\right),\, \mu\left(B_x^2 + B_y^2\right) =$$

$$\left\{-\frac{B_xB_z + \lambda B_y}{\sqrt{\left(B_x^2 + B_y^2\right)\left(B^2 + \lambda^2\right)}},\, -\frac{B_yB_z - \lambda B_x}{\sqrt{\left(B_x^2 + B_y^2\right)\left(B^2 + \lambda^2\right)}},\, \frac{B_x^2 + B_y^2}{\sqrt{\left(B_x^2 + B_y^2\right)\left(B^2 + \lambda^2\right)}}\right\},$$

$$(3.90)$$

$$\mathbf{I}_{y'}^{(-)} = \{-\mu\left(B_xB_z + \lambda B_y\right),\, -\mu\left(B_yB_z - \lambda B_x\right),\, -\mu\left(B_x^2 + B_y^2\right) =$$

$$\left\{-\frac{B_xB_z + \lambda B_y}{\sqrt{\left(B_x^2 + B_y^2\right)\left(B^2 + \lambda^2\right)}},\, -\frac{B_yB_z - \lambda B_x}{\sqrt{\left(B_x^2 + B_y^2\right)\left(B^2 + \lambda^2\right)}},\, \frac{B_x^2 + B_y^2}{\sqrt{\left(B_x^2 + B_y^2\right)\left(B^2 + \lambda^2\right)}}\right\}.$$

$$(3.91)$$

Substituting λ in accordance with (3.88) and taking into account two signs, we get

The components of these expressions are:

$$\frac{B_xB_z + \lambda B_y}{\sqrt{\left(B_x^2 + B_y^2\right)\left(B^2 + \lambda^2\right)}} = \frac{B_xB_z}{B\sqrt{B_y^2 + B_y^2}}\cos\varepsilon \pm \frac{B_y}{\sqrt{B_x^2 + B_y^2}}\sin\varepsilon; \quad (3.92)$$

$$\frac{B_yB_z - \lambda B_x}{\sqrt{\left(B_x^2 + B_y^2\right)\left(B^2 + \lambda^2\right)}} = \frac{B_yB_z}{B\sqrt{B_x^2 + B_y^2}}\cos\varepsilon \pm \frac{B_x}{\sqrt{B_x^2 + B_y^2}}\sin\varepsilon; \quad (3.93)$$

$$\frac{B_x^2 + B_y^2}{\sqrt{\left(B_x^2 + B_y^2\right)\left(B^2 + \lambda^2\right)}} = \frac{\sqrt{B_x^2 + B_y^2}}{B}\cos\varepsilon. \quad (3.94)$$

Two signs in these expressions in combination with two signs in (3.90) and (3.91) correspond to two possible directions normal to the plane MOx' and two possible configurations of the Cartesian coordinates – right and left. For a unique choice of the right coordinate system, axis Ox' which when $\varepsilon = 0$ coincides with the line OA, and the axis Oy' with the direct line OE, we can consider a special case of the orientation of the system $Ox'y'z'$ regarding the $Oxyz$ system, in which the choice of the proper ones in expressions (3.90)–(3.94) would be obvious. Such a particular case can be the rotation of the $Ox'y'z'$ system around the spatial diagonal of a cubic crystallographic cell. With this the Oz' is oriented along the unit vector

$$\mathbf{l}^{[111]} = \left\{ \frac{\sqrt{3}}{3}, \frac{\sqrt{3}}{3}, \frac{\sqrt{3}}{3} \right\}, \tag{3.95}$$

and the orientation of the axes Ox' and Oy' should be selected corresponding to two characteristic values of the angle ε: zero and 90°. Comparison of the signs included in expressions (3.90)–(3.94) with the signs of the components of the vector $\mathbf{l}_{y'}$ resulting in this particular case, allows unambiguously to choose signs (3.90)–(3.94). As a result, we obtain the components of the vector $\mathbf{l}_{y'}$ as:

$$\left(\mathbf{l}_{y'}\right)_x = \frac{B_x B_z}{B\sqrt{B_x^2 + B_y^2}} \cos \varepsilon - \frac{B_y}{\sqrt{B_x^2 + B_y^2}} \sin \varepsilon =$$

$$= \frac{1}{B\sqrt{B_x^2 + B_y^2}} \left(B_x B_z \cos \varepsilon - B B_y \sin \varepsilon \right); \tag{3.96}$$

$$\left(\mathbf{l}_{y'}\right)_y = \frac{B_y B_z}{B\sqrt{B_x^2 + B_y^2}} \cos \varepsilon + \frac{B_x}{\sqrt{B_x^2 + B_y^2}} \sin \varepsilon =$$

$$= \frac{1}{B\sqrt{B_x^2 + B_y^2}} \left(B_y B_z \cos \varepsilon + B B_x \sin \varepsilon \right); \tag{3.97}$$

$$\left(\mathbf{l}_{y'}\right)_z = \frac{\sqrt{B_x^2 + B_y^2}}{B} \cos \varepsilon. \tag{3.98}$$

A direct check can be made to ensure that the ratio:

$$\left(\mathbf{l}_{y'}\right)_x^2 + \left(\mathbf{l}_{y'}\right)_y^2 + \left(\mathbf{l}_{y'}\right)_z^2 = 1, \tag{3.99}$$

that is, the normalization condition on the unit is completely satisfied.

Vector $l_{x'}$. Now we find the vector $\mathbf{l}_{x'}$. As perpendicular to the vectors $\mathbf{C}$ and $\mathbf{l}_z$, using task No. 3:

$$\mathbf{l}_{x'} = \frac{\left[\mathbf{l}_{y'} \times \mathbf{l}_{z'} \right]}{\left[\left[\mathbf{l}_{y'} \times \mathbf{l}_{z'} \right] \right]}. \tag{3.100}$$

The vector product of vectors $\mathbf{l}_{y'}$ and $\mathbf{l}_{z'}$ has the appearance

$$[\mathbf{l}_{y'} \times \mathbf{l}_{z'}] = \begin{vmatrix} \mathbf{i} & \mathbf{j} & \mathbf{k} \\ (\mathbf{l}_{y'})_x & (\mathbf{l}_{y'})_y & (\mathbf{l}_{y'})_z \\ (\mathbf{l}_{z'})_x & (\mathbf{l}_{z'})_y & (\mathbf{l}_{z'})_z \end{vmatrix} = \mathbf{i} \cdot \left\{ (\mathbf{l}_{y'})_y (\mathbf{l}_{z'})_z - (\mathbf{l}_{y'})_z (\mathbf{l}_{z'})_y \right\} +$$

$$+ \mathbf{j} \cdot \left\{ (\mathbf{l}_{y'})_z (\mathbf{l}_{z'})_x - (\mathbf{l}_{y'})_x (\mathbf{l}_{z'})_z \right\} + \mathbf{k} \cdot \left\{ (\mathbf{l}_{y'})_x (\mathbf{l}_{z'})_y - (\mathbf{l}_{y'})_y (\mathbf{l}_{z'})_x \right\}.$$

$$(3.101)$$

The module of a vector product is

$$\left| \left[\mathbf{l}_{y'} \times \mathbf{l}_{z'} \right] \right| = 1. \tag{3.102}$$

Substituting (3.96)–(3.98) and (3.73), we obtain the components of the vector $\mathbf{l}_{x'}$ as

$$\left(\mathbf{l}_{x'} \right)_x = \frac{1}{B\sqrt{B_x^2 + B_y^2}} \left(BB_y \cos\varepsilon + B_x B_z \sin\varepsilon \right); \tag{3.103}$$

$$\left(\mathbf{l}_{x'} \right)_y = \frac{1}{B\sqrt{B_x^2 + B_y^2}} \left(BB_x \cos\varepsilon + B_y B_z \sin\varepsilon \right); \tag{3.104}$$

$$*\left(\mathbf{l}_{x'} \right)_z = \frac{\sqrt{B_x^2 + B_y^2}}{B} \sin\varepsilon. \tag{3.105}$$

It can be seen that here the normalization condition is also satisfied.

3.4.6. Set of unit vectors of rotated coordinate system

So, in accordance with (3.103)–(3.105), (3.96)–(3.98) and (3.73), we obtain a set of unitary vectors of the system $Ox'y'z'$, whose components are expressed in the $Oxyz$ system, in the form:

$$\left(\mathbf{l}_{x'} \right)_x = \frac{1}{B\sqrt{B_x^2 + B_y^2}} \left(BB_y \cos\varepsilon + B_x B_z \sin\varepsilon \right); \tag{3.106}$$

$$\left(\mathbf{l}_{x'} \right)_y = \frac{1}{B\sqrt{B_x^2 + B_y^2}} \left(BB_x \cos\varepsilon + B_y B_z \sin\varepsilon \right); \tag{3.107}$$

$$\left(\mathbf{l}_{x'} \right)_z = -\frac{\sqrt{B_x^2 + B_y^2}}{B} \sin\varepsilon; \tag{3.108}$$

$$\left(\mathbf{l}_{y'} \right)_x = -\frac{1}{B\sqrt{B_x^2 + B_y^2}} \left(B_x B_z \cos\varepsilon - BB_y \sin\varepsilon \right); \tag{3.109}$$

$$\left(\mathbf{1}_{y'}\right)_y = -\frac{1}{B\sqrt{B_x^2 + B_y^2}}\left(B_y B_z \cos\varepsilon - BB_x \sin\varepsilon\right);$$

(3.110)

$$\left(\mathbf{1}_{y'}\right)_z = -\frac{\sqrt{B_x^2 + B_y^2}}{B}\cos\varepsilon.$$

(3.111)

$$\left(\mathbf{1}_{z'}\right)_x = \frac{B_x}{B};$$

(3.112)

$$\left(\mathbf{1}_{z'}\right)_y = \frac{B_y}{B};$$

(3.113)

$$\left(\mathbf{1}_{z'}\right)_z = \frac{B_z}{B};$$

(3.114)

where, in accordance with (3.63)

$$B = \sqrt{B_x^2 + B_y^2 + B_z^2}.$$

(3.115)

3.4.7. Full view of direct transition matrix

We reduce the components of the vectors $\mathbf{1}_{x'}$, $\mathbf{1}_{y'}$, $\mathbf{1}_{z'}$ in a single table:

	x'	y'	z'
x	$\left(\mathbf{1}_{x'}\right)_x$	$\left(\mathbf{1}_{y'}\right)_x$	$\left(\mathbf{1}_{z'}\right)_x$
y	$\left(\mathbf{1}_{x'}\right)_y$	$\left(\mathbf{1}_{y'}\right)_y$	$\left(\mathbf{1}_{z'}\right)_y$
z	$\left(\mathbf{1}_{x'}\right)_z$	$\left(\mathbf{1}_{y'}\right)_z$	$\left(\mathbf{1}_{z'}\right)_z$

(3.116)

In this case, the transition matrix takes the form:

$$\overset{\leftrightarrow}{A} = \begin{pmatrix} \left(\mathbf{1}_{x'}\right)_x & \left(\mathbf{1}_{y'}\right)_x & \left(\mathbf{1}_{z'}\right)_x \\ \left(\mathbf{1}_{x'}\right)_y & \left(\mathbf{1}_{y'}\right)_y & \left(\mathbf{1}_{z'}\right)_y \\ \left(\mathbf{1}_{x'}\right)_z & \left(\mathbf{1}_{y'}\right)_z & \left(\mathbf{1}_{z'}\right)_z \end{pmatrix}.$$

(3.117)

Writing the matrix using the component notation:

$$\overset{\leftrightarrow}{A} = \begin{pmatrix} A_{11} & A_{12} & A_{13} \\ A_{21} & A_{22} & A_{23} \\ A_{31} & A_{32} & A_{33} \end{pmatrix},$$

(3.118)

we get:

$$A_{11} = (\mathbf{1}_{x'})_x = \frac{1}{B\sqrt{B_x^2 + B_y^2}}(BB_y \cos\varepsilon + B_x B_z \sin\varepsilon); \qquad (3.119)$$

$$A_{12} = (\mathbf{1}_{y'})_x = \frac{1}{B\sqrt{B_x^2 + B_y^2}}(B_x B_z \cos\varepsilon - BB_y \sin\varepsilon); \qquad (3.120)$$

$$A_{13} = (\mathbf{1}_{z'})_x = \frac{B_x}{B}; \qquad (3.121)$$

$$A_{21} = (\mathbf{1}_{x'})_y = -\frac{1}{B\sqrt{B_x^2 + B_y^2}}(BB_x \cos\varepsilon - B_y B_z \sin\varepsilon); \qquad (3.122)$$

$$A_{22} = (\mathbf{1}_{y'})_y = \frac{1}{B\sqrt{B_x^2 + B_y^2}}(B_y B_z \cos\varepsilon + BB_x \sin\varepsilon); \qquad (3.123)$$

$$A_{23} = (\mathbf{1}_{z'})_y = \frac{B_y}{B}; \qquad (3.124)$$

$$A_{31} = (\mathbf{1}_{x'})_z = -\frac{\sqrt{B_x^2 + B_y^2}}{B}\sin\varepsilon; \qquad (3.125)$$

$$A_{32} = (\mathbf{1}_{y'})_z = -\frac{\sqrt{B_x^2 + B_y^2}}{B}\cos\varepsilon; \qquad (3.126)$$

$$A_{33} = (\mathbf{1}_{z'})_z = \frac{B_z}{B}. \qquad (3.127)$$

3.4.8. Inverse transition matrix

The inverse transitiom from the system $Oxyz$ to the system $Ox'y'z'$ takes place when using the inverse transition matrix A^{-1} and the equation

$$\mathbf{a'} = \overset{\leftrightarrow}{A}^{-1}\cdot\mathbf{a} \qquad (3.128)$$

where the the matrix $\overset{\leftrightarrow}{A}^{-1}$ is obtained from the direct transition matrix (3.117) by replacing rows with columns:

$$\overset{\leftrightarrow}{A}^{-1} = \begin{pmatrix} (\mathbf{1}_{x'})_x & (\mathbf{1}_{x'})_y & (\mathbf{1}_{x'})_z \\ (\mathbf{1}_{y'})_x & (\mathbf{1}_{y'})_y & (\mathbf{1}_{y'})_z \\ (\mathbf{1}_{z'})_x & (\mathbf{1}_{z'})_y & (\mathbf{1}_{z'})_z \end{pmatrix}. \qquad (3.129)$$

$$\overset{\leftrightarrow}{A}^{-1} = \begin{pmatrix} A_{11}^{-1} & A_{12}^{-1} & A_{13}^{-1} \\ A_{21}^{-1} & A_{22}^{-1} & A_{23}^{-1} \\ A_{31}^{-1} & A_{32}^{-1} & A_{33}^{-1} \end{pmatrix}, \qquad (3.130)$$

and:

$$A_{11}^{-1} = (1_{x'})_x = \frac{1}{B\sqrt{B_x^2 + B_y^2}}\left(BB_y \cos\varepsilon + B_x B_z \sin\varepsilon\right);$$

(3.131)

$$A_{12}^{-1} = (1_{x'})_x = \frac{1}{B\sqrt{B_x^2 + B_y^2}}\left(BB_x \cos\varepsilon + B_y B_z \sin\varepsilon\right);$$

(3.132)

$$A_{13}^{-1} = (1_{x'})_z = -\frac{\sqrt{B_x^2 + B_y^2}}{B}\sin\varepsilon;$$

(3.133)

$$A_{21}^{-1} = (1_{y'})_z = -\frac{1}{B\sqrt{B_x^2 + B_y^2}}\left(B_x B_z \cos\varepsilon - BB_y \sin\varepsilon\right);$$

(3.134)

$$A_{22}^{-1} = (1_{y'})_y = \frac{1}{B\sqrt{B_x^2 + B_y^2}}\left(B_y B_z \cos\varepsilon - BB_x \sin\varepsilon\right);$$

(3.135)

$$A_{23}^{-1} = (1_{y'})_z = \frac{\sqrt{B_x^2 + B_y^2}}{B}\cos\varepsilon;$$

(3.136)

$$A_{31}^{-1} = (1_{z'})_x = \frac{B_x}{B},$$

(3.137)

$$A_{32}^{-1} = (1_{z'})_y = \frac{B_y}{B},$$

(3.138)

$$A_{33}^{-1} = (1_{z'})_z = \frac{B_z}{B},$$

(3.139)

By direct verification, we can verify that the product of the forward and reverse transition matrices is equal to the identity matrix.

3.4.9. Transition matrices in spherical coordinate system

In the review, the direction of the axis Oz' specified in the Cartesian coordinate system $Oxyz$ through the projections of the segment OM on the axes of this system, is equal to B_x, B_y and B_z. The same direction can be set in the spherical coordinate system $Or\theta\varphi$ through the polar and azimuth angles θ and φ, as shown in Fig. 3.4.

Here

$$B_x = B \sin\theta \, \cos\varphi$$ (3.140)
$$B_y = B \sin\theta \, \sin\varphi$$ (3.141)
$$B_z = B \cos\theta$$ (3.142)

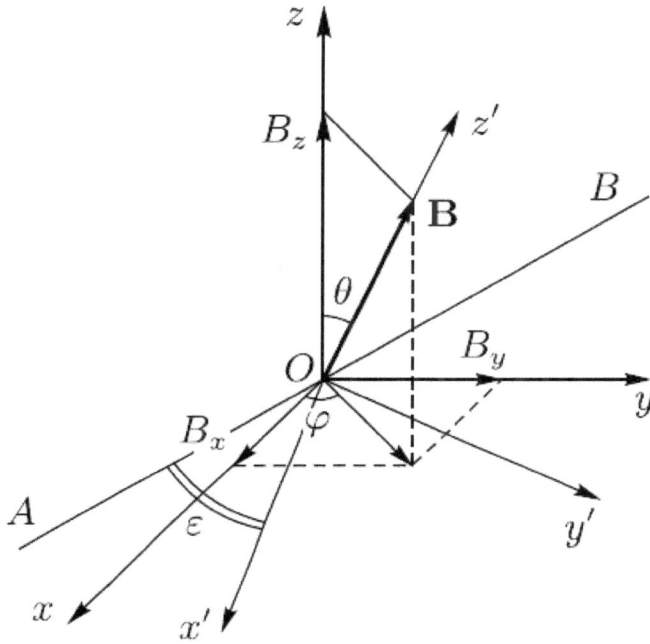

Fig. 3.4. The geometry of the problem in a spherical coordinate system.

Substituting these expressions into (3.1190–(3.127), we obtain the components of the transition matrix in the form:

$$A_{11} = \cos\theta\cos\varphi\ \sin\varepsilon + \sin\varphi\ \cos\varepsilon \tag{3.143}$$
$$A_{12} = \cos\theta\cos\varphi\ \cos\varepsilon - \sin\varphi\ \sin\varepsilon \tag{3.144}$$
$$A_{13} = \sin\theta\cos\varphi \tag{3.145}$$
$$A_{21} = \cos\theta\sin\varphi\ \sin\varepsilon - \cos\varphi\ \cos\varepsilon \tag{3.146}$$
$$A_{22} = \cos\theta\sin\varphi\ \cos\varepsilon + \cos\varphi\ \sin\varepsilon \tag{3.147}$$
$$A_{23} = \sin\theta\sin\varphi \tag{3.148}$$
$$A_{31} = -\sin\theta\sin\varepsilon \tag{3.149}$$
$$A_{32} = -\sin\theta\cos\varepsilon \tag{3.150}$$
$$A_{33} = \cos\theta. \tag{3.151}$$

The components of the inverse matrix, respectively, have the form:

$$A_{11}^{-1} = \cos\theta\cos\varphi\sin\varepsilon + \sin\varphi\cos\varepsilon; \tag{3.152}$$
$$A_{12}^{-1} = \cos\theta\sin\varphi\sin\varepsilon - \cos\varphi\cos\varepsilon; \tag{3.153}$$
$$A_{13}^{-1} = -\sin\theta\sin\varepsilon; \tag{3.154}$$
$$A_{21}^{-1} = \cos\theta\cos\varphi\cos\varepsilon - \sin\varphi\sin\varepsilon; \tag{3.155}$$

$$A_{22}^{-1} = \cos\theta\sin\varphi\cos\varepsilon + \cos\varphi\sin\varepsilon; \tag{3.156}$$

$$A_{31}^{-1} = -\sin\theta\cos\varepsilon; \tag{3.157}$$

$$A_{31}^{-1} = \sin\theta\cos\varphi; \tag{3.158}$$

$$A_{32}^{-1} = \sin\theta\sin\varphi; \tag{3.159}$$

$$A_{33}^{-1} = \cos\theta; \tag{3.160}$$

We write these matrices in the form of tables. So from (3.143) - (3.151) we get the matrix of the direct transition:

$$\overleftrightarrow{A} = \begin{pmatrix} \cos\theta\cos\varphi\sin\varepsilon + & \cos\theta\cos\varphi\cos\varepsilon - & \sin\theta\cos\varphi \\ + \sin\varphi\cos\varepsilon & - \sin\varphi\sin\varepsilon & \\ \cos\theta\sin\varphi\sin\varepsilon - & \cos\theta\sin\varphi\cos\varepsilon + & \sin\theta\sin\varphi \\ - \cos\varphi\cos\varepsilon & + \cos\varphi\sin\varepsilon & \\ - \sin\theta\sin\varepsilon & - \sin\theta\cos\varepsilon & \cos\theta \end{pmatrix}. \tag{3.161}$$

Accordingly, from (3.152)–(3.160) we get the inverse transition matrix:

$$\overleftrightarrow{A}^{-1} = \begin{pmatrix} \cos\theta\cos\varphi\sin\varepsilon + & \cos\theta\sin\varphi\sin\varepsilon - & - \sin\theta\sin\varepsilon \\ + \sin\varphi\cos\varepsilon & - \cos\varphi\cos\varepsilon & \\ \cos\theta\cos\varphi\cos\varepsilon - & \cos\theta\sin\varphi\cos\varepsilon + & - \sin\theta\cos\varepsilon \\ - \sin\varphi\sin\varepsilon & + \cos\varphi\sin\varepsilon & \\ \sin\theta\cos\varphi & \sin\theta\sin\varphi & \cos\theta \end{pmatrix}. \tag{3.162}$$

By direct verification, we can verify that the product of the forward and reverse transition matrices is equal to the identity matrix.

3.5. Cubic anisotropy

One of the most common types of anisotropy is cubic, so consider this type of anisotropy in more detail. Before proceeding to obtain the rotation matrices, we first turn to the general geometric properties of such anisotropy, first of all, a cubic cell, as well as to the main variants of orientation of the [111] axes relative to the plane of the magnetic plate.

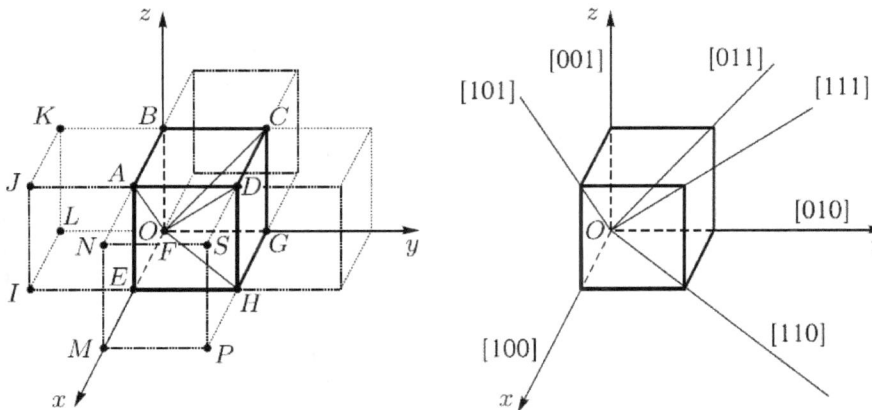

Fig. 3.5. Directional pattern in a cubic cell.

3.5.1. Designation of cubic cell axes

Due to the wide applicability in practice of materials with cubic anisotropy, a certain system of indices has been developed to designate the axes of such anisotropy [261, 273, 274, 294]. Consider this system on the example of a cubic cell, for which we turn to Fig. 3.5.

Figure 3.5 *a* and shows a cubic cell *ABCD–EFGH*, placed in the Cartesian coordinate system *Oxyz* in such a way that the beginning of this system, point *O*, coincides with one of the vertices of the cube *F*, and the axes *Ox, Oy, Oz* – with the edges coming out of this vertex *FE, FG, FB*. The marked cell is surrounded by all other similar cubic cells, for example, *NADS-MEHP, JKBA-ILFE* and others. The position of any point in space is determined by three numbers corresponding to the projections of this point on the coordinate axes *Ox, Oy, Oz*. As a unit of length, the cube cell edge length is used. In this case, the nodal points of the full structure corresponding to the vertices of the cubic cell structure will be determined by a set of three integers.

For example, point *E* has coordinates $x_E = 1$, $y_E = 0$, $z_E = 0$, point *A* – coordinates $x_A = 1$, $y_A = 0$, $z_A = 1$, point *K* – $x_K = 0$, $y_K = -1$, $z_K = 1$, and so on. In the notation adopted in analytic geometry [292, 293], these points are written as: *E*(1, 0, 0), *A*(1, 0, 1), *K*(0, −1, 1).

3.5.2. Crystallographic directions

However, in crystallography, the main task is not so much to determine the coordinates of individual points (although this is important when analyzing the internal structure of the cell), but rather on certain directions passing through the lattice sites. Such directions are given as straight lines, passing through two points, one of which is usually the origin of coordinates coordinated with the cell node (point O), and the other is the point of the nearest node through which the given direction passes. Since all three coordinates of point O are always equal to zero, to specify the direction, it is sufficient to specify the three coordinates of another point of a given direction. These coordinates are the name of the indexes of directions and are written in the order of the axes Ox, Oy, Oz, and then enclosed in square brackets. For points lying in the region of negative coordinates, the indices are denoted by a minus sign, located above the corresponding number.

For a more detailed explanation of the rules of indexation of directions, we note some of them, the most characteristic (Fig. 3.5, a). For example, the direction corresponding to the edge of the cube FE, coinciding with the axis Ox, is denoted as [100]. The direction corresponding to the other edge FG is denoted by [010], the direction of the edge FB is [001]. The spatial diagonal direction of the face of $EOGH$, that is, OH, is denoted by [110], the direction FC is [011], the direction FA is [101]. The direction corresponding to the spatial diagonal FD is designated [111]. The directions passing through the points lying in the region of the negative parts of the coordinate frames have the form: $FK - [0\bar{1}1]$, $FI - [1\bar{1}0]$, $FJ - [1\bar{1}1]$, and so on.

Quite common in practice, for example, in the CMD technique [38–41], is the direction passing through the spatial diagonal of the parallelepiped composed of two semi-adjacent cubic cells, for example, passing through the points F and S. Such direction is denoted as [211]. Accordingly, the directions of the diagonals of the faces of the parallelepiped $NBFM$ and $MFGP$ are denoted by [201] and [210].

If it is necessary to work with other directions that do not pass through point F, they are to be directed to the direction passing through this point by parallel transfer, after which the designations take the same form. For example, the designation of the direction passing through the points B and D has the same appearance as the designation of the direction passing through the points F and H, that is, [110]. The direction passing through the points E and G is [110] and so on. The scheme of designations of the most commonly used directions of a cubic cell is shown in Fig. 3.5 b. The directions most commonly used are [100], [110], and [111], since they sufficiently reflect the anisotropic properties of a cubic crystal. Of the other directions often used are also [211].

3.5.3. Crystallographic planes

Another important characteristic of the anisotropic properties of a crystal are the planes passing through the lattice sites.

To denote planes, the same index system is used, as for directions. However, there are two slightly different interpretations in this question.

Thus, according to [274, p. 11], a plane perpendicular to this direction is denoted by the same set of indices as the direction itself. In this case, the designation of the plane, in contrast to the designation of direction, is enclosed the in parentheses. For example, perpendicular to the direction of *FE* (Fig. 3.5 *a*), that is [100], the *FBCG* plane is located, so such a plane is denoted by (100).

Perpendicular to the direction of *FD*, that is, [111] is a plane passing through the points *A*, *C* and *H*, that is (111). For more complex planes, the same rule is maintained. So, the plane passing through the points *J*, *C*, *G*, *I*, has a normal direction *FP*, corresponding to [210], so that is denoted as (210). The plane passing through the points *M*, *N*, *C*, *G* is denoted (120) and so on.

On the other hand, according to [273, 294], the planes are characterized by the lengths of the segments, cut off on the coordinate axes. In this case, the unit of measurement is the length of the cube edge, and the indices are calculated as the inverse of these cut lengths. In the case of fractional values, the entire system of indices is multiplied by the minimum number, leading to a whole number. The indices obtained in this way are called 'Miller indices' and are also written in the order the coordinate coordinates are enclosed in the parentheses.

In order to avoid division by zero when taking inversely, the planes are drawn so that they do not pass through the origin of coordinates, but they are at least one atomic distance.

So to designate a crystallographic plane parallel to the *Oyz* coordinate plane, here we can no longer take the *FBCG* plane, but one should take the *EADH* plane parallel to it, which cuts a single segment on the *Ox* axis, and does not cut off on the *Oy* and *Oz* axes due to the parallelness of the *Oyz* plane. The inverse values of the corresponding sections will be 1, 0 and 0 (since the absence of clipping is equivalent to dividing by an infinitely large amount).

In this case, the designation of this plane in the Miller indexes will be (100). For the planes passing through the points G and B parallel to the planes Oxz and Oxy respectively, designations of the Miller indexes will be (010) and (001). It can be seen that in these three-cuts, Miller's notations coincide with those accepted in the first interpretation [274]. For a plane passing through the points E, B, G, Miller's notation has the form (111). It can be seen that this designation also coincides with that adopted in the first interpretation [274], thanks to that the plane passing through the points E, B, G is parallel to the plane passing through the points A, C, H used in the first interpretation.

Not so easy, however, is the case for a number of other planes. Thus, the $JCGI$ plane cuts on the axes of the coordinates Ox, Oy, Oz the segments 1/2, 1 and ∞, respectively. The inverse of these cuts give Miller indices in the form (210). Some 'complication' here is the need to multiply the lengths of segments by 2 to bring them to integer values. Similarly, for a plane passing through the points M, N, C, G, we obtain Miller indices in the form (120). It can be seen that in this case the designations of the same planes in both interpretations coincide. Such a coincidence becomes obvious if we turn to two types of recording the equation of the plane, which follows from analytical geometry [292, 293]. So the plane perpendicular to the vector n = $\{A, B, C\}$, is described by the equation:

$$\frac{x}{\dfrac{D}{A}} + \frac{y}{\dfrac{D}{B}} + \frac{z}{\dfrac{D}{c}} + 1 = 0, \tag{3.163}$$

where in the case of a plane not passing through the origin, has the place $D \neq 0$.

Separating this equation as a whole into D and each of the terms contained in the coefficient contained in it, we obtain:

$$\frac{x}{a} + \frac{y}{b} + \frac{z}{c} + 1 = 0, \tag{3.164}$$

whence, introducing the notation $a = D/A$, $b = D/B$, $c = D/AC$, we get:

$$\frac{x}{a} + \frac{y}{b} + \frac{z}{c} + 1 = 0, \tag{3.165}$$

This expression is the traditionally known 'the equation of a plane in segments' [292,293], where *a, b, c* are segments cut off by this plane on the corresponding coordinates.

Thus, it can be seen that both of the above interpretations of the characteristics of the crystallographic planes using Miller indices are completely equivalent, that is, they lead to the same results. In the present monograph the following will be mainly used.

the first interpretation defines the plane through the normal to a given direction, which, however, only reflects the personal tastes of the authors, without in any way preventing the use of another equivalent interpretation.

Comment. For cells of a different symmetry other than cubic, various systems for designating indexes of directions were also proposed, however, as noted in [274, p. 11], none of these systems found widespread use comparable to the Miller indices, so each use of such systems requires an accompanying explanation.

3.5.4. Cubic cell layout

Now that the basic notation has been established, let us turn directly to the study of cubic symmetry, which is based on an elementary cubic cell, the scheme of which is shown in Fig. 3.6. We note some important directions for the practice and angles. We assume that the cell has a unit size, that is, the length of each edge of the cube is equal to one. The length of the diagonals of the face of the cube, for example, *AC* and *BD* in Fig. 3.6 *b*, is equal to $\sqrt{2}$, and the length of the spatial diagonals of the cube, for example, *AG* and *BH* in Fig. 3.6 *c* is equal to $\sqrt{3}$. On the side face of the cube *ABCD* (Fig. 3.6 *b*) the length of the edges $AB = BC = CD = DA = 1$, the length of the diagonals: $AC = BD = \sqrt{2}$.

The angles between the diagonals AC and BD are equal to 90°, these diagonals make angles of 45° with the edges of the cube *AB, BC, CD, DA*.

On the *ABGH* cube's diagonal plane, the sides *AB* and *GH* are 1, the diagonals of the faces *AH* and *BG* are $\sqrt{2}$. The length of the spatial diagonals *AG* and *BH* is equal to $\sqrt{3}$.

The angles between the spatial diagonals and the diagonals of the faces are equal:

$$\angle GAH = \angle GBH = \angle BHA = \angle BGA = \arctan \sqrt{2}/2 \approx 35.2644°.$$

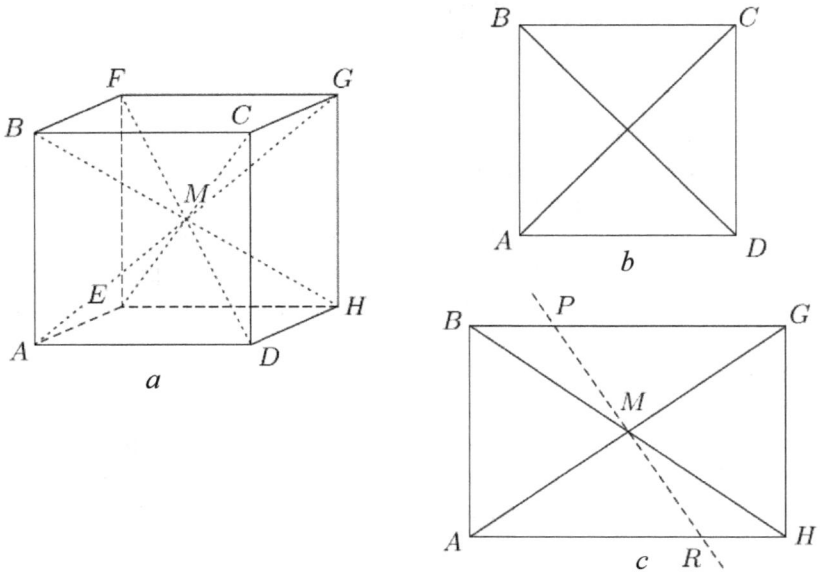

Fig. 3.6. Diagram of a cubic crystallographic cell. *a* – scheme of the cell cubic perspective; *b* – scheme of the lateral face of the cube; *c* – diagram of the cube's diagonal plane.

The angles between spatial diagonals and edges are:

$$\angle BAG = AGBHG = \angle ABH = \angle AGH = \arctan \sqrt{2} \approx 54.7356°.$$

The angles between spatial diagonals are equal:

$$\angle AMB = \angle HMG = 2 \arctan \sqrt{2}/2 \approx 70.5288°; \quad \angle AMH = \angle BMG = 2 \operatorname{arctg} \sqrt{2} \approx 109.4712°.$$

Direct *PMR* is perpendicular to the diagonal *AG*. Moreover, the angles between this straight line and the other diagonal *BH* are equal:

$$\angle RMH = \angle BMP = \pi/2 - 2 \arctan \sqrt{2}/2 \approx 19.4712,4°;$$
$$\angle RMB = \angle HMP = \pi/2 + 2 \arctan \sqrt{2}/2 \approx 160.5288°.$$

3.5.5. Different orientations of the cubic cell relative to the plane of the magnetic plate

We will consider the important for practice case of a magnetic plate, the material of which has cubic symmetries. At the given stage, the plate demagnetizing factor will not be taken into account (this will be done later), and the plate will be considered only as the basis for the corresponding orientation of the Cartesian coordinate system $Oxyz$. That is, the plate plane will correspond to the coordinate plane Oxy, and the normal to it will be the coordinate axis Oz.

Figure 3.7 shows the three main directions of the axes of the cubic cell [001], [011] and [111] along the normal to the plane of the magnetic plate.

It can be seen from the figure that the orientations of the axes (for example, for the YIG of the corresponding easy magnetization) of the type [111] in all cases are different. Consider the arrangement of the axes of the [111] type for these main orientations in more detail.

Orientation [001]. Figure 3.8 shows the layout of the axes of the [111] type (above) and their projections onto the Oxy plane (below). The coordinate axes Ox, Oy, Oz coincide with the edges of the cubic cell [100], [010], [001].

In the lower figure, the solid thick lines correspond to the location of the halves of the [111] axes above the Oxy plane. It can be seen from the figure that the angles between the projections of the [111] axes on the plate plane are $90°$.

Orientation [011]. Figure 3.9 shows the layout of the [111] type axes (above) and their projections onto the Oxy plane (below). The Oz coordinate axis coincides with the diagonal of the face of the cubic cell [011], the Ox axis coincides with the edge of the cubic cell [100], the Oy axis coincides with the diagonal of the face of the cube [011]. In the lower figure, solid thick lines correspond to the location of the halves of the [111] axes above the Oxy plane, the double solid lines – the arrangement of the [111] axes in the Oxy plane.

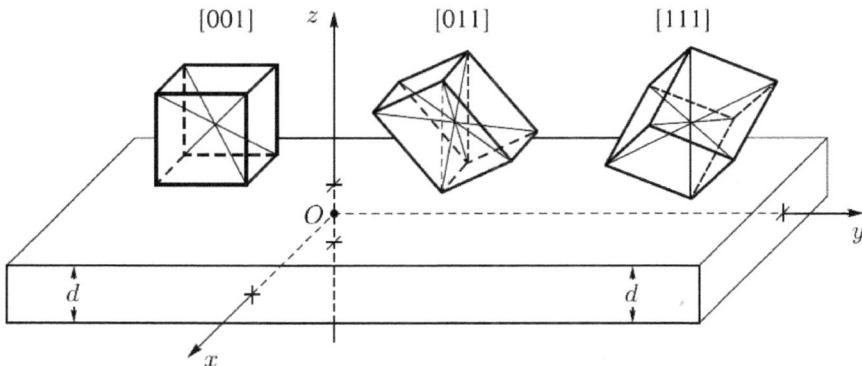

Fig. 3.7. The general scheme of the main orientations of the cubic cell relative to the plane of the magnetic plate.

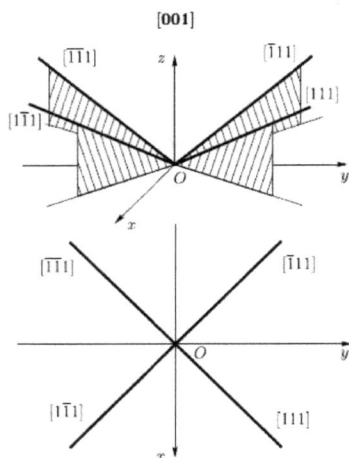

Fig. 3.8. Orientation [001]

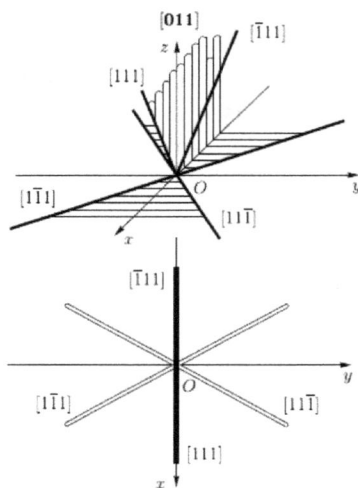

Fig. 3.9. Orientation [011]

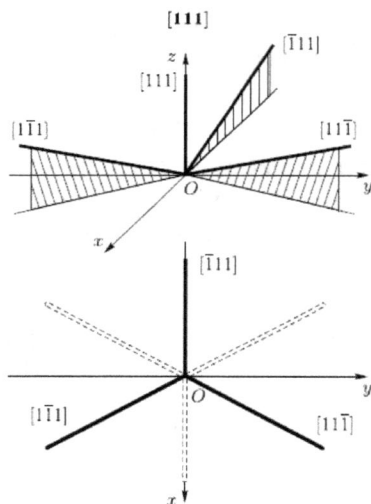

Fig. 3.10. Orientation [111]

It can be seen from the figure that the two projections of the [111] axes on the plate plane are oriented along the axis Ox, and the other two make angles along $\pm$ arctg $\sqrt{2}/2$ with the axis Oy, which corresponds to $\pm35.2644°$.

Orientation [111]. Figure 3.10 shows the layout of the axes of the type [111] (above) and their projections on the plane Oxy (below). The coordinate axis Oz coincides with the diagonal of the face of the cubic cell [111], the Ox axis lies in the plane passing through the vertices of the cube of the unit cell (0, 0, 0), (1, 0, 0), (1, 1, 1) and perpendicular to the [111] axis, the Oy axis is perpendicular to the Ox and Oz axes, making up with them the right three. The angles between the $[11\bar{1}]$, $[1\bar{1}1]$, $[\bar{1}11]$ axes and the Oxy plane are ($\pi/2-2$ arctg $\sqrt{2}/2$) = 19.4712°.

In the lower figure, solid thick lines correspond to the location of the halves of the [111] axes above the Oxy plane, double dashed lines indicate the location of the halves of the [111] axes below the Oxy plane.

It can be seen from the figure that the projections of [111] axes of the type on the plate plane that are closest to each other form angles of 60° between each other, and the angles between the projections of technical halves of [111] axes, which are located above or below the Oxy plane, are equal to 120°.

3.6. Transformation matrices in case of cubic anisotropy

Now, after the basic properties of the cubic cell have been established, we proceed to consider the transformation matrices in the case of cubic anisotropy. We confine ourselves to the simplest, most important for practice, variants of the rotation of a cubic cell – around axes of the type [001], [011] and [111].

3.6.1. Transformation matrix for axis [001]

The task is to find the transformation matrix from the coordinate system $Oxyz$, whose axes are directed along the edges of the cube, to the coordinate system $Ox'y'z'$, the axis Oz' of which is directed along the edge of a cube of type [001], and the axis Ox' and Oy' rotated around this axis at an arbitrary angle.

General geometry of the problem. The overall geometry of the problem is illustrated in Fig. 3.11. Here the dotted line shows a cubic cell, along the edges of which in the directions [100], [010] and [001] are oriented to the axis of the original coordinate $Oxyz$ system. The Oz' axis of the rotated coordinate system is directed along the edge of the cube [001], denoted by the segment OM. The line AB passes through the point O and lies in the plane Oxy, coinciding with the axis Ox. The line CE passes through the point O and lies in the plane Oxy, coinciding with the axis Oy. The CME plane passes through the Oz and Oz' axes. By construction, the $ACBE$ plane is perpendicular to the Oz and Oz' axes. so the axes of the

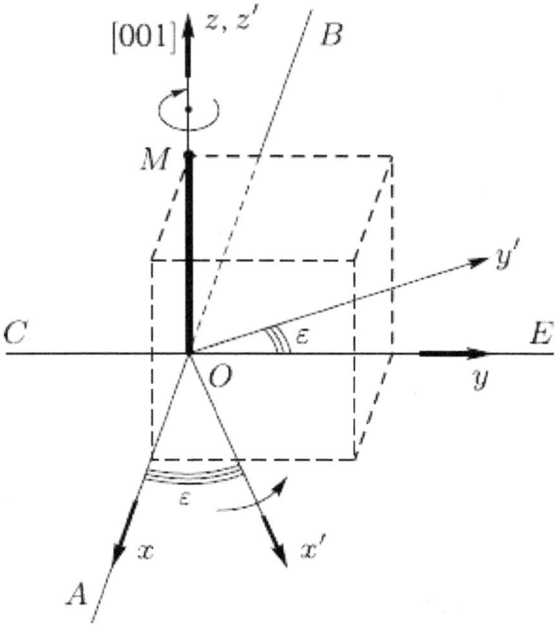

Fig. 3.11. General geometry of the problem for orientation [001].

rotated system are Ox' and Oy lie in the ACBE plane. In this case, the rotation of the system $Ox'y'z'$ around the Oz' axis is determined by the angle ε, given by the mutual arrangement of the Ox and Ox' axes. By virtue of the mutual perpendicularity of the axes Ox' and $Oxyz$, as well as direct OA and OC, the angle between the axes Oy and Oy' is also equal to ε.

Note on the direction of reference of the angle of rotation. In Fig. 3.11 the angle ε is represented as the angle between the straight line OA and the axis Ox'. We will consider as positive the direction of reference of the angle ε from the straight line OA to the axis Ox', That is, from the positive part of the axis Ox to the positive part of the axis Oy, as indicated by the arrow on the Ox' axis. Thus, the angle ε is measured from the straight line OA, which coincides with the positive part of the axis Ox. This choice of reference direction is convenient if we rotate the coordinate system $Ox'y'z'$ around the axis Oz'. At the same time, if we consider the system $Ox'y'z'$ if the cubic cell with the $Oxyz$ system associated with it is stationary and relative to it, then with the same reference direction of the angle ε

the cell will rotate in the opposite direction, as shown by the arrow on the axis Oz near point M.

Thus, if we look from the direction of the positive direction of the axis Oz, then the angle ε is counted counterclockwise, and the cell rotates clockwise. If we look with the negative direction of the axis Oz, then the angle ε is counted clockwise, and the cell is rotated counterclockwise.

The direction of reference of the angle ε adopted here corresponds to the classical choice of the direction of rotation of the coordinate system adopted in mathematics [292, 293]. That is, the cell remains stationary, and the system of the coordinates rotates. However, in the experiment, the opposite situation is often observed: the laboratory coordinate system remains stationary, and the sample under study (that is, the cell) rotates. In this case, for correct comparison with experiment, in all the resulting expressions it is necessary to change the sign of the angle ε to the opposite.

The overall structure of the matrix of the direct transition. The transformation of the vector components from one system to another is carried out using the transition matrix using the formula:

$$\mathbf{a} = \ddot{A}\mathbf{a}', \tag{3.166}$$

where $\mathbf{a}$ is a vector in the original coordinate system, $\mathbf{a}'$ is the same vector in the rotated coordinate system, $\ddot{A}$ is the transition matrix from the rotated system to the initial one.

The components of the matrix $\ddot{A}$ are the reference cosines of unitary vectors $\mathbf{l}_{x'}$, $\mathbf{l}_{y'}$, $\mathbf{l}_{z'}$ of the axes of the $Ox'y'z'$ coordinate system, relative to the $Oxyz$ system, that is, the first task is to find these different factors.

Obtaining the unit vectors of the axes of the rotated coordinate system. Components of unitary axes of the rotated coordinate system $\mathbf{l}_{x'}$, $\mathbf{l}_{y'}$, $\mathbf{l}_{z'}$ are determined by the rules of analytic geometry [292, 293].

Obviously, the unitary vector of the axis Oz' coincides with the unitary vector of the axis Oz. Getting single in $\mathbf{l}_{x'}$ and $\mathbf{l}_{y'}$ is explained by Fig. 3.12, which shows a view of the Oxy plane from the positive direction of the Oz axis when the coordinate system $Ox'y'$ is rotated relative to Oxy.

From the figure one can see that

$$\mathbf{l}_{x'} = \mathbf{i}(\cos\varepsilon) + \mathbf{j}(\sin\varepsilon); \tag{3.167}$$
$$\mathbf{l}_{y'} = \mathbf{i}(-\sin\varepsilon) + \mathbf{j}(\cos\varepsilon); \tag{3.168}$$

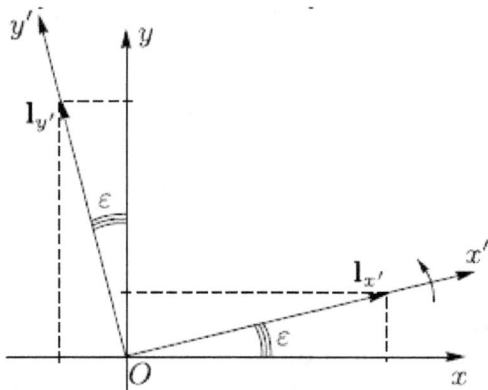

Fig. 3.12. Rotation of the coordinate system *Ox'y'* relative to *Oxy*.

and:
$$\mathbf{l}_{z'} = \mathbf{kl}, \tag{3.169}$$
where **i, j, k** are the unit vectors of the coordinate axes of the *Oxyz* system. So, in accordance with (3.167), (3.168), (3.169), the set of unitary vectors of the *Ox'y'z'* system, expressed in the *Oxyz* system, has the form:

$$\mathbf{l}_{x'} = \{\cos \varepsilon, \sin \varepsilon, 0\}; \tag{3.170}$$

$$\mathbf{l}_{y'} = \{-\sin \varepsilon, \cos \varepsilon, 0\}; \tag{3.171}$$

$$\mathbf{l}_{z'} = \{0; 0; 1\}. \tag{3.172}$$

The explicit view of the direct transition matrix. We reduce the components of the vectors $\mathbf{l}_{x'}$, $\mathbf{l}_{y'}$, $\mathbf{l}_z$ in a single table:

	x'	y'	z'
x	$\cos\varepsilon$	$-\sin\varepsilon$	0
y	$\sin\varepsilon$	$\cos\varepsilon$	0
z	0	0	1

(3.173)

In this case, the transition matrix takes the form:

$$\ddot{A}_{001}(\varepsilon) = \begin{pmatrix} \cos\varepsilon & -\sin\varepsilon & 0 \\ \sin\varepsilon & \cos\varepsilon & 0 \\ 0 & 0 & 1 \end{pmatrix} \tag{3.174}$$

Special cases of angles of rotation. We give the form of the transition matrix in some useful for the practice of partial cases. When $\varepsilon = 0$, the transition matrix has the form

$$\ddot{A}_{001}(0) = \begin{pmatrix} 1 & 0 & 0 \\ 0 & 1 & 0 \\ 0 & 0 & 1 \end{pmatrix}. \tag{3.175}$$

When $\varepsilon = 45°$, the transition matrix has the form

$$\ddot{A}_{001}(45°) = \begin{pmatrix} \dfrac{\sqrt{2}}{2} & \dfrac{\sqrt{2}}{2} & 0 \\ \dfrac{\sqrt{2}}{2} & \dfrac{\sqrt{2}}{2} & 0 \\ 0 & 0 & 1 \end{pmatrix} \tag{3.176}$$

When $\varepsilon = 90°$, the transition matrix has the form

$$\ddot{A}_{001}(90°) = \begin{pmatrix} 0 & -1 & 0 \\ 1 & 0 & 0 \\ 0 & 0 & 1 \end{pmatrix} \tag{3.177}$$

Reverse transition from the source to rotated system. Reverse transition from system $Oxyz$ to system $Ox'y'z'$ performed using the matrix $\ddot{A}^{-1}$ according to the formula

$$\mathbf{a} = \ddot{A}\mathbf{a}'. \tag{3.178}$$

The components of the matrix $\ddot{A}^{-1}$ can be found given that

$$\mathbf{l} = \mathbf{i}'(\cos \varepsilon) + \mathbf{j}'(-\sin \varepsilon) \tag{3.179}$$

$$\mathbf{l'}_y = \mathbf{i}'(\sin \varepsilon) + \mathbf{j}'(\cos \varepsilon) \tag{3.180}$$

$$\mathbf{l}_z = \mathbf{k}'\mathbf{l}, \tag{3.181}$$

where $\mathbf{i}'$, $\mathbf{j}'$, $\mathbf{k}'$ are the unit vectors of the coordinate axes of the system $Ox'y'z'$. Thus, the inverse transition matrix is

$$\ddot{A}_{001}^{-1}(\varepsilon) = \begin{pmatrix} \cos\varepsilon & \sin\varepsilon & 0 \\ -\sin\varepsilon & \cos\varepsilon & 0 \\ 0 & 0 & 1 \end{pmatrix} \tag{3.182}$$

By direct verification, we can verify that the product of the forward and reverse transition matrices (3.174) and (3.182) is equal to the identity matrix.

3.6.2. Transformation matrix for axis [011]

The task is to find the transformation matrix from the coordinate system $Oxyz$, whose axes are directed along the edges of the cube, to the coordinate system $Ox'y'z'$, the Oz' axis of which is directed along the diagonal of the face of a cube of the type [011], and the axis Ox' and Oy' rotated around this axis at an arbitrary angle.

General geometry of the problem. The overall geometry of the problem is illustrated in Fig. 3.13. Here the dotted line shows a cubic cell, the length of the edges of which is assumed to be equal to one. Along the edges of the cell in the directions [100], [010] and [001] the axes of the original system are oriented to the axis of the coordinates $Oxyz$. The Oz' axis of the rotated coordinate system is directed diagonally to the face of the cube [011], denoted by the segment OM. The line AB passes through the point O and lies in the plane Oxy, coinciding with the axis Ox. Points R and P lie on this line on both sides of point O at unit distance from it. The $OCME$ plane passes through the Oz and Oz' axes, coinciding with the Oyz plane. Direct line CE passes through point O, lies in the $OCME$ plane and makes up with the axis Oz' angles $\angle MOC = \angle MOE = 90°$. Point D on the line OC corresponds to the intersection of this line with the continuation of the edge of the cube MN. By virtue of construction, the $ACBEO$ plane is perpendicular to the axis Oz'. The axes of the rotated system Ox' and Oy' lie in the $ACBEO$ plane. Turn of the $Ox'y'z'$ system around the Oz axis is determined by the angle ε. This angle is determined by the relative position of the axis Ox' and straight OA and is counted from straight OA to straight OE. By virtue of the mutual perpendicularity of the axes Ox' and Oy', as well as direct OA and OE, the angle between direct OE and axis Oy' is also equal to ε.

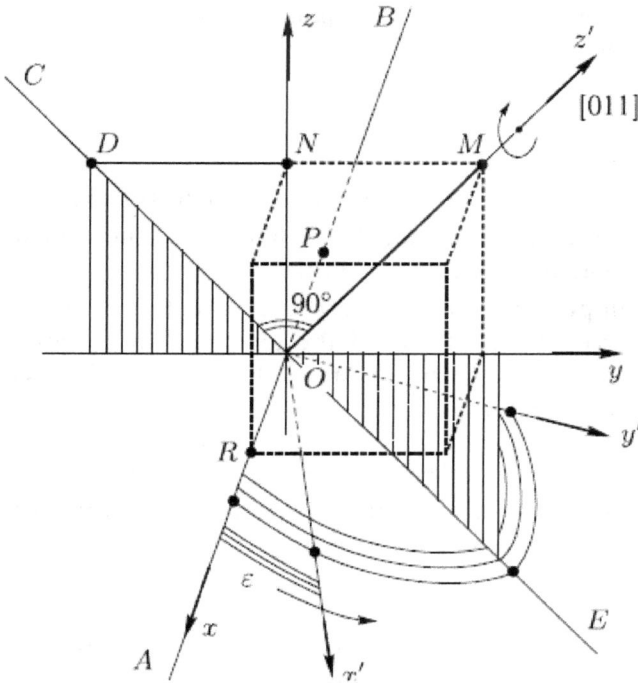

Fig. 3.13. General geometry of the problem for orientation [011].

Note on the direction of reference of the angle of rotation. Figure 3.13 shows the angle ε represented as the angle between the axis Ox' and straight line OA. We will consider as positive the direction of reference of the angle ε from the straight line OA to the straight line OE, as indicated by the arrow on the axis Ox'. This choice of reference direction is convenient if we rotate the coordinate system $Ox'y'z'$ around the axis Oz'. At the same time, if we consider the $Ox'y'z'$ system as fixed and turned relative to it a cubic cell together with its associated $Oxyz$ system, then with the same direction of reference of the angle ε, the cell will rotate in the opposite direction, as indicated by the arrow on the Oz' axis. Thus, if we look from the direction of the positive direction of the axis Oz', then the angle ε is counted counterclockwise, and the cell rotates clockwise. If we look with the sides of the negative direction of the axis Oz', then the angle ε is counted clockwise, and the cell rotates counterclockwise.

The overall structure of the matrix of the direct transition. The transformation of the vector components from one system to another is carried out using the transition matrix and the formula:

$$\mathbf{a} = \ddot{A}\mathbf{a}'.$$

(3.183)

where $\mathbf{a}$ is a vector in the original coordinate system, $\mathbf{a}'$ is the same vector in the rotated coordinate system, $\ddot{A}$ is the transition matrix from the rotated system to the initial one.

The components of the matrix $\ddot{A}$ are the reference cosines of unitary vectors $\mathbf{l}_{x'}$, $\mathbf{l}_{y'}$, $\mathbf{l}_{z'}$ of the axes of the coordinate system $Ox'y'z'$, relative to the $Oxyz$ system, that is, the first task is to find these different factors.

Obtaining the unitary vectors of the axes of the rotated coordinate system. Components of the unitary axes of the rotated coordinate system $\mathbf{l}_{x'}$, $\mathbf{l}_{y'}$, $\mathbf{l}_{z'}$ are determined according to the rules of analytical geometry, with the help of auxiliary problems Nos. 1–9 (section 3.3.1).

Vector $\mathbf{l}_{z'}$. We first find the unitary vector of the Oz axis, that is, the vector $\mathbf{l}_{z'}$. In the direction this vector coincides with the segment OM, whose components are equal:

$$(OM)_x = 0$$

(3.184)

$$(OM)_x = (OM)_z = 1.$$

(3.185)

Its length is equal to:

$$OM = \sqrt{(OM)_x + (OM)_y + (OM)_z} = \sqrt{2}.$$

(3.186)

Normalizing in accordance with problem No. 1, we get:

$$\mathbf{l}_{z'} = \left\{0; \frac{\sqrt{2}}{2}; \frac{\sqrt{2}}{2}\right\}.$$

(3.187)

The independence of this vector from the angle of rotation of the $Ox'y'z'$ system around the Oz' axis is consistent with the general formulation of the problem.

Vector $\mathbf{l}_{y'}$. We now find the vector $\mathbf{l}_{y'}$, which depends on the angle ε of rotation of the system $Ox'y'z'$ around the axis Oz'. We define this angle as corresponding to the intersection line of the plane MOx', passing through the axis Oz' and Ox', with the $AOBM$ plane passing through the Ox axis and the OM line. This definition follows from the perpendicularity of both these planes to the $ACBEO$ plane, which in turn follows from the passage of these two planes through the OM line, that is, the Oz' axis. In this case, ε is measured in the $ACBEO$ plane from the Ox axis towards the OE line. The task of building a vector $\mathbf{l}_y$ consists in expressing the components of this vector through the angle ε. Next is the vector $\mathbf{l}_{y'}$ determined through the normal vector to the plane MOx', therefore, as a first step, we associate the equation of this plane with the angle ε. For such a connection, we use problem No. 8, according to which the cosine of the angle between the planes is determined through the normal vectors to these planes. The angle ε is between two planes: $AOBM$ and MOx'.

We can find the equation of the $AOBM$ plane using problem No. 6 from the condition of passing the plane through three points: $R(1, 0, 0)$, $M(0, 1, 1)$ and $P(-1, 0, 0)$. So we get

$$\begin{vmatrix} x & y & z & 1 \\ 1 & 0 & 0 & 1 \\ 0 & 1 & 1 & 1 \\ -1 & 0 & 0 & 1 \end{vmatrix} = 0. \tag{3.188}$$

from where we get the $AOBM$ plane equation in the form

$$y-z = 0. \tag{3.189}$$

To find the equation of the plane MOx' we will construct a bundle of planes passing through the axis Oz', that is, the line OM, after which we will select the desired plane from this bundle.

As the reference planes of the beam, we select the $AOBM$ and $MODN$ planes intersecting along the OM line.

The equation for the $AOBM$ plane is (3.189).

The equation of the plane $MODN$ we will find using the problem No. 6, given that this plane passes through three points $O(0, 0, 0)$, $M(0, 1, 1)$ and $N(0, 0, 1)$, that is, from ratios

$$\begin{vmatrix} x & y & z & 1 \\ 0 & 0 & 0 & 1 \\ 0 & 1 & 1 & 1 \\ 0 & 0 & 0 & 1 \end{vmatrix} = 0. \tag{3.190}$$

from where we get the equation of the plane *MODN* in the form

$$x = 0. \tag{3.191}$$

Thus, to build a beam of planes passing through the *Oz'* axis, we obtain two equations of the *AOBM* and *MODN* planes:

$$y - z = 0; \tag{3.192}$$

$$x = 0. \tag{3.193}$$

Using problem No. 9, we obtain the beam equation in the form

$$(y-z) + \lambda(x) \tag{3.194}$$

where λ is a parameter that must be determined from the condition of equality of the angle ε between the *AOBM* plane and another, not yet known *MOx'* plane, chosen from the beam. Using problem No. 8, based on (3.189) and (3.194), we find the cosine of the angle between the *AOBM* and *MOx'* planes:

$$\cos \varepsilon = \frac{\sqrt{2}}{\sqrt{\lambda^2 + 2}}. \tag{3.195}$$

From this we get

$$\sin \varepsilon = \frac{\lambda}{\sqrt{\lambda^2 + 2}}. \tag{3.196}$$

Using problem No. 7, we find the vector $\mathbf{l}_{y'}$ as the normal vector to the plane *MOx'*, whose equation has the form (3.194). The normalizing factor is

$$\mu = \frac{1}{\sqrt{\lambda^2 + 2}}. \tag{3.197}$$

We obtain two possible normals:

$$\mathbf{l}_{y'}^{(+)} = \left\{ -\frac{\lambda}{\sqrt{\lambda^2 + 2}}, -\frac{1}{\sqrt{\lambda^2 + 2}}, \frac{1}{\sqrt{\lambda^2 + 2}} \right\}; \tag{3.198}$$

$$\mathbf{l}_{y'}^{(+-)} = \left\{ -\sin \varepsilon, -\frac{\sqrt{2}}{2} \cos \varepsilon, \frac{\sqrt{2}}{2} \cos \varepsilon \right\}. \tag{3.199}$$

Replacing λ through the sine and cosine of the angle ε, as well as taking into account the two sine signs (3.196), for the vector $\mathbf{l}_{y'}$ we get four possible expressions:

$$\mathbf{l}_{y'}^{(+-)} = \left\{ -\sin\varepsilon, -\frac{\sqrt{2}}{2}\cos\varepsilon, \frac{\sqrt{2}}{2}\cos\varepsilon \right\}. \tag{3.200}$$

$$\mathbf{l}_{y'}^{(++)} = \left\{ \sin\varepsilon, -\frac{\sqrt{2}}{2}\cos\varepsilon, \frac{\sqrt{2}}{2}\cos\varepsilon \right\}; \tag{3.201}$$

$$\mathbf{l}_{y'}^{(-+)} = \left\{ \sin\varepsilon, \frac{\sqrt{2}}{2}\cos\varepsilon, -\frac{\sqrt{2}}{2}\cos\varepsilon \right\}; \tag{3.202}$$

$$\mathbf{l}_{y'}^{(--)} = \left\{ -\sin\varepsilon, \frac{\sqrt{2}}{2}\cos\varepsilon, -\frac{\sqrt{2}}{2}\cos\varepsilon \right\}; \tag{3.203}$$

These four expressions correspond to the two possible directions normal to the plane MOx' and two possible configurations of the Cartesian coordinate system – right and left. For unambiguous choice of the right coordinate system, which coincides with $\varepsilon = 0°$ with the original one, consider two characteristic values of the angle ε: zero and 90°.

When $\varepsilon = 0°$ from (3.200)–(3.203) we get:

$$\mathbf{l}_{y'}^{(+-)} = \left\{ 0, -\frac{\sqrt{2}}{2}, \frac{\sqrt{2}}{2} \right\}; \tag{3.204}$$

$$\mathbf{l}_{y'}^{(++)} = \left\{ 0, -\frac{\sqrt{2}}{2}, \frac{\sqrt{2}}{2} \right\}; \tag{3.205}$$

$$\mathbf{l}_{y'}^{(-+)} = \left\{ 0, \frac{\sqrt{2}}{2}, -\frac{\sqrt{2}}{2} \right\}; \tag{3.206}$$

$$\mathbf{l}_{y'}^{(--)} = \left\{ 0, \frac{\sqrt{2}}{2}, -\frac{\sqrt{2}}{2} \right\}. \tag{3.207}$$

When $\varepsilon = 90°$ from (3.200)–(3.203) we get:

$$\mathbf{l}_{y'}^{(+-)} = \left\{ -1, 0, 0 \right\}; \tag{3.208}$$

$$\mathbf{l}_{y'}^{(++)} = \left\{ 1, 0, 0 \right\}; \tag{3.209}$$

$$\mathbf{l}_{y'}^{(-+)} = \{1,0,0\};$$ (3.210)

$$\mathbf{l}_{y'}^{(--)} = \{-1,0,0\};$$ (3.211)

From Fig. 3.13 one can see that:

$$\mathbf{l}_{y'}(\varepsilon = 0) = \left\{0, \frac{\sqrt{2}}{2}, -\frac{\sqrt{2}}{2}\right\};$$ (3.212)

$$\mathbf{l}_{y'}(\varepsilon = 90°) = \{-1,0,0\}.$$ (3.213)

Comparing these expressions with (3.204)–(3.211) shows that only (3.207) and (3.211) satisfy the required conditions, which corresponds to $\mathbf{l}_{y'}^{(--)}$, that is, the vector $\mathbf{l}_{y'}$ has the form:

$$\mathbf{l}_{y'} = \left\{-\sin\varepsilon, \frac{\sqrt{2}}{2}\cos\varepsilon, -\frac{\sqrt{2}}{2}\cos\varepsilon\right\}.$$ (3.214)

Vector $\mathbf{l}_{x'}$. Now, since the vectors are $\mathbf{l}_{y'}$ and $\mathbf{l}_{z'}$ were obtained, it remains to find the direction vector $\mathbf{l}_{x'}$ of the Ox' axis. We define this vector as perpendicular to the vectors $\mathbf{l}_{y'}$ and $\mathbf{l}_{z'}$, using problem No. 3.

The vector product of the vectors $\mathbf{l}_{y'}$ and $\mathbf{l}_{z'}$, taking into account (3.214) and (3.187) has the appearance

$$\left[\mathbf{l}_{y'} \times \mathbf{l}_{z'}\right] = \mathbf{i}(\cos\varepsilon) + \mathbf{j}\left(\frac{\sqrt{2}}{2}\sin\varepsilon\right) + \mathbf{k}\left(-\frac{\sqrt{2}}{2}\sin\varepsilon\right).$$ (3.215)

The module of a vector product is

$$\left\|\left[\mathbf{l}_{y'} \times \mathbf{l}_{z'}\right]\right\| = 1.$$ (3.216)

So we get

$$\mathbf{l}_{x'} = \left\{\cos\varepsilon, \frac{\sqrt{2}}{2}\sin\varepsilon, -\frac{\sqrt{2}}{2}\sin\varepsilon\right\}.$$ (3.217)

A set of unitary vectors of a rotated coordinate system. So, in accordance with (3.217), (3.214), (3.187), we obtain a set of unitary vectors of the $Ox'y'z'$ system, the components of which are distributed in the $Oxyz$ system, in the form:

$$\mathbf{l}_{x'} = \left\{\cos\varepsilon, \frac{\sqrt{2}}{2}\sin\varepsilon, -\frac{\sqrt{2}}{2}\sin\varepsilon\right\}.$$ (3.218)

$$\mathbf{l}_{y'} = \left\{ -\sin\varepsilon, \frac{\sqrt{2}}{2}\cos\varepsilon, -\frac{\sqrt{2}}{2}\cos\varepsilon \right\}. \tag{3.219}$$

$$\mathbf{l}_{z'} = \left\{ 0, \frac{\sqrt{2}}{2}, \frac{\sqrt{2}}{2} \right\}. \tag{3.220}$$

The explicit view of the direct transition matrix. We reduce the components of the vectors $\mathbf{l}_{x'}$, $\mathbf{l}_{y'}$ and $\mathbf{l}_{z'}$ in a single table:

	x'	y'	z'
x	$\cos\varepsilon$	$-\sin\varepsilon$	0
y	$\dfrac{\sqrt{2}}{2}\sin\varepsilon$	$\dfrac{\sqrt{2}}{2}\cos\varepsilon$	$\dfrac{\sqrt{2}}{2}$
z	$-\dfrac{\sqrt{2}}{2}\sin\varepsilon$	$-\dfrac{\sqrt{2}}{2}\cos\varepsilon$	$\dfrac{\sqrt{2}}{2}$

(3.221)

In this case, the transition matrix takes the form

$$\overset{\leftrightarrow}{A}_{011}(\varepsilon) = \begin{pmatrix} \cos\varepsilon & -\sin\varepsilon & 0 \\ \dfrac{\sqrt{2}}{2}\sin\varepsilon & \dfrac{\sqrt{2}}{2}\cos\varepsilon & \dfrac{\sqrt{2}}{2} \\ -\dfrac{\sqrt{2}}{2}\sin\varepsilon & -\dfrac{\sqrt{2}}{2}\cos\varepsilon & \dfrac{\sqrt{2}}{2} \end{pmatrix} \tag{3.222}$$

Special cases of angles of rotation. We give the form of the transition matrix in some partial cases useful for practice.

At $\varepsilon = 0°$ the transition matrix has the form

$$\overset{\leftrightarrow}{A}_{011}(0) = \begin{pmatrix} 1 & 0 & 0 \\ 0 & \dfrac{\sqrt{2}}{2} & \dfrac{\sqrt{2}}{2} \\ 0 & -\dfrac{\sqrt{2}}{2} & \dfrac{\sqrt{2}}{2} \end{pmatrix}. \tag{3.223}$$

where $\varepsilon = \text{arctg}\,\dfrac{\sqrt{2}}{2}$ the transition matrix has the form

$$\ddot{A}_{011}\left(\operatorname{arctg}\frac{\sqrt{2}}{2}\right) = \begin{pmatrix} \dfrac{\sqrt{6}}{3} & -\dfrac{\sqrt{3}}{3} & 0 \\ \dfrac{\sqrt{6}}{3} & \dfrac{\sqrt{3}}{3} & \dfrac{\sqrt{2}}{2} \\ \dfrac{\sqrt{6}}{3} & \dfrac{\sqrt{3}}{3} & \dfrac{\sqrt{2}}{2} \end{pmatrix} \qquad (3.224)$$

At $\varepsilon = \operatorname{arctg}\sqrt{2}$ the transition matrix has the form:

$$\ddot{A}_{011}\left(\operatorname{arctg}\sqrt{2}\right) = \begin{pmatrix} \dfrac{\sqrt{3}}{3} & -\dfrac{\sqrt{6}}{3} & 0 \\ \dfrac{\sqrt{3}}{3} & \dfrac{\sqrt{6}}{3} & 0 \\ \dfrac{\sqrt{3}}{3} & \dfrac{\sqrt{6}}{3} & \dfrac{\sqrt{2}}{2} \end{pmatrix} \qquad (3.225)$$

At $\varepsilon = 90°$ the matrix has the form

$$\ddot{A}_{111}(90°) = \begin{pmatrix} 0 & -1 & 0 \\ \dfrac{\sqrt{2}}{2} & 0 & \dfrac{\sqrt{2}}{2} \\ -\dfrac{\sqrt{2}}{2} & 0 & \dfrac{\sqrt{2}}{2} \end{pmatrix} \qquad (3.226)$$

Reverse transition matrix. The reverse transition from the *Oxyz* system to the *Ox'y'z'* system takes place using the $\ddot{A}^{-1}$ reverse transition matrix and the formula:

$$\mathbf{a'} = \ddot{A}^{-1}\mathbf{a}. \qquad (3.227)$$

where the matrix $\ddot{A}^{-1}$ is obtained from the direct transition matrix by replacing with the columns:

$$\ddot{A}_{011}^{-1}(\varepsilon) = \begin{pmatrix} \cos\varepsilon & \dfrac{\sqrt{2}}{2}\sin\varepsilon & \dfrac{\sqrt{2}}{2}\sin\varepsilon \\ -\sin\varepsilon & \dfrac{\sqrt{2}}{2}\cos\varepsilon & -\dfrac{\sqrt{2}}{2}\cos\varepsilon \\ 0 & \dfrac{\sqrt{2}}{2} & \dfrac{\sqrt{2}}{2} \end{pmatrix} \qquad (3.228)$$

By direct verification, we can verify that the product of the forward and reverse transition matrices (3.222) and (3.228) is equal to the identity matrix.

3.6.3. Transformation matrix for axis [111]

The task is to find the transformation matrix from the coordinate system $Oxyz$, whose axes are directed along the edges of the cube, to the coordinate system $Ox'y'z'$, the axis Oz' which is directed along the spatial diagonal of the cube of the [111] type, and the Ox' and Oy' axes rotated around this axis at an arbitrary angle.

General geometry of the problem. The overall geometry of the problem is illustrated in Fig. 3.14. Here the dotted line shows a cubic cell along the edges of which in the directions [100], [010] and [001] the axes of the initial coordinate system $Oxyz$. The Oz axis of the rotated coordinate system is directed along the spatial diagonal

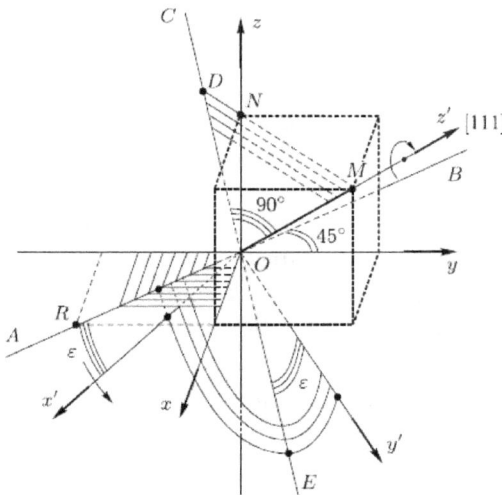

Fig. 3.14. General geometry of the problem for orientation [111].

of the cube [111], denoted by the segment *OM*. The straight line *AB* passes through the point *O* and lies in the plane *Oxy*, making angles of 45° with the axes *Ox* and *Oy*. Due to the symmetry of the cubic cell, ∠*MOA* = ∠*MOB* = 90°. The *MOC* plane passes through the *Oz* and *Oz'* axes. Direct *COE* lies in the *MOC* plane and makes up with the axis *Oz'* the angles ∠*MOC* = ∠*MOE* = 90°. Point *D* on the line *OC* corresponds to the intersection of this line with the continuation of the diagonal of the face of the cube *MN*. Due to the perpendicularity of straight lines *AB* and *CE* to the axis *Oz'* the plane *ACBEO* is perpendicular to the axis *Oz'*. So the axes of the rotated system are *Ox'* and *Oy'* lie in the *ACBEO* plane. In this case, the rotation of the *Ox'y'z'* system around the *Oz'* is determined by the angle ε. This angle is determined by the relative position of the axis *Ox'* and straight *OA* and is counted from straight *OA* to straight *OE*. By virtue of the mutual perpendicularity of the axes *Ox'* and *Oy'*, as well as the straight lines *OA* and *OE*, the angle between the straight line *OE* and the axis *Oy'* are also equal to ε.

Note on the direction of reference of the angle of rotation. In Fig. 3.14 the angle ε is represented as the angle between the axis *Ox'* and direct *OA*. We will consider as positive the direction of reference of the angle ε from the straight line *OA* to the straight line *OE*, as indicated by the arrow on the axis *Ox'*. This choice of reference direction is convenient if we rotate the coordinate system *Ox'y'z'* around the axis *Oz'*. At the same time, if we consider the *Ox'y'z'* system as a fixed and the cubic cell together with the associated *Oxyz* system rotate to its relatively, then with the same direction of reference of the angle ε the cell will rotate in the opposite direction, as indicated by the arrow on the axis *Oz'*. Thus, if we look from the positive direction of the axis *Oz*, then the angle ε is counted clockwise, and the cell rotates clockwise. If we look from the negative direction of the *Oz* axis, then the angle ε is counted clockwise, and the cell rotates counterclockwise. The overall structure of the matrix of the direct transition. The transformation of the vector components from one system to another is carried out using the transition matrix using the formula (3.10):

$$\mathbf{a} = \ddot{A}\mathbf{a}', \qquad (3.229)$$

where **a** is a vector in the original coordinate system, **a'** is the same vector in the rotated coordinate system, $\ddot{A}$ is the transition matrix from the rotated system to the initial one.

The components of the matrix $\ddot{A}$ are the direction cosines of the unit vectors of the $l_{x'}$, $l_{y'}$ and $l_{z'}$ of the axes of the coordinate system $Ox'y'z'$, relative to the system $Oxyz$, that is, the first task is to find these different factors.

Obtaining the unitary vectors of the axes of the rotated coordinate system. Components of unitary axes of rotated coordinate system $l_{x'}$, $l_{y'}$ and $l_{z'}$ are determined according to the rules of analytical geometry, with the help of auxiliary numbers Nos. 1–9.

Vector $l_{z'}$. We first find the unitary vector of the axis Oz', that is vector $l_{z'}$. In the direction this vector coincides with the segment OM, whose components are equal

$$(OM)_x = (OM)_y = (OM)_z = 1. \qquad (3.230)$$

Its length is equal to

$$OM = \sqrt{(OM)_x + (OM)_y + (OM)_z} = \sqrt{3}. \qquad (3.231)$$

Normalizing in accordance with task number 1, we get

$$l_{z'} = \left\{ \frac{\sqrt{3}}{3}, \frac{\sqrt{3}}{3}, \frac{\sqrt{3}}{3} \right\}, \qquad (3.232)$$

The independence of this vector from the angle of rotation of the $Ox'y'z'$ system around the Oz' axis is consistent with the general formulation of the problem.

Vector $l_{y'}$. We now find the vector $l_{y'}$, which depends on the angle ε of rotation of the $Ox'y'z'$ system around the axis Oz'. We define this angle as the corresponding line of intersection of the plane MOx' passing through the axis Oz' and Ox', with the $AOBM$ plane passing through the straight line AB and the line OM. This definition follows from the perpendicularity of both these planes to the plane $ACBEO$, which in turn follows from the passage of these planes through the line OM, that is, the axis Oz'. In this case, ε is measured in the $ACBEO$ plane from the Ox axis towards the OE line. The task of building the vector $l_{y'}$ consists in expressing the components of this vector through the angle ε. Next, the vector $l_{y'}$ will be determined through the normal vector to the plane MOx', therefore, as a first step, we associate the equation of this plane with the angle ε. For such a connection, we use problem No. 8, according to which the cosine of the angle between the planes is determined through the normal vectors to these planes. The angle ε is between two planes: $AOBM$ and MOx'.

We can find the equation of the $AOBM$ plane using problem No. 6 from the condition of passing the plane through three points: $O(0, 0, 0)$, $M(1, 1, 1)$ and $R(1, -1, 0)$. So we get

$$\begin{vmatrix} x & y & z & 1 \\ 0 & 0 & 0 & 1 \\ 1 & 1 & 1 & 1 \\ 1 & -1 & 0 & 1 \end{vmatrix} = 0. \tag{3.233}$$

from where we get the $AOBM$ plane equation in the form

$$x + y - 2z = 0. \tag{3.234}$$

To find the equation of the plane MOx' we will construct a bundle of planes passing through the axis Oz', that is, the line OM, after which we will select the desired plane from this bundle.

As the reference planes of the beam, we select the $AOBM$ and $MODN$ planes intersecting along the OM line. The equation for the $AOBM$ plane is (3.234).

The equation of the plane $MODN$ we will find using the problem No. 6, given that this plane passes through three points $O(0, 0, 0)$, $M(1, 1, 1)$ and $N(0, 0, 1)$, that is, from the relationship

$$\begin{vmatrix} x & y & z & 1 \\ 0 & 0 & 0 & 1 \\ 1 & 1 & 1 & 1 \\ 0 & 0 & 1 & 1 \end{vmatrix} = 0, \tag{3.235}$$

from where we get the equation of the plane $MODN$ in the form

$$x - y = 0. \tag{3.236}$$

Thus, to build a beam of planes passing through the Oz' axis, we obtain two equations of the $AOBM$ and $MODN$ planes:

$$x + y - 2z = 0. \tag{3.237}$$

$$x - y = 0. \tag{3.238}$$

Using problem No. 9, we obtain the beam equation in the form:

$$(x + y - 2z) + \lambda(x - y) = 0, \tag{3.239}$$

or

$$(1 + \lambda)x + (1 - \lambda)y - 2z = 0. \tag{3.240}$$

where λ is a parameter that must be determined from the condition of equality of the angle ε between the $AOBM$ plane and another, not yet known MOx' plane, chosen from the beam. Using problem No. 8, based on (3.234) and (3.240), we find the cosine of the angle between the $AOBM$ and MOx' planes:

$$\cos \varepsilon = \frac{\sqrt{3}}{\sqrt{\lambda^2 + 3}}. \tag{3.241}$$

From this we get

$$\sin \varepsilon = \pm \frac{\lambda}{\sqrt{\lambda^2 + 3}}. \tag{3.242}$$

Using problem No. 7, find the vector $\mathbf{l}_{y'}$ as the normal vector to the plane MOx', whose equation has the form (3.240). The normalizing factor is

$$\mu = -\frac{\sqrt{2}}{2\sqrt{\lambda^2 + 3}}. \tag{3.243}$$

We obtain two possible normals:

$$\mathbf{l}_{y'}^{(+)} = \left\{ -\frac{\sqrt{2}(1+\lambda)}{2\sqrt{\lambda^3 + 3}}, -\frac{\sqrt{2}(1-\lambda)}{2\sqrt{\lambda^2 + 3}}, \frac{\sqrt{2}}{2\sqrt{\lambda^3 + 3}} \right\}; \tag{3.244}$$

$$\mathbf{l}_{y'}^{(-)} = \left\{ -\frac{\sqrt{2}(1+\lambda)}{2\sqrt{\lambda^3 + 3}}, \frac{\sqrt{2}(1-\lambda)}{2\sqrt{\lambda^2 + 3}}, -\frac{\sqrt{2}}{2\sqrt{\lambda^3 + 3}} \right\}; \tag{3.245}$$

Replacing λ through the sine and cosine of the angle ε according to (3.241) and (3.242), and also taking into account the two sine signs in (3.242), for the vector $\mathbf{l}_{y'}$ we get four possible expressions:

$$\mathbf{l}_{y'}^{(++)} = \left\{ -\frac{\sqrt{6}}{6}\cos\varepsilon - \frac{\sqrt{2}}{2}\sin\varepsilon, \; -\frac{\sqrt{6}}{6}\cos\varepsilon + \frac{\sqrt{2}}{2}\sin\varepsilon, \; \frac{\sqrt{6}}{3}\cos\varepsilon \right\},$$

(3.246)

$$\mathbf{l}_{y'}^{(+-)} = \left\{ -\frac{\sqrt{6}}{6}\cos\varepsilon + \frac{\sqrt{2}}{2}\sin\varepsilon, \; -\frac{\sqrt{6}}{6}\cos\varepsilon - \frac{\sqrt{2}}{2}\sin\varepsilon, \; \frac{\sqrt{6}}{3}\cos\varepsilon \right\},$$

(3.247)

$$\mathbf{l}_{y'}^{(-+)} = \left\{ \frac{\sqrt{6}}{6}\cos\varepsilon + \frac{\sqrt{2}}{2}\sin\varepsilon, \; \frac{\sqrt{6}}{6}\cos\varepsilon - \frac{\sqrt{2}}{2}\sin\varepsilon, \; -\frac{\sqrt{6}}{3}\cos\varepsilon \right\},$$

(3.248)

$$\mathbf{l}_{y'}^{(--)} = \left\{ \frac{\sqrt{6}}{6}\cos\varepsilon - \frac{\sqrt{2}}{2}\sin\varepsilon, \; \frac{\sqrt{6}}{6}\cos\varepsilon + \frac{\sqrt{2}}{2}\sin\varepsilon, \; -\frac{\sqrt{6}}{3}\cos\varepsilon \right\},$$

(3.249)

These four expressions correspond to the two possible directions normal to the plane *MOx'* and two possible configurations of the Cartesian coordinate system - right and left. For unambiguous

Choosing the right coordinate system, axis *Ox'* which when at $\varepsilon = 0°$ coincides with the straight line *OA*, and the axis *Oy'* – with straight line *OE*, we consider two characteristic values of the angle ε: zero and 90°.

When $\varepsilon = 0°$ from (3.246)–(3.249) we get:

$$\mathbf{l}_{y'}^{(++)} = \left\{ -\frac{\sqrt{6}}{6}, -\frac{\sqrt{6}}{6}, \frac{\sqrt{6}}{3} \right\};$$

(3.250)

$$\mathbf{l}_{y'}^{(+-)} = \left\{ -\frac{\sqrt{6}}{6}, -\frac{\sqrt{6}}{6}, \frac{\sqrt{6}}{3} \right\};$$

(3.251)

$$\mathbf{l}_{y'}^{(-+)} = \left\{ \frac{\sqrt{6}}{6}, \frac{\sqrt{6}}{6}, -\frac{\sqrt{6}}{3} \right\};$$

(3.252)

$$\mathbf{l}_{y'}^{(--)} = \left\{ \frac{\sqrt{6}}{6}, \frac{\sqrt{6}}{6}, -\frac{\sqrt{6}}{3} \right\};$$

(3.253)

When $\varepsilon = 90°$ from (3.246)–(3.249) we get:

$$\mathbf{l}_{y'}^{(++)} = \left\{ -\frac{\sqrt{2}}{2}\sin\varepsilon, \frac{\sqrt{2}}{2}\sin\varepsilon, 0 \right\}; \tag{3.254}$$

$$\mathbf{l}_{y'}^{(+-)} = \left\{ \frac{\sqrt{2}}{2}\sin\varepsilon, -\frac{\sqrt{2}}{2}\sin\varepsilon, 0 \right\}; \tag{3.255}$$

$$\mathbf{l}_{y'}^{(-+)} = \left\{ \frac{\sqrt{2}}{2}\sin\varepsilon, -\frac{\sqrt{2}}{2}\sin\varepsilon, 0 \right\}; \tag{3.256}$$

$$\mathbf{l}_{y'}^{(--)} = \left\{ -\frac{\sqrt{2}}{2}\sin\varepsilon, \frac{\sqrt{2}}{2}\sin\varepsilon, 0 \right\}. \tag{3.257}$$

From Fig. 3.14 it can be seen that:

$$\mathbf{l}_{y'}\left(\varepsilon = 0°\right) = \left\{ \frac{\sqrt{6}}{6}, \frac{\sqrt{6}}{6}, -\frac{\sqrt{6}}{3} \right\}; \tag{3.258}$$

$$\mathbf{l}_{y'}\left(\varepsilon = 90°\right) = \left\{ -\frac{\sqrt{2}}{2}, \frac{\sqrt{2}}{2}, 0 \right\}. \tag{3.259}$$

Comparing these expressions with (3.250)–(3.257) shows that only (3.253) and (3.257) satisfy the required conditions, which corresponds to $\mathbf{l}_{y'}^{(--)}$, that is, the vector $\mathbf{l}_{y'}$ has the appearance

$$\mathbf{l}_{y'} = \left\{ \frac{\sqrt{6}}{6}\cos\varepsilon - \frac{\sqrt{2}}{2}\sin\varepsilon, \frac{\sqrt{6}}{6}\cos\varepsilon + \frac{\sqrt{2}}{2}\sin\varepsilon, -\frac{\sqrt{6}}{3}\cos\varepsilon \right\}, \tag{3.260}$$

or

$$\mathbf{l}_{y'} = \left\{ \frac{\sqrt{6}}{6}\left(\cos\varepsilon - \sqrt{3}\sin\varepsilon\right), \frac{\sqrt{6}}{6}\left(\cos\varepsilon - \sqrt{3}\sin\varepsilon\right), -\frac{\sqrt{6}}{3}\cos\varepsilon \right\}, \tag{3.261}$$

Vector $\mathbf{l}_{x'}$. Now determine the vector $\mathbf{l}_{x'}$ as a vector perpendicular to the vectors $\mathbf{l}_{y'}$ and $\mathbf{l}_{z'}$, using problem No. 3.

The vector product of the vectors $\mathbf{l}_{y'}$ and $\mathbf{l}_{z'}$ taking into account (3.261) and (3.232) has the appearance

$$\left[\mathbf{l}_{y'} \times \mathbf{l}_{z'} \right] = \left\{ \frac{\sqrt{6}}{6}\left(\sqrt{3}\cos\varepsilon + \sin\varepsilon\right), -\frac{\sqrt{6}}{6}(\ ) \right\} \tag{3.262}$$

The module of a vector product is

$$\left[\left[\mathbf{1}_{y'} \times \mathbf{1}_{z'}\right]\right] = 1 \tag{3.263}$$

So we get

$$\mathbf{1}_{x'} = \left\{ \frac{\sqrt{6}}{6}\left(\sqrt{3}\cos\varepsilon + \sin\varepsilon\right), -\frac{\sqrt{6}}{6}\left(\sqrt{3}\cos\varepsilon - \sin\varepsilon\right), -\frac{\sqrt{6}}{3}\sin\varepsilon \right\}, \tag{3.264}$$

A set of unitary vectors of a rotated coordinate system. So, in accordance with (3.264), (3.261), (3.232), we obtain a set of unitary vectors of the *Ox'y'z'* system of components which are distributed in the system Oxyz, in the form:

$$\mathbf{1}_{x'} = \left\{ \frac{\sqrt{6}}{6}\left(\sqrt{3}\cos\varepsilon + \sin\varepsilon\right), -\frac{\sqrt{6}}{6}\left(\sqrt{3}\cos\varepsilon - \sin\varepsilon\right), -\frac{\sqrt{6}}{3}\sin\varepsilon \right\}; \tag{3.265}$$

$$\mathbf{1}_{y'} = \left\{ \frac{\sqrt{6}}{6}\left(\cos\varepsilon - \sqrt{3}\sin\varepsilon\right), \frac{\sqrt{6}}{6}\left(\cos\varepsilon + \sqrt{3}\sin\varepsilon\right), -\frac{\sqrt{6}}{3}\cos\varepsilon \right\}; \tag{3.266}$$

$$\mathbf{1}_{z'} = \left\{ \frac{\sqrt{3}}{3}, \frac{\sqrt{3}}{3}, \frac{\sqrt{3}}{3} \right\}. \tag{3.267}$$

The explicit view of the direct transition matrix. We reduce the components of the vectors $\mathbf{1}_{x'}$, $\mathbf{1}_{y'}$ and $\mathbf{1}_{z'}$ in a single table:

	x'	y'	z'
x	$\frac{\sqrt{6}}{6}\left(\sqrt{3}\cos\varepsilon + \sin\varepsilon\right)$	$\frac{\sqrt{6}}{6}\left(\cos\varepsilon - \sqrt{3}\sin\varepsilon\right)$	$\frac{\sqrt{3}}{3}$
y	$\frac{\sqrt{6}}{6}\left(\sqrt{3}\cos\varepsilon - \sin\varepsilon\right)$	$\frac{\sqrt{6}}{6}\left(\cos\varepsilon + \sqrt{3}\sin\varepsilon\right)$	$\frac{\sqrt{3}}{3}$
z	$-\frac{\sqrt{6}}{3}\sin\varepsilon$	$-\frac{\sqrt{6}}{3}\cos\varepsilon$	$\frac{\sqrt{3}}{3}$

$$\tag{3.268}$$

In this case, the transition matrix takes the form

$$\ddot{A}_{111}(\varepsilon) = \left(\begin{array}{cc|c} \dfrac{\sqrt{6}}{6}\left(\sqrt{3}\cos\varepsilon + \sin\varepsilon\right) & \dfrac{\sqrt{6}}{6}\left(\sqrt{3}\cos\varepsilon - \sqrt{3}\sin\varepsilon\right) & \dfrac{\sqrt{3}}{3} \\ \hline \dfrac{\sqrt{6}}{6}\left(\sqrt{3}\cos\varepsilon - \sin\varepsilon\right) & \dfrac{\sqrt{6}}{6}\left(\cos\varepsilon + \sqrt{3}\sin\varepsilon\right) & \dfrac{\sqrt{3}}{3} \\ \hline -\dfrac{\sqrt{6}}{3}\sin\varepsilon & -\dfrac{\sqrt{6}}{3}\cos\varepsilon & \dfrac{\sqrt{3}}{3} \end{array} \right) \quad (3.269)$$

Special cases of angles of rotation. We give the form of the transition matrix useful for practice.

When $\varepsilon = 0$, the transition matrix has the form

$$\ddot{A}_{111}(0) = \left(\begin{array}{c|c|c} \dfrac{\sqrt{2}}{2} & \dfrac{\sqrt{6}}{6} & \dfrac{\sqrt{3}}{3} \\ \hline -\dfrac{\sqrt{2}}{2} & \dfrac{\sqrt{6}}{6} & \dfrac{\sqrt{3}}{3} \\ \hline 0 & \dfrac{\sqrt{6}}{3} & \dfrac{\sqrt{3}}{3} \end{array} \right) \quad (3.270)$$

When $\varepsilon = 30°$, the transition matrix has the form

$$\ddot{A}_{111}(30°) \left(\begin{array}{c|c|c} \dfrac{\sqrt{6}}{3} & 0 & \dfrac{\sqrt{3}}{3} \\ \hline -\dfrac{\sqrt{6}}{6} & -\dfrac{\sqrt{2}}{2} & \dfrac{\sqrt{3}}{3} \\ \hline -\dfrac{\sqrt{6}}{6} & -\dfrac{\sqrt{2}}{2} & \dfrac{\sqrt{3}}{3} \end{array} \right) \quad (3.271)$$

When $\varepsilon = 60°$, the transition matrix has the form

$$\ddot{A}_{111}(60°) = \left(\begin{array}{c|c|c} \dfrac{\sqrt{2}}{2} & -\dfrac{\sqrt{6}}{6} & \dfrac{\sqrt{3}}{3} \\ \hline 0 & \dfrac{\sqrt{6}}{3} & \dfrac{\sqrt{3}}{3} \\ \hline -\dfrac{\sqrt{2}}{2} & -\dfrac{\sqrt{6}}{6} & \dfrac{\sqrt{3}}{3} \end{array} \right) \quad (3.272)$$

When $\varepsilon = 90°$, the transition matrix has the form

$$\ddot{A}_{111}(90°) = \begin{pmatrix} \dfrac{\sqrt{6}}{6} & -\dfrac{\sqrt{2}}{2} & \dfrac{\sqrt{3}}{3} \\[2mm] \dfrac{\sqrt{6}}{6} & \dfrac{\sqrt{2}}{2} & \dfrac{\sqrt{3}}{3} \\[2mm] \dfrac{\sqrt{6}}{3} & 0 & \dfrac{\sqrt{3}}{3} \end{pmatrix} \qquad (3.273)$$

Reverse transition matrix. The reverse transition from the *Oxyz* to the *Ox'y'z'* system by using the inverse transition matrix $\ddot{A}^{-1}$ using the formula:

$$\mathbf{a}' = \ddot{A}^{-1}\mathbf{a}, \qquad (3.274)$$

where the matrix $\ddot{A}^{-1}$ is obtained from the direct transition matrix (3.269) by replacing the rows with columns:

$$\ddot{A}_{111}^{-1}(\varepsilon) = \begin{pmatrix} \dfrac{\sqrt{6}}{6}\begin{pmatrix}\sqrt{3}\cos\varepsilon + \\ +\sin\varepsilon\end{pmatrix} & -\dfrac{\sqrt{6}}{6}\begin{pmatrix}\sqrt{3}\cos\varepsilon - \\ -\sin\varepsilon\end{pmatrix} & \dfrac{\sqrt{3}}{3}\sin\varepsilon \\[4mm] \dfrac{\sqrt{6}}{6}\begin{pmatrix}\cos\varepsilon - \\ -\sqrt{3}\sin\varepsilon\end{pmatrix} & \dfrac{\sqrt{6}}{6}(\cos\varepsilon + \\ +\sqrt{3}\sin\varepsilon) & -\dfrac{\sqrt{6}}{3}\cos\varepsilon \\[4mm] \dfrac{\sqrt{3}}{3} & \dfrac{\sqrt{3}}{3} & \dfrac{\sqrt{3}}{3} \end{pmatrix} \qquad (3.275)$$

By direct verification, we can verify that the product of the forward and reverse transition matrices (3.269) and (3.275) is equal to the identity matrix.

3.6.4. Variety of options for obtaining conversion matrices

In the previous section, the variants of obtaining transformation matrices for various orientations with coordinate systems are given. It should be noted that the matrices obtained here are far from the only ones. The specific nature of the matrices is related both to the orientation of the original coordinate system and the axes around which the cubic cell rotates. So for example, if we turn to the rotation around the [111] axis shown in Fig. 3.14, it can be seen that the equivalent rotation can be carried out around any other spatial diagonal of the cube with the same result of changing symmetry, however, the rotation matrix will be somewhat different. The multiplicity of the form of the rotation matrices here is similar to the multiplicity of the choice of Euler angles, the main of which can be 12, and taking into account the sign and the left axis of the coordinate systems, significantly more [261, footnote on p. 451].

Thus, the matrices presented here should not be regarded as something universal, suitable for all cases of life. This is nothing more than an example corresponding to the specific geometry of the problem. However, since, for different types of rotation matrices, the transformation will in any case rely on the symmetry of the cubic cell, the immediate elements of the matrices will be on top of each other. to some extent similar. That is, numerical coefficients of the type $\sqrt{3}/3$, $\sqrt{6}/6$ or the like, as well as cosines and sine of the angles of rotation multiplied by these coefficients will often be encountered. in various classes with different signs. At the same time, each specific task will require its own consideration, as applied to the given conditions, and the mathematical methods presented here should be considered nothing more than a convenient working tool to obtain the necessary matrices.

Conclusions for chapter 3

This chapter provides a brief overview of the basic mathematical apparatus used in solving problems related to working with anisotropic crystals. The main results obtained in this chapter are as follows.

1. The main types of lattice symmetry are given. The geometry of cells corresponding to cyclical, monoclinic, orthorhombic (orthorhombic), tetragonal (square), trigonal, hexagonal and cubic structures. The features of magnetic symmetry are given. The main types of magnetic anisotropy, such as uniaxial and cubic, are noted. Other types of magnetic anisotropy are briefly mentioned, such as uniaxial high-order, mixed uniaxial-cubic, helical (spiral). The structural features of several materials are considered: iron, hematite, orthoferrites, spinel ferrites, garnets ferrites, iron yttrium garnet and barium hexaferrite.

156

2. The apparatus of transition matrices between coordinate systems of different spatial orientation is considered as the main mathematical technique for working with anisotropic crystals. Definitions of direct and inverse transformation matrices are given, their interconnection and the possibility of becoming each other. Examples of direct and inverse transformations concerning the consideration of ferromagnetic resonance in the case of dnoaxial and cubic anisotropy are qualitatively discussed. The transition matrix in general form is given. It is noted that in the case of transition from one Cartesian coordinate system to another, the elements of the transition matrices are the cosines of the angles between the axes of the one and the other coordinate systems. It is noted that when solving specific problems it is convenient to first write down the anisotropy energy density in the system where its form is the simplest, and then make the transition to the system directly required conditions of the problem. Auxiliary geometric problems used to obtain the components of the coordinate transformation matrices are given. Matrices of direct and inverse transformations are obtained for the case of an arbitrary orientation of one Cartesian coordinate system relative to another. The same transition matrices in the spherical coordinate system are given.

3. The main features of cubic anisotropy are considered. The main practical options for the designation of axes, crystallographic directions and planes of cubic symmetry are given. The scheme of an elementary cubic cell is shown and its main geometrical parameters are noted. Different types of orientation of the cubic cell relative to the Cartesian coordinate system are considered. The orientations of crystallographic systems of the [001], [011], and [111] types, for which their specific position is relative to the axes of the Cartesian system, are identified. The obtained transformation matrices correspond to the rotation of the cell with respect to the axes of the type [001], [011] and [111] by an arbitrary angle. Examples of transformation matrices for a number of important angles of rotation are given. The multiplicity of possible variations of the form of transformation matrices is noted, which is associated with the multiplicity of choice of Euler angles options, which determine the rotation of two-Cartesian systems relative to each other.

4

Energy density of magnetic anisotropy

This chapter is devoted to the use of the mathematical apparatus of the transition matrices described in the previous chapter for calculating the magnetic anisotropy energy density in various cases. The main attention is paid to uniaxial and cubic anisotropy for a number of the most characteristic orientations. The material of the chapter does not so much represent a significant novelty, as it is intended for the fullest possible generalization of the more or less well-known results of the article, scattered about a variety of articles made in various ways with different designations and auxiliary functions. The consideration begins with a description of various symmetry arrangements, with the help of which the energy density is calculated first, first uniaxial and then cubic anisotropy, with different orientations of the cubic cell. The results presented here rely for the most part on the same literary sources as given in the previous chapter. For new sources not mentioned there, the corresponding references are given in the text.

4.1. Symmetry operations

Any solid body is a set of atoms bound together by certain interatomic interactions. The periodic structure of the crystal, manifested in its symmetry, imposes certain restrictions on the energy density of such interactions. Within the framework of this monograph, we will be interested only in magnetic interactions. Other types of interactions including elastic and magnetoelastic are supposed to be considered in other monographies of this series. We show how to apply the symmetry operations to obtain transformation matrices for such interactions.

4.1.1. Reflection in coordinate planes

In the previous section it is mentioned that the operations of both crystallographic and magnetic symmetry represent a reflection in certain planes and a turn at a given angle around a certain axis. Mathematical description of such transformations is conveniently performed in the matrix form, that is, when the operation of symmetry is specified using a specific matrix. Let us show the formation of such a matrix by the example of the reflection of the magnetization vector $\mathbf{m}$ in the plane Oxz, for which we turn to Fig. 4.1.

It can be seen from the figure that the vector $\mathbf{m}$, given in the right quadrant of the $Oxyz$ system, is converted into the vector $\mathbf{m}'$, located in the left quadrant of the same system. In this case, the components of the vector $\mathbf{m}$, such as m_x and m_z, remain the same, and the component m_y is replaced with an equal component in its

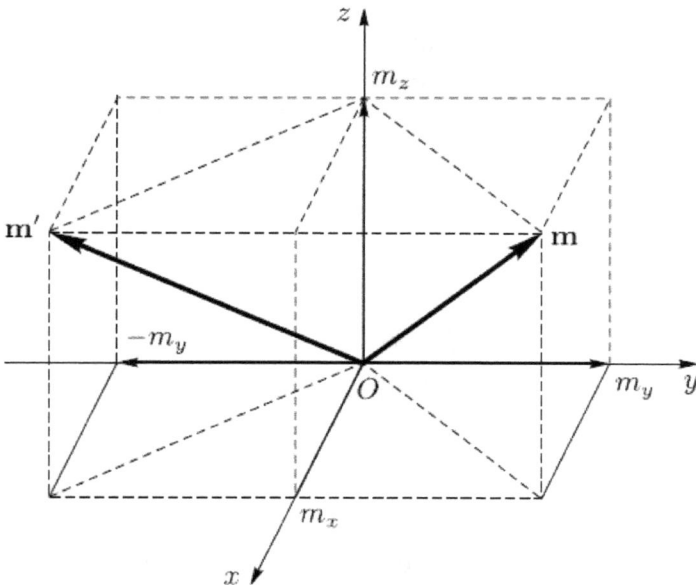

Fig. 4.1. The reflection pattern of the vector $\mathbf{m}$ in the Oxz plane.

magnitude, but opposite in sign, $-m_y$. Such a conversion can be written in the following form:

$$\mathbf{m}' = \ddot{A}\mathbf{m}, \tag{4.1}$$

where $\ddot{A}$ is the matrix describing the reflection of the vector $\mathbf{m}$ in the Oxz plane, that is, performing the conversion of the vector $\mathbf{m}$ into the vector $\mathbf{m}'$. Writing this ratio in the matrix form, we obtain:

$$\begin{pmatrix} m_x \\ -m_y \\ m_z \end{pmatrix} = \begin{pmatrix} A_{11} & A_{12} & A_{13} \\ A_{21} & A_{22} & A_{23} \\ A_{31} & A_{32} & A_{33} \end{pmatrix} \begin{pmatrix} m_x \\ m_y \\ m_z \end{pmatrix} =$$

$$= \begin{pmatrix} A_{11}m_x + A_{12}m_y + A_{13}m_z \\ A_{21}m_x + A_{22}m_y + A_{23}m_z \\ A_{31}m_x + A_{32}m_y + A_{33}m_z \end{pmatrix},$$

$$\tag{4.2}$$

whence, equating the first and the last of this chain of equalities, we obtain the matrix $\ddot{A}$ in the form

$$\ddot{A}_{xz} = \begin{pmatrix} 1 & 0 & 0 \\ 0 & -1 & 0 \\ 0 & 0 & 1 \end{pmatrix} \tag{4.3}$$

where the subscript xz means that the matrix belongs to the reflection in the Oxz plane.

Similarly, we obtain the reflection matrices in the Oyz and Oxy planes:

$$\ddot{A}_{yz} = \begin{pmatrix} -1 & 0 & 0 \\ 0 & 1 & 0 \\ 0 & 0 & 1 \end{pmatrix} \tag{4.4}$$

$$\overset{\leftrightarrow}{A}_{xy} = \begin{pmatrix} 1 & 0 & 0 \\ 0 & 1 & 0 \\ 0 & 0 & -1 \end{pmatrix}$$ (4.5)

Of particular interest may be matrices describing the reflection in two planes passing through the Oz axis in the middle (at an angle of 45°) to the Oxz and Oyz planes.

The first of these planes intersects the Oxy plane between the positive directions of the Ox and Oy axes, so that when reflected in it, the components m_x and m_y change in lengths, each retaining its sign.

The second of these points crosses the Oxy plane between the positive and negative directions of each of the Ox and Oy axes, so that when reflected in it, the components m_x and m_y change in lengths, but the signs of both and at the same time change to the opposite.

The corresponding matrices are:

$$\overset{\leftrightarrow}{A}_{+45} = \begin{pmatrix} 0 & 1 & 0 \\ 1 & 0 & 0 \\ 0 & 0 & 1 \end{pmatrix}$$ (4.6)

$$\overset{\leftrightarrow}{A}_{-45} = \begin{pmatrix} 0 & -1 & 0 \\ -1 & 0 & 0 \\ 0 & 0 & 1 \end{pmatrix}$$ (4.7)

where the indices '±45' mean the angle that these planes form with the positive direction of the Ox axis.

4.1.2. Rotation by some characteristic angles

Another important symmetry operation is a 90° rotation around a particular coordinate axis. Obtaining the rotation matrices in this case is trivially simple and similar to the one considered above, so the matrix for the main coordinates we give without output:

$$\ddot{A}_{x90} = \begin{pmatrix} 1 & 0 & 0 \\ 0 & 0 & -1 \\ 0 & 1 & 0 \end{pmatrix} \qquad (4.8)$$

$$\ddot{A}_{y90} = \begin{pmatrix} 0 & 0 & 1 \\ 0 & 1 & 0 \\ -1 & 0 & 0 \end{pmatrix} \qquad (4.9)$$

$$\ddot{A}_{z90} = \begin{pmatrix} 0 & -1 & 0 \\ 1 & 0 & 0 \\ 0 & 0 & 1 \end{pmatrix} \qquad (4.10)$$

These matrices describe a rotation of 90°, observed from the corresponding axis (then towards it) in the counterclockwise direction. Turns in the opposite direction or when looking against the corresponding axis are described by similar matrices with other signs, however, for the implementation of symmetry operations of such a reflection, as a rule, threefold parameters are quite enough here are the matrices.

For practice, it may be important to turn around the same axes by 180°. The corresponding matrices are listed below:

$$\ddot{A}_{x180} = \begin{pmatrix} 1 & 0 & 0 \\ 0 & -1 & 0 \\ 0 & 0 & -1 \end{pmatrix} \qquad (4.11)$$

$$\ddot{A}_{y180} = \begin{pmatrix} -1 & 0 & 0 \\ 0 & 1 & 0 \\ 0 & 0 & -1 \end{pmatrix} \qquad (4.12)$$

$$\ddot{A}_{z180} = \begin{pmatrix} -1 & 0 & 0 \\ 0 & -1 & 0 \\ 0 & 0 & 1 \end{pmatrix} \qquad (4.13)$$

In some cases, it may be of interest to rotate around the axis corresponding to the spatial diagonal of the cubic cell. The orientation of such an axis can be clearly seen from the view of cell No. 7 in Fig. 3.1, where this axis corresponds to a straight line

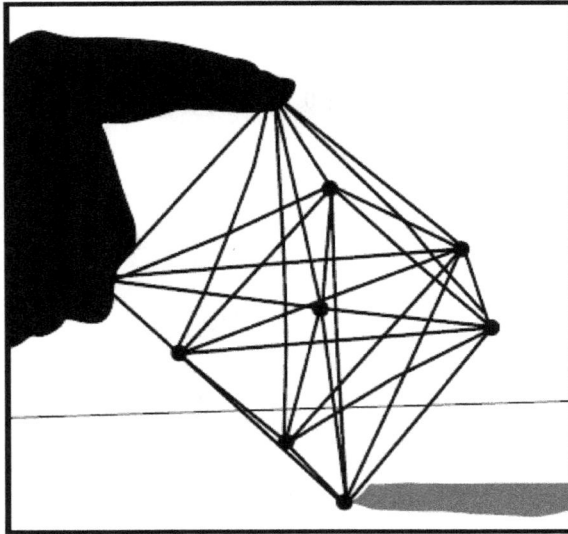

Fig. 4.2. Photo of a cubic cell spatial model.

passing through the vertices of the cube F and D. There, on the left, it shows a view of a cubic cell along this axis from the side of top D. For the sake of clarity of transformations, it is convenient to use a mechanical model of a cubic cell, made for example of wire, similar to that shown in Fig. 4.2.

Using this model, it is easy to see that turning the cell 120° in the counterclockwise direction transforms the components of the magnetization vector

$$\begin{pmatrix} m_x \\ \hline m_y \\ \hline m_z \end{pmatrix} \rightarrow \begin{pmatrix} m_z \\ \hline m_x \\ \hline m_y \end{pmatrix}, \tag{4.14}$$

and turning it clockwise is the same to transform

$$\begin{pmatrix} m_x \\ \hline m_y \\ \hline m_z \end{pmatrix} \rightarrow \begin{pmatrix} m_y \\ \hline m_z \\ \hline m_x \end{pmatrix} \tag{4.15}$$

The corresponding rotation matrices obtained using a transformation similar to (4.2) are:
counterclockwise

$$\overset{\leftrightarrow(+)}{A}_{c120} = \begin{pmatrix} 0 & 0 & 1 \\ 1 & 0 & 0 \\ 0 & 1 & 0 \end{pmatrix};$$

(4.16)

clockwise

$$\overset{\leftrightarrow(-)}{A}_{c120} = \begin{pmatrix} 0 & 1 & 0 \\ 0 & 0 & 1 \\ 1 & 0 & 0 \end{pmatrix}.$$

(4.17)

By direct multiplication it is easy to verify that the product of these matrices is equal to one.

4.1.3. Rotate at an arbitrary angle

The most common types of magnetic anisotropy are 'easy axis' (easy-axis or uniaxial) and 'easy plane' (easy-plane). Such types of anisotropy imply the invariance of the energy density of the magnetic medium when rotating around a given axis at an arbitrary angle.

Find the transformation matrix for such a rotation. As earlier, we will consider turning counterclockwise when lookingfrom the positive direction of the corresponding axis (here Oz).

Consider Fig. 4.3, where the scheme of rotation of the vector **m** around the centre of the Oxy plane by the angle φ is shown. In the original position, this vector makes an angle α with the axis Ox and has components

$$m_x = m_0 \cos \alpha,$$ (4.18)
$$m_y = m_0 \sin \alpha,$$ (4.19)

where m_0 is the length of the vector **m**. After turning by the angle φ,

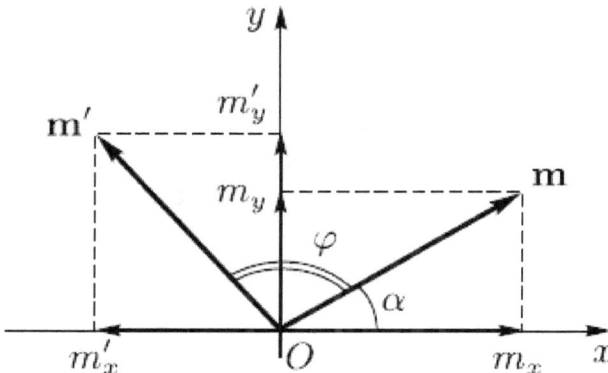

Fig. 4.3. The rotation scheme of the vector **m** through arbitrary angle φ.

it is converted into a vector **m**′ of the same length having components

$$m'_x = m_0 \cos(\alpha + \varphi), \qquad (4.20)$$
$$m'_y = m_0 \sin(\alpha + \varphi) \qquad (4.21)$$

We omit the coefficient m_0, as it does not change when turning (or we set it equal to one) and write down the rotation operation in the matrix form:

$$\mathbf{m}' = \ddot{A}\mathbf{m}, \qquad (4.22)$$

where $\ddot{A}$ is the desired rotation matrix. Writing the same in the components, we get

$$\left(\begin{array}{c} \cos(\alpha + \varphi) \\ \hline \sin(\alpha + \varphi) \end{array} \right) = \left(\begin{array}{c|c} A_{11} & A_{12} \\ \hline A_{21} & A_{22} \end{array} \right) \left(\begin{array}{c} \cos\alpha \\ \hline \sin\alpha \end{array} \right). \qquad (4.23)$$

Performing multiplication in the right part, we obtain

$$\left(\begin{array}{c} \cos(\alpha + \varphi) \\ \hline \sin(\alpha + \varphi) \end{array} \right) = \left(\begin{array}{c} A_{11}\cos\alpha + A_{12}\sin\alpha \\ \hline A_{21}\cos\alpha + A_{22}\sin\alpha \end{array} \right). \qquad (4.24)$$

Expanding the components of the vector on the left side according to the rules of trigonometry [270] and equating the components of both of the studied vectors line by line, after bringing similar terms, we get the system of equations:

$$\left(A_{11} - \cos\varphi \right)\cos\alpha + \left(A_{12} + \sin\varphi \right)\sin\alpha = 0; \qquad (4.25)$$

$$\left(A_{21} - \sin\varphi \right)\cos\alpha + \left(A_{22} - \cos\varphi \right)\sin\alpha = 0. \qquad (4.26)$$

Due to the arbitrariness of the angle α, this system can only be satisfied if all coefficients of $\cos\alpha$ and $\sin\alpha$ are zero. Using this condition as a way of finding A_{ik}, where $i, k = 1, 2$, we obtain the desired rotation matrix in the plane Oxy at an arbitrary angle φ in the form

$$\vec{A}_\varphi = \left(\begin{array}{c|c} \cos\varphi & -\sin\varphi \\ \hline \sin\varphi & \cos\varphi \end{array} \right). \qquad (4.27)$$

4.1.4. Rotation in cases of trigonal and hexagonal symmetry

Special cases of rotary symmetry are trigonal (rhombohedral) and hexagonal. To describe such symmetry types, rotation is performed around the same axis parallel to the vertical edge of the corresponding cell (for example, AD for cell No. 5 or AG for cell No. 6 in Fig. 3.1) passing through the centre of symmetry of the base of the cell. Usually such an axis is designated as Oz. In this case, the plane Oxy containing the two other axes of the Cartesian coordinate system Ox and Oy, being perpendicular to Oz, turns out to be parallel to the horizontal plane of the cell: ABC or DEF for the trigonal cell (No. 5 in Fig. 3.1) or $ABCDEF$ or $GHIJKL$ for the hexagonal cell (No. 6 in Fig. 3.1).

The trigonal or hexagonal nature of the symmetry of the cell requires its alignment with the previous position when rotated by 120° or 60°. Obviously, such turns are special cases of turning by an arbitrary angle described by the matrix (4.27). From this matrix, setting the angle of rotation to be 60° and 120°, we obtain the corresponding matrices, as special cases:

$$\overset{\leftrightarrow}{A}_{60} = \left(\begin{array}{c|c} \dfrac{1}{2} & -\dfrac{\sqrt{3}}{2} \\ \hline \dfrac{\sqrt{3}}{2} & \dfrac{1}{2} \end{array} \right), \tag{4.28}$$

and also:

$$\overset{\leftrightarrow}{A}_{120} = \left(\begin{array}{c|c} -\dfrac{1}{2} & -\dfrac{\sqrt{3}}{2} \\ \hline \dfrac{\sqrt{3}}{2} & -\dfrac{1}{2} \end{array} \right) \tag{4.29}$$

For a three-dimensional vector, the component along the Oz axis of which does not change during rotation, the same matrices must be supplemented with a row and a column, at the intersection of which the unit is located. That is, the rotation matrix at an arbitrary angle around the Oz axis is:

$$\overset{\leftrightarrow}{A}_{\varphi} = \left(\begin{array}{c|c|c} \cos\varphi & -\sin\varphi & 0 \\ \hline \sin\varphi & \cos\varphi & 0 \\ \hline 0 & 0 & 1 \end{array} \right). \tag{4.30}$$

To rotate at angles of 60° and 120° we get:

$$
\overleftrightarrow{A}_{60} = \begin{pmatrix} \dfrac{1}{2} & -\dfrac{\sqrt{3}}{2} & 0 \\ \dfrac{\sqrt{3}}{2} & \dfrac{1}{2} & 0 \\ 0 & 0 & 1 \end{pmatrix}; \qquad (4.31)
$$

and also:

$$
\overleftrightarrow{A}_{120} = \begin{pmatrix} -\dfrac{1}{2} & -\dfrac{\sqrt{3}}{2} & 0 \\ \dfrac{\sqrt{3}}{2} & -\dfrac{1}{2} & 0 \\ 0 & 0 & 1 \end{pmatrix}. \qquad (4.32)
$$

4.2. Uniaxial magnetic anisotropy energy density

We will assume that the energy density of magnetic anisotropy in the most general case can be represented as a series expansion in powers of the normalized magnetization components (i.e. $m_i = M_i/M_0$):

$$
\begin{aligned}
U = &\sum_{i=1}^{3} a_i m_i + \sum_{i,k=1}^{3} a_{ik} m_i m_k + \sum_{i,k,l=1}^{3} a_{ikl} m_i m_k m_l + \\
&+ \sum_{i,k,l,m=1}^{3} a_{iklm} m_i m_k m_l m_m + \sum_{i,k,l,m,n=1}^{3} a_{iklmn} m_i m_k m_l m_m m_n + \\
&+ \sum_{i,k,l,m,n,p=1}^{3} a_{iklmnp} m_i m_k m_l m_m m_n m_p + ...,
\end{aligned} \qquad (4.33)
$$

where a_i, a_{ik}, a_{ikl}, a_{iklm}, a_{iklmn}, a_{iklmnp}... – constant coefficients (anisotropy constants), the form of which will be determined by the condition consisting in the conservation of energy density during coordinate transformations corresponding to the given crystallographic symmetry.

The simplest type of symmetry is uniaxial, another, somewhat more complicated, but also quite common is cubic. Consider these types of anisotropy consistently.

4.2.1. Second order anisotropy in magnetization

The symmetry of the uniaxial anisotropy requires the invariance of the energy density in the case of two-variable transformations: 1. when the sign of the projection of the magnetization vector on the anisotropy axis changes, 2. when the component of the magnetization vector normal to the anisotropy axis is rotated by an arbitrary angle around this axis. Consider these requirements sequentially. We introduce the Cartesian coordinate system $Oxyz$ in such a way that the coordinate axis Oz coincides with the anisotropy axis. In this case, the plane perpendicular to the anisotropy axis will be Oxy. The first requirement essentially means that the energy density remains unchanged when the magnetization vector is reflected in the Oxy plane. The transformation matrix is given by (4.5). The transformation of the magnetization vector in such a reflection, in accordance with the general rule (4.1), has the form:

$$\begin{pmatrix} m'_x \\ m'_y \\ m'_z \end{pmatrix} = \begin{pmatrix} 1 & 0 & 0 \\ 0 & 1 & 0 \\ 0 & 0 & -1 \end{pmatrix} \begin{pmatrix} m_x \\ m_y \\ m_z \end{pmatrix} = \begin{pmatrix} m_x \\ m_y \\ -m_z \end{pmatrix}. \tag{4.34}$$

It can be seen that the components of the magnetization vector m_x and m_y remain unchanged, while the component m_z changes the sign. This means that all the terms of the energy density that have m_z in the odd number of penalties change the sign during the transformation. This circumstance is unacceptable, therefore, all coefficients of terms in expression (4.33), containing m_z in odd degrees, must be set equal to zero. Expression (4.33), up to second order terms, takes the form

$$U = a_x m_x + a_y m_y + a_{xx} m_x m_x + \left(a_{xy} + a_{yx}\right) m_x m_y + \\ + a_{yy} m_y m_y + a_{zz} m_z m_z. \tag{4.35}$$

The second requirement means that the energy density remains unchanged when the magnetization vector is rotated around the Oz axis by an arbitrary angle φ. The transformation matrix is given by (4.27). Conversion of the magnetization vector at such a rotation, in accordance with the general rule (4.1), has the form:

$$\begin{pmatrix} m'_x \\ m'_y \\ m'_z \end{pmatrix} = \begin{pmatrix} \cos\varphi & \sin\varphi & 0 \\ -\sin\varphi & \cos\varphi & 0 \\ 0 & 0 & 1 \end{pmatrix} \begin{pmatrix} m_x \\ m_y \\ m_z \end{pmatrix} = \begin{pmatrix} m_x \cos\varphi + m_y \sin\varphi \\ -m_x \sin\varphi + m_y \cos\varphi \\ m_z \end{pmatrix}. \tag{4.36}$$

Included in (4.35) combinations of the components of the magnetization vector take the form:

$$m_x \rightarrow m_x \cos\varphi - m_y \sin\varphi; \qquad (4.37)$$

$$m_y \rightarrow m_x \sin\varphi + m_y \cos\varphi; \qquad (4.38)$$

$$m_x m_x \rightarrow \left(m_x \cos\varphi - m_y \sin\varphi\right)^2; \qquad (4.39)$$

$$m_x m_y \rightarrow \left(m_x \cos\varphi - m_y \sin\varphi\right)\left(m_x \sin\varphi + m_y \cos\varphi\right); \qquad (4.40)$$

$$m_y m_y \rightarrow \left(m_x \sin\varphi + m_y \cos\varphi\right)^2; \qquad (4.41)$$

$$m_z m_z \rightarrow m_z^2. \qquad (4.42)$$

Since the angle φ is arbitrary, it follows that all combinations, except the last (4.42), change as this angle changes. So, to keep the energy density unchanged, all coefficients a_i and a_{ik}, except a_{zz}, must be set equal to zero. As a result, expression (4.35) takes the form

$$U = a_{zz}m_z^2, \qquad (4.43)$$

which is the required expression for the energy density of uniaxial anisotropy with a light axis that coincides with the axis Oz.

In practice, in some cases it is convenient to use the expression of the energy density through the components of magnetization perpendicular to the anisotropy axis. For this, one should use the condition of preservation of the length of the magnetization vector, in normalized components having the form

$$m_x^2 + m_y^2 + m_z^2 = 1, \qquad (4.44)$$

whence, expressing m_z and substituting in (4.43), up to a constant term, we get

$$U = -a_{zz}\left(m_x^2 + m_y^2\right) \qquad (4.45)$$

Next, for the uniaxial anisotropy, we introduce the subscript a, for the a_{zz} constant, the designation K. In such records from (4.43) and (4.45) we get

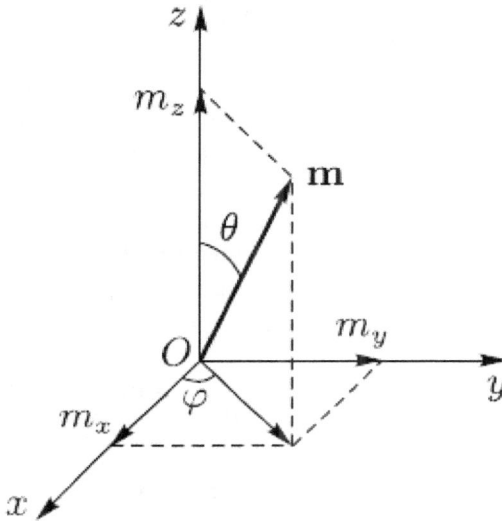

Fig. 4.4. Normalized vector of magnetization in Cartesian and spherical system of coordinates.

$$U_a = Km_z^2 = -K\left(m_x^2 + m_y^2\right). \tag{4.46}$$

We write the obtained expressions in the spherical coordinate system shown in Fig. 4.4. The components of the normalized vector of magnetization are:

$$m_x = \sin\theta\cos\varphi; \tag{4.47}$$
$$m_y = \sin\theta\sin\varphi \tag{4.48}$$
$$m_z = \cos\theta \tag{4.49}$$

Substituting these components into the first and second equations (4.46), we get:

$$U_a^{(1)} = K\cos^2\theta; \tag{4.50}$$
$$U_a^{(2)} = -K\sin^2\theta. \tag{4.51}$$

With an accuracy of the sign of a constant, the energy density of uniaxial anisotropy in the form of the second of these averages is used in the CMD technique [40, p. 23, forms. (1.14)]. Consider this question in the next section.

4.2.2. Easy axis and easy plane anisotropy

It is important to note that the nature of anisotropy depends on the sign of its constant. All the above arguments did not concern this sign, that is, the 'default' anisotropy was assumed to be simply uniaxial without taking into account the equilibrium position of the magnetization vector. However, it is easy to see from formula (4.46) that with a positive sign of this constant, the potential has a minimum as $m_z \to 0$ or $(m_x^2 + m_y^2) \to + \infty$. This means that in equilibrium the vector of magnetization is located in a plane perpendicular to the anisotropy axis. Moreover, the azimuthal orientation of the vector is indifferent. This anisotropy is called 'easy plane'. With a negative constant sign, on the contrary, the case is realized when the equilibrium position of the magnetization is oriented along the anisotropy axis, that is, the anisotropy 'easy plane' or 'easy axis' has a place.

In most practical applications, for example, in the technique of cylindrical magnetic domains (CMD) [35–42], the anisotropy of the easy axis type takes place, as a rule, oriented perpendicular to the film plane. At the same time, based on the convenience to work with positive values, the constant K can be assumed to be positive, but then in the formula (4.46) the signs must be changed. So

$$U_a = -Km_z^2 = K(m_x^2 + m_y^2), \qquad (4.52)$$

where for the realization of easy-axis anisotropy it is assumed that $K > 0$. Recall that the formula given here includes the normalized components of the magnetization vector $m_{x,y,z}$, that is, to obtain the full components of $M_{x,y,z}$, and multiply them by the saturation magnetization M_0. In this case, the formula (4.52) takes the form

$$U_a = -\frac{K}{M_0^2} M_z^2 = \frac{K}{M_0^2} \left(M_x^2 + M_y^2 \right). \qquad (4.53)$$

A convenient characteristic of easy-axis anisotropy is the effective 'anisotropy field', determined in accordance with the usual rule for obtaining effective fields [7, 8], without taking into account the exchange interaction, having the form (1.2)

$$H_e = -\frac{\partial U}{\partial M}. \qquad (4.54)$$

In this case, the anisotropy field obtained by differentiating the first equality of formula (4.53) with the subsequent subsequent replacement of M_z by M_0 is obtained equal to

$$H_a = \frac{2K}{M_0},$$ (4.55)

In this form, the anisotropy field, minus the demagnetization of the form and some influence of the domains, determines the magnitude of the external field required to 'lay' the magnetization vector in the plane of the uniaxial film used for the CMD. The qualitative picture of the phenomenon here is such that in the absence of an external field the magnetization vector is oriented exactly along the anisotropy axis, where the field H_a 'holds' it. As the external field increases in the film plane, the magnetization vector rotates in the direction of this field, so that there is an orientational transition, ending with the complete alignment of the magnetization along the external field. This alignment corresponds to the phase transition, discussed in more detail in Chapter 5. This phase transition just happens when an external field is equal to the anisotropy field with an accuracy of up to the additives mentioned above (demagnetization of the form and domains). If the anisotropy field is so much higher than these additives that they can be neglected, then it just corresponds to the phase transition field.

4.2.3. Uniaxial high order anisotropy

In addition to the two previous sections, it should be added that the review was limited to members in the expansion (4.33) not higher than the second order. Accounting for members of higher orders can be carried out similarly. In this case, after taking into account the symmetry operations, only members with an even order in the mz component remain in the expansion, that is,

$$U_A = a_{zz}m_z^2 + a_{zzzz}m_z^4 + a_{zzzzzz}m_z^6 + a_{zzzzzzzz}m_z^8 + ...,$$ (4.56)

However, it should be noted that, in most real materials, the values of the constants of higher orders tend to decrease, the faster the order, so that in most practical cases in the decomposition (4.56) it is quite possible to confine ourselves only to the first term

4.2.4. Uniaxial magnetic anisotropy energy density with an arbitrary direction of the axis

In the previous sections, the situation was considered when the axis of magnetic anisotropy coincides with one of the axes of the

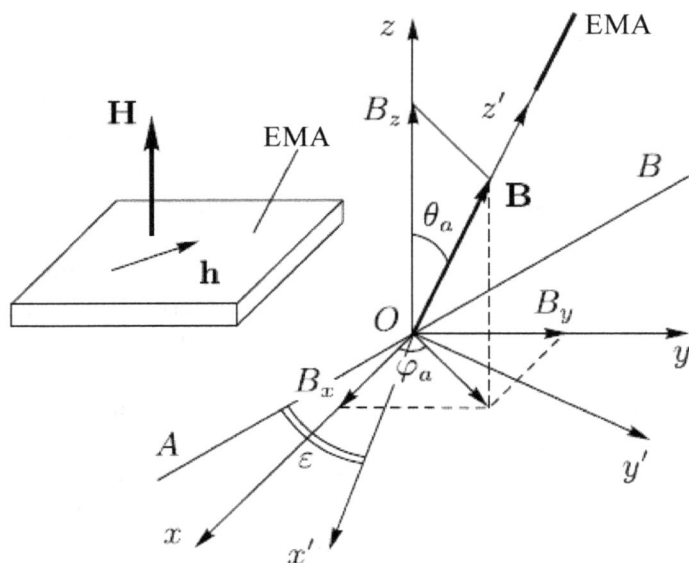

Fig. 4.5. Geometry of the uniaxial anisotropy problem with an arbitrary direction of the axis.

laboratory Cartesian coordinate systems (in this case with the Oz axis). Now we turn to the important case for the practice of the same anisotropy, the orientation of which axis relative to the laboratory coordinate system can be arbitrary.

The geometry of the problem is illustrated in Fig. 4.5. The figure on the left shows the relative position of the constant and variable fields, as well as the anisotropy axis relative to the plane of the ferrite plate.

The initial (laboratory) coordinate system is $Oxyz$. The Oxy plane coincides with the plane of the ferrite plate, the axis Oz is perpendicular to it. The constant field **H** is parallel to the axis Oz.

At this stage, the demagnetizing factor of the plate will not be explicitly taken into account in order to more clearly reveal the role of anisotropy in its pure form. A detailed account of the demagnetizing factor will be carried out further when referring to the real anisotropy of films of mixed ferrites–garnets (Chapter 7). Here we will assume that anisotropy can be described using a single axis, which we will call the 'easy magnetization axis' (EMA). In the figure, its guide vector is denoted by **B**.

The direction of the axis of easy magnetization with respect to the $Oxyz$ system is arbitrary, and in spherical coordinates, it is determined by the polar and azimuthal angles θ_a and φ_a. The $Ox'y'z'$ coordinate system is oriented so that its axis Oz' directed along the axis of anisotropy of the EMA. The straight line AB, passing through the point O, lies in the Oxy plane perpendicular to the axis Oz'. The angle of rotation of the $Ox'y'z'$ system around the axis Oz', measured from the line OA, is denoted by ε.

We assume that the energy density of uniaxial anisotropy recorded in the $Ox'y'z'$ system has the form

$$U_a^{(1)} = K(m_{x'}^2 + m_{y'}^2),\qquad(4.57)$$

where $m_{x',y'}$ are the components of the normalized magnetization vector, K – constant, for the case of easy-axis anisotropy it is positive: $K > 0$. The introduction of the index (1) at the energy density at this stage is not of fundamental importance and is explained in more detail below. The problem is to obtain an expression for the energy density depending on the direction of orientation of the anisotropy axis in the $Oxyz$ system, that is, on the polar and azimuthal angles θ_a and φ_a.

Since the expression for the energy density of uniaxial anisotropy is given in the $Ox'y'z'$ system, then the components of the magnetization vector $m_{x'}$ and $m_{y'}$ defined in the $Ox'y'z'$ system are expressed in the components of the same vector m_x, m_y, m_z in the $Oxyz$ system. Substituting these components into (4.33) will give the desired expression for the energy density in the $Oxyz$ system, expressed as m_x, m_y, m_z. Thus, it is necessary to find the vector ratio

$$\mathbf{m}' = \ddot{B}\mathbf{m}.\qquad(4.58)$$

where $\overleftrightarrow{B}$ is the transition matrix of the $Ox'y'z'$ system to the $Oxyz$ system. From the general rule of working with transition matrices, it follows that the desired matrix is inverse to the matrix of an arbitrary transition from the $Oxyz$ system to the $Ox'y'z'$ system. It is technically easier to find first direct matrix $\overleftrightarrow{A}$ providing communication

$$\mathbf{m} = \ddot{A}\mathbf{m}',\qquad(4.59)$$

since the components of this matrix are the coordinates of the guides of the axes of the $Ox'y'z'$ system regarding the $Oxyz$ system. Components of the direct matrix can be found through the components vectors $\mathbf{B}$ using θ_a, φ_a, ε as the Euler angles with the line of nodes AB [261].

In the general case, the direct matrix has the components given in formulas (3.143)–(3.151), that is, being written as a table, it takes the form (3.161) (the index a is omitted for simplicity of recording)

$$\overrightarrow{A} = \begin{pmatrix} \cos\theta\cos\varphi\sin\varepsilon + & \cos\theta\cos\varphi\cos\varepsilon - & \\ + \sin\varphi\cos\varepsilon & - \sin\varphi\sin\varepsilon & \sin\theta\cos\varphi \\ \cos\theta\sin\varphi\sin\varepsilon - & \cos\theta\sin\varphi\cos\varepsilon + & \\ - \cos\varphi\cos\varepsilon & + \cos\varphi\sin\varepsilon & \sin\theta\sin\varphi \\ - \sin\theta\sin\varepsilon & - \sin\theta\cos\varepsilon & \cos\theta \end{pmatrix}. \qquad (4.60)$$

The components of the inverse matrix correspond to formulas (3.152)–(3.160), that is, being written as a table, this matrix has the form (3.162)

$$\overleftrightarrow{A}^{-1} = \begin{pmatrix} \cos\theta\cos\varphi\sin\varepsilon + & \cos\theta\sin\varphi\sin\varepsilon - & \\ + \sin\varphi\cos\varepsilon & - \cos\varphi\cos\varepsilon & - \sin\theta\sin\varepsilon \\ \cos\theta\cos\varphi\cos\varepsilon - & \cos\theta\sin\varphi\cos\varepsilon + & \\ - \sin\varphi\sin\varepsilon & + \cos\varphi\sin\varepsilon & - \sin\theta\cos\varepsilon \\ \sin\theta\cos\varphi & \sin\theta\sin\varphi & \cos\theta \end{pmatrix}. \qquad (4.61)$$

It can be seen that (4.61) is also obtained from (4.60) by replacing the rows with columns. By direct verification, we can verify that the product (4.60) and (4.61) is equal to the identity matrix.

Shown in Fig. 4.5 the geometry in the choice of rotation of the $Ox'y'z'$ system around the Oz' axis has no limit. We now consider a simpler particular case. The Ox' axis is compatible? with the line OA, that is, according to Euler, with the line of nodes [261], perpendicular to the projection of the anisotropy axis on the Oxy plane. The angle ε is obtained equal to zero, so that the inverse matrix (4.61) takes the form

$$\overleftrightarrow{A}^{-1} = \begin{pmatrix} \sin\varphi & - \cos\varphi & 0 \\ \cos\theta\cos\varphi & \cos\theta\sin\varphi & - \sin\theta \\ \sin\theta\cos\varphi & \sin\theta\sin\varphi & \cos\theta \end{pmatrix}. \qquad (4.62)$$

This matrix in the relation (4.58) plays the same role as the matrix $\overleftrightarrow{B}$. given there. When substituting (4.62), relation (4.58) takes the form

$$\begin{pmatrix} m_{x'} \\ m_{y'} \\ m_{z'} \end{pmatrix} = \begin{pmatrix} \sin\varphi & - \cos\varphi & 0 \\ \cos\theta\cos\varphi & \cos\theta\sin\varphi & - \sin\theta \\ \sin\theta\cos\varphi & \sin\theta\sin\varphi & \cos\theta \end{pmatrix} \begin{pmatrix} m_x \\ m_y \\ m_z \end{pmatrix}. \qquad (4.63)$$

Writing down by the components, we get:

$$m_{x'} = m_x \sin\varphi - m_y \cos\varphi; \qquad (4.64)$$

$$m_{y'} = m_x \cos\theta\cos\varphi + m_y \cos\theta\sin\varphi - m_z \sin\theta; \qquad (4.65)$$

$$m_{z'} = m_x \sin\theta\cos\varphi + m_y \sin\theta\sin\varphi + m_z \cos\theta. \qquad (4.66)$$

Substituting these expressions into (4.57), we obtain the energy density of uniaxial anisotropy in the $Oxyz$ system as

$$U_a^{(1)} = K\left\{ m_x^2\left(\cos^2\theta\cos^2\varphi + \sin^2\varphi\right) + m_y^2\left(\cos^2\theta\sin^2\varphi + \cos^2\varphi\right) + \right.$$
$$+ m_z^2\sin^2\theta - 2m_x m_y \sin^2\theta\sin\varphi\cos\varphi -$$
$$\left. -2m_x m_z \sin\theta\cos\theta\cos\varphi - 2m_y m_z \sin\theta\cos\theta\sin\varphi \right\}. \qquad (4.67)$$

Here, the angles θ and φ characterize the orientation of the EMA with respect to the $Oxyz$ system, that is, as before, they can be marked with the index a and introduce the notations θ_a and φ_a:

$$U_a^{(1)} = K\left\{ m_x^2\left(\cos^2\theta_a \cos^2\varphi_a + \sin^2\varphi_a\right) + m_y^2\left(\cos^2\theta_a \sin^2\varphi_a + \cos^2\varphi_a\right) + \right.$$
$$+ m_z^2\sin^2\theta_a - 2m_x m_y \sin^2\theta_a \sin\varphi_a \cos\varphi_a -$$
$$\left. -2m_y m_z \sin\theta_a \cos\theta_a \sin\varphi_a - 2m_z m_x \sin\theta_a \cos\theta_a \cos\varphi_a \right\}. \qquad (4.68)$$

Introducing the traditional designation for the anisotropy field [7,8]

$$H_a = \frac{2K}{M_0}, \qquad (4.69)$$

we write (4.68) in the form

$$U_a^{(1)} = \frac{1}{2} M_0 H_0 \left\{ m_x^2\left(\cos^2\theta_a \cos^2\varphi_a + \sin^2\varphi_a\right) + \right.$$
$$+ m_y^2\left(\cos^2\theta_a \sin^2\varphi_a + \cos^2\varphi_a\right) +$$
$$+ m_z^2\sin^2\theta_a - 2m_x m_y \sin^2\theta_a \sin\varphi_a \cos\varphi_a -$$
$$\left. -2m_y m_z \sin\theta_a \cos\theta_a \sin\varphi_a - 2m_z m_x \sin\theta_a \cos\theta_a \cos\varphi_a \right\}. \qquad (4.70)$$

For a number of tasks of considerable interest are effective anisotropy fields, which are determined by the formula (1.2) or (4.54) (taking into account the fact that $\mathbf{m} = \mathbf{M}/M_0$)

$$\mathbf{H}_a = -\frac{1}{M_0}\frac{\partial U}{\partial \mathbf{m}}. \qquad (4.71)$$

The result is:

$$H_{ax}^{(1)} = -H_a\left\{m_x\left(\cos^2\theta_a\cos^2\varphi_a + \sin^2\varphi_a\right) = \right.$$
$$\left. -m_y\sin^2\theta_a\sin\varphi_a\cos\varphi_a - m_z\sin\theta_a\cos\theta_a\cos\varphi_a\right\}; \tag{4.72}$$

$$H_{ay}^{(1)} = -H_a\left\{m_y\left(\cos^2\theta_a\sin^2\varphi_a + \cos^2\varphi_a\right) - \right.$$
$$\left. -m_x\sin^2\theta_a\sin\varphi_a\cos\varphi_a - m_z\sin\theta_a\cos\theta_a\sin\varphi_a\right\}; \tag{4.73}$$

$$H_{az}^{(1)} = -H_a\left\{m_z\sin^2\theta_a - \right.$$
$$\left. -m_y\sin\theta_a\cos\theta_a\sin\varphi_a - m_x\sin\theta_a\cos\theta_a\cos\varphi_a\right\}. \tag{4.74}$$

We now give an expression for the energy density (4.67) in a spherical coordinate system. In accordance with Fig. 4.4, the components of the normalized magnetization vector **m** have the form (4.47)–(4.49):

$$m_x = \sin\theta\cos\varphi; \tag{4.75}$$

$$m_y = \sin\theta\sin\varphi; \tag{4.76}$$

$$m_z = \cos\theta. \tag{4.77}$$

Substituting (4.75)–(4.77) into (4.68), we obtain

$$U_a^{(1)} = K\left\{\left(\cos^2\theta_a\cos^2\varphi_a + \sin^2\varphi_a\right)\sin^2\theta\cos^2\varphi + \right.$$
$$+\left(\cos^2\theta_a\cos^2\varphi_a + \cos^2\varphi_a\right)\sin^2\theta\sin^2\varphi + $$
$$+\sin^2\theta_a\cos^2\theta - 2\sin^2\theta_a\sin\varphi\cos\varphi_a\sin^2\theta\sin\varphi\cos\varphi - $$
$$-2\sin\theta_a\cos\theta_a\sin\varphi_a\sin\theta\cos\theta\sin\varphi - $$
$$\left. -2\sin\theta_a\cos\theta_a\cos\varphi_a\sin\theta\cos\theta\cos\varphi\right\}. \tag{4.78}$$

If the orientation of the spherical coordinate system is not specified by any specific conditions of the problem, then in order to simplify the recording, it is convenient to choose it so that the origin line of the azimuthal coordinate falls on the projection of the EMA with the plane perpendicular to the polar axis (i.e. along the Ox axis in Fig. 4.5). This gives $\varphi_a = 0$, so that (4.68) takes the form:

$$U_a^{(1)} = K\left(m_x^2 \cos^2\theta_a + m_y^2 + m_z^2 \sin^2\theta_a - 2m_x m_z \sin\theta_a \cos\theta_a\right). \tag{4.79}$$

Substituting into this expression (4.75) - (4.77), we get

$$U_a^{(1)} = K\left(\cos^2\theta_a \sin^2\theta \cos^2\varphi + \sin^2\theta \sin^2\varphi + \sin^2\theta_a \cos^2\theta - 2\sin\theta_a \cos\theta_a \sin\theta \cos\theta \cos\varphi\right). \tag{4.80}$$

↔ 4.2.5. Azimuth diagram of the energy density of magnetic uniaxial anisotropy with an arbitrary direction of the axis

The dependence of the normalized energy density on the azimuth angle φ for an arbitrary direction of the axis is illustrated by the azimuth diagram shown in Fig. 4.6. As an example, we put $\varphi_a = 0°$, $\theta_a = 60°$, $\theta = 60°$. Curves are constructed according to the formula (4.80). The solid line corresponds to $m_z > 0$, the dotted line to $m_z < 0$.

It can be seen from the figure that the minimum of the energy density for $m_z > 0$, that is, when the EMA is inclined from the positive part of the Oz axis to the positive part of the Ox axis, it comes at an angle $\varphi = 0°$. When $m_z < 0$ the minimum is at $\varphi = 180°$, which corresponds to the EMA slope from the negative part of the Oz axis to the negative part of the Ox axis.

4.2.6. Another kind of expression for the energy density of uniaxial magnetic anisotropy with an arbitrary direction of the axis

The above consideration is based on the expression for the energy density of the form (4.57):

$$U_a^{(1)} = K(m_{x'}^2 + m_{y'}^2). \tag{4.81}$$

According to the general position of theoretical physics, the potential energy is determined with an accuracy to the constant term [295], that is, in its records, in addition to (4.81), other options are possible. Consider one of them, which has great practical value.

So, from the condition of constancy of the length of the normalized vector of magnetization, having the form

$$m_{x'}^2 + m_{y'}^2 + m_{z'}^2 = 1, \tag{4.82}$$

we get

$$U_a^{(1)} = K - Km_{z'}^2. \tag{4.83}$$

We introduce the notation:

$$U_z^{(2)} = -Km_{z'}^2, \tag{4.84}$$

so that

$$U_a^{(1)} = K + U_a^{(2)}. \tag{4.85}$$

Since the potential energy is determined to within a constant term, it can be assumed that the energy densities $U_a^{(1)}$ and $U_a^{(2)}$ are equivalent to each other. The representation of the energy density of uniaxial anisotropy in the form of $U_a^{(2)}$ can be useful in a number of problems, so we give it here in the technical notation as $U_a^{(1)}$. In this case, instead of (4.68), we get:

$$U_a^{(2)} = -\frac{1}{2}M_0 H_a \{ m_x^2 \sin^2\theta_a \cos^2\varphi_a + m_y^2 \sin^2\theta_a \sin^2\varphi_a +$$
$$+ m_z^2 \cos^2\theta_a + 2m_x m_y \sin^2\theta_a \sin\varphi_a \cos\varphi_a +$$
$$+ 2m_x m_z \sin\theta_a \cos\theta_a \cos\varphi_a + 2m_y m_z \sin\theta_a \cos\theta_a \sin\varphi_a \}. \tag{4.86}$$

Entering the anisotropy field in accordance with (4.69), we obtain

$$\tag{4.87}$$

Effective fields take the form:

$$H_{ax}^{(2)} = H_a \{ m_x \sin^2\theta_a \cos^2\varphi_a +$$
$$+ m_y \sin^2\theta_a \sin\varphi_a \cos\varphi_a + m_z \sin\theta_a \cos\theta_a \cos\varphi_a \}; \tag{4.88}$$

$$H_{ay}^{(2)} = H_a \{ m_y \sin^2\theta_a \sin^2\varphi_a +$$
$$+ m_x \sin^2\theta_a \sin\varphi_a \cos\varphi_a + m_z \sin\theta_a \cos\theta_a \sin\varphi_a \}; \tag{4.89}$$

$$H_{az}^{(2)} = H_a \{ m_z \cos^2 \theta_a +$$
$$+ m_y \sin \theta_a \cos \theta_a \sin \varphi_a + m_x \sin \theta_a \cos \theta_a \cos \varphi_a \}. \tag{4.90}$$

Expressing in (4.86) the components of magnetization through the angles θ and φ in accordance with (4.75)–(4.77), we obtain

$$U_a^{(2)} = -K \{ \sin^2 \theta_a \cos^2 \varphi_a \sin^2 \theta \cos^2 \varphi +$$
$$+ \sin^2 \theta_a \sin^2 \varphi_a \sin^2 \theta \sin^2 \varphi +$$
$$+ \cos^2 \theta_a + 2 \sin^2 \theta_a \sin \varphi_a \cos \varphi_a \sin^2 \theta \sin \varphi \cos \varphi +$$
$$+ 2 \sin \theta_a \cos \theta_a \cos \varphi_a \sin \theta \cos \theta \cos \varphi +$$
$$+ 2 \sin \theta_a \cos \theta_a \sin \varphi_a \sin \theta \cos \theta \sin \varphi \}. \tag{4.91}$$

If, like the previous case, in the expression (4.86) we $\varphi_a = 0$, then we get

$$U_a^{(2)} = -K \left(m_x^2 \sin^2 \theta_a + m_z^2 \cos^2 \theta_a + 2 m_x m_z \sin \theta_a \cos \theta_a \right), \tag{4.92}$$

Substituting the components of the magnetization vector in the form (4.75)–(4.77), we get:

$$U_a^{(2)} = -K \{ \sin^2 \theta_a \sin^2 \theta \cos^2 \varphi + \cos^2 \theta_a \cos^2 \theta +$$
$$+ 2 \sin \theta_a \cos \theta_a \sin \theta \cos \theta \cos \varphi \}. \tag{4.93}$$

The dependence of the normalized energy density on the azimuth angle φ in this case, taking into account relation (4.85), completely coincides with that given in Fig. 4.6, which is a consequence of the equivalence of energy densities $U_a^{(1)}$ and $U_a^{(2)}$.

Comment. Above, when discussing formula (4.85), it is noted that the energy density $U_a^{(1)}$ and $U_a^{(2)}$ differs only by a constant value, that is, in terms of the potential, expressions (4.70) and (4.87) are equivalent. However, derived from these expressions the effective fields (4.72–(4.74) and (4.88)–(4.90) are not equivalent. For a qualitative consideration, it can be assumed that the physical meaning of the easy-axis type anisotropy corresponds to the fields (4.88)–(4.90), whereas the fields of the form (4.72)–(4.74) correspond to anisotropy of the 'easy plane' type. However, it must be said that such a definition is not indisputable. Therefore, the use of fields of

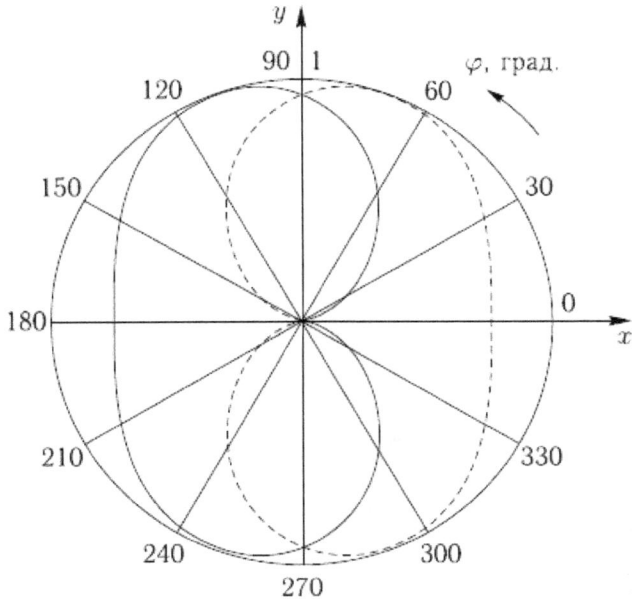

Fig. 4.6. Azimuthal diagram of the energy density of uniaxial magnetic anisotropy with the direction of the axis determined by the angles $\varphi_a = 0°$, $\theta_a = 60°$, at $\theta = 60°$

one kind or another should be carried out with full consideration of the specific nature of the anisotropy of the considered problem. Moreover, in some cases, for example, with a change in temperature, the sign of the constant K may change, so that the easy-axis anisotropy will turn into the easy-plane anisotropy or vice versa. When solving problems of this type, when passing through a point of change of sign, the constants and expressions for the fields should replace each other accordingly. The continuity of the change of the effective field will not be violated in this case, since the constant at the point of change of sign is equal to zero, so that all the fields at this point are also equal to zero.

4.3. Energy density of cubic magnetic anisotropy

We first consider the energy density of cubic anisotropy for the simplest case – the orientation [001]. This case is the base case for any distributions, for which the energy density is obtained from the base case by rotating the crystal cell, as will be shown later.

4.3.1. Energy density of cubic anisotropy for orientation [001]

The symmetry of cubic anisotropy requires an invariance of the energy density during two-dimensional transformations: when the sign of the projection of the magnetization vector changes to any of the axes coinciding with the edges of the cubic cell; when turning 90° around any of the technical axes. Consider these requirements sequentially. We introduce the Cartesian coordinate system $Oxyz$ so that the coordinate axes Ox, Oy and Oz coincide with the three edges of the cubic cell originating from the same vertex. In this case, the planes of the cubic cell formed by two neighbouring edges will coincide with the coordinate planes Oxy, Oxz and Oyz. The first requirement essentially means the invariance of the energy density in the reflection of the magnetization vector in these coordinate planes. The transformation matrices are given by the formulas (4.3), (4.4), (4.5). Transformations of the magnetization vector for such a discharge, performed similarly to (4.34), show that in all the cases, the two components of the magnetization remain unchanged, and the third changes the sign. This means that all the terms of the energy density that have any of the components of magnetization to an odd degree, change the sign. In order to maintain unchanged energy density for such transformations, all coefficients of terms in expression (4.33) containing m_x, m_y or m_z to an odd degree must be set equal to zero.

Expression (4.33) with the accuracy of the fourth order terms, takes the form

$$U = a_{xx}m_x^2 + a_{yy}m_y^2 + a_{zz}m_z^2 + a_{xxxx}m_x^4 + a_{yyyy}m_y^4 + a_{zzzz}m_z^4 +$$
$$\left(a_{xxyy} + a_{yyzz}\right)m_x^2 m_y^2 + \left(a_{yyzz} + a_{zzyy}\right)m_y^2 m_z^2 + \left(a_{zzxx} + a_{xxzz}\right)m_z^2 m_x^2. \quad (4.94)$$

Rotation by 90° around the Ox axis, performed using the A_{x90} matrix (4.8) leads to the following transformation of the magnetization vector:

$$\begin{pmatrix} m_{x'} \\ m_{y'} \\ m_{z'} \end{pmatrix} = \begin{pmatrix} 1 & 0 & 0 \\ 0 & 0 & -1 \\ 0 & 1 & 0 \end{pmatrix} \begin{pmatrix} m_x \\ m_y \\ m_z \end{pmatrix}. \quad (4.95)$$

Included in (4.95) the combinations of the components of the magnetization vector take the form:

$$m_x^2 \rightarrow m_x^2; \quad (4.96)$$

$$m_y^2 \rightarrow m_z^2;$$ (4.97)

$$m_z^2 \rightarrow m_y^2;$$ (4.98)

$$m_x^4 \rightarrow m_x^4;$$ (4.99)

$$m_y^4 \rightarrow m_z^4;$$ (4.100)

$$m_z^4 \rightarrow m_y^4;$$ (4.101)

$$m_x^2 m_y^2 \rightarrow m_x^2 m_z^2;$$ (4.102)

$$m_y^2 m_z^2 \rightarrow m_z^2 m_y^2;$$ (4.103)

$$m_z^2 m_x^2 \rightarrow m_y^2 m_x^2.$$ (4.104)

Substituting these combinations into (4.95), comparing with the original expression and doing a similar procedure of turning by 90° around the Oy and Oz axes using the matrices (4.9), (4.10), we get:

$$a_{xx} = a_{yy} = a_{zz} = a_0;$$ (4.105)

$$a_{xxxx} = a_{yyyy} = a_{zzzz} = a_1;$$ (4.106)

$$\left(a_{xxyy} + a_{yyxx}\right) = \left(a_{yyzz} + a_{zzyy}\right) = \left(a_{zzxx} + a_{xxzz}\right) = a_2;$$ (4.107)

where a_0, a_1, a_2 are auxiliary notations introduced to shorten the record.

For such types, (4.94) takes the form

$$u = a_0\left(m_x^2 + m_y^2 + m_z^2\right) + a_1\left(m_x^4 + m_y^4 + m_z^4\right) +$$
$$a_2\left(m_x^2 m_y^2 + m_y^2 m_z^2 + m_z^2 m_x^2\right),$$ (4.108)

which is the required expression for the energy density of cubic magnetic anisotropy in the Cartesian coordinate system, the axes of which coincide with the edges of the cubic cell. A significant simplification of the resulting expression suggests the preservation of the length of the magnetization vector:

$$m_x^2 + m_y^2 + m_z^2 = 1.$$ (4.109)

So, substituting this expression in (4.108), from the first term we get a_0 (1), that is, a constant value, which, given the definition of the potential with an accuracy of an arbitrary constant, can be omitted.

Next, squaring (4.109), we get:

$$\left(m_x^4 + m_y^4 + m_z^4\right) + 2\left(m_x^2 m_y^2 + m_y^2 m_z^2 + m_z^2 m_x^2\right) = 1.$$

(4.110)

Multiplying this expression by $-a_1$ and adding to (4.108), citing similar terms and omitting the constant terms, by the definition of the potential with an accuracy of an arbitrary constant, we get:

$$U = \left(a_2 - 2a_1\right)\left(m_x^2 m_y^2 + m_y^2 m_z^2 + m_z^2 m_x^2\right).$$

(4.111)

Similarly, multiplying (4.110) by $-a_2/2$, adding to (4.108), omitting the constant terms and giving similar terms, we get

$$U = \left(a_1 - a_2/2\right)\left(m_x^4 + m_y^4 + m_z^4\right).$$

(4.112)

By virtue of equality (4.111) and (4.112) up to arbitrary constants, one can see that the expression for the energy density of cubic anisotropy has two equivalent values (4.111) and (4.112), differing in the values of constants.

Apparently, the more common is the form (4.111), where to denote the coefficient in the first round brackets in cubic anisotropy constant, is equal to

$$K_1 = (a_2 - 2a_1)$$

(4.113)

At the same time, the energy density of cubic anisotropy, for which we often use the 'c' index, takes the form:

$$U_c = K_1 \left(m_x^2 m_y^2 + m_y^2 m_z^2 + m_z^2 m_x^2\right).$$

(4.114)

This type is used, for example, in [259, p. 160, in the forms. (4.7)]. From the formula (4.112) we can get another type of energy density, equivalent to the first:

$$U_c = -\frac{K_1}{2}\left(m_x^4 + m_y^4 + m_z^4\right).$$

(4.115)

where to get the coefficient in front of the bracket, it suffices to express a_2 through K_1 from (4.113) and then substitute it into (4.112). This type is used for example in [7, p. 90, forms. (2.2.24')]

Here, the formulas (4.114) and (4.115) are usually sufficient to account for cubic anisotropy in most real materials. However, in some rays it is required to take into account in the expansion of the energy density (4.33) the terms of higher order. Moreover, for cubic anisotropy, the following order after the fourth will be the sixth. As a result of taking into account the invariance of the energy density upon reflection of the magnetization vector in the coordinate planes and rotated by 90° around the coordinate axes, the sixth order energy density component remaining from (4.33) takes the form

$$U_c^{(6)} = K_2 m_x^2 m_y^2 m_z^2. \tag{4.116}$$

The total energy density is the sum of (4.114) or (4.115) and (4.116), so it takes the form

$$U_c = K_1 \left(m_x^2 m_y^2 + m_y^2 m_z^2 + m_z^2 m_x^2 \right) + K_2 m_x^2 m_y^2 m_z^2, \tag{4.117}$$

or

$$U_c = -\frac{K_1}{2} \left(m_x^4 + m_y^4 + m_z^4 \right) + K_2 m_x^2 m_y^2 m_z^2, \tag{4.118}$$

In the form (4.117), the expression for the energy density of cubic magnetic anisotropy is used, for example, in [7, p. 83, formula (2.2.3)].

Comment. It should be noted, however, that in most real materials, the constant K_2, as a rule, in absolute value is much less than K_1. So in the iron–yttrium garnet the difference between them reaches an order and more. Especially clearly for this material, the ratio of the constants in a wide temperature range is presented in [8, p. 62, Fig. 2.12]. At the same time, when accounting for K_2, the mathematical calculations are noticeably more complicated. Apparently, for this reason, in most of the known to the second of this monograph the work of consideration is limited only to the constant K_1, that is, the energy density of the cubic anisotropy is used in the form (4.114) or (4.115). However, for example, in papers [131, 204], devoted to the excitation of hypersound in a plate with cubic anisotropy, the K_2 constant is present in the analytic calculations, although its role is not considered in further detail.

4.3.2. Cubic anisotropy field

To characterize cubic anisotropy, a 'cubic anisotropy field' is often introduced, equal to [7, p. 85, forms. (2.2.10)]

$$H_c = \frac{K_1}{M_0}. \tag{4.119}$$

Note, however, that this field is not obtained from the energy density by differentiation, similar to that performed when calculating the field of uniaxial anisotropy. Nevertheless, this expression presents certain conveniences, especially when analyzing the orientational dependences of ferromagnetic resonance [6–8] (in the case of IYG, one must take into account the sign of the constant in the interpretation given below).

4.3.3. Features of cubic anisotropy of iron yttrium garnet

The iron–yttrium garnet (IYG), for which two values of the anisotropy field are found in the literature, has the greatest interest for the further presentation. So in [7, p. 95, the caption to Fig. 2.2.8] the value of $H_c = -36$ Oe is given, and according to [296, p. 307, Table 15] the same value is equal to -45 Oe. Considering that the saturation magnetization M_0 for IYG is usually assumed to be 140 G (or $4\pi M_0 = 1750$ G), the anisotropy constants in both these cases are respectively -5046 erg×cm^{-3} and -6300 erg×cm^{-3}.

At the same time, the negative sign of the constant means that the light axes of anisotropy are spatial diagonals of the cube. In this case, for clarity, we can assume that the negative anisotropy field (4.119) 'repels' the magnetization from the coordinate axes with rotation around them, that is, from the edges of the cubic cell.

The interpretation of the sign of the anisotropy constant given here is quite common [6–8, 296]. However, in practical examination of the anisotropy of an iron–yttrium garnet is not always convenient to have a constant negative. Therefore, to facilitate the calculations, we can set the constant to be positive, that is, assume $K_1 > 0$, but the signs of the expressions (4.114) and (4.115) must be changed. Similarly, in the yttrium garnet, the constant K_2 is also negative [8, p. 62, Fig. 2.12], so that, assuming for convenience that $K_2 > 0$, the sign of expression (4.116) and the signs before the second terms in expressions (4.117), (4.118) should also be changed. As a result of this change, the expressions for the energy density (4.117) and (4.118) take the form:

$$U_c = -K_1\left(m_x^2 m_y^2 + m_y^2 m_z^2 + m_z^2 m_x^2\right) - K_2 m_x^2 m_y^2 m_z^2, \qquad (4.120)$$

and:

$$U_c = \frac{K_1}{2}\left(m_x^4 + m_y^4 + m_z^4\right) - K_2 m_x^2 m_y^2 m_z^2, \qquad (4.121)$$

where it is assumed that $K_1 > 0$ and $K_2 > 0$.

With such a change of the sign of the K_1 constant, the 'anisotropy field' (4.119) becomes positive, which also facilitates working with it.

Within the framework of this monograph for the energy density of cubic anisotropy, the form (4.120) is used everywhere under the assumption $K_1 > 0$ and $K_2 > 0$.

4.3.4. Cubic anisotropy energy density for orientation [001] in the spherical coordinate system

In the previous sections, the expressions for the magnetic anisotropy energy density for uniaxial and cubic cases were obtained. The conclusion was based on the use of symmetry operations, such as the reflection in the plane and rotation around the axis. To perform such operations, the most convenient is the Cartesian coordinate system, which contains basically coordinate planes and straight coordinate axes. However, for some problems, especially with the participation of uniaxial anisotropy, the spherical coordinate system is more convenient. Such a situation takes place, for example, in the technique of cylindrical magnetic domains [35–42] or in finding the orientational dependences of the frequency of ferromagnetic resonance [6–8, 31]. To record the energy density in a spherical coordinate system, the Cartesian components of the magnetization vector in spherical coordinates should be expressed, just as it was done in deriving expressions (4.75)–(4.77) or in more detail when deriving (4.47)–(4.49) based on Fig. 4.4. Further, the components of the magnetization vector thus obtained should be substituted into the expressions for the energy density obtained in the Cartesian coordinate system. Such a transformation for uniaxial anisotropy has already been performed in Section 4.2.4. We now show the implementation of a similar transformation for the case of cubic anisotropy, and, for the sake of generality, we consider both the constants K_1 and K_2.

So, substituting (4.75)–(4.77) into (4.120), taking into account the remark about the signs as constants, we get:

$$U_c = -K_1\left(\sin^4\theta\sin^2\varphi\cos^2\varphi + \sin^2\theta\cos^2\theta\right) - $$
$$-K_2\sin^4\theta\cos^2\theta\sin^2\varphi\cos^2\varphi. \qquad (4.122)$$

In a similar form, up to the signs of the constants and taking into account the trigonometric formulas for a double angle, the energy density of cubic anisotropy is given, for example, in [7, p. 96, forms. (2.2.34)]. Similarly, from (4.121) we get:

$$U_c = \frac{K_1}{2}\left(\sin^4\theta\cos^4\varphi + \sin^4\theta\sin^4\varphi + \cos^4\theta\right) - $$
$$-K_2\sin^4\theta\cos^2\theta\sin^2\varphi\cos^2\varphi. \qquad (4.123)$$

By simple, although somewhat cumbersome trigonometric transformations, it can be verified that the difference between the first terms of these expressions is equal to one. It means that, despite the difference in the recording forms, the potentials defined by these two formulas differ by a constant value, that is, they describe the same potential energy.

The formulas given here refer to the orientation of a [001] type cubic cell. The formulas for other orientations in the spherical coordinate system are obtained from the formulas of the Cartesian system in a similar way using the substitution of the components of the magnetization vector in the form (4.75)–(4.77).

4.3.5. Azimuth diagram of energy density of magnetic anisotropy for orientation [001]

Let us go back a little to formulas (4.114), (4.115) and give a convenient reception of a visual image of the distribution of the energy density of cubic anisotropy in three-dimensional space. So a rather convenient way here is the construction of an azimuthal diagram, that is, the representation of the energy density depending on the azimuthal orientation angle of the magnetization vector. Similar the diagram was constructed for uniaxial anisotropy in Fig. 4.6. Here we perform the same construction of the case for cubic

188

anisotropy with orientation of the [001] type. A common technique here is the expression of the components of the magnetization vector m_x, m_y, m_z in a spherical coordinate system with subsequent substitution into the energy density written through these components. Such a substitution gives the energy density expressed in terms of the polar and azimuth angles θ and φ. In the considered case of orientation [001], a similar procedure leads to formulas (4.122) and (4.123). The azimuthal diagram is obtained as the dependence of the energy density on the azimuthal angle of the magnetization vector φ for a given value of the polar angle θ. This value can be selected based on the visibility of the picture. Practice shows that a fairly convenient value is θ = 60°. Such an azimuthal diagram, obtained on the basis of formulas (4.114), (4.115) and normalized to one, is presented in Fig. 4.7. Here the lines are solid and dashed correspond to different signs of the first anisotropy constant K_1. Solid line – $K_1 > 0$ (case of IYG or nickel), dotted line – $K_1 < 0$ (case of iron). In the two steps, the second constant K_2, due to its smallness, is not taken into account. It can be seen from the figure that the azimuthal dependence of the energy density on the angle φ has a fourth-order symmetry, that is, it repeats every 90°. Such symmetry corresponds exactly to the symmetry of the cube, which is set by the face on the

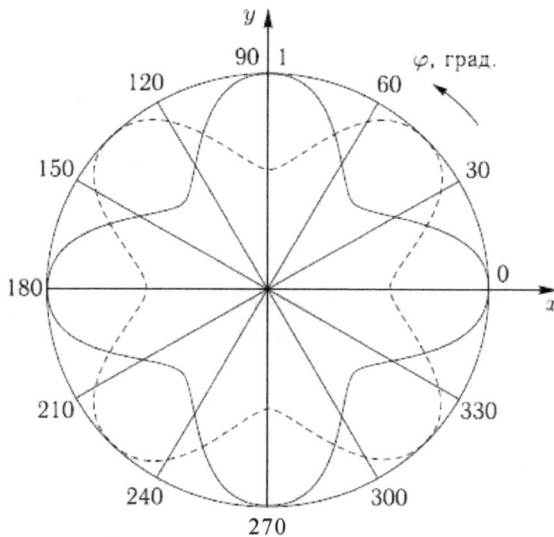

Fig. 4.7. Azimuth diagram of the magnetic anisotropy energy density for the [001] orientation. The polar angle is θ = 60°. The solid line is $K_1 > 0$, the dotted line is $K_1 < 0$.

plane corresponding to $\theta = 90°$. When $K_1 > 0$, the energy density minima occurs at angles of 45°, 135°, 225°, 315°, that is, they pass through planes corresponding to projections of [111] axes on the Oxy plane in the system whose Cartesian axes coincide with the edges of the cube. Energy density maxima occur between these planes. This kind of azimuth diagram is due to the fact that with a positive sign of the K_1 constant the [111] type axis are easy magnetization axes.

The opposite sign of the anisotropy constant, as can be seen from the nature of the dotted curve, maxima and minima changes

This exactly corresponds to the fact that, with a positive sign of the same constant, the axes of the [111] type represent the axes of a difficult magnetization.

4.3.6. Energy density of cubic anisotropy for orientation [011]

In Section 4.3.4, the energy density of cubic anisotropy was obtained for the [001] orientation. However, in a number of practical tasks (for example, in the CMD technique) other orientations are important. We now give further results calculated in accordance with the matrices obtained in Section 3.6, as the expression for the energy density of a cubic magnetic anisotropy with a constant K_1 with different crystallographic orientations. Let's start with the orientation [011].

General formulation of the problem. There is an expression for the energy density of the cubic magnetic anisotropy in the Cartesian coordinate system $Oxyz$ for the case when the edges of the cubic crystallographic cell, that is, the [001], [010] and [100] crystallographic axes, are oriented along the axes Ox, Oy and Oz.

As shown in Section 4.3.3, this expression, up to the fourth-order terms, has two equivalent forms (4.120) and (4.121). We restrict ourselves to the consideration of the form (4.120). The task is to record the energy density in the same system, but with the direction of the crystallographic axis [011] along the coordinate axis Oz, and the crystallographic cell itself can be rotated around this axis at an arbitrary angle.

The transformation of magnetization vectors. To transform the magnetization vectors, we use the transition matrix method, according to which the transformation of the vector components when the coordinate system is rotated in accordance with formula (3.10)

$$\mathbf{a} = \ddot{A}\mathbf{a}', \qquad (4.124)$$

where **a** is a vector in the original coordinate system, **a'** is a vector in rotated coordinate system, $\vec{A}$ is the transition matrix from rotated $Ox'y'z'$ system to the initial $Oxyz$ system.

The matrix of such a transition has the form (3.222)

$$\vec{A}_{011}(\varepsilon) = \begin{pmatrix} \cos\varepsilon & -\sin 3 & 9 \\ \dfrac{\sqrt{2}}{2}\sin\varepsilon & \dfrac{\sqrt{2}}{2}\cos\varepsilon & \dfrac{\sqrt{2}}{2} \\ -\dfrac{\sqrt{2}}{2}\sin 2 & -\dfrac{\sqrt{2}}{2}\cos e & \dfrac{\sqrt{2}}{2} \end{pmatrix} \tag{4.125}$$

Let us now consider the transformation of the magnetization vector. According to the transformation formula (4.124), we find that the components of the magnetization vector normalized to M_0 are transformed as follows:

$$\begin{pmatrix} m_x \\ m_y \\ m_z \end{pmatrix} = \vec{A}_{011} \begin{pmatrix} m_{x'} \\ m_{y'} \\ m_{z'} \end{pmatrix}, \tag{4.126}$$

from which we get:

$$m_x = \cos\varepsilon\, m_{x'} - \sin\varepsilon\, m_{y'}; \tag{4.127}$$

$$m_y = \frac{\sqrt{2}}{2}\sin\varepsilon\, m_{x'} + \frac{\sqrt{2}}{2}\cos\varepsilon\, m_{y'} + \frac{\sqrt{2}}{2} m_{z'}; \tag{4.128}$$

$$m_z = \frac{\sqrt{2}}{2}\sin\varepsilon\, m_{x'} - \frac{\sqrt{2}}{2}\cos\varepsilon\, m_{y'} + \frac{\sqrt{2}}{2} m_{z'}; \tag{4.129}$$

Energy density conversion. The energy density of magnetic anisotropy in the $Oxyz$ system, whose axes are directed along the edges of the cube, has the form (4.120):

$$U_c = -K_1\left(m_x^2 m_y^2 + m_y^2 m_z^2 + m_z^2 m_x^2\right) - K_2 m_x^2 m_y^2 m_z^2, \tag{4.130}$$

where K_1, K_2 are the constants of cubic anisotropy (in the case of IYG, they are positive). We restrict ourselves to considering only the first anisotropy constant, setting $K_2 = 0$. At the same time, we obtain

$$U_c = -K_1 \left(m_x^2 m_y^2 + m_y^2 m_z^2 + m_z^2 m_x^2 \right).$$

(4.131)

To obtain the energy density expressed in terms of the components of the magnetization vectors in the system $Ox'y'z'$ it is necessary to substitute into (4.131) the components of vector magnetization determined by the expressions (4.75)–(4.77),.(4.132)

To reduce the cumbersome form of the calculations, it is useful to group the expressions (4.131) in the form:

$$U_c = -K_1 \left[m_x^2 \left(m_y^2 + m_z^2 \right) + \left(m_y m_z \right)^2 \right].$$

(4.132)

A special case. The general expression is rather cumbersome, so consider the special case corresponding to $\varepsilon = 0$. In this case, the axis Oz' is oriented along the diagonal of the face of the cube, the Ox' axis – along the edge of the cube coinciding with the line OA, and the axis Oy' – along the OE line.

The transition matrix (4.125) for this case takes the form:

$$\ddot{A}_{011}(0) = \begin{pmatrix} 1 & 0 & 0 \\ 0 & \dfrac{\sqrt{2}}{2} & \dfrac{\sqrt{2}}{2} \\ 0 & \dfrac{\sqrt{2}}{2} & \dfrac{\sqrt{2}}{2} \end{pmatrix}$$

(4.133)

The components of the magnetization vector (4.127)–(4.129) take the form:

$$m_x = m_{x'};$$

(4.134)

$$m_y = \frac{\sqrt{2}}{2} m_{y'} + \frac{\sqrt{2}}{2} m_{z'};$$

(4.135)

$$\# m_z = -\frac{\sqrt{2}}{2} m_{y'} + \frac{\sqrt{2}}{2} m_{z'};$$

(4.136)

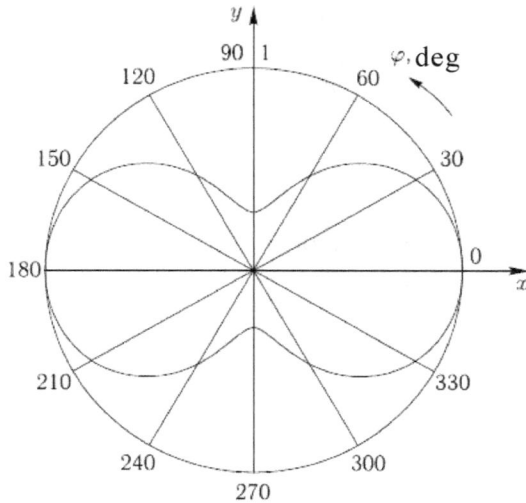

Fig. 4.8. Azimuth diagram of the magnetic anisotropy energy density for orientation [011]. Polar angle θ = 60°.

Substituting (4.134)–(4.136) into (4.132), we obtain the energy density of magnetic anisotropy in the form

$$U_c = -\frac{K_1}{4}\left(m_{y'}^4 + m_{z'}^4 + 4m_{x'}^2 m_{y'}^2 + 4m_{x'}^2 m_{z'}^2 - 2m_{y'}^2 m_{z'}^2\right). \tag{4.137}$$

If further the task will be solved in the rotated coordinate system $Ox'y'z'$, that is, the law of motion will be recorded in this system, it is possible to rename this system to $Oxyz$ for convenience and remove strokes from the magnetization components system – rotated). In this case, the energy density (4.137) takes the form:

$$U_c = -\frac{K_1}{4}\left(m_y^4 + m_z^4 + 4m_x^2 m_y^2 + 4m_x^2 m_z^2 - 2m_y^2 m_z^2\right). \tag{4.138}$$

Two of the four axes of the [111] type lie in the plane $Ox'y'$ (plane $ACBEO$ in Fig. 3.13), forming with the axis Ox' (OA line in the same figure) azimuth angles: 60°, 120°, 240°, 300°. Two other axes lie in the plane $Ox'z'$ (the $AOBM$ plane in the same figure), that is, their action on the plane $Ox'y'$ occurs at angles of 0° and 180°.

4.3.7. Azimuth diagram of energy density of magnetic anisotropy for orientation [011]

The dependence of the normalized energy density on the azimuth angle is illustrated in Fig. 4.8. The method of construction of this figure is similar to that adopted for the case of orientation [001] considered above, that is, the polar angle of the magnetization vector is known considered to be constant $\theta = 60°$. In this case, the energy density with respect to m_z is quadratic; therefore, the form of the diagram shown does not depend on the sign of this component (unlike in the case of the [111] orientation considered below). It can be seen from the figure that the minima of the energy density occur at angles of 90° and 270°, that is, they are located in the middle of the [111] axes lying in the $Ox'y'$ plane. The diagram has a two-petal character, corresponding to the location of a two-axis [111] type in a plane perpendicular to the polar axis of the spherical coordinate system, and two more in a plane perpendicular to the first one. The figure is constructed with a positive sign of the anisotropy constant, when the [111] axes are light. Changing the sign of the constant to the opposite results in a rotation of the diagram by 90°, is like the previous case.

4.3.8. Energy density of cubic orientation for for orientation [111]

Let us now consider a more complex case, corresponding to the [111] orientation, that is, when one of the spatial coordinates of the cubic cell is directed along the Oz axis of the Cartesian coordinate system.

General formulation of the problem. There is an expression for the energy density of the cubic magnetic anisotropy in the Cartesian coordinate system $Oxyz$ for the case when the edges of the cubic crystallographic cell, that is, the [001], [010] and [100] crystallographic axes are oriented along the axes Ox, Oy and Oz (this expression is given in section 4.3.4). The task is to record the energy density in the same system, but with the direction of the crystallographic axis [111] along the coordinate axis Oz, and the crystallographic cell itself can be rotated around this axis at an arbitrary angle.

Transformation of magnetization vectors. To convert of the magnetization vectors we will use the transition matrix method, according to which the transformation of the components of the vector when the coordinate system is rotated is in accordance with the formula (3.10)

$$\mathbf{a} = \vec{A}\mathbf{a}',$$

(4.139)

where $\mathbf{a}$ is a vector in the original coordinate system, $\mathbf{a}'$ is a vector in rotated coordinate system, $\vec{A}$ is a transition matrix from rotated system $Ox'y'z'$ to the initial system $Oxyz$.

The transition matrix has the form (3.269)

$$\vec{A}_{111}(\varepsilon) = \begin{pmatrix} \frac{\sqrt{6}}{6}(\sqrt{3}\cos\varepsilon + \sin\varepsilon) & \frac{\sqrt{6}}{6}(\cos\varepsilon - \sqrt{3}\sin\varepsilon) & \frac{\sqrt{3}}{3} \\ -\frac{\sqrt{6}}{6}(\sqrt{3}\cos\varepsilon - \sin\varepsilon) & \frac{\sqrt{6}}{6}(\cos\varepsilon + \sqrt{3}\sin\varepsilon) & \frac{\sqrt{3}}{3} \\ -\frac{\sqrt{6}}{3}\sin\varepsilon & -\frac{\sqrt{6}}{3}\cos\varepsilon & \frac{\sqrt{3}}{3} \end{pmatrix}.$$

(4.140)

Let us now consider the transformation of the magnetization vector. According to the transformation formula (4.139), we find that the components of the magnetization vector normalized to M_0 are transformed as follows:

$$\begin{pmatrix} m_x \\ m_y \\ m_z \end{pmatrix} = \vec{A}_{111} \begin{pmatrix} m_{x'} \\ m_{y'} \\ m_{z'} \end{pmatrix},$$

(4.141)

where do we get:

$$m_x = \frac{\sqrt{6}}{6}\left(\sqrt{3}\cos\varepsilon + \sin\varepsilon\right)m_{x'} + \frac{\sqrt{6}}{6}\left(\cos\varepsilon - \sqrt{3}\sin\varepsilon\right)m_{y'} + \frac{\sqrt{3}}{3}m_{z'};$$

(4.142)

$$m_y = \frac{\sqrt{6}}{6}\left(\sqrt{3}\cos\varepsilon - \sin\varepsilon\right)m_{x'} +$$
$$+ \frac{\sqrt{6}}{6}\left(\cos\varepsilon + \sqrt{3}\sin\varepsilon\right)m_{y'} + \frac{\sqrt{3}}{3}m_{z'};$$

(4.143)

$$m_z = -\frac{\sqrt{6}}{3}\sin\varepsilon\, m_{x'} - \frac{\sqrt{6}}{3}\cos\varepsilon\, m_{y'} + \frac{\sqrt{3}}{3}m_{z'}.$$

(4.144)

Energy density conversion. The magnetic anisotropy energy density in the $Oxyz$ system, whose axes are directed along the edges of the cube, has the form (4.120), (4.130):

$$U_c = -K_1\left(m_x^2 m_y^2 + m_y^2 m_z^2 + m_z^2 m_x^2\right) - K_2 m_x^2 m_y^2 m_z^2,$$

(4.145)

where K_1, K_2 are the constants of cubic anisotropy (in the case of IYG both are assumed positive).

As in the previous case, we restrict ourselves to considering only the first anisotropy constant, setting $K_2 = 0$. At the same time, according to (4.131), we have:

$$U_c = -K_1\left(m_x^2 m_y^2 + m_y^2 m_z^2 + m_z^2 m_x^2\right).$$ (4.146)

To obtain the energy density, expressed in terms of the components of the magnetization vectors in the system $Ox'y'z'$, it is necessary to substitute the components of the magnetization vectors defined by the expressions (4.142)–(4.144) into expression (4.146).

To reduce the cumbersome computation, the terms of the expressions (4.146) should be grouped in the form:

$$U_c = -K_1\left[m_x^2\left(m_y^2 + m_z^2\right)+\left(m_y m_z\right)^2\right].$$ (4.147)

A special case. The general expression is quite cumbersome, so consider the special case corresponding to $\varepsilon = 0$. In this case, the axis Oz' is oriented along the spatial diagonal of the cube, the axis Ox' – along the line OA (Fig. 3.14), and the axis Oy' along the OE line (same place). The transition matrix for this case is

$$A_{111}(0) = \begin{pmatrix} \dfrac{\sqrt{2}}{2} & \dfrac{\sqrt{6}}{6} & \dfrac{\sqrt{3}}{3} \\ \dfrac{\sqrt{2}}{2} & \dfrac{\sqrt{3}}{6} & \dfrac{\sqrt{3}}{3} \\ 0 & -\dfrac{\sqrt{6}}{3} & \dfrac{\sqrt{3}}{3} \end{pmatrix}$$ (4.148)

The components of the magnetization vector take the form:

$$m_x = \frac{\sqrt{2}}{2}m_{x'} + \frac{\sqrt{6}}{6}m_{y'} + \frac{\sqrt{3}}{3}m_{z'};$$ (4.149)

$$m_y = -\frac{\sqrt{2}}{2}m_{x'} + \frac{\sqrt{6}}{6}m_{y'} + \frac{\sqrt{3}}{3}m_{z'};$$ (4.150)

$$m_z = -\frac{\sqrt{6}}{3}m_{y'} + \frac{\sqrt{3}}{3}m_{z'};$$

(4.151)

Substituting (4.149)–(4.151) into (4.147), we obtain the energy density of the magnetic anisotropy in the form

$$U_c = -\frac{K_1}{12}\left(3m_{x'}^4 + 3m_{y'}^4 + 4m_{z'}^4 + 6m_{x'}^2 m_{y'}^2 - \right.$$
$$\left. -12\sqrt{2}m_{x'}^2 m_{y'}m_{z'} + 4\sqrt{2}m_{x'}^3 m_{z'}\right).$$

(4.152)

If further the task will be solved in the rotated coordinate system $Ox'y'z'$, that is, the law of motion will be recorded in this system, it is possible to rename this system to $Oxyz$ for convenience and remove strokes from the magnetization components (not forgetting that this system is rotated). In this case, the energy density (4.152) takes the form

$$U_c = -\frac{K_1}{12}\left(3m_x^4 + 3m_y^4 + 4m_z^4 + 6m_x^2 m_y^2 - \right.$$
$$\left. -12\sqrt{2}m_x^2 m_y m_z + 4\sqrt{2}m_x^3 m_z\right).$$

(4.153)

Projections of the [111] axes on the plane $Ox'y'$ (*ACBEO* plane in Fig. 3.14) are in azimuth angles: 30°, 90°, 150°, 210°, 270°, 330°. With this positive direction of the axis Oz' correspond to the angles 90°, 210°, 330°.

4.3.9. Azimuth diagram of energy density of magnetic anisotropy for orientation [111]

The dependence of the normalized energy density on the azimuth angle is illustrated in Fig. 4.9. The solid line corresponds to $m_z > 0$, the dotted line $m_z < 0$.

It can be seen from the figure that when $m_z > 0$, the energy density minima is at angles of 90°, 210°, 330°, and for $m_z < 0$ at angles of 30°, 150°, 270°, which corresponds to orientations of projections of [111] axes on the Oxy plane, taking into account the direction of the Oz axis (i.e. in the upper and lower half-spaces relative to the plane Oxy). The figure is constructed with a positive sign of the anisotropy constant, when the [111] axes are light. Changing the sign of a constant to the opposite leads to the replacement of the

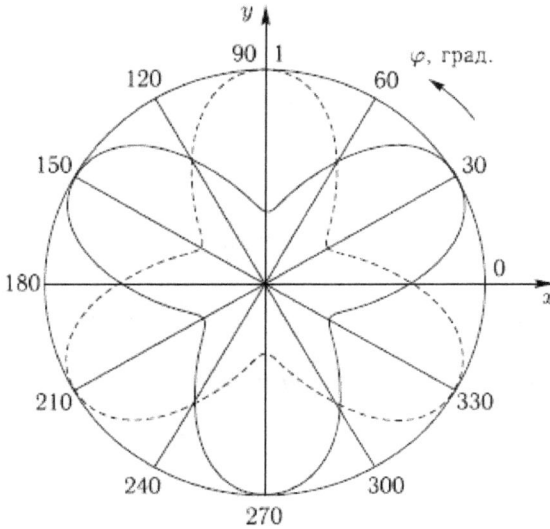

Fig. 4.9. Azimuth diagram of the magnetic anisotropy energy density for the [111] orientation. The polar angle is $\theta = 60°$. The solid line is $m_z > 0$, the dotted line is $m_z < 0$

maxima of the two maximally corresponding minima and vice versa, that is, to the rotation of the initial curves by 60°. That is, we can assume that the diagram is antisymmetric with respect to a plane perpendicular to the polar axis.

4.3.10. Energy density with some other rotation angles about the axis [111]

The obtained expression (4.153) corresponds to $\varepsilon = 0$, that is, it is not the only possible one. As an example, we will give some expressions corresponding to the energy density of cubic anisotropy for other angles of cell rotation with respect to an axis of the [111] type.

These expressions are:

$$U_c(111) = -\frac{K}{12}\left(3m_x^4 + 3m_y^4 + 4m_z^4 + 6m_x^2m_y^2 - \right.$$

$$\left. -12\sqrt{2}m_xm_y^2m_z + 4\sqrt{2}m_x^3m_z\right),$$

$$(4.154)$$

$$U_c(111) = -\frac{K}{12}\left(3m_x^4 + 3m_y^4 + 4m_z^4 + 6m_x^2 m_y -\right.$$
$$\left. -12\sqrt{2}m_x^2 m_y m_z + 4\sqrt{2}m_y^3 m_z\right), \tag{4.155}$$

$$U_c(111) = -\frac{K}{12}\left(3m_x^4 + 3m_y^4 + 4m_z^4 + 6m_x^2 m_y^2 +\right.$$
$$\left. +12\sqrt{2}m_x m_y^2 m_z - 4\sqrt{2}m_x^3 m_z\right), \tag{4.156}$$

$$U_c(111) = -\frac{K}{12}\left(3m_x^4 + 3m_y^4 + 4m_z^4 + 6m_x^2 m_y^2 +\right.$$
$$\left. +12\sqrt{2}m_x^2 m_y m_z - 4\sqrt{2}m_y^3 m_z\right), \tag{4.157}$$

$$U_c(111) = -\frac{K}{12}\left(3m_x^4 + 3m_y^4 + 4m_z^4 + 6m_x^2 m_y^2 - 4m_x^3 m_z +\right.$$
$$\left. +12m_x^2 m_y m_z + 12m_x m_y^2 m_z - 4m_y^3 m_z\right). \tag{4.158}$$

These expressions correspond to the successive rotation of the cubic cell from the initial position, corresponding to $\varepsilon = 0$ at angles corresponding to the location of the projections of [111] axes on the film plane. The expression (4.154) corresponds to $\varepsilon = 0$ and coincides with (4.153), the expression (4.155) corresponds to $\varepsilon = 30°$, (4.156) to $\varepsilon = 60°$, (4.157) to $\varepsilon = 90°$, (4.158) to $\varepsilon = 120°$. It can be seen that expression (4.158) coincides with (4.154), as it should be, since the cell is is in a position completely equivalent to the initial one. The structure of all the expressions given is almost the same, the only difference is in the signs and in the permutation of the indices x and y in the last term. Such a small difference is determined by the high degree of symmetry of the cubic cell with respect to the [111] axis. However, when deviating from angles that are multiples of 30°, the expression for the energy density becomes more complicated. As an example, we give the energy density for a position that differs from a multiple of 30° at an angle equal to half of this multiplicity, that is, when $\varepsilon = 45°$:

$$U_c(111) = -\frac{K}{12}\left(3m_x^4 + 3m_y^4 + 4m_z^4 + 6m_x^2 m_y^2 - 4m_x^3 m_z +\right.$$
$$\left. +12m_x^2 m_y m_z + 12m_x m_y^2 m_z - 4m_y^3 m_z\right). \tag{4.159}$$

It can be seen that the structure of the first four terms is preserved here, but the others change to a rather significant degree. With an even greater deviation from the multiplicity of 30°, the structure of the expression for the energy density of cubic anisotropy is even more complicated, so that in these cases one should use the full rotation matrix of the form (4.140).

4.4. Energy density, effective fields and their time derivatives for cubic magnetic anisotropy at different orientations of the cubic cell

In view of the importance for further exposition, we present a summary of the main expressions relating to cubic anisotropy. As such, we choose the energy density, the effective fields, and also the derivatives of the effective particles over time. The need for similar expressions occurs, for example, when solving the problem of exciting a powerful hypersound [130,131,264–266,297], in problems related to reorientation of magnetization [297–301], precession of the equilibrium state of magnetization [302–332], and in a number of others. For an effective field, we will use its traditional definition [4,7,8]

$$H_i = -\frac{\partial U_c}{\partial M_i} = -\frac{1}{M_0}\frac{\partial U_c}{\partial m_i},$$

(4.160)

where U_c is the anisotropy energy density.

4.4.1. Orientation [001] (cube edge)

Energy density:

$$U_c(001) = -K_1\left(m_x^2 m_y^2 + m_y^2 m_z^2 + m_z^2 m_x^2\right).$$

(4.161)

Effective fields:

$$H_{cx} = \frac{2K_1}{M_0}m_x\left(m_y^2 + m_z^2\right);$$

(4.162)

$$H_{cy} = \frac{2K_1}{M_0} m_y \left(m_z^2 + m_x^2 \right);$$

$$(4.163)$$

$$H_{cz} = \frac{2K_1}{M_0} m_z \left(m_x^2 + m_y^2 \right);$$

$$(4.164)$$

Derivatives from effective time series:

$$\frac{\partial H_{cx}}{\partial t} = \frac{2K_1}{M_0} \left\{ \left(m_y^2 + m_z^2 \right) \frac{\partial m_x}{\partial t} + 2m_z m_y \frac{\partial m_y}{\partial t} + 2m_x m_z \frac{\partial m_z}{\partial t} \right\};$$

$$(4.165)$$

$$\frac{\partial H_{cy}}{\partial t} = \frac{2K_1}{M_0} \left\{ 2m_y m_x \frac{\partial m_x}{\partial t} + \left(m_z^2 + m_x^2 \right) \frac{\partial m_y}{\partial t} + 2m_y m_z \frac{\partial m_z}{\partial t} \right\};$$

$$(4.166)$$

$$\frac{\partial H_{cz}}{\partial t} = \frac{2K_1}{M_0} \left\{ 2m_z m_x \frac{\partial m_x}{\partial t} + 2m_z m_y \frac{\partial m_y}{\partial t} + \left(m_x^2 + m_y^2 \right) \frac{\partial m_z}{\partial t} \right\}.$$

$$(4.167)$$

4.4.2. Orientation [011] (diagonal cube face)

Energy density:

$$U_c(011) = -\frac{K_1}{4} \left(m_y^4 + m_z^4 + 4m_x^2 m_y^2 + 4m_x^2 m_z^2 - 2m_y^2 m_z^2 \right).$$

$$(4.168)$$

Effective fields:

$$H_{cx} = \frac{K_1}{M_0} m_x \left(m_y^2 + m_z^2 \right);$$

$$(4.169)$$

$$H_{cy} = \frac{K_1}{M_0} m_y \left(2m_x^2 + m_y^2 - m_z^2 \right);$$

$$(4.170)$$

$$H_{cz} = \frac{K_1}{M_0} m_z \left(2m_x^2 - m_y^2 - m_z^2 \right).$$

$$(4.171)$$

Derivatives from effective time series:

$$\frac{\partial H_{cx}}{\partial t} = \frac{K_1}{M_0} \left\{ \left(m_y^2 + m_z^2 \right) \frac{\partial m_x}{\partial t} + 2m_x m_y \frac{\partial m_y}{\partial t} + 2m_x m_z \frac{\partial m_z}{\partial t} \right\};$$

$$(4.172)$$

$$\frac{\partial H_{cx}}{\partial t} = \frac{K_1}{M_0} \left\{ 4m_x m_y \frac{\partial m_x}{\partial t} + \left(2m_x^2 + 3m_y^2 - m_z^2 \right) \frac{\partial m_y}{\partial t} - 2m_y m_z \frac{\partial m_z}{\partial t} \right\};$$

$$(4.173)$$

$$\frac{\partial H_{cz}}{\partial t} = \frac{K_1}{M_0} \left\{ 4m_x m_z \frac{\partial m_x}{\partial t} - 2m_y m_z \frac{\partial m_y}{\partial t} + \left(2m_x^2 + m_y^2 - 3m_z^2 \right) \frac{\partial m_z}{\partial t} \right\}.$$

$$(4.174)$$

4.4.3. Orientation [111] (spatial diagonal of the cube)

Energy density ($\varepsilon = 30°$):

$$U_c(111) = -\frac{K_1}{12}\left(3m_x^4 + 3m_y^4 + 4m_z^4 + 6m_x^2 m_y^2 - 12\sqrt{2}m_x^2 m_y m_z + 4\sqrt{2}m_y^3 m_z\right).$$

(4.175)

Effective fields:

$$H_{cx} = \frac{K_1}{M_0}\left(m_x^3 + m_x m_y^2 - 2\sqrt{2}m_x m_y m_z\right);$$

(4.176)

$$H_{cy} = \frac{K_1}{M_0}\left(m_y^3 + m_x^2 m_y - \sqrt{2}m_x m_y m_z\right);$$

(4.177)

$$H_{cz} = \frac{K_1}{M_0}\left(\frac{\sqrt{2}}{3}m_y^3 + \frac{4}{3}m_z^3 - \sqrt{2}m_x^2 m_y\right);$$

(4.178)

Derivatives from effective time series:

$$\frac{\partial H_{cx}}{\partial t} = \frac{K_1}{M_0}\left\{\left(3m_x^2 + m_y^2 - 2\sqrt{2}m_y m_z\right)\frac{\partial m_x}{\partial t} + \right.$$

$$\left. +2m_x\left(m_y - \sqrt{2}m_z\right)\frac{\partial m_y}{\partial t} - 2\sqrt{2}m_x m_y\frac{\partial m_z}{\partial t}\right\},$$

(4.179)

$$\frac{\partial H_{cy}}{\partial t} = \frac{K_1}{M_0}\left\{2m_x\left(m_y - \sqrt{2}m_z\right)\frac{\partial m_x}{\partial t} + \right.$$

$$\left. +\left(m_x^2 + 3m_y^2 + 2\sqrt{2}m_y m_z\right)\frac{\partial m_y}{\partial t} - \sqrt{2}\left(m_x^2 - m_y^2\right)\frac{\partial m_z}{\partial t}\right\},$$

(4.180)

$$\frac{\partial H_{cz}}{\partial t} = \frac{K_1}{M_0}\left\{-2\sqrt{2}\,m_x m_y\frac{\partial m_x}{\partial t} - \right.$$

$$\left. -\sqrt{2}\left(m_x^2 - m_y^2\right)\frac{\partial m_y}{\partial t} + 4m_z^2\frac{\partial m_z}{\partial t}\right\}.$$

(4.181)

4.5. The physical meaning of different types of expressions for magnetic anisotropy energy density

In conclusion of this chapter, we make some remarks concerning the physical meaning of the expressions obtained for the anisotropy energy. Let us consider different types of anisotropy separately.

4.5.1. Uniaxial anisotropy

In Section 4.2.1, an expression for the energy density of uniaxial anisotropy (4.45) with an axis along Oz is obtained based on the application of symmetry operations, with an accuracy up to the notation, having the form

$$U_a^{(1)} = K\left(m_x^2 + m_y^2\right). \tag{4.182}$$

At the same time, based on the condition of preservation of the length of the magnetization vector (4.44)

$$m_x^2 + m_y^2 + m_z^2 = 1, \tag{4.183}$$

can be written (4.182) as

$$U_a^{(2)} = -Km_z^2. \tag{4.184}$$

It can be seen that these two expressions differ only by a constant term, so that

$$U_a^{(1)} = KU_a^{(2)} \tag{4.185}$$

In this case, since in the general case the equation of motion is determined by the derivative of the energy density by the generalized coordinate [295], in the process of differentiation $U_a^{(1)}$ and $U_a^{(2)}$ by the components the magnetization constant term disappears in any case. Therefore, from the point of view of potential, expressions (4.182) and (4.184) can be considered equivalent.

However, this is not the case when we want to find the derivatives of this potential. Thus, the calculation of the effective values using the formula [7, 8]

$$\mathbf{H}_e = -\frac{1}{M_0}\frac{\partial U}{\partial \mathbf{m}}, \tag{4.186}$$

taking into account the designation of the 'anisotropy field' [7,8]

$$H_a = \frac{2K}{M_0}, \tag{4.187}$$

for the energy density of the form (4.182) leads to the expression:

$$H_{ax}^{(1)} = -H_a m_x;$$ (4.188)

$$H_{ay}^{(1)} = -H_a m_y;$$ (4.189)

$$H_{az}^{(1)} = 0.$$ (4.190)

At the same time, from (4.184) we get:

$$H_{ax}^{(2)} = 0;$$ (4.191)

$$H_{ay}^{(2)} = 0;$$ (4.192)

$$H_{az}^{(2)} = H_a m_z.$$ (4.193)

Obviously, the groups of expressions (4.188)–(4.190) and (4.191)–(4.193) are not equivalent. In this case, from the qualitative side, based on (4.188)–(4.190), one can see that when the magnetization vector **m** deviates from the Oz axis, the fields arising along the Ox and Oy axes tend to return the magnetization vector to the Oz axis the stronger, the larger the deviation. Proceeding from (4.191)– (4.193), it is seen that with a similar deviation of **m** from the Oz axis, that is, when m_z decreases, the field that arises, although it also tends to return m to the Oz axis, but as the deviation increases (there is a decrease in m_z) the field in turn decreases.

Therefore, it can be assumed that the expression of energy density in the form of (4.182) is more consistent with the physical meaning of the problem, which consists in describing uniaxial anisotropy with a positive constant, that is, an easy-axis anisotropy. If the anisotropy constant is negative $K < 0$, then the position is reversed. Fields (4.188)–(4.190) with magnetization deviation from the axis Oz tend to attract magnetization to the axes Ox and Oy, but the force of such attraction weakens as the deviation increases (since this force is proportional to the cosine of the angle between the field and the magnetization). At the same time, the field (4.193) tends to push away the magnetization from the Oz axis, that is, to put it in the Oxy plane, the stronger, the more the magnetization is deflected from this plane. So, as one moves away from the Oxy plane, the effect of the anisotropy fields (4.188)–(4.190) on the magnetization weakens, and the fields (4.191)–(4.193) increase.

Thus, it can be assumed that in the case of a negative anisotropy constant, that is, when describing an anisotropy of the type light plane, the physical sense is more consistent with the expression of the energy density in the form (4.184).

Equivalence of two expressions for energy density.

Despite the difference in the effective values, the initial expressions for the energy density of the form (4.182) and (4.184) are equivalent in terms of potential. We explain this in a simplified example of the energy density depending only on the two magnetization magnitudes m_x and m_z. That is, we assume:

$$U_a^{(1)} = Km_x^2;$$
(4.194)

$$U_a^{(2)} = -Km_z^2.$$
(4.195)

Consider first (4.194). Differentiating by m_x we get

$$\frac{\partial U_a^{(1)}}{\partial m_x} = 2Km_x.$$
(4.196)

The second differentiation gives

$$\frac{\partial^2 U_a^{(1)}}{\partial m_x^2} = 2K.$$
(4.197)

Equality to zero of expression (4.196) gives the value $m_x = 0$, for which $U_a^{(1)}$ a has an extremum. When $K > 0$ from (4.197), we find that this extremum is a minimum. That is, when choosing the expression for the energy of the form $U_a^{(1)}$ (4.194), the equilibrium position of magnetization is the alignment of its vector along the Oz axis

Carrying out similar actions with (4.195), we obtain

$$\frac{\partial U_a^{(2)}}{\partial m_z} = -2Km_z.$$
(4.198)

The second differentiation gives

$$\frac{\partial^2 U_a^{(2)}}{\partial m_z^2} = -2K.$$
(4.199)

When $K > 0$, the second derivative (4.199) is negative, that is, the condition $m_z = 0$, obtained from (4.198), corresponds to the maximum energy, so the equilibrium position is the alignment of the magnetization perpendicular to the axis Oz, that is, again along the axis Ox.

Thus, returning to the three-dimensional case, it can be assumed that both expressions (4.194) and (4.195), despite the difference in appearance, for $K > 0$ lead to the same condition – the equilibrium magnetization with its orientation along the axis Ox.

4.5.2. Anisotropy of the form (demagnetization)

In the case of limited dimensions of the sample under study, along with the magnetic anisotropy determined by the crystal structure of the material constituting the sample, the anisotropy of the shape of the sample as a whole is also important.

The anisotropy energy form, also called the demagnetization energy, in the general case has the form (2.5):

$$U_p = -\frac{1}{2}\mathbf{MH}_m,$$
(4.200)

where $\mathbf{H}_m$ is the demagnetization field. In general, the field $\mathbf{H}_m$ is non-uniform, that is, it depends on the coordinates of the observation point inside the body.

However, its homogeneity is preserved if the sample has the shape of an ellipsoid, special cases of which are sphere, thin cylinder or thin film. In these cases, the demagnetizing field takes the form [4,6–8]

$$\mathbf{H}_m = -\overset{\leftrightarrow}{N}\mathbf{M} = -M_0\overset{\leftrightarrow}{N}\mathbf{m},$$
(4.201)

where $\mathbf{m}$ is the normalized magnetization vector, and $\overset{\leftrightarrow}{N}$ is the tensor of the demagnetizing ellipsoid factors:

$$\overset{\leftrightarrow}{N} = \begin{pmatrix} 0 & 0 & 0 \\ 0 & N_y & 0 \\ 0 & 0 & N_z \end{pmatrix}$$
(4.202)

In the case of a thin film oriented perpendicular to the Ox axis, the demagnetizing factor tensor has the form

$$\vec{\vec{N}} = \begin{pmatrix} 0 & 0 & 0 \\ 0 & 0 & 0 \\ 0 & 0 & 0 \end{pmatrix}$$

(4.203)

Substituting (4.203) into (4.201), performing multiplication and substituting the resulting expression into (4.200), we obtain the demagnetization energy density in the form:

$$U_m^{(1)} = 2\pi M_0^2 m_x^2.$$

(4.204)

Taking into account the condition of preserving the length of the magnetization vector

$$m_x^2 + m_y^2 + m_z^2 = 1.$$

(4.205)

This expression can be written up to a constant term.

$$U_m^{(2)} = -2\pi M_0^2 \left(m_y^2 + m_z^2 \right).$$

(4.206)

It can be seen that (4.204) and (4.206) coincide in form with (4.184) and (4.182) respectively. Thus, in the demagnetization energy density, the coefficient $-2\pi M_0^2$ plays the same role as the uniaxial anisotropy constant K in the anisotropy energy density. However, unlike the anisotropy energy density, the sign of this coefficient is negative, that is, the expressions (4.204) and (4.206) describe anisotropy of the 'easy plane' type. With this equilibrium position the magnetization corresponds to alignment precisely in this plane. Guided by approximately the same considerations as in the choice between the formulas (4.182) and (4.184), one can conclude that formula (4.204) corresponds better to the physical meaning, which will mainly be used further

4.5.3. External field anisotropy

The external field has a certain direction, so we can conditionally speak of its 'anisotropy', which manifests itself in the anisotropic dependence of the density of the interaction energy of the field with magnetization at different orientations of the latter.

The energy density of the interaction of magnetization with external field U_h, according to (2.4), has the form

$$U_h = -\mathbf{MH}. \tag{4.207}$$

Recording the magnetization and the external field by components, we obtain:

$$U_h = -M_0 H_0 \left(m_x h_x + m_y h_y + m_z h_z \right). \tag{4.208}$$

Here all the magnetization components are included only in the first degree, therefore the condition of preserving the length of the magnetization vector does not lead to the simplification of the expression, and the effective fields obtained by differentiating between the components of magnetization, are obtained simply equal to the components of the external field H_0 and $h_{x,y,z}$.

4.5.4. Cubic anisotropy

In Section 4.3.3, two expressions (4.120) are obtained for the energy density of cubic anisotropy with cell orientation like [001] and (4.121), with $K_2 = 0$, taking the form:

$$U_c^{(1)} = -K_1 \left(m_x^2 m_y^2 + m_y^2 m_z^2 + m_z^2 m_x^2 \right). \tag{4.209}$$

$$U_a^{(2)} = \frac{K_1}{2} \left(m_x^4 + m_y^4 + m_z^4 \right). \tag{4.210}$$

equivalent to the constant term.

In actually existing magnetic materials with cubic symmetry, two types of orientations of the axes of easy magnetization are most common: along spatial and diagonal lines and along the edges of the cubic cell. The first type at room temperature occurs in an iron–yttrium garnet [7, 8], from metals in nickel and cobalt [4]. In this case, the anisotropy constant is positive ($K_1 > 0$). The second kind of axis orientation at room temperature is observed in iron [4], where the anisotropy constant in the interpretation of signs inversion adopted here (see section 4.3.3) is negative ($K_1 < 0$).

When the temperature changes, the anisotropy constant may change the sign, as is the case, for example, in nickel, where below 100°C this the constant is negative, and above 200°C it is positive [4] (in [4], the temperature interval was characterized only by its boundaries, since the anisotropy inside it changes with temperature extremely slowly). The anisotropy constant may vary particularly in alloys with a change in the percentage composition of the components. For example, in a permalloy (iron alloy with nickel) with a nickel content equal to 70%, the anisotropy constant, being positive, is 9000 erg cm^{-3}, and with a content of 80% becomes negative equal to -3000 erg cm^{-3} [4]. The mutual relation between the first K_1 and the second K_2 constants of cubic anisotropy (in accordance with (4.120) or (4.121)) can play a significant role in the character of the anisotropy of cubic agnetics. So, according to [134], with $K_1 < 0$ the light axes are cube edges (axes of type [100]), for any value of K_2, while for $K_1 > 0$, to realize easy axes of type [111], the inequality $K_1 > K_2/3$ is required, where the constant K_2 is assumed positive. However, in the majority of real magnetics, the second constant in absolute value is much smaller than the first, so that the reduced inequality is not particularly critical. For example, for a IYG at room temperature, K_1 is about 6000 erg cm^{-3}, and K_2 is about 300 erg cm^{-3}. At helium temperatures, the values approach 25 000 erg cm^{-3} and 10 000 erg cm^{-3}, so the ratio $K_1 > K_2/3$ is always deliberately fulfilled (the figures given are taken from the overview monograph [8, p. 62, Fig. 2.12], where the figure is given without reference to the primary source).

Returning to the formulas (4.209) and (4.210), we note that in this in this case from calculating the effective values of the forces apparently acting on the magnetization vector, a clear picture is not obtained as for uniaxial anisotropy. That is, can not give preference one or the other of these formulas to describe the orientation of the EMA with respect to axes of the type [100] or [111]. As noted earlier in section 4.3.1 regarding the interpretation of the formulas (4.120) and (4.121), in the literature, both expressions of the type (4.209) and (4.210) are used, therefore, in this monograph, the authors do not give preference to one or the other, and following their own favorite tradition, they use the form (4.209).

However, in any case, when using the formulas (4.209) or (4.210), you should not just rely on the resulting mathematical formulas, but by all means analyze in detail the physical meaning of the resulting solutions.

4.6. Some other types of anisotropy

In section 3.1.4, a variety of different anisotropy types was noted. We give here some examples concerning more complex anisotropy types than the simple uniaxial and cubic ones discussed above.

4.6.1. High order anisotropy

First of all, we note that uniaxial anisotropy, except for the second one, can have higher orders of decomposition in magnetization (like (4.56)). That is, when the anisotropy axis is oriented, for example, along the coordinate axis Ox, decomposition takes place (hereinafter, the designations of the constants K_i corresponding to different numbers 'i' are conditional)

$$U_a = K_1 m_x^2 + K_2 m_x^4 + K_3 m_x^6 + ... K_m m^{2n} + ... \qquad (4.211)$$

The axes of uniaxial anisotropy of different order can be oriented in different directions. Thus, for example, in [333] it is reported about the existence of a second-order uniaxial anisotropy in some metal panels of an iron–yttrium garnet with an axis perpendicular to the film plane, which is accompanied by a fourth-order uniaxial anisotropy, whose axis lies in the film plane.

It should be noted, however, that the uniaxial anisotropy constants of high orders, as a rule, are much smaller than the first second order constant, and this decrease is manifested the more strongly the higher the order of anisotropy. As in the case of uniaxial anisotropy, in the expansion of the potential in the components of magnetization (4.33) for cubic anisotropy higher orders can be considered. Thus, taking into account the sixth order, as a result of applying the operations of symmetry, leads, up to a sign, to an expression similar to (4.116):

$$U_a = K_2 m_x^2 m_y^2 m_z^2. \qquad (4.212)$$

where it is taken into account that the sign of the constant K_2 must be positive, as well as for the constant K_1 in the formulas (4.120) and (4.121). Here the anisotropy 'repels' the magnetization vector from the edges of the cubic cell coinciding with the axes of the Cartesian coordinate system, that is, it acts in the same direction as the fourth order anisotropy. In this case, the full expression for the energy density in the expansion to the sixth order has the form (4.120) and (up to a sign) is given in [259, p. 160, forms. (4.7)]:

$$U_c = -K_1 \left(m_x^2 m_y^2 + m_y^2 m_z^2 + m_z^2 m_x^2 \right) - K^2 m_x^2 m_y^2 m_z^2. \qquad (4.213)$$

In [7, p. 90, forms. (2.2.24?)] (also with the accuracy up to the sign) another type is used (4.121), equivalent to the first:

$$U_c = \frac{K_1}{2}\left(m_x^4 + m_y^4 + m_z^4\right) - K^2 m_x^2 m_y^2 m_z^2. \qquad (4.214)$$

Similarly, expressions for the cubic anisotropy of higher orders can be obtained from expression (4.33). To do this, use the same symmetry operations as in the derivation of the expressions (4.212)–(4.214). Thus, the expression for the energy density of the eighth-order cubic anisotropy in the coordinate system, the axis which are oriented along the edges of the cube, has the form:

$$U_c = K_3\left(m_x^4 + m_y^4 + m_y^4 m_z^4 + m_z^4 m_x^4\right), \qquad (4.215)$$

and a similar expression for the tenth order is

$$U_c = K_4\left(m_x^4 m_y^4 m_z^2 + m_y^4 m_z^4 m_x^2 + m_z^4 m_x^4 m_y^2\right). \qquad (4.216)$$

Similarly, expressions of even higher orders can be obtained. It should be noted, however, that for cubic anisotropy, as well as for uniaxial, in most real materials, the values of the constants of higher orders as a rule decrease the sooner the order is higher. So for an iron–yttrium garnet, the K_2 constant at room temperature is less than one tenth of the K_1 constant [8, p. 62, Fig. 2.12]. Thus, in most practical cases, in the expansions (4.213) or (4.214) it is quite possible to confine ourselves only to the first term.

4.6.2. Rhombic anisotropy

In addition to uniaxial and cubic, many types of magnetic anisotropy are found in many real materials. So, in the widely studied as a medium for memory devices on CMD films, mixed ferrite garnet was noted, in addition to uniaxial and rhombic anisotropy [40, p. 25], as well as [276]. The use of rhombic symmetry operations (reflection in all three-dimensional planes) in terms of the second order decomposition (4.33) leads to the expression

$$U_r = K_1 m_x^2 + K_2 m_y^2 + K_1 m_z^2, \qquad (4.217)$$

where, given the preservation of the length of the magnetization vector

$$m_x^2 + m_y^2 + m_z^2 = 1, \qquad (4.218)$$

up to a renormalization of the constants and the constant term, we get

$$U_r = K_1 m_x^2 + K_2 m_y^2, \qquad (4.219)$$

That is, two anisotropy constants are needed to describe the rhombic anisotropy of even the lowest (second) order.

4.6.3. Oblique uniaxial anisotropy

A special place is occupied by the uniaxial 'induced' anisotropy observed in technical films of mixed ferrite–garnets, whose axis is deviated from the normal to the film plane. Such anisotropy is discussed in more detail later in Chapter 7. We also note here only the simplest form of the energy density of such anisotropy and compare it with the energy density of the rhombic anisotropy. Let's execute consideration in the geometry presented on Fig. 4.5. Let us proceed from the form of the energy density of uniaxial anisotropy (4.57)

$$U_a = K\left(m_{x'}^2 + m_{y'}^2\right), \qquad (4.220)$$

where $m_{x',y'}$ is the components of the normalized vector in the rotated coordinate system $Ox'y'z'$

For simplicity, we assume that the deviation of the anisotropy axis from the normal takes place in the plane Oxz. In this case, in accordance with Fig. 4.5, we set $\varepsilon = 0$ and $\varphi_a = 0$, so the transition matrix (4.61) takes the form

$$\ddot{A}^{-1} = \begin{pmatrix} 0 & -1 & 0 \\ \sin\theta_a & 0 & \sin\theta_a \\ \sin\theta_a & 0 & \cos\theta_a \end{pmatrix} \qquad (4.221)$$

The required transformation is determined by the formula of the type (4.58)

$$\mathbf{m}' = \ddot{A}^{-1}\mathbf{m}. \qquad (4.222)$$

We write (4.222) in expanded form:

$$\begin{pmatrix} m_x \\ m'_y \\ m'_z \end{pmatrix} = \begin{pmatrix} 0 & -1 & 9 \\ \cos\theta_a & 0 & -\sin\theta_a \\ \sin\theta_a & 0 & \cos\theta_a \end{pmatrix} \begin{pmatrix} m_x \\ m_y \\ m_z \end{pmatrix} \qquad (4.223)$$

Performing multiplication, we get:

$$m_{x'} = -m_y; \qquad (4.224)$$

$$m_{y'} = m_x \cos\theta_a + m_z \sin\theta_a; \qquad (4.225)$$

$$m_{z'} = m_x \sin\theta_a + m_z \cos\theta_a. \qquad (4.226)$$

Substituting (4.224)–(4.226) into (4.220), we obtain:

$$U_a = m_x^2 \cos^2\theta_a + m_y^2 + m_z^2 \sin^2\theta_a - 2m_x m_z \sin\theta_a \cos\theta_a. \qquad (4.227)$$

Eliminate m_z from this expression using the relation (4.218), which we write in the form:

$$m_z = \sqrt{1 - m_x^2 - m_y^2}. \qquad (4.228)$$

Substituting this expression into (4.227) and citing similar terms, we get:

$$U_a = \sin^2\theta_a + m_x^2 \left(\cos^2\theta_a - \sin^2\theta_a\right) + m_y^2 \cos^2\theta_a - $$
$$- 2m_x \sqrt{1 - m_x^2 - m_y^2} \sin\theta_a \cos\theta_a. \qquad (4.229)$$

We introduce the notation:

$$K_1 = \cos^2\theta_a - \sin^2\theta_a; \qquad (4.230)$$

$$K_2 = \cos^2\theta_a. \qquad (4.231)$$

Substituting these designations into (4.229) and omitting the constant term, which is not important for potential energy, we get:

$$U_a = K_1 m_x^2 + K_2 m_y^2 - 2\sqrt{K_2\left(K_2 - K_1\right)} m_2 \sqrt{1 - m_x^2 - m_y^2}. \qquad (4.232)$$

This is up to the notation of constants, an expression for the energy density of uniaxial anisotropy recorded in the $Oxyz$ system, provided that the anisotropy axis, being in the Oxz plane, is deflected from the Oz axis by the angle θ_a. Unlike the expression for uniaxial anisotropy of the type (4.220), here two constants K_1 and K_2 are required, instead of one K in (2.220). Two constants are required because in addition to the magnitude of the anisotropy, the orientation of its axis is also specified here. If we set the angle θ_a equal to zero, then from (4.230) and (4.231) we can see that the constants are compared to each other, the root in the last term of the expression (4.232) vanishes, so that the expression (4.232) takes the form similar to (4.220) with a single constant before the sum of the squares of the transverse magnetization components. It is interesting to compare the oblique uniaxial anisotropy with the rhombic one. So, from a comparison of expressions (4.232) and (4.219), one can see that the uniaxial anisotropy differs from the rhombic additional term, the latter in expression (4.232). That is, when arbitrary value of the angle θa does not have a complete correspondence between these two types of anisotropy.

Comment. It should be noted that in the literature known to the authors of this monograph, a significant distinction between the effects of the manifestation of rhombic and oblique uniaxial anisotropy is revealed not enough. A rather detailed consideration of this question is given in [276], however, the angles of deviation of the anisotropy axis from the normal to the film plane are chosen rather small, not exceeding several degrees. At the same time, it was shown in [22–25, 30, 240, 275, 277, 278] that these angles can be up to ten degrees or more, and they are determined by strengthening the angle of deflection of the [111] axis of the substrate from the normal to its surface.

The difficulty in distinguishing between two types anisotropy is apparently due to the fact that both anisotropy, in determining the anisotropy field by the method of domain nucleation, leads to azimuthal dependencies of a very similar form. In the consideration carried out later in this monograph (Chapter 7), only the oblique uniaxial anisotropy was taken into account, which gave an interpretation of the experimental results of ferromagnetic resonance with a fairly high accuracy (no worse than 10%), so that taking rhombic anisotropy does not seem to was required. However, in the opinion of the authors, the question of the role of both anisotropy in the formation of both the nucleation field of the domains and the frequency of ferromagnetic resonance is of particular interest, that is, open to future research.

Conclusions for chapter 4

In this chapter, the described mathematical apparatus of transition matrices is applied to calculate the energy density of the magnetic anisotropy of uniaxial and cubic nature. The main results obtained in this chapter are as follows.

1. Various symmetry operations are considered that are necessary for calculating the anisotropy energy density in various crystal structures. As the main operations, the reflection is given in coordinate planes and rotated at certain angles around the coordinates. Matrices of transformations of the magnetization vector for such operations are presented. The rotation matrices for characteristic angles corresponding to uniaxial, cubic, trigonal, and hexagonal symmetries are obtained.

2. The energy density of magnetic anisotropy is considered in the most general form as a series in powers of the normalized components of magnetization. The terms of the series corresponding to uniaxial and cubic anisotropy are selected. Based on the requirement of maintaining tight energy when the sign of the projection of the magnetization vector on the anisotropy axis changes, as well as when the component of the magnetization vector normal to the anisotropy axis is rotated around this axis, a general expression is obtained for the energy density of uniaxial anisotropy. Two equivalent forms of the energy density record are given, passing one into the other with allowance for the conservation of the length of the magnetization vector. The same types of records in the spherical coordinate system are obtained. The types of anisotropy are considered of the type 'easy axis' and 'easy plane', the transition of one source to another is demonstrated when the sign of the anisotropy constant is changed. The singularities of the uniaxial anisotropy of higher order are noted. An expression for the energy density of uniaxial anisotropy with an arbitrary direction of the axis relative to the Cartesian coordinate system is obtained. Transition matrices corresponding to rotation are given.

The magnetization vector at arbitrary polar and azimuthal angles of the spherical coordinate system, as well as at an arbitrary angle around the anisotropy axis. Effective fields obtained corresponding to the given orientations. An azimuthal energy density diagram of the uniaxial magnetic anisotropy is constructed, reflecting the rotation of the magnetization vector around the anisotropy axis for a given value of the polar angle. The single-petal nature of the diagram is revealed, which corresponds to the deviation of the anisotropy axis from the polar axis of the spherical coordinate system.

3. The energy density of the cubic anisotropy for the [001] orientation is considered. Based on the general definition of the magnetic anisotropy energy density as a series expansion in powers of the normalized components of magnetization, the terms of the series corresponding to uniaxial and cubic anisotropy are distinguished. Proceeding from the requirement of conservation of energy density when changing the sign of the projection of the magnetization vector on any of the axes coinciding with the edges of the cubic cell, as well as turning 90° around any of the technical axes, a general expression is obtained for the energy density of cubic anisotropy up to fourth order terms by magnetization. Two equivalent distributions of any other type of energy density are given. which, when taking into account the conservation of the length of the magnetization vector, transform one into another with an accuracy of a constant term. The definition of a cubic anisotropy field is introduced, and the features of such anisotropy for an iron yttrium garnet are considered. The cubic anisotropy of high orders is considered, the expression for the sixth-order energy density in magnetization is given, and the sixth-order anisotropy constant is small compared to the fourth-order constant. The transition to a spherical coordinate system is performed and two equivalent expressions for the energy density are obtained, containing the polar and azimuth angles of the magnetization vector. It is shown that these expressions differ by a constant value, that is, they are equivalent to each other. An azimuthal energy density diagram of the cubic magnetic anisotropy of the [001] orientation is constructed. Its four-'petal' character was discovered, which corresponds to projections of [111] axes on a plane perpendicular to the polar axis of the spherical coordinate system. The rotation of the diagram by 90∘ is noted when the sign of the anisotropy constant changes.

4. The energy density of the cubic anisotropy for the orientation [011] is considered. The transition matrices were obtained and the previously obtained expression for the energy density in the [001] orientation to the [011] orientation was converted. Two equivalents of each other are recorded. The expressions for the energy density for orientation [011], obtained from the corresponding expressions for orientation [001]. An azimuthal diagram of the energy density of cubic magnetic anisotropy in orientation [011] is constructed. Revealed is her two-petal character, corresponding to the location of

a biaxial axis of the [111] type in a plane perpendicular to the polar axis of the spherical coordinate system, and two more in a plane perpendicular to the first one. The symmetry of the diagram relative to the plane perpendicular to the polar axis is noted, as well as its rotation by 90° with a change in the sign of the constant anisotropy.

5. The energy density of the cubic anisotropy for the [111] orientation is considered. The transition matrices were obtained and the previously obtained expression for the energy density in the [001] orientation was converted to orientation [111]. An azimuthal diagram of the energy density of cubic magnetic anisotropy in the [111] orientation is constructed. Revealed is its three-petal character, corresponding to the location of the projections of the [111] axes on a plane perpendicular to the polar axis of the spherical coordinate system. Antisymmetry of the diagram with respect to a plane perpendicular to the polar axis, as well as its rotation by 60° when the sign of the anisotropy constant changes. Expressions are given for the energy density of the cubic anisotropy of the [111] orientation with rotation around the polar axis at angles that are multiples of 30°. It is noted the structure of the obtained expressions is marked, due to the symmetry of the cubic cell with respect to the [111] axes.

6. A general summary of expressions for the energy density, effective pixels and their derivatives about time for the cubic cell orientations of the following type is given: [001] (cube edge), [011] (cube face diagonal), [111] (cube spatial diagonal).

7. The features, physical meaning and equivalence of permissible energy density records for various types of anisotropy are considered: uniaxial, cubic, shape anisotropy (demagnetization), as well as the anisotropic nature of the external field. Shown is the equivalence of different recording types, as well as their physical difference. Recommendations for the use of other types of recording are given in accordance with the physical meaning of the solved cottages.

8. Some special types of magnetic anisotropy are considered: uniaxial and cubic high rows, rhombic and uniaxial inclined. Expressions for the energy density listed in the anisotropy types are given. It has been shown that both rhombic and oblique uniaxial types of anisotropy require two-points for their description. Comparison of the orthorhombic and oblique uniaxial anisotropy was made, the fundamental difference between them was noted, the reasons for the difficulties of their separate determination in the experiment are discussed.

5

Orientational transition in a magnetic medium

This chapter is devoted to the description of orientational transitions under certain specific conditions. The main provisions of the physics of orientational transitions and their relationship with physics are considered phase transitions. A model of a simple orientational transition based on minimization of the energy density, as well as the conditions of its 'pulling in' and 'smearing' in more complex rays, is considered in detail. Attention is paid to numerical methods for calculating orientational transitions — simple and dynamic establishment. The features of orientational transitions in real materials – films of mixed garnets – garnet with uniaxial anisotropy, whose axis is inclined from the normal to the film plane. The results of experiments on the study of anisotropy in such plates are presented. When considering the surveys listed in the polls, we will mainly follow the works [4.132–139, 258], the remaining references are indicated in the text.

5.1. General ideology of orientational transition

As noted above, the main subject of this work is the study of ferromagnetic resonance and magnetostatic waves under the conditions of the orientational transition. Therefore, given the terminological remarks, given in Section 1.2.1, briefly dwell on the concept of an orientational transition and consider its main features.

The orientational transition is understood to be a nonstationary state, in which the magnet is located during the transition from one stationary state to another, and these states differ equilibrium orientation of the magnetization inside the magnetic. The stationary state of a magnet at a given temperature is determined by the equilibrium between the three fields: the anisotropy field, the demagnetization field and the field applied to the magnet from the outside. In this case, the magnetization vector at each point of the magnetic is in a state determined by the minimum of the potential energy at this point, and in the absence of thermal motion rests. The potential energy of a magnetic in space can have several minima of different depths separated by potential barriers of a certain height. That is, the magnetization vector at each point of the magnet is located, as it were, in a potential well, with a deviation from the lowest point of which a force arises that returns the magnetization to this point. Temperature movement is the oscillation of magnetization inside such a well. At a sufficiently high temperature, the amplitude of oscillations may increase to such an extent that their energy will exceed the potential barrier at the edge of the well. As a result such an excess, the magnetization vector may jump from the origin of one well to the next, where its orientation, determined by the bottom of the well, will change. If the depth of the new pit exceeds the depth the initial one, the vector of magnetization can remain in it, that is, it will begin to oscillate around a new stationary state. The excess kinetic energy acquired during the fall to the bottom is more a deep well will be transferred to the crystal lattice, due to which it will heat up slightly, but being an open system will quickly transfer this excess to the surrounding space (thermostat). Thus, the jump of the magnetization vector from a single potential well in the other, it will occur with a change in its orientation (since the orientations of all are different). Such a change in the orientation of the magnetization is an orientational transition. Temperature is not the only factor that can stimulate an orientational transition. Of the three factors mentioned above: anisotropy, demagnetization, and field, the external magnetic field is the most easily susceptible to change. So even at a constant temperature, under the influence of the field, the configuration of the potential well in which the magnetization vector is located may vary. In the simplest case, the mutual depth of the two-axis sediments varies, up to the point that one of them turns back to the inflection surface, so that the magnetization vector freely rolls from it into the next hole, thus making an orientational transition. A typical example of such a transition is the magnetization of a uniaxial magnet in the direction perpendicular to the anisotropy axis. In the absence of a field, the magnetization vector is oriented exactly along the anisotropy axis (we do not take into account the demagnetization), and as the field increases, it gradually deviates from this axis, turning to the field direction. The transition ends when the magnetization vector is aligned exactly along the field, after which a further increase in the field of its orientation does not change. If the magnet has several anisotropy axes, for example, cubic symmetry of the [111] type, like IYG (or, for example, nickel), then the application of the field causes the magnetization to rotate in its direction, and, due to its multi-axis (ie, the presence of several spatially separated potentialities), the motion of the vector Magnetization can be quite complex and occur along a curved line. Ultimately, the magnetization vector with a sufficiently large field, it is still oriented along its direction, deviating in the limit to infinitely small angles. The described picture of the orientational transition suggests that the spatially bounded sample of the magnetic material as a whole is uniformly magnetized, that is, the directions of the magnetization vector are all the same at all points. However, due to the demagnetization of the shape, these directions in different sample currents, especially near the surface and different angles may vary. In this case, the orientational transition in different sample points will occur differently, but ultimately the field will still force the magnetization in all sample line up along its direction.

With a sufficient sample size, the magnetization can no longer be oriented in one direction in the entire sample, since in this case a significant external field arises in the outer space, which has a large energy. The desire to reduce this energy causes the sample to be divided into domains, and in the case of a sufficiently large uniaxial anisotropy, the directions of the magnetization vectors in the neighboring domains acquire mutually opposite directions, so that the distribution of magnetization becomes nonuniform. The application of an external field along the anisotropy axis leads to the destruction of the domains of an unfavorable sign, so that the sample again finds a uniform state. Each of these states of a magnet is called a 'phase', so that in the absence of a field, the magnet is in a 'non-uniform phase' – with domains, and with a sufficient field it passes into a 'uniform phase' – without domains. Such a transition from one phase to another under the action of a field is a phase transition. If the field is removed in a homogeneous phase, the magnet does not immediately go into a non-uniform phase, since a certain energy is required for the nucleation of the domains. This energy is required to overcome the exchange interaction, forcing all the spins of the magnetic to line up in one direction. For the nucleation of domains, it is necessary to deploy some of the spins relative to the remaining ones, that is, to overcome a certain potential barrier. Such overcoming of the barrier is usually carried out at the expense of small numbers of rings of the new phase, which are formed due to thermal fluctuations of the magnetization in different points of the sample. Since the transition from a homogeneous phase to a non-uniform phase does not occur immediately, but through the embryo, a complete reversal of the magnetic in the forward and reverse directions occurs with hysteresis, that is, with a certain loss of energy. Thus, the transition from a uniform phase to a non-uniform phase under the action of a field applied along the anisotropy axis is a first-order phase transition. Imagine now that sample magnetization occurs field applied perpendicular to the anisotropy axis. In this case, the magnetization vectors inside each of the domains will turn to the field to the same extent, making the same angles with it. In this case, the mutual width of the domains, determined by the orientationof the magnetization vectors at the edges of the sample will not change. The change will comprehend only the energy density of the domain domains (since the magnetization vector inside them will rotate at angles less than 180 degrees), as a result of which the period of the domain structure will change, however, in a completely symmetrical way. When the field reaches a value sufficient to align the orientation of the magnetization along its direction, the domains will disappear altogether and the entire sample will be magnetized uniformly. Thus, here as the field increases, a transition from the non-uniform to the homogeneous phase will occur. However, now when the field decreases, the two directions in which the magnetization vector may deviate during the inhomogeneous phase nucleation will be equivalent, as a result of which the nucleation will occur immediately in both directions, so that the uniform magnetization distribution will be distorted in the form of a sinusoidal wave, the period of which will be equal to the period of the emerging domain structure . The reason for the excitation of such a wave will be thermal fluctuation, and the energy barrier here will not have to be overcome, since there will be no need to reorient the direction of magnetization to the opposite. Thus, a transition from a homogeneous phase to a non-uniform one will occur here, and without loss of energy and without any hysteresis, that is, it will be a second-order phase transition.

In the absence of a field, the magnetization vectors will be oriented in two opposite directions along the anisotropy axis. In the field above the transition, the magnetization vectors in the entire sample volume will be oriented in a single direction - along the field, which is perpendicular to the anisotropy axis. Thus, within each domain, the magnetization vector will be reoriented from the old direction (along the axis) to the new (along the field). That is, in this case, an orientational transition will take place, ending with an increase in the field by a second-order phase transition. The directions of the magnetization vectors in the domains of various signs in the process of transition will be between themselves a certain angle smaller than 180 degrees. This state of magnetization is called the 'angular phase'. In the absence of a field, the magnetization vectors in the neighboring domains will be antiparallel to each other, i.e. make an angle of 180 degrees between each other. This state of magnetization is called the 'collinear phase'. The transition between the collinear and angular phases can also be considered a second-order phase transition, since it occurs without any energy expenditure in an infinitely small, but non-zero field. Thus, it can be assumed that the magnetization of a uniaxial magnet by a field perpendicular to the anisotropy axis occurs through an orientational transition, the beginning and end of which are two second-order phase transitions - one from the collinear phase to the angular one, the second from the angular to the homogeneous phase.

Comment. In the formation of the equilibrium state of magnetization, in addition to those listed in the energy IDs (anisotropy, demagnetization, and external field), other types of energy, such as elastic or magnetoelastic, can also be involved, and their significant. In [4, pp. 320–322], it was shown that the magnetoelastic interaction prevents the formation of domains with a 90-degree orientation of the magnetization, and the difference between the magnetization of the neighbouring domains is 180 degrees. In [47, 48], the existence of a significant frequency gap of a magnetoelastic origin in the spectrum of a magnetoelastic crystal caused by spontaneous striction was revealed. However, in this monograph, with the exception of some specific rays, we confine ourselves to the three types of energy mentioned, and we will leave more complex options for the formation of the energy potential for separate consideration.

5.1.1. Simple orientational transition

Let us consider the simplest type of orientational transition using the

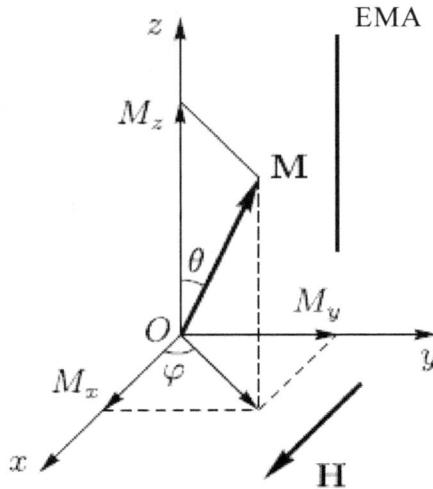

Fig. 5.1. Task geometry to illustrate simple orientation transition. EMA – easy magnetization axis.

example of magnetization of an unlimited medium with a uniaxial anisotropy field perpendicular to the anisotropy axis. Task geometry shown in Fig. 5.1. The Cartesian coordinate system $Oxyz$ is chosen in such a way that the axis Oz is parallel to the axis of magnetic anisotropy.

Comment. Here anisotropy is implied uniaxial, such as 'easy axis' therefore, following the tradition established in the CMD technique ([22–25, 30], as well as [38, p. 14 and further]), for the axis of similar anisotropy, the name 'easy magnetization axis' is used or in the form of the abbreviation 'EMA'.

The field **H** is perpendicular to such an anisotropy axis, and the axis Ox is directed along this field. The axis Oy is perpendicular to the axes Ox and Oz, due to the symmetry of the problem, its direction in the plane Oxy is arbitrary. In addition to the Cartesian system $Oxyz$, a spherical system with the the coordinates $Or\theta\varphi$, the origin of which falls on the same point O as the Cartesian system, the polar axis $O\theta$ coincides with the axis Oz, that is, with the EMA, and the azimuthal $O\varphi$ with the axis Ox, that is, with the direction of the field. The magnetization vector **M** has a length M_0, the components M_x, M_y, M_z and makes an angle θ with the axis Oz, and its projection on the plane Oxy makes the angle φ with the axis Ox.

In the absence of a field, the magnetization vector is oriented along the Oz axis, and when turned on, the field rotates in its direction what exactly is the orientational transition. It can be assumed that due to the symmetry of the problem with respect to the Oxz plane, the magnetization vector will always remain in this plane during its rotation, that is, the angle φ will always remain zero [4].

Assuming that, in the absence of a field, the magnetization vector is oriented in the positive direction of the Oz axis, we obtain:

$$M_x = M_0 \sin \theta; \tag{5.1}$$
$$M_y = 0 \tag{5.2}$$
$$M_z = M_0 \cos \theta, \tag{5.3}$$

i.e

$$\mathbf{M} = \{M_0 \sin \theta, \ 0, \ M_0 \cos \theta\} \tag{5.4}$$

The field, oriented in the positive direction of the axis Ox, is:

$$H_x = H_0; \tag{5.5}$$
$$H_y = 0; \tag{5.6}$$
$$H_z = 0, \tag{5.7}$$

i.e

$$\mathbf{H} = \{H_0, \ 0, \ 0\}. \tag{5.8}$$

First, we restrict ourselves to considering the transition in the interval of angles $0 \leq \theta° \leq 90°$, since this is quite sufficient for preliminary identification of the physical picture. The case of the initial orientation of the magnetization vector in the negative direction of the Oz axis is completely similar to symmetric reflection in the Oxy plane.

We will search for the equilibrium position of the magnetization vector, based on the minimum energy density of the magnetic material as a whole.

The energy density of uniaxial anisotropy (with $M_y = 0$) is equal to [4, 40]:

$$U_a = K\left(\frac{M_x}{M_0}\right)^2 = \frac{K}{M_0^2}M_x^2. \tag{5.9}$$

It can be seen that for $K > 0$, the energy minimum corresponds to $M_x = 0$, that is, the alignment of the magnetization along the Oz axis. The energy density of the interaction of magnetization with the external field is [4, 40]:

$$U_m = -\mathbf{MH}, \tag{5.10}$$

whence it can be seen that, due to its negativity, as well as the fact that the angle between the $\mathbf{M}$ and $\mathbf{H}$ vectors is $90° - \theta$, it is minimal when the magnetization is oriented along the field (i.e. at $\theta = 90°$). The total energy density is equal to the sum (5.9) and (5.10):

$$U = U_a + U_m = \frac{K}{M_0^2} - \mathbf{MH}. \tag{5.11}$$

Writing in a spherical coordinate system, we get

$$U = K \sin^2 \theta - M_0 H_0 \sin \theta, \tag{5.12}$$

Determine the effective anisotropy field [6–8, 40]

$$H_a = \left.\frac{\partial U_a}{\partial M_x}\right|_{M=M_0} = \frac{2K}{M_0}. \tag{5.13}$$

We introduce the normalized external field

$$h = \frac{H_0}{H_a} = \frac{M_0 H_0}{2K}. \tag{5.14}$$

Expressing $M_0 H_0$ from (5.14) and substituting in (5.12), we obtain the energy density in the form

$$U = K\left(\sin^2 \theta - 2h\sin \theta\right). \tag{5.15}$$

The equilibrium position of the magnetization vector will be sought from the condition of the minimum energy density (5.15) with respect to the angle θ. Such a condition is the equality to zero of the first derivative U with respect to θ and the positivity of the second one [268].

Find these derivatives:

$$\frac{dU}{d\theta} = 2K\cos\theta(\sin\theta - h);$$ (5.16)

$$\frac{d^2U}{d\theta^2} = 2K(\cos^2\theta - \sin^2\theta + h\sin\theta).$$ (5.17)

Equating (5.16) to zero, we obtain the condition for the presence of an extremum:

$$\cos\theta\ (\sin\theta\ - h) = 0.$$ (5.18)

Since this equation contains two factors, for the left to be equal to zero, it is sufficient that at least one of the factors be equal to zero. That is, two cases are possible. Consider them sequentially.

The first case.

$$\cos\theta\ = 0,$$ (5.19)

$$\theta = \frac{\pi}{2}.$$ (5.20)

Substituting (5.20) into (5.17), we find the second derivative:

$$\frac{d^2U}{d\theta^2} = 2K(-1 + h).$$ (5.21)

The extremum is the minimum in the case of positivity of the second derivative [268]. Since $K > 0$, the minimum occurs when $(-1 + h) > 0$, that is, $h > 1$ or, taking into account (5.13) and (5.14), when $H_0 > H_a$, that is, when the external field is greater than the anisotropy field.

Second case:

$$\sin\theta - h = 0,$$ (5.22)

i.e

$$\sin\theta\ = h,$$ (5.23)

or

$$\theta = \arcsin h.$$ (5.24)

Substituting (5.24) into (5.17) we find the second derivative:

$$\frac{d^2U}{d\theta^2} = 2K(1 - h^2).$$ (5.25)

The extremum is the minimum in the case of positivity of the second derivative [268]. Since $K > 0$, the minimum occurs when $(1 - h^2) > 0$, that is, $|h| < 1$ or, in view of (5.13) and (5.14), with $H_0 < H_a$, that is, with an external field less than the anisotropy field. In the case of $H_0 = H_a$, that is, $h = 1$, the second derivatives (5.21) and (5.25) are zero, that is, at this point the function (5.15) has a bend [268], which corresponds to the joining of two cases considered. In this case, the function (5.15) at the junction point remains continuous, which corresponds to a second-order phase transition.

Figure 5.2 shows how the energy density (5.15), normalized to K, changes as the angle θ changes between the magnetization vector and the axis Oz. Curves are constructed at different levels of the normalized field h. The equilibrium orientations of the magnetization vector for a given value of h correspond to the minima of the curves marked by dots. Due to symmetry with respect to the Oxy plane, the magnetization vector can have two orientations equivalent in energy, corresponding to the angles θ and $180° - \theta$, which manifests itself as the presence of two minima symmetric with respect to $90°$ on the curves.

It can be seen that as the field h increases (that is, during the transition from curve 1 to curve 6) the equilibrium orientation of

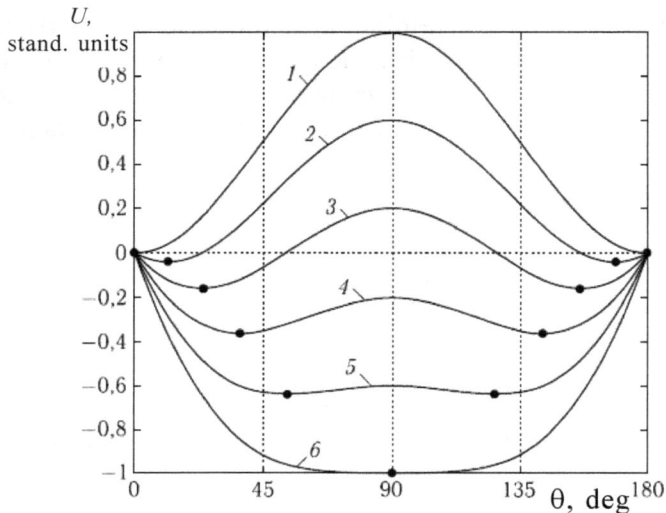

Fig. 5.2. The dependence of the normalized energy density on the orientation of the magnetization vector at different values of the activated field h: 1 – 0.0; 2 – 0.2; 3 – 0.4; 4 – 0.6; 5 – 0.8; 6 – 1.0

the magnetization tends to $\theta = 90°$, that is, the magnetization rotates in the direction of the field. In this case, the absolute energy level decreases, that is, the application of the field leads to a general decrease in the energy of the system. With further increase of the field above $h = 1$, the energy continues decrease in accordance with (5.17), but the position of the minimum remains in place in accordance with (5.20).

Thus, a phase transition occurs at $h = 1$, since the symmetry of the structure changes (instead of the two equilibrium positions of the magnetization vector, only one remains), and the energy passes through this point without any jump in a continuous manner, that is, the transition is a transition of the second kind.

Turn now to Fig. 5.3, which shows the dependence of the equilibrium value of the angle θ (a), as well as the normalized magnetization components $M_{x,z}$ (b) on the field h. The equality of

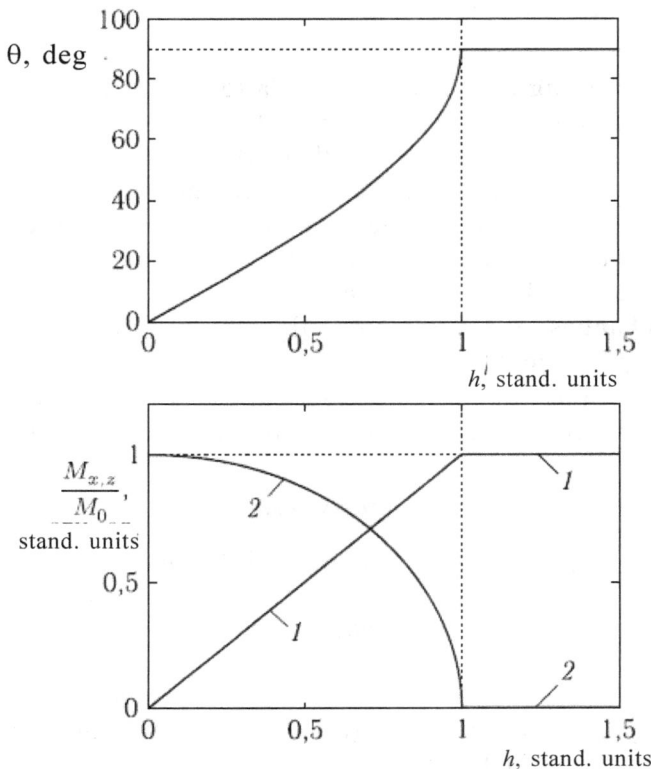

Fig. 5.3. The dependence of the equilibrium value of the angle θ (a), as well as the normalized magnetization components $M_{x,z}$ (b), on the field h. $1 - M_x/M_0$; $2 - M_z/M_0$

the external field H_0 to the anisotropy field H_a corresponds to $h = 1$, as indicated by the vertical dotted line.

In Fig. 5.3 *a* the curve at $0 \leq h \leq 1$ is constructed by the formula (5.24), and when $h \geq 1$ according to the formula (5.20).

In Fig. 5.3 *b* the curve 1 at $0 \leq h \leq 1$ is built according to the formula (5.1) subject to (5.23), which corresponds to

$$M_x = M_0 h = M_0 \frac{H_0}{H_a}, \qquad (5.26)$$

and when $h > 1$ according to the same formula (5.1), but with allowance for (5.20), which corresponds to $M_x = M_0$.

Curve 2 at $0 \leq h \leq 1$ is constructed by the formula (5.3) with (5.23), which corresponds to

$$M_z = M_0 \sqrt{1 - h^2} = M_0 \sqrt{1 - \left(\frac{H_0}{H_a}\right)^2}, \qquad (5.27)$$

and $h \geq 1$ according to the same formula (5.3), but with allowance for (5.20), which corresponds to $Mz = M_0$.

It can be seen from the figure that all the above dependences smoothly change to the point $h = 1$, and the orientation of the magnetization vector it also changes smoothly, that is, an orientational transition occurs. After the point $h = 1$, all dependencies do not change further, so that the orientation of the magnetization vector no longer changes, and it remains aligned along the Ox axis, i.e. along the applied field.

At the point $h = 1$, all curves undergo a kink, but remain continuous, which corresponds to a second-order phase transition.

5.1.2. Lengthy orientational transition

The above description of the calculation of the simplest orientational transition is intended to give no more than the basic outline of the calculation methodology. That is, first, the energy density as a function of the components of magnetization is recorded in one or another coordinate system, after which, based on considerations of symmetry or some other factors, two of the components are expressed in terms of the third, which allows us to represent the initial energy density as a function of the only remaining component of magnetization. Then, the energy density is differentiated by this component, the positions of extremes are determined from the equality of zero of the first derivative, and the positivity of the second derivative allows identifying extremes as minima of the original function, which gives the equilibrium position of the magnetization.

Such an approach is more or less general and is used in most cases. However, such simple mathematical calculations, as in the previous section, are far from being always obtained. So, already the deviation of the field direction from the perpendicular to the anisotropy axis, for example, the introduction of the H_z field component (Fig. 5.1), leads to asymmetry of the problem, as a result of which expression (5.12) becomes much more complicated. The solution of such an equation (usually carried out by numerical methods) leads to a 'tightening' of the alignment of the magnetization along the field, so that with its infinite increase the magnetization vector only infinitely tends to the direction of the field, never reaching it.

Let us consider such an inhibition of the orientational transition on the simplest example of the deviation of the field direction from the normal to the anisotropy axis. For simplicity, we will assume that the magnetization vector in the process of rotation moves in a plane passing through the EMA and the direction of the field, that is, in a spherical coordinate system (Fig. 5.1) the angle φ can be set to zero. In this case, it is sufficient to consider only the Oxz plane, where the EMA and both vectors – magnetization and fields – will be located. The orientation of these curtains and angle designations are shown in Fig. 5.4. It can be seen from the figure that in the adopted geometry the magnetization vector still corresponds to the formula (5.4), that is,

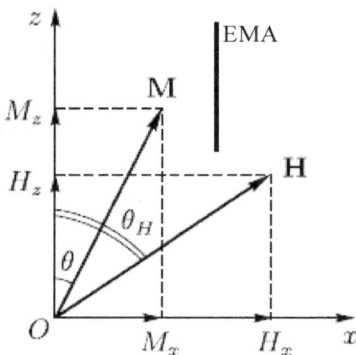

Fig. 5.4. Problem of geometry when the direction of the field does not coincide with the perpendicular to the axis of easy magnetization.

$$\mathbf{M} = \{M_0 \sin\theta, 0, M_0 \cos\theta\}, \tag{5.28}$$

and the field vector has the form

$$\mathbf{H} = \{H_0 \sin\theta_H, 0, H_0 \cos\theta_H\}. \tag{5.29}$$

The energy density, in accordance with (5.11), takes the form

$$U = K \sin^2\theta - M_0 H_0 (\sin\theta_H \sin\theta + \cos\theta_H \cos\theta). \tag{5.30}$$

Similar to the previous case, we introduce an anisotropy field:

$$H_a = \frac{2K}{M_0}, \tag{5.31}$$

as well as the normalized external field:

$$h = \frac{H_0}{H_a} = \frac{M_0 H_0}{2K}. \tag{5.32}$$

Using these transpositions, we bring (5.30) to the form

$$U = K\left(\sin^2\theta - 2h\sin\theta_H \sin\theta - 2h\cos\theta_H \cos\theta\right). \tag{5.33}$$

The minimum energy density is obtained by equating to zero the derivative with respect to θ from this expression. Excluding $\cos\theta$ from the resulting equation due to $\sin\theta$, squaring and arranging in powers of $\sin\theta$, we get:

$$\sin^4\theta - 2h\sin\theta_H \sin^3\theta + \left(h^2 - 1\right)\sin^2\theta +$$
$$+ 2h\sin\theta_h \sin\theta - h^2 \sin^2\theta_H = 0. \tag{5.34}$$

This is a complete fourth degree equation for $\sin\theta$. To determine the solvability of this equation, a control check was performed, which consisted in a graphical construction of the dependence of the left side of this equation on θ. With such a construction, the solution corresponded to the dependency passing through zero. In the real-time range of field $0 \le h \le 10$ and its orientation relative to the EMA in the range of $0 \le \theta_H \le 90°$, verification showed that this equation has a unique solution that provides the minimum energy in a quarter of the Oxz plane corresponding to $M_x \geq 0$ and $M_z \geq 0$.

Due to uniqueness, it can be assumed that the solution of this equation in an analytical form is quite possible, for example, by the Ferrari method, for which the Cardano equation will be resolving [261, 262]. The scheme of such a solution is presented in this monograph in Section 2.4. Some variants of similar solutions can be found, for example, in [297, 334, 335]. However, if a special saving of computer time is not required, then, as a rule, the numerical solution performed by the zero search method is more convenient. Some examples of the use of numerical solutions are given in the subsequent lavas of this monograph. There is also a limit we do not consider the particular case, convenient for illustration physical picture of the phenomenon.

Comment. For simplicity, we restrict ourselves to considering the case when both vectors – fields and magnetization – are in the positive quarter of the Oxz plane, so what is $0 \leq \theta_H \leq 90°$, as well as $M_x \geq 0$ and $M_z \geq 0$. Cases of other field orientations and magnetization, taking into account the bidirectionality of the EMA, are symmetric to the one under consideration, that is, completely analogous to it. Some features of the initial orientations of the magnetization vector, leading to its spasmodic reorientation, are discussed further in Chapter 7, devoted to FMR in films with the EMA slope.

Turn to Fig. 5.5, where the dependences of the polar angle of the magnetization vector θ on the normalized field value are shown for various orientations. The numbers on the curves correspond to the values of the polar angle of the field vector is θ_H in degrees.

From the figure it can be seen that the magnetization vector at a field greater than the anisotropy field (that is, $h > 1$), is aligned exactly along of this field only at $\theta_H = 90°$, that is, when the field is exactly perpendicular to the EMA. This case is just as described above. It is realized because of the symmetry of the equilibrium orientation of the magnetization vector with respect to the Ox axis, so that the minimum of total energy falls on this axis. However, when the field vector deviates from the Ox axis, the minimum of the total energy shifts in the same direction as the field. Therefore, for all other angles except $\theta_H = 90°$, the magnetization vector with increasing field only tends to its direction, and from the figure it can be seen that even with $h = 4$ norm. units the difference between the angles θ and θ_H remains fairly significant. The magnitude of the

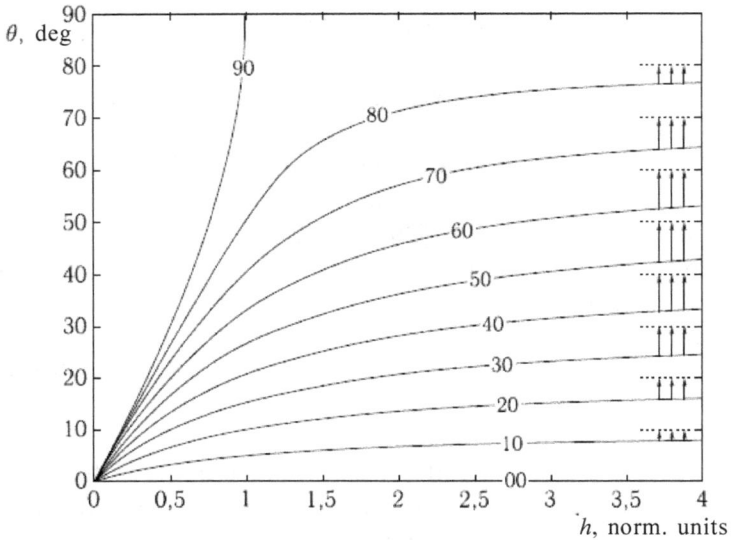

Fig. 5.5. Dependences of the polar angle of the magnetization vector on the normalized field value for various orientations. The numbers on the curves are the values of the angle θ_H in degrees

difference can be estimated from the intervals between the solid curves shown with arrows, corresponding to the angle θ and dotted horizontal lines corresponding to the angle θ_H. The difference in the figure is maximal (reaches ~10°) near $\theta_H = 50°$, which is close to the average position of the field vector between the axes Ox and Oz, that is, $\theta_H = 45°$. At $\theta_H = 90°$, the difference is absent for the above reason, and for $\theta_H = 0°$, the difference is also absent, since in this case the direction of the field coincides with the direction of the EMA, and the magnetization vector is oriented along Oz axis for any field size.

Thus, when the field orientation differs from the direction of the normal to the EMA, an orientational transition occurs, which in a field equal to the anisotropy field does not end, but continues for much larger field values, that is, in the field is 'smeared'. The change in the components of the magnetization vector at such an orientational transition is illustrated in Fig. 5.6, which shows the dependences of these components on the normalized field value at different orientations.

It can be seen from the figure that when $\theta_H = 90°$, the change in the components with increasing field corresponds to the simplest case considered above. In this case, the orientational transition ends with a phase transition when the magnetization is aligned exactly along

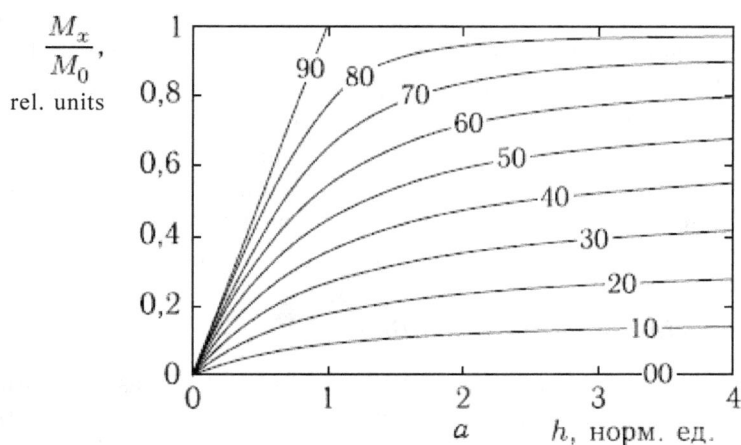

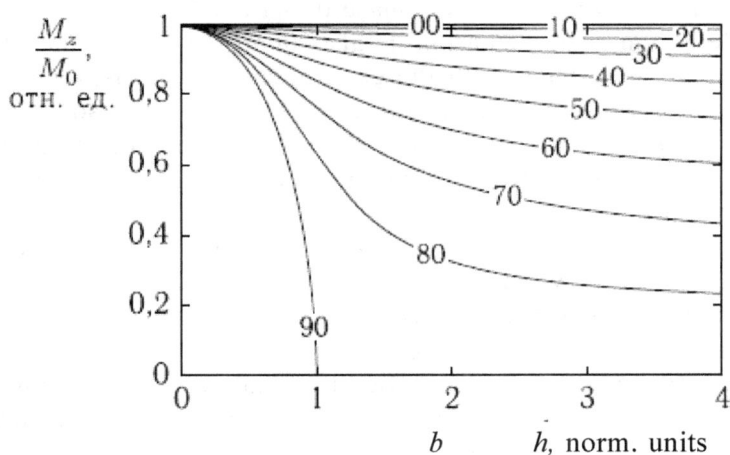

Fig. 5.6. Dependences of normalized magnetization vector components on the normalized field value at different orientations. a – component m_x; b – component m_z. The numbers on the curves are the angle θ_H in degrees

the field. In all other cases, the magnetization only tends to line up closer to the field, and the values of its components are determined by formulas (5.1)–(5.3).

Thus, in the case of $\theta_H = 90°$, with a field smaller than the anisotropy field, there are two equivalent energy orientations of the magnetization vector on opposite sides of the axis Ox (up and down in Fig. 5.4), and for a field larger than the anisotropy field only one along the axis Ox. In limited samples, these two equivalent

234

orientations due to demagnetization can lead to the formation of a domain structure. In the case of a deviation of the field vector from the normal to the EMA, the two possible orientations of the magnetization vector become nonequivalent and are realized from them more favourable. However, in the limited samples, in some interval of fields, a domain structure may also take place, and the possibility of its existence in the first place determined by the magnitude of the field components along the EMA. Some of these topics are discussed in more detail later in Section 5.2.

5.1.3. Smearing of phase transition

So, from the review we can see that the second-order phase transition point, which ends with a simple orientational transition when the field vector deviates from the perpendicular to the direction of the EMA, is now not a point, so the phase transition is 'smeared'. With a small deviation such 'smearing' is also small, so when comparing the results of calculation with the experiment, it can, as a rule, be neglected. So, when measuring the normal uniaxial anisotropy in films of mixed garnet ferrites used in CMD devices, the method of magneto-optical observation of domains is often used when the film is magnetized by a field in its plane.

In this case, the value of the external field is taken as the anisotropy field, at which the domains disappear with increasing field or nucleation when it decreases [25,30,38,40,140,229–231,235]. In this case, the accuracy of measuring the anisotropy field does not exceed 10–15% due to the difficulty of taking into account the influence of the domain structure, as well as reducing the contrast of the observed domain pattern near the transition [25, 230, 235]. In this case, the deviation of the field in the range of degrees (up to ~5°) affects the measurement accuracy slightly. On the other hand, the deviation of the anisotropy axis from the normal to the film plane already within the degrees dinits, even if the field is sufficiently accurate in its plane, leads to a strong dependence of the field of disappearance (or nucleation) of the domains on the direction of magnetization of the film in its plane, and the anisotropy of this dependence can reach an order or more [25, 30, 32].

Another factor that significantly complicates the calculation of the orientational transition is cubic anisotropy, which is usually present in most of the compositions of mixed garnets [24, 25, 38, 40, 276, 277]. At the same time, even when magnetizing the film exactly in its plane, the magnetization vector during the orientational transition remains in the same plane only if the field is precisely oriented along certain crystallographic directions. In other cases, about the time of its movement from the normal to the plane of the film to the direction of the field in its plane, the magnetization vector can describe a rather complex curve, significantly extending from the plane between the normal and the direction of the field. However, in some cases, cubic anisotropy can also be studied by observing domains when the film is magnetized in its plane [276, 277], although ferromagnetic resonance can also give fairly accurate results [24, 25, 35, 337].

An additional complication of the nature of the orientational transition can be provided by the magnetoelastic interaction that is present in many garnets [24, 25, 38, 40, 338]. An interesting variant of the orientational transition study is the excitation elastic vibrations under the action of an alternating field applied to the film, causing the movement of domains. In this case, the magnetization of the film by a constant field in its plane, up to the destruction of the domains, makes it possible to record the phase transition by the appearance or the disappearance of elastic frivolations. The amplitude of the excited elastic oscillations is maximal near the phase transition, as a result of which the sensitivity of the method is so high that it can successfully determine not only uniaxial but also cubic anisotropy [237, 238]. Due to the difficulty of analyzing the nature of the change magnetization in the orientational transition, numerical methods are important, among which, first of all, the establishment method, briefly discussed in the following section.

5.1.4. The basic scheme of the establishment method

The establishment method is widely used in calculating the structures of domain structures, magnetization distribution in domains and in periodic scattering lines [339–349]. Comparison of this method with some analytical models is given, for example, in [350]. Since the basis of such cottages is the determination of the equilibrium position of the magnetization vector, the method can also be successfully applied to calculate the orientational transitions. The method of establishment in the classical version is the implementation of successive approximations, consisting in a sequence of successive steps followed by steps of the same type.

At the first step, some, presumably more or less close to the desired, the initial direction of the magnetization vector is set. Based on this direction, the components of magnetization are determined, which are used to find the anisotropy field.

From the external field and the anisotropy field obtained in this way, the total field is determined, the direction of which is determined. The magnetization vector is oriented along such a total field, and from its components the anisotropy field is again calculated.

This anisotropy field is again added to the external field, the orientation of the total field is found, along which the magnetization vector is directed again. The components of such a new magnetization vector are again used to calculate the anisotropy field, and so on. Thus, there is a gradual approximation of the direction of the magnetization vector to the equilibrium one, which, after reaching the required accuracy, is taken as the desired one. In this form, the establishment method is essentially an iteration method widely used in computational mathematics [261, 351, 352]. The convergence of the establishment process is ensured by the uniqueness of the minimum energy density, from which the anisotropy field is determined. If the energy density has several minima, as is the case for cubic anisotropy, then the initial orientation of the magnetization vector should be chosen so that there are no potential arresters between the desired minimum and the initial position of the magnetization. Otherwise, the magnetization vector may 'fall' not into the minimum that needs to be determined, but into the next or even more distant. That is, before setting the initial orientation of the magnetization, one should first qualitatively study the configuration of the surface of the energy density and choose the initial position as close as possible to the desired one.

5.1.5. Convergence of the establishment method

An illustration of the convergence of the method of establishment is Fig. 5.7, where for the case of uniaxial anisotropy the dependences of the components of the magnetization on the field are shown, obtained by the establishment method with a different number of iterations N.

The geometry of the problem corresponds to that shown in Fig. 5.1, that is, the anisotropy axis corresponds to the coordinate axis Oz, and the field is applied along Ox axis, so the field direction is perpendicular to the anisotropy axis. The figure on a larger

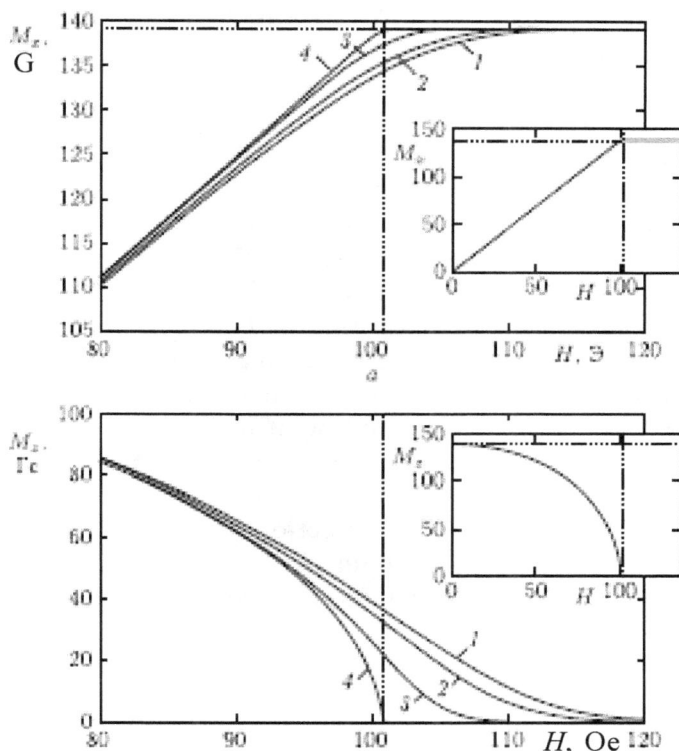

Fig. 5.7. The dependences of the components of the magnetization on the field at the orientational transition, obtained by the method of establishing with a different number of iterations N: $1 - 1$, $2 - 2$; $3 - 10$, $4 - 100$. a - component of magnetization M_x; b is the component of magnetization M_z. On the inset, the same curves for $\N = 100$ in the full interval along the horizontal axis, starting from $H = 0$ Oe. Vertical dotted lines correspond to the field of the phase transition $H_a = 100.53$ Oe, the horizontal ones - to the full magnetization $M_0 = 139.26$ G.

scale shows the areas of the curves corresponding to the end of the orientational transition, that is, the neighborhood of the phase transition point. For comparison, the inset shows the same curves with $N = 100$ in full scale along both axes. When constructing the curves, the following parameters were used: the anisotropy constant $K = 7000$ erg×cm^{-3}, which corresponds to the anisotropy field $H_a = 2K/M_0 = 100.53$ Oe, magnetization: $4\pi M_0 = 1750$ G ($M_0 = 139.26$ G).

The figure shows that the general course of the curves corresponds to that shown in Fig. 5.3 b. The sharp drop in dependences corresponds to curve 4, obtained for the number of iterations $N = 100$. With a smaller number iterations, the curves far from the phase transition point are still quite close to each other; however, in the region of the order of 10% of the transition field, the magnetization vector is not quite to equilibrium direction. So, in the field $H = 100.53$ Oe, equal to the anisotropy field H_a, that is, the one at which the magnetization should line up along the field, the component of magnetization M_z (along the EMA) at one iteration is 38 G (curve 1), at two Gs (curve 2), with ten it is 20 G (curve 3) and only with one hundred iterations it reaches zero, as the theory requires at the phase transition point. This deterioration in the accuracy of the method of establishing near the end point of the transition occurs because as the field approaches this point, the magnetization vector rotates closer and closer to the direction of the field, so that the force tending to tighten it to the direction of the field decreases, tending to zero at the transition point. Thus, we can conclude that when using the establishment method to achieve sufficient accuracy near the phase transition point, the number of iterations should be chosen sufficiently large, as required by the conditions of the task. The solution of some problems by the establishment method is given in this monograph in Chapters 6–7.

5.1.6. Dynamic setting method

Another possibility of the numerical finding of the equilibrium orientation of the magnetization is the method of dynamic establishment. Actually, this method made the majority of works cited above [339–349], with a particularly vivid and detailed description of the method contained in the synthesis paper [342].

Despite the similarity of the names, this method has a fundamental difference from the simple establishment method discussed in the previous section. Here, for a given value of the external field, the equilibrium position of the magnetization is not sought by successive radios, and after the initial orientation is set, the magnetization is released into the free precession, the law of motion of which is given by the classical Landau–Lifshitz equation. The equation is solved numerically in a complete nonlinear form (three first order equations of the type (2.97)–(2.99)), and the precession damping is specified using the dissipative term Hubert [6–9]. With such a free precession, the amplitude of oscillations of the magnetization vector due to attenuation gradually decreases, with the result that at the end of the relaxation the vector stops at the position corresponding to the equilibrium state. This final position is determined more precisely, the smaller the attenuation and the step of numerical calculation of the precession.

An important advantage of the method is the possibility of determining the establishment of magnetization in two or even three dimensions without first assuming that the magnetization vector moves along a certain plane, which is often present in the simple method setting. The disadvantage of the method is a longer required machine time, which can be critical when calculating complex structures. Thus, for the calculation using a simple method of establishing, several hundreds of iterations are sufficient, whereas for calculating the precession of the magnetization under technical conditions, already for one circle of precession approximately the same number of machine steps is required. That is the dynamic method setting, compared with the method of simple setting, requires machine time at least an order of magnitude longer. Therefore, the advantage of the dynamic establishment method is especially clearly revealed in the case of a plurality of minima of energy density, primarily in two-dimensional structures [341–343], and also when calculating the magnetization distribution in periodic domain structures [340, 343, 345, 349].

5.2. Orientational transition in ribbons of mixed garnet–ferrites

The orientational transition in films of mixed ferrite–garnets has some peculiarities that distinguish it from the usual transition considered above in an unbounded medium or in an EMA plate oriented along the normal to the plane of the plate. The difference is due to the fact that, in such films the EMA, as a rule, is deviated from the above normal by angles that make up units and even tens of degrees. In this case, the orientational dependence of the saturation of the film by the field in its plane, recorded by the disappearance of the domains, becomes strongly anisotropic. That is, the disappearance of the domains during magnetization of the film along the projection of the EMA on its plane occurs when the field is much smaller than with magnetization in the direction perpendicular to the same projection.

An important feature of mixed garnet films, which makes them very attractive for CMD realization, is high uniaxial anisotropy, the EMA of which is perpendicular (or almost perpendicular) to the plane of the film itself. Moreover, the field of such uniaxial anisotropy H_a is, as a rule, an order of magnitude or more exceeds the magnetization field $4\pi M_0$. Thus, the typical value of $4\pi M_0$ is 100–200 G, while the anisotropy field H_a lies within 5–20 kOe. For orthoferrites which when used showed for the first time a high applied value of CMD [35], this ratio is even greater and reaches two orders [40]. Thus, when considering orientational transitions occurring at fields comparable to the anisotropy field, in the first approximation the addition of the demagnetization field of the plate to the field of uniaxial anisotropy can be neglected, which is done further.

5.2.1. Deviation of the axis of easy magnetization from normal to film plane

So, since the EMA in films of mixed ferrite garnet tends to strongly deviate from the normal to the film plane, we will carry out further consideration in two stages. Consider the first mechanism of deviation of the EMA from the normal to the film plane, and then the orientational dependences of the saturation field when the film is magnetized by a field in its plane.

The mechanism of deviation. Regarding films of mixed ferrite–garnets there are two experimental methods:

1. The presence of strong uniaxial anisotropy, without noticeable cubic traces (anisotropy field – hundreds and thousands of Oersted).

2. The only axis of such anisotropy is oriented close to the normal to the plane of the film, but it is often deviated from this normal by a small angle (as a rule, units of degrees).

Consider these facts consistently. First of all, we dwell on the strong uniaxial anisotropy. In the general case, thanks to the unit cell arrangement, which has cubic symmetry, single crystals of mixed ferrites possess cubic magnetic anisotropy. The field of such anisotropy is relatively small and amounts to no more than several tens of Oersteds.

However, the epitaxial films, grown on substrates and gadolinium-gallium garnet substrates, first, have a lattice constant that is somewhat different from the substrate and, second, they have a rather small thickness, which to some extent interferes with the ordering crystal lattice properly. In this connection, two mechanisms for the formation of uniaxial anisotropy have been proposed in the literature, a brief discussion of which is given in [40, pp. 24-25]. Following this work, we note the basic features. The first mechanism is magnetostrictive, that is, it consists in the formation of uniaxial anisotropy due to the mechanical stresses of the film, due to the difference in the lattice parameters of the film and the substrate [353]. The second mechanism is 'growth anisotropy', that has noncubic ordering associated with the unequal colonization by magnetic ions of the permissible lattice nodes signs [354].

According to [40], in real films the role of the first mechanism is relatively small, but the main reason for the formation of uniaxial anisotropy is the second mechanism. It is supported by the fact that in a number of cases in the process of annealing the film lattice is gradually ordered, the corresponding atoms take their proper places and the anisotropy disappears. However, the unconditionality of such ordering in [40] is not approved (the exact link is missing). Moreover, the nature of the second mechanism to the end, apparently, remains unclear. In favor of such an assertion, one can quote from [40, p. 24]: "Identifying the microscopic mechanism by which the symmetry of the growth process causes the appearance of uniaxial anisotropy turned out to be a difficult task. This mechanism may be associated with single-ion magnetocrystalline anisotropy, the energy of the dipole interaction, or with the anisotropy of the exchange interaction. There are no references to the role of the listed mechanisms in [40].

The experimental results, stating the presence of strong uniaxial anisotropy in ferrite–garnet films of various compositions, were contained in a significant number of papers, of which, without pretending for completeness, one can specify [25, 42, 275, 278.355–358].

We now turn to the interpretation of the second experimental fact – the deviation of the axis of uniaxial anisotropy from the normal to the film plane. For films of the [111] type in experiments it became established that the main cause of such a deviation is a certain misorientation of the [111] crystallographic axis of the substrate relative to the normal to the film plane. In this case, the deviation of the axis of the magnetic (growth) anisotropy, as a rule, significantly exceeds the deviation of the crystallographic axis of the substrate. That is, in the film, compared with the substrate, the axis deviation is greatly enhanced. Among the works where such enhancement was observed, it can be noted [25,30,240,359]. Apparently, the most

consistent phenomenological description of this effect is given in
[240,359]. The task of their work is to show the possibility of the
formation of uniaxial (orthorhombic) anisotropy in the most general
case due to the misorientation of the growth of the film relative
to cubic axes. It is stated that with a general record of the energy
density of the orthorhombic anisotropy

$$E = A_{ij}\alpha_i\,\alpha_j \tag{5.35}$$

where A_{ij} are constant coefficients such that

$$A_{xx} + A_{yy} + A_{zz} = 0, \tag{5.36}$$

and

$$A_{ij} = A_{ji} \tag{5.37}$$

moreover, $\alpha_{i,j}$ are the direction cosines of the magnetization vector,
and in the further assumption that

$$A_{ij} = A_{ij}\,(\boldsymbol{\beta}) \tag{5.38}$$

where β is the growth direction of the film, the magnetic anisotropy
energy density can be expressed by decomposing into spherical
harmonics of the lowest order, which gives it the following
expression:

$$E = A\left(\alpha_x^2\beta_x^2 + \alpha_y^2\beta_y^2 + \alpha_z^2\beta_z^2\right) +$$
$$+B\left(\alpha_x\alpha_y\beta_x\beta_y + \alpha_y\alpha_z\beta_y\beta_z + \alpha_z\alpha_y\beta_z\beta_x\right). \tag{5.39}$$

That is, it is argued that for a phenomenological description of
the amplification of the deviations of the magnetic anisotropy axis
as compared with the axis of crystallographic anisotropy, only two
parameters A and B are sufficient.

Replacing $\alpha_{x,y,z}$ with the components of the normalized
magnetization vector

$$\mathbf{m} = \left\{m_x, m_y, m_z\right\}, \tag{5.40}$$

as well as introducing the direction vector of the axis of epitaxial
growth, that is, the [111] crystallographic axis of the substrate

$$\mathbf{p} = \{p_x, p_y, p_z\},\tag{5.41}$$

that is, replacing $\beta_{x,y,z}$ with $p_{x,y,z}$, we bring the formula (5.39) to the form

$$E = A\left(m_x^2 p_x^2 + m_y^2 p_y^2 + m_z^2 p_z^2\right) +$$
$$B\left(m_x m_y p_x p_y + m_y m_z p_y p_z + m_z m_x p_z p_x\right).\tag{5.42}$$

Unfortunately, in the references cited, a sufficiently detailed derivation of this formula is absent, and the numerical analysis was performed only for some of the partial rays.

As an example, we can assume that the vector $\mathbf{p}$ is deviated from the Oz axis in the direction of the Ox axis, that is, in the Oxz plane, and the deviation of the vector $\mathbf{m}$ occurs in the same plane. Turning to a spherical coordinate system and minimizing the expression for the energy density by the polar angle of the magnetization vector, it can be shown that the equilibrium position of the magnetization is determined by the biquadratic equation with respect to the sine of this polar angle (or, in the general case, a complete fourth-degree equation). The resolution of such an equation gives the dependence of the orientation of the magnetic anisotropy axis on the angle between the vector $\mathbf{p}$ and the normal to the film plane. The condition of exceeding the deviation of the anisotropy axis over the deviation of the vector $\mathbf{p}$ allows us to find the required values of the constants A and B. Approximately this procedure is done in the articles cited above. Here the authors of the monograph do not give it in view of a certain cumbersome. References to the above formula (5.39) are also found in [275, 355–358]. The effect of the axis tilt on some dynamic properties, including FMR and domain mobility, was noted in [22–25, 278]. The study of oblique uniaxial anisotropy by the magneto-optical method was performed in [25, 30, 42, 229, 231, 235, 276, 277, 360], and also by the method of ferromagnetic resonance in [22–25].

In [43], the effect of inclined anisotropy (mainly cubic type [111]) on the nonreciprocal alignment of a single domain wall formed by a gradient field [361] along an ethylene axis was noted. In [278], the effect of the inclination of the axis of light magnetization on the mobility of domain boundaries in films of garnets–ferrites was noted.

Comment. It should be noted that the general structure of the formula (5.42) is very similar to the structure of the formula for the energy density of the magnetoelastic interaction (2.14):

$$U_{me} = B_1 \left(m_x^2 u_{xx} + m_z^2 u_{zz} \right) +$$
$$+ 2B_2 \left(m_x m_y u_{xy} + m_y m_z u_{yz} + m_z m_x u_{zx} \right), \qquad (5.43)$$

where u_{ik} is the strain tensor:

$$u_{ik} = \frac{1}{2} \left(\frac{\partial u_i}{\partial x_k} + \frac{\partial u_k}{\partial x_i} \right), \qquad (5.44)$$

u_i are the components of the elastic displacement, B_1, B_2 are the 'first' and 'second' constants of the magnetoelastic interaction. The analogy consists in replacing the components of the strain tensor u_{ik} by the product of the components $p_i p_k$ of the vector **p**. However, formula (5.43) corresponds to the orientation [001], whereas for the [111] orientation, which corresponds to formula (5.42), the form of the energy density changes substantially. In the case of an isotropic medium, the constants of the magnetoelastic interaction are equal to each other: $B_1 = B_2$. That is, the structure of formula (5.43), apparently, is the simplest of all possible. It can be assumed that with the help of formula (5.43) one can describe the first of the considered ways of the formation of uniaxial anisotropy – magnetostriction, including the tilt of the axis, it is only necessary to properly determine the components of the strain tensor, which can be done on the basis of the lattice parameters of the substrate material and the bulk single crystal of ferrite of a given composition. However, it should be taken into account that in the vast majority of cases epitaxial films of garnet ferrites are grown on substrates with the [111] orientation, for which the expression for the density of magnetoelastic energy will be more complicated than (5.43). That is, the marked analogy should be approached with some caution.

5.2.2. The effect of axis tilt on film saturation

Let us now consider how the above-described deviation of the easy magnetization axis from the normal affects the saturation field of the film when magnetized in its plane.

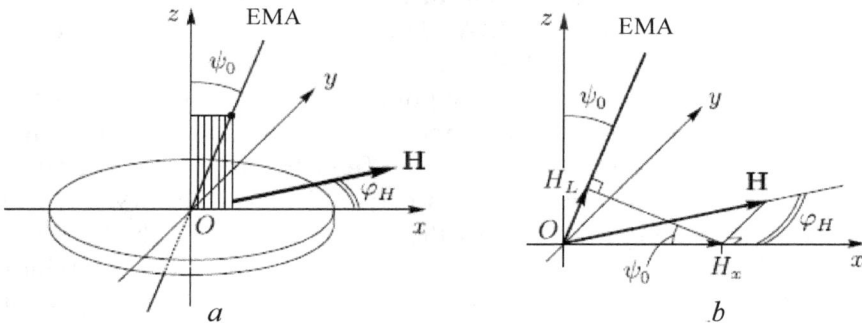

Fig. 5.8. The scheme of formation of the component field, parallel to the EMA.

As is known [4], the magnetization of a specimen with domains occurs in accordance with two mechanisms: the displacement of the domain boundaries and the rotation of the magnetization vector in the direction of the field. The first due to the component of the magnetic field along the EMA, the second across. Consider these mechanisms separately.

Domain boundary displacement. The first mechanism of saturation of the film is due to the displacement of the boundaries of the domains when it is magnetized along EMA. As the field of the domain increases, the orientation of the magnetization in which it coincides with the direction of the field, increases, and the domains with the opposite orientation of the magnetization are decreasing. As a result, with a sufficient field value, the domains of the unfavourable sign completely disappear, and the entire film becomes uniformly magnetized along the field. The disappearance of the disadvantageous domains occurs when the field component, parallel to the EMA, reaches the saturation field $H_{sat}^{(0)}$.

The diagram of the formation of the component field, parallel to the EMA, is illustrated in Fig. 5.8.

Figure 5.8 *a* shows the overall geometry of the problem. The *Oxy* plane of the Cartesian coordinate system coincides with the plane of the film, the *Oz* axis is perpendicular to it. In this case, the EMA is deviated from the *Oz* axis (that is, it is normal to the film plane) at an angle of ψ_0. The *Ox* axis of the coordinate system is parallel to the projection of the EMA on the film plane. The field **H** is applied in the film plane and makes the angle φ_H with the *Ox* axis.

Figure 5.8 *b* shows a diagram of the formation of the component of the field **H** along the EMA. This component is obtained by twofold design. First, the field vector **H** is projected onto the axis *Ox*, as a result of which its component H_x is obtained. From the general geometry of the problem, it can be seen that only this component gives the projection of the field **H** onto the EMA, since the component along the *Oy* axis is perpendicular to the EMA (since the *Oy* axis is perpendicular to the *Oxz* plane in which the EMA lies), and the component along the *Oz* axis is generally zero (since the **H** field lies in the *Oxy* plane). Further, the H_x field component is projected on the EMA, giving the H_L component. This projection occurs in the same plane *Oxz*, since the field H_x and the EMA both lie in this plane. From the geometry of two-sided triangles formed by the vectors **H** and **H**$_x$, as well as **H**$_x$ and **H**$_L$, one can see that $H_x = H \cos \varphi$ and $H_L = H_x \sin \psi_0$. Thus, we obtain $H_L = H \cos \varphi \sin \psi_0$.

Assuming now that the saturation of the film occurs at the field $H_s^{(d)}$, when the field component H_L reaches the value $H_{sat}^{(0)}$, corresponding to the complete disappearance of the domains of an unfavorable sign, i.e. saturation along the EMA, we get

$$H_s^{(d)} = \frac{H_{sat}^{(0)}}{\sin \psi_0 \cos \varphi_H}, \qquad (5.45)$$

From Fig. 5.8 it can be seen that when the sign of $\cos \varphi_H$ is reversed (which takes place in a certain range of angles), the magnetization vector also changes its direction along the EMA to the opposite. However, the sign of the field $H_s^{(d)}$ should not change, since both directions of saturation of the sample along the EMA are equal. Therefore, in the formula (5.45) $\cos \varphi_H$ must be taken modulo, with the result that it takes the form:

$$H_s^{(d)} = \frac{H_{sat}^{(0)}}{\sin \psi_0 |\cos \varphi_H|}. \qquad (5.46)$$

From the formula obtained, we can see that for $\varphi_H = 0° + 180°n$, where n is integer, the field $H_s^{(d)}$ has a minimum, and for $\varphi_H = 90° + 180°n$ it tends to infinity.

Rotation of the magnetization vector. We now consider the second mechanism by which the saturation of the film occurs by rotating the magnetization vector in the direction of the field. The action of this mechanism is due to the field component perpendicular to the EMA. In this case, as the field increases, the boundaries of the domains, if shifted, are insignificant, but the mutual width of the domains of either sign does not change, that is, the domain structure remains symmetric. The magnetization of the film occurs by rotating the magnetization vectors in both fields to the direction of the field completely symmetrically with respect to the *Oxy* plane. That is, an orientational transition occurs within each domain, which ends with a complete (or almost complete) alignment of all magnetization vectors along the field, so that the domain structure disappears and the entire film becomes uniformly magnetized. Thus, at the moment of disappearance of the domain structure, a change occurs in the symmetry of the magnetic state of matter: instead of two phases, corresponding to the directions of magnetization in the domains in mind, one corresponding to homogeneous magnetization. Such a change corresponds to a second-order phase transition, described in detail in Section 5.1.1. This transition occurs when the field $H_r^{(d)}$ the formation of which is illustrated in Fig. 5.9.

Fig. 5.9 *a* shows the general geometry of the problem. The orientation of the coordinate system *Oxyz*, EMA and field **H** relative to the film fully correspond to those shown in Fig. 5.8. In the rotation of the magnetization vector, initially oriented along the EMA, the field component perpendicular to the EMA is involved. The rotation ends when the value of this component becomes equal to the anisotropy field H_a (with an accuracy of the demagnetization field, where we ignore the pattern of the field component perpendicular to the EMA of smallness). The locus of the ends of such a component is a cylinder whose axis coincides with the EMA and the radius is H_a. Since the external field is applied in the *Oxy* plane, the transition occurs on the sectional line of the cylinder of this by plane. According to the general rules of analytical geometry [292], a section of a circular cylinder with a plane not parallel to the axis of the cylinder is an ellipse, denoted by the letters *ABCD*. The general structure of this ellipse is shown in Fig. 5.9 *b*. Its semi-axes along *Ox* and *Oy* are, respectively, *a* and *b*. From the geometry of Fig. 5.9 *a* it can be seen that $a = H_a/\cos\psi_0$, $b = H_a$. The components of the field **H**, which play the role of variables, are respectively equal: $H_x = \cos\varphi_H$, $H_y = \sin\varphi_H$ (Fig. 5.9 *c*). The classical equation of an ellipse has the form [292]

$$\frac{x^2}{a^2}+\frac{y^2}{b^2}=1. \tag{5.47}$$

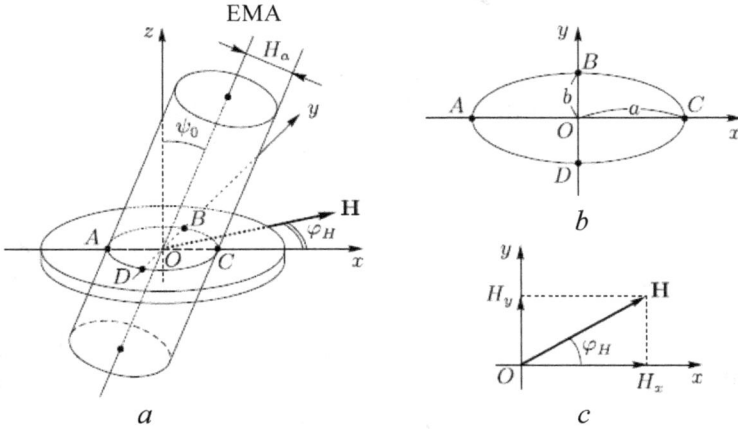

Fig.5.9. Formation of components of the field perpendicular to EMA.

Substituting the values of a and b into this equation, as well as $x = H_x$ and $y = H_y$

$$\frac{H^2 \cos^2 \varphi_H \sin^2 \psi_0}{H_a^2} + \frac{H^2 \sin^2 \varphi_h}{H_a^2} = 1.$$

$$(5.48)$$

From this equation, we find the field required to saturate the film:

$$H = \frac{H_a}{\sqrt{1 - \cos^2 \varphi_H \sin^2 \psi_0}}.$$

$$(5.49)$$

Assuming that the EMA is deviated from the normal by a small angle, so that $\sin^2 \psi_0 \ll 1$, and also using the approximate ratio

$$\frac{1}{\sqrt{1 + \pi}} \approx 1 - \frac{1}{2} x,$$

$$(5.50)$$

representing the first term of the Taylor expansion in the function $y = (1 + x)^{-1/2}$ in a neighborhood of the point $x = 0$, we get

$$H_r^{(d)} = H_a \left(1 + \frac{1}{2} \sin^2 \psi_0 \cos^2 \varphi_H \right).$$

$$(5.51)$$

This expression gives the orientational dependence of the saturation field of the film when magnetized in the plane.

5.2.3. The general orientational dependence of the film saturation field

The resulting expression (5.51), together with the above (5.46), describe the complete saturation of the film in accordance with two mechanisms: domain displacement and rotation of the magnetization vector. Real saturation at a given orientation of the field **H** occurs when its value H_S reaches the smaller of two values: $H_s^{(d)}$ or $H_r^{(d)}$.

Figure 5.10 shows the dependence of the saturation field of the film H_S on the angle φ_H, which determines the orientation of the field applied in its plane. Curve 1 is constructed by the formula (5.46) and corresponds to the saturation of the film due to the displacement of the boundaries of the domains. Curve 2 is constructed by the formula (5.51) and corresponds to the saturation of the film due to the rotation of the magnetization vector. Real saturation, determined by the minimum of the fields $H_s^{(d)}$ or $H_r^{(d)}$, shown in solid line. The dashed parts of curves 1 and 2 correspond to unrealizable values of technical fields.

It can be seen from the figure that the dependence is periodic in nature, repeating through 180°. The angles 0° and 180° correspond to the magnetization along the projection of the EMA on the film

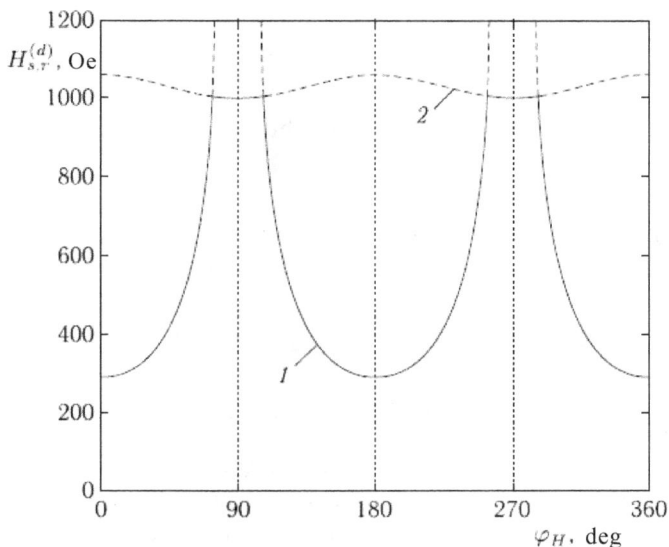

Fig. 5.10 Dependence of the saturation field of films and the slope of the EAM on the angle φ_H, defining the orientation of the field in the plane. Curve 1 is constructed using formula (5.46), curve 2, using formula (5.51). Construction parameters: H_a = 1000 Oe; $H_{sat}^{(0)}$ = 100 Oe; ψ_0 = 20°

plane, the angles 90° and 270° – to the magnetization perpendicular to this projection.

Below the solid parts of curve 1, the domain boundaries shift, mainly without noticeable rotation of the magnetization vector. Some rotation takes place only on the approach of these curves to the increasing portions of the solid curve. Below is a solid stack. curve 2, an orientational transition takes place without a noticeable displacement of domain boundaries.

In this case, solid portions of curve 1 correspond to the point of phase transition of the first kind, taking place with increasing field. When the field decreases from a value greater than saturation, the domains may not appear immediately, but when the corresponding nuclei take effect. That is, there may be a certain hysteresis. The corresponding curve of nucleation of the domain structure will pass slightly below curve 1. The solid portions of curve 2 correspond to the second-order phase transition point. In this case as the field decreases from high magnitudes, the nucleation of domains occurs immediately from the sinusoidal instability of the orientation of the magnetization, which is determined by the soft spin-wave mode [230]. In this case, there is no hysteresis.

5.3. Experimental study of films with inclination of the anisotropy axis

In this section, we briefly describe the results of experiments on the study of magnetic anisotropy in films of mixed ferrite garnet, which have a significant deviation of the axis of uniaxial anisotropy from the normal to the film plane. The experiments were carried out in 1972–1980 with the participation of one of the authors of the present monographs and are described in detail in [22–25,30]. Here we present their results, without going into unnecessary details.

5.3.1. Anisotropy field and domain structure period

An experimental study of magnetic anisotropy was carried out on films, the main characteristics of which are given in Table 5.1. More than 40 samples were studied (grown by I.G. Avaeva and V. B. Kravchenko), so that only the most common properties are listed in the table.

Table 5.1

Composition	Anisotropy type	$\underline{H}_a$, Oe	M_0, G
$(EuEr)_3(FeGa)_5O_{12}$	uniaxial	5000...9000	5...12
$(EuY)_3(FeGa)5O_{12}$	uniaxial	1000...2000	12...20
$(YSm)_3(FeGa)_5O_{12}$	uniaxial	500...800	20...30
$(YGdYb)_3(FeGa)_5O_{12}$	cubic	100...200	15...20
$(YGdYbBi)_3(FeAl)_3O_{12}$	uniaxial	2000...7000	10...20
$(YTmSmBi)_3(FeAl)_5O_{12}$	uniaxial	2000...7000	10...20

The first column of the table shows the nominal compositions of the films, determined by the ratio of the components in the mixture during their growth. The first four compositions were intended primarily for the study of CMD. Bismuth was included in the last two compositions, which ensured their high magneto-optical properties – strong rotation of the plane of polarization of light propagating along the direction of magnetization (the Faraday effect) [146].

The second column shows the nature of the anisotropy. In all films with the uniaxial anisotropy, the cubic magnetic anisotropy was not observed. In the films with the composition $(YGdYb)_3$ $(FeGa)5O_{12}$, the uniaxial anisotropy was relatively small, but the cubic hard enough.

The third column shows the values of the uniaxial anisotropy field. In the fourth, the saturation magnetization values. Can see that larger values of the anisotropy field mainly correspond to lower values of magnetization. Let us first consider the nature of the orientational transition with tangential magnetization of films, for which we follow the period domain structure observed by the magneto-optical method according to the Faraday effect. It should be noted that in all the flaks, regardless of the composition and type of the domain structure, when the field is applied magnetized in the film plane perpendicular to the projection of the EMA on the film plane ($\varphi_H = 90 + 180n$), a monotonic decrease in the period of the domain structure by two to three times was observed, similar to that noted earlier in [160, 161.229–231, 235, 236].

Figure 5.11 shows a typical dependence of the period of a strip domain structure on the field in the film plane. The film of composition $Y_{0.76}Gd_{0.76}Yb_{0.68}Bi_{0.8}Fe_{3.9}Al_{1.1}O_{12}$ was studied; it had a field of uniaxial anisotropy $H_a = 2500$ Oe, the saturation magnetization was about 34 G, and the width of the strip domains in the absence of a field was about 22 μm. The [111] crystallographic axis of the

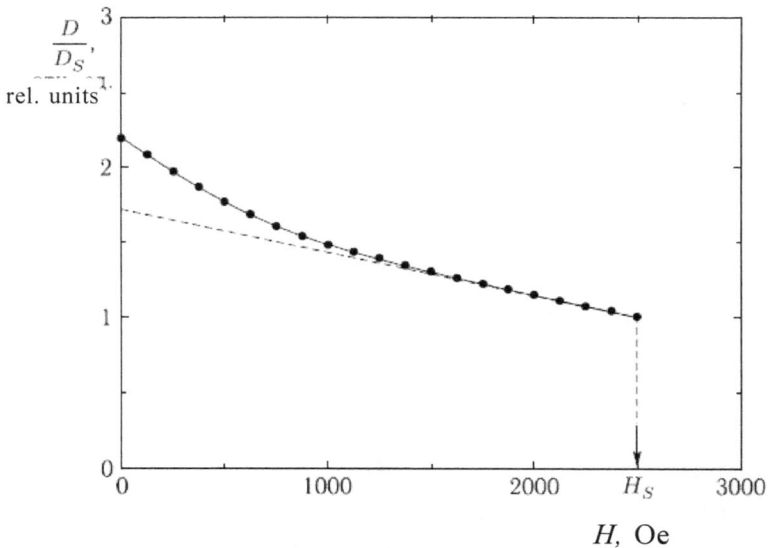

Fig. 5. 11. Dependence of the normalized period of the stripe domain structure D/D_S on the field in the film plane. D_S is the observed period of the domain structure at the phase transition point

substrate was deflected from the normal to the film plane by 4°, which gave the EMA deviation from the normal of 15°. Due to the presence of bismuth, the film had a high magneto-optical activity: in red light, the rotation of the polarization plane was about 20 000 deg/cm. With a film thickness of about 20 μm, this ensured the rotation of the plane of polarization of light in neighbouring domains at angles of about 40°, so that the domain structure was traced quite contrastingly until it disappeared completely. From the figure it can be seen that the period of the domain structure when approaching the phase transition point ($H_a = 2500$ Oe) tends not to zero, as predicted in [160, 161, 229, 231, 235], but to the final non-zero value (about 10 μm).

Such behaviour of the period near the phase transition was theoretically interpreted in [230], where it was shown that at the transition point the structure period should be equal to the length of the zero-frequency spin wave during the formation of the soft mode [133,134]. The length of such a wave is deliberately different from zero and is determined by the energy balance between the demagnetization field and the exchange interaction (since the anisotropy here is completely compensated by the external field).

According to [230], the field of the phase transition H_p is related to the anisotropy field H_a by the relation:

$$H_p = H_a - 2\pi M_0 \frac{D_S^2}{b^2}, \qquad (5.52)$$

where D_S is the period of the domain structure at the transition point, b is the film thickness. The dependence of the period on the field at the transition point is determined by the angular coefficient of the form:

$$k_s = \frac{d}{dH} B(H) \bigg|_{H=H_p} = -\frac{b}{16\pi^2 M_0 D_S}. \qquad (5.53)$$

With real values of the parameters for the studied film $D_S = 10$ μm, $H_a = 2500$ Oe, $b = 20$ μm, $M_0 = 34$ G, formula (5.52) shows that the addition to the anisotropy field (the second term) is about 8 Oe. At the level $H_a = 2500$ Oe such a small additive can be neglected, that is, $H_p = H_a$. In this case, from the formula (5.53) we obtain the slope $k_S = -37 \cdot 10^{-5}$ Oe^{-1}. From the experimental relationship shown in Fig. 5.11, it can be seen that the angular coefficient of the tangent to the solid curve at the transition point (that is, the dotted line) is $k_S = -29 \cdot 10^{-5}$ Oe^{-1}. This value quite closely (with an accuracy of about 20%) corresponds to the calculated value.

Summarizing what has been said, it can be concluded that the finiteness of the period of the domain structure at the phase transition point is quite close. The agreement between the experimental values of the derivative of the period of the domain structure over the field and the theoretical value is in favour of a theory that takes into account the non-uniformity of the magnetization distribution film thickness [230].

5.3.2. Orientation dependences of the saturation field

We now turn to the consideration of the orientation dependences of the saturation field upon magnetization of a film with a strong inclination of the EMA field applied in the plane of the film. The theoretical consideration of this issue is made in Chapter 7, but now we turn to the experimental results, based on the materials of technical works [25, 30]. The measurements were carried out on technical samples, the parameters of which are given in Table 5.1. The saturation field for a given angle with a certain step was determined by the fact of the disappearance of the domain structure observed by the magneto-optical method (according to the Faraday effect). In all the cases, ultata were similar, i.e. orientation dependences looked like that shown in Fig. 5.10. For a more visual illustration of the entire set samples were selected five of the most characteristic, the composition of which is given in Table 5.2. The obtained dependences of the saturation field of the films on the angle ϕH, which determines the orientation of the field in the plane, are shown in Fig. 5.12. The curve numbers correspond to the sample numbers given in Tables 5.2 and 5.3.

Table 5.2

No.	Composition
1	$Eu_1Er_2Fe_{4.3}Ga_{0.7}O_{12}$
2	$Eu_1Er_2Fe_{13}Ga_1O_{12}$
3	$Eu_{0.65}Y_{12.35}Fe_{4.4}Ga_{0.6}O_{11}$
4	$Y_{2.52}Sm_{0.38}Fe_{3.85}Ga_{1.15}O_{12}$
5	$Y_{1.3}Gd_1Yb_{0.7}Fe_{4.1}Ga_{10.9}O_{12}$

Table 5.3

No.	$H_S^{(max)}$, Oe	$H_S^{(min)}$, Oe	$\psi_{[111]}$, deg	ψ_0, deg
1	7000 ± 500	160 ± 10	4 ± 1	14 ± 2
2	1300 ± 100	$450\pm$	4 ± 1	20 ± 2
3	1000 ± 100	$250\pm$	4 ± 1	17 ± 2
4	500 ± 100	$300\pm$	4 ± 1	15 ± 2
5	90 ± 10	$90\pm$	4 ± 1	0 ± 2

The origin and character of the dependencies obtained in the experiment fully corresponds to that shown in Fig. 5.10, therefore, is not discussed in detail here. We note only some remarks concerning the specific parameters of the samples studied, for which we turn to Table 5.3.

Here, the first column corresponds to the numbers of samples, the compositions of which are given in Table 5.2. In the second and third columns Table 5.3 shows the values of the maximum $H_S^{(max)}$ and minimum $H_S^{(min)}$ fields at which the domain structure disappears.

The measurements of these fields were carried out at angles of $0°$, $90°$, $180°$, $270°$ and $360°$, respectively. It can be seen that values correspond to extremums of curves in fig. 5.12. Given the smallness of the addition to the anisotropy field of the second term in formula

(5.52), it can be assumed that with a high degree of accuracy the field $H_S^{(max)}$ and the anisotropy field H_a are equal to each other: $H_S^{(max)} = H_a$, that provides a convenient way to measure the anisotropy field.

The fourth column of the table shows the angles $\psi_{[111]}$ corresponding to the deviation of the [111] axis of the substrate from the normal to the film plane. All substrates were cut from a single rod of a gadolinium–gallium garnet grown by zone melting, and cutting was carried out perpendicular to the axis of the rod. X-ray diffraction analysis performed on an already grown rod revealed a deviation of the [111] crystallographic axis from the rod axis by 4°, which gave the corresponding deviation of the [111] axis of the substrate from the normal to the cutting plane of the rod by the same 4°.

However, the deviation of the magnetic anisotropy axis from the normal did not work out on films grown on such substrates. The deviation was measured by observing the domains in a tangential field close to the anisotropy field, with the film being rotated around axis perpendicular to the one that corresponded to the minimum of the saturation field. Such a rotation was carried out before the technical support, while the saturation field was not compared with the field that corresponded to the maximum saturation value. The resulting angle of deviation of the normal to the plane of the film from its initial position parallel to the axis of observation was taken as the angle of deviation of the EMA from the normal to the plane of the film. The angles obtained by this way from the normal to the plane of the film are given in the fifth column of the table. From the table it can be seen that, in the first of four samples, the angle of deviation of the EMA from the normal exceeds that for the [111] axis by three to five times, and there is no correlation with the anisotropy field. Such an increase in the deviation corresponds to that observed in [25, 43, 275, 278, 355–358], discussed in detail in Section 5.2.1. In the last sample (No. 5) such a deviation is absent, moreover, the axis of magnetic anisotropy is almost exactly oriented along the normal. In connection with this curve 5 in Fig. 5.12 has the form of a straight horizontal line without any noticeable extremes. It can be assumed that the absence of a deviation is caused by the absence of a non-cubic ordering mechanism for films of such a composition, which was discussed in Section 5.2.1. However, we note that although the magneto-optical method does not reveal a noticeable anisotropy in this sample, nevertheless, a sufficiently

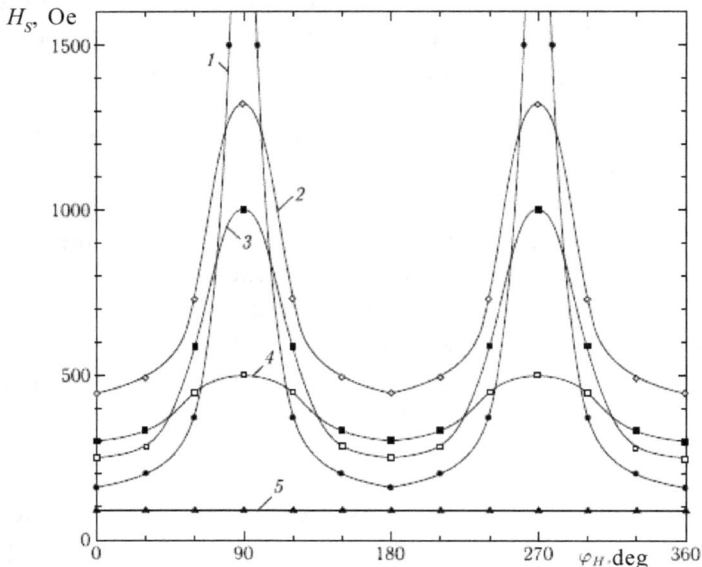

Fig. 5.12. Dependences of the saturation field of films with an EMA tilt on the angle φ_H, which determines the field orientation in the plane. Curve numbers correspond to sample numbers given in Tables 5.2 and 5.3

noticeable anisotropy of a cubic character can be detected by the method of ferromagnetic resonance [22–25], discussed in detail in Chapter 7 (Section 7.6).

Conclusions for chapter 5

This chapter is devoted to the description of orientational transitions in various words, including the issues of numerical analysis of the process of establishing magnetization in real materials. The main results obtained in this chapter are as follows.

1. The general ideology of the orientational transition is considered. The definition of an orientational transition is presented as an intermediate state in which a magnet is located when passing from one stationary state to another, and these extreme states are distinguished by an equilibrium orientation of the magnetization inside the magnet. It is noted that the equilibrium state of magnetization is determined by at least the potential formed by the anisotropy, demagnetization and external field energies. Examples of orientational transitions, primarily a smooth rotation of the

magnetization vector during magnetization of a uniaxial magnet by a field perpendicular to the axis of magnetic anisotropy, are given. It is noted that the desire to reduce the potential energy of the demagnetizing field leads to the splitting of the limited sample into domains. The definitions of homogeneous, inhomogeneous, collinear and angular phases are given. It is noted that the nucleation of the domain structure at the phase transition point of the second

It occurs through a zero-frequency spin wave, the length of which is equal to the period of the emerging domain structure.

2. A simple orientational transition in the magnetization geometry of a uniaxial magnetic field perpendicular to the anisotropy axis is considered. Based on the minimization of the total energy density of the anisotropy and the external field, the dependences of the longitudinal and transverse components of the magnetization on the field are obtained. The fact of the orientation transition is revealed when the orientation of the external field deviates from the normal to the anisotropy axis. It is shown that the minimization of density energy in this case requires the resolution of a complete fourth-degree algebraic equation with respect to the sine of the polar angle of the magnetization vector. Using the method of numerical solution of such an equation, the dependences of the polar angle of the magnetization vector on the normalized magnitude of the field are obtained for different orientations. It is shown that only when the field is oriented parallel or perpendicular to the anisotropy axis, the orientational transition ends with a clear second-order phase transition at one point, and for all intermediate orientations the magnetization vector only infinitely approaches the field direction, never reaching it. Thus, a kind of 'smearing' of a phase transition across the field occurs and becomes stronger with increasing field direction from the normal to the anisotropy axis.

3. We considered the method of establishment, which allows by successive approximations to calculate the dependence spring position from the field in cases of a forced orientation transition. The implementation of the method consists in setting some initial direction of magnetization, on the basis of which is determined minimum density of total energy. This position of the minimum is then taken as the position of the magnetization vector, on the basis of which the minimum energy density is again found. Sequential implementation of similar values (iterations) in the case of convergence of the method allows us to find the equilibrium position of the magnetization vector with any predetermined accuracy. The

conditions of convergence are considered and it is shown that the accuracy of the method as it approaches to the phase transition point gradually decreases, which, however, does not interfere with its successful implementation in the majority of cases required in practice. The dynamic establishment method is briefly characterized, which differs from the simple determination method by taking into account the vector precession magnetization at each iteration. Its advantages and disadvantages compared with the simple method of establishing.

4. The features of the orientational transition in real films of mixed garnet ferrites, where the magnetic anisotropy axis deviates from the normal to the film plane by a considerable angle, are considered. A sharply anisotropic dependence of the saturation field is noted on the orientation of the external field applied in the film plane. It is shown that the maxima of this dependence correspond to the orientational transition of the magnetization vector, similar to that at magnetization of a uniaxial magnet by a field perpendicular to the anisotropy axis. The minima of the same dependence correspond to the saturation of the film due to the destruction of the domains of the unfavourable sign of the field component parallel to the anisotropy axis.

5. Possible mechanisms for the deviation of the anisotropy axis in ferrite-garnet films from the normal to the film plane are considered. It is noted that significant angles of deflection of the magnetic anisotropy axis of the film from the normal are the result of increased deflection [111] crystallographic axis of the substrate from the same normal. Two possible mechanisms for such amplification are mentioned: the magnetoelastic interaction, provided that the lattice parameters of the film and the substrate are different, and the non-cubic ordering of magnetic ions in the nodes of the figure

6. The calculation of the mechanisms of saturation of the film with oblique anisotropy when it is magnetized in the plane is performed. Considered saturation by shifting the boundaries of domains, as well as by rotating the vector magnetization in the direction of the field. Analytical expressions are obtained, allowing to calculate the dependence of the saturation field on the angle of orientation of the external field, based on the anisotropy field values and the saturation field of the film by magnetizing it along the anisotropy axis. The resulting dependence was constructed, the maxima of which fall on the directions in which the external field is oriented perpendicular to the projection of the anisotropy axis on the film plane, and the

minima on the direction when the external field is parallel to this projection.

7. The results of experiments on the study of anisotropy in films of mixed ferrites–garnets with a tilt of the axis of easy magnetization are presented. The dependence of the period of the domain structure of films on the field under the conditions of a second-order phase transition was studied. It is shown that in the process of increasing the field, the period of the domain structure decreases, and at the phase transition point it always remains finite. It is noted that the final value of the period of the domain structure at the phase transition point is in favour of the theory that assumes a nonuniform distribution of the magnetization across the film thickness. The results of an experimental study of the dependence of the saturation field on the orientation of the external field in the film plane are presented. A good agreement is noted with the theoretical mechanisms of saturation formation, which take into account the displacement of the domain boundaries and the rotation of the magnetization vector to the field direction.

6

Ferromagnetic resonance in plates with uniaxial and cubic anisotropy

This chapter is devoted to the description of the phenomenon of ferromagnetic resonance in an anisotropic medium, including the conditions of the orientational transition. The ferromagnetic resonance in a plate with uniaxial and cubic anisotropy at different directions along the cube axes. A dynamic susceptibility tensor is obtained for ferromagnetic resonance in a plate with normal uniaxial anisotropy magnetized by a field, perpendicular to the anisotropy axis. As an example, which is important for further discussion, we consider ferromag   `netic resonance in a plate with uniaxial and cubic magnetic anisotropy, with the EMA and the [111] axis perpendicular to the plane of the plate. In such a geometry, in view of the asymmetric arrangement of the axes of the [111] type relative to the plane of the plate, cases of tangent and normal magnetization are fundamentally different, so they are considered separately. When considering the surveys listed in the survey, we will mainly follow the works [6–8], the remaining references are indicated in the text.

6.1. Tangential magnetization

We first consider the case of tangential magnetization, that is, when the field is applied in the plane of the plate. We assume that the applied field is larger than the anisotropy field, that is, the equilibrium orientation of the magnetization is rather close to the field direction. So here we consider the orientational transition in the tightened state, that is, when a further increase in the field only approximates the orientation of the magnetization to the direction of the field (Section 5.1.2).

6.1.1. Energy density

The geometry of the problem is shown in Fig. 6.1.

The coordinate system $Oxyz$ is oriented in such a way that the plane Oyz is parallel to the plane of the plate, and the axis Ox is perpendicular to it. The anisotropy axis is perpendicular to the plane of the plate, that is, directed along the axis Ox. The axis of cubic anisotropy [111] will be assumed to be directed also perpendicular to the plane of the plate, that is, along the same axis Ox. The choice of such an orientation of the [111] axis for this problem is not accidental, but is due to the most common technological practice, according to which garnet ferrite films are grown on substrates of c gadolinium–gallium garnet with this orientation [38, 40].

Comment. The choice of the name of the coordinate axes given here differs from that adopted in Fig. 5.1. This is due to the fact that the geometry of Fig. 5.1 is traditionally used in problems related to the domain structure, primarily the CMD, where the Oz axis is oriented perpendicular to the film plane [38, 40]. It also uses an orientation that is more common in the description of wave processes, primarily for problems concerning the propagation of surface magnetostatic waves (SMSW) in the film plane [1, 2]. In

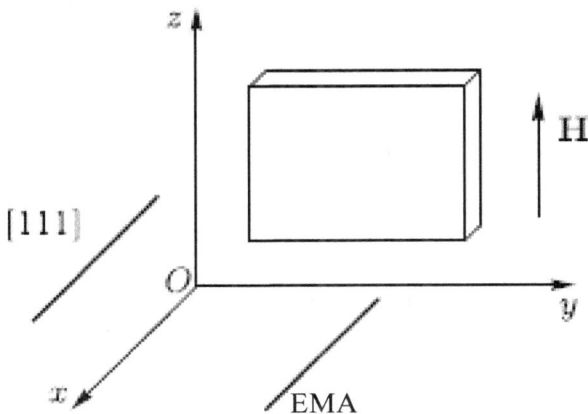

Fig. 6.1. The geometry of the problem with a tangent magnetization of the magnetic plate. EMA – easy magnetization axis.

this case, the field orientation coincides with the film plane, i.e., it is directed along Oz (which is generally traditional for FMR [6–8]), and the SMSW propagation occurs in the Oyz plane, so that the Ox axis is perpendicular to the film plane. In the adopted geometry, the field vector has the form

$$\mathbf{H} = \{0,0,H_0\} \tag{6.1}$$

The energy density without cubic anisotropy is

$$U_a = -\frac{K}{M_0^2}M_x^2 + 2\pi M_x^2 - M_z H_0, \tag{6.2}$$

where the first term is the energy density of the uniaxial anisotropy ($K > 0$), the second is the demagnetization energy density of the plate and the third is the energy density of the interaction of magnetization with a constant field. To record the energy density of the cubic anisotropy, it is more convenient to introduce a coordinate system associated with the edges of the cubic cell. Denote such a system by $Ox'y'z'$, so that the energy density of the cubic anisotropy in this system is

$$U_c = -\frac{K_1}{M_0^4}\left(M_{x'}^2 M_{y'}^2 + M_{y'}^2 M_{z'}^2 + M_{z'}^2 M_{x'}^2\right). \tag{6.3}$$

Here it is assumed that $K_1 > 0$, so the axes of the [111] type are light. The constant K_2 is assumed to be significantly smaller than K_1, therefore we do not consider it.

The total energy density is

$$U = U_a + U_c = \frac{2\pi M_0^2 - K}{M_0^2}M_x^2 - M_z H -$$

$$-\frac{K_1}{M_0^4}\left(M_{x'}^2 M_{y'}^2 + M_{y'}^2 M_{z'}^2 + M_{z'}^2 M_{x'}^2\right). \tag{6.4}$$

Since it is assumed that the field is so large that the magnetization is oriented along the field, to ensure the possibility of linearization of the equations of motion, the problem should be considered in the system $Oxyz$, so expression (6.4) should be brought to this system. For such a cast, we use the apparatus of the transition matrix (Chapter 3).

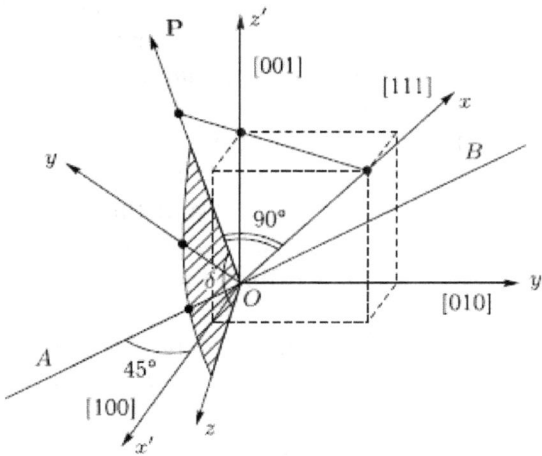

Fig. 6.2. Geometry in which the transition matrix is obtained.

Transition matrix. To obtain the transition matrix, we turn to the geometry shown in Fig. 6.2.

The coordinate system $Oxyz$ is associated with the plane of the plate and with the direction of the field. The main task will be solved in this system. System $Ox'y'z'$ is associated with the cubic cell, so that its axes are oriented along the edges of the cube. It is in this system that the energy density of cubic anisotropy (6.3) is written, which represents the last term in expression (6.4). That is, it is necessary to make a transition from the magnetization, having the components $M_{x'y'z'}$ to the magnetization having the components $M_{x,y,z}$, that is, to express the components $M_{x'y'z'}$ through $M_{x,y,z}$. Like section 3.6.3, we use the apparatus of Euler angles [261]. The straight line AB (line of nodes) lies in the plane $O_{x'y'}$ and is up with the Ox' axis angle of 45°. The Ox axis runs along the spatial diagonal of the cell, that is, coincides with the [111] axis. The vector **P** lies in the plane passing through the [001] and [111] axes of the cell, that is, through the coordinate axes Ox and Oz'. This vector makes an angle of 90° with the axis Ox, that is, lies in the plane of the plate. Line AB, being perpendicular to the axis Ox, also lies in the plane of the plate.

Since the plane containing the vector **P** axis [111] runs through the diagonal of the face of the cube extending from point O, it can be seen that the vector **P** is perpendicular to the straight line AB. In the same plane, in addition to the [111] axis, which coincides with the Ox axis, lies another axis of the [111] type, more precisely the $[\bar{1}\bar{1}1]$ axis, whose projection on the $Oxyz$ plane coincides with the vector **P**. The shaded plane passes through the vector **P** and the straight line AB, and since the axis Ox is perpendicular to AB, the axis Oz also lies in the shaded plane, that is, this plane coincides with the plane of the plate Oyz. The angle between the axis Oz and the vector **P** is denoted by δ. It is this angle that determines the rotation of the cubic cell around the axis Ox, that is, the [111] axis. With such a turn, the straight line AB and the vector **P** they also rotate around this axis, simultaneously with the rotation of all three axes of the the $Ox'y'z'$, as they follow the rotation of the edges of the cubic cell.

Thus, the angle δ determines the orientation of the crystallographic cells with respect to the $Oxyz$ system associated with the plate, that is, with the field as well.

When δ = 0°, the Oz axis coincides with the projection of one of the [111] axes (axes $\left[\overline{1}\,\overline{1}1\right]$) onto the Oyz plane (that is, onto the plane of the plate). When δ = 30°, the Oz axis is perpendicular to this projection, that is, it is located midway between two projections of [111] axes, which together form an angle of 60° (in accordance with Fig. 3.10 in Section 3.5.5). The initial expression for the energy density of cubic anisotropy (6.3) is written in the system $Ox'y'z'$. According to the accepted conditions of the problem (that is, to solve the problem in the system associated with the plate), this expression must be converted to the $Oxyz$ system. Formula (6.3) contains the components of the magnetization vector $M_{x',y',z'}$. To transform such a vector, we use formula (3.14)

$$\mathbf{a}' = \ddot{A}^{-1}\mathbf{a}, \tag{6.5}$$

where $\ddot{A}^{-1}$ is the transition matrix whose components are determined by the cosines of the angles between the axes of the coordinate systems $Oxyz$ and $Ox'y'z'$. Further, the substitution of the components $M_{x',y',z'}$ obtained in terms of $M_{x,y,z}$, will give an expression for the energy density expressed in $M_{x,y,z}$, that is, in the $Oxyz$ system, where the main problem will be solved .

As shown in Section 3.2.6, when solving a problem in a system associated with a plate, that is, $Oxyz$, the matrix $\ddot{A}^{-1}$ must be the inverse transition matrix.

To obtain the required matrix, we find the cosines of the angles between the axes of the systems $Oxyz$ and $Ox'y'z'$ using technical techniques of analytic geometry, as in Section 3.6.

As a result, we obtain the following table of the components of the transition matrix:

	x	y	z
x'	$\dfrac{\sqrt{3}}{3}$	$-\dfrac{\sqrt{6}}{6}\left(-\dfrac{\sqrt{6}}{6}\sin\delta+\sqrt{3}\cos\delta\right)$	$\dfrac{\sqrt{6}}{6}\left(\sin\delta-\sqrt{3}\cos\delta\right)$
y'	$\dfrac{\sqrt{3}}{3}$	$\dfrac{\sqrt{6}}{6}\left(\sin\delta-\sqrt{3}\cos\delta\right)$	$\dfrac{\sqrt{6}}{6}\left(\sin\delta+\sqrt{3}\cos\delta\right)$
z'	$\dfrac{\sqrt{3}}{3}$	$\dfrac{\sqrt{6}}{3}\sin\varepsilon$	$\dfrac{\sqrt{6}}{3}\cos\varepsilon$

$$(6.6)$$

That is, the required matrix is

$$\ddot{A}_{111}(\delta)=\begin{pmatrix} \dfrac{\sqrt{3}}{3} & -\dfrac{\sqrt{6}}{6}(\sin\delta+\sqrt{3}\,\cos\delta) & (\dfrac{\sqrt{6}}{6}\left(\sqrt{3}\sin\gamma-\cos\delta\right) \\[2mm] \dfrac{\sqrt{3}}{3}\cos\delta & -\dfrac{\sqrt{6}}{6}(\sin\delta-\sqrt{3}\,\cos\delta) & \dfrac{\sqrt{6}}{6}\left(\sqrt{3}\sin\delta+\cos\delta\right) \\[2mm] \dfrac{\sqrt{3}}{3} & \dfrac{\sqrt{6}}{3}\sin\varepsilon & \dfrac{\sqrt{6}}{3}\cos\varepsilon \end{pmatrix}$$

$$(6.7)$$

A note on the equivalence of matrices. Note that the matrix obtained here is different in form from the similar matrix (3.275) given in Section 3.6.3. This happened for the reason that in Figs. 6.2 and Fig. 3.14 several different Euler angle systems are adopted. Indeed, as noted in [261], there may exist 12 different systems of Euler angles differing in the orientation of the line of nodes and the direction of reference of the angles with respect to the axes. Moreover, all systems are equivalent, and on the physics of phenomena this does not affect, and the choice of a particular system, as a rule, is due to the convenience of conducting mathematical calculations. Here, the system of Euler angles, somewhat different from that adopted in Section 3.6.3, was chosen, which was done in order to avoid further dashed notation when the cubic cell was rotated around the [111] axis. Let us show how the matrix (6.7) obtained here can be reduced to the form (3.275).

First of all, we note that according to the tradition established in analytic geometry [292, 293], the axes of the original coordinate system are usually denoted by letters without strokes, and the axes rotated are marked with strokes. The axes in Fig. 3.14 are makred like this fere. Here, the problem is solved in the *Oxyz* system, which rotates relative to the $Ox'y'z'$ system, therefore the notation is taken backwards.

At the same time from the comparison of Fig. 6.2 with Fig. 3.14 it can be seen that the notation used here for the axes Ox, Oy, Oz, and also Ox', Oy', Oz' are replaced by Fig. 3.14 to the following (given in the same order): Oz', Ox', Oy', as well as Ox, Oy, Oz. That is, writing (6.6) in the notation with the system in Fig. 3.14, we get

(6.8)

	x'	y'	z'
x	$-\frac{\sqrt{6}}{6}(\sin\delta+\sqrt{3}\cos\delta)$	$\frac{\sqrt{6}}{6}(\sqrt{3}\sin\delta-\cos\delta)$	$\frac{\sqrt{3}}{3}$
y	$-\frac{\sqrt{6}}{6}(\sin\delta-\sqrt{3}\cos\delta)$	$-\frac{\sqrt{6}}{6}(\sqrt{3}\sin\delta+\cos\delta)$	$\frac{\sqrt{3}}{3}$
z	$\frac{\sqrt{6}}{3}\sin\delta$	$\frac{\sqrt{6}}{3}\cos\delta$	$\frac{\sqrt{3}}{3}$

. (6

We present the columns of this table in the order of following the designations of the axes x', y', z':

	x'	y'	z'
x	$\frac{\sqrt{3}}{3}$	$-\frac{\sqrt{6}}{6}\left(\sin\delta+\sqrt{3}\cos\delta\right)$	$-\frac{\sqrt{6}}{6}\left(\sqrt{3}\sin\delta-\cos\delta\right)$
y	$\frac{\sqrt{3}}{3}$	$-\frac{\sqrt{6}}{6}\left(\sin\delta-\sqrt{3}\cos\delta\right)$	$-\frac{\sqrt{6}}{6}\left(\sqrt{3}\sin\delta+\cos\delta\right)$
z	$\frac{\sqrt{3}}{3}$	$\frac{\sqrt{6}}{3}\sin\delta$	$\frac{\sqrt{6}}{3}\cos\delta$

(6.9)

Since here the shaded and non-hatched systems are reversed, we should take the inverse matrix, that is, replace the rows with columns:

	x	y	z
x'	$-\frac{\sqrt{6}}{6}(\sin\delta+\sqrt{3}\cos\delta)$	$-\frac{\sqrt{6}}{6}(\sin\delta-\sqrt{3}\cos\delta)$	$\frac{\sqrt{6}}{3}\sin\delta$
y'	$\frac{\sqrt{6}}{6}(\sqrt{3}\sin\delta-\cos\delta)$	$-\frac{\sqrt{6}}{6}(\sqrt{3}\sin\delta+\cos\delta)$	$\frac{\sqrt{6}}{3}\cos\delta$
z'	$\frac{\sqrt{3}}{3}$	$\frac{\sqrt{3}}{3}$	$\frac{\sqrt{3}}{3}$

(6.10)

We now take into account that from the comparison of Fig. 6.2 and Fig. 3.14 it can be seen that the angles ε and δ differ by 180°, i.e.

$$\delta = \varepsilon + 180°,$$

(6.11)

so that

$$\sin \delta = -\sin \varepsilon ; \qquad (6.12)$$

$$\cos \delta = -\cos \varepsilon. \qquad (6.13)$$

Making corresponding substitutions in (6.10), we obtain:

	x	y	z
x'	$\dfrac{\sqrt{6}}{6}(\sin \varepsilon + \sqrt{3}\cos \varepsilon)$	$\dfrac{\sqrt{6}}{6}(\sin \varepsilon - \sqrt{3}\cos \varepsilon)$	$-\dfrac{\sqrt{6}}{3}\sin \varepsilon$
y'	$-\dfrac{\sqrt{6}}{6}(\sqrt{3}\sin \varepsilon - \cos \varepsilon)$	$\dfrac{\sqrt{6}}{6}(\sqrt{3}\sin \varepsilon + \cos \varepsilon)$	$-\dfrac{\sqrt{6}}{3}\cos \varepsilon$
z'	$\dfrac{\sqrt{3}}{3}$	$\dfrac{\sqrt{3}}{3}$	$\dfrac{\sqrt{3}}{3}$

$$(6.14)$$

It can be seen that the table obtained in this way, up to a change in the order of the terms in brackets, completely coincides with the table of the matrix (3.275) obtained in the system adopted in Fig. 3.14, which proves the equivalence of both systems.

6.1.2. The transformation matrix when rotating around the [111] axis

Thus, according to the accepted geometry (Fig. 6.2), the rotation of the cubic cell by the angle δ around the [111] axis, which coincides with the normal to the plate plane, is described by a matrix of the form (6.7)

$$\vec{A}_{111}(\delta) = \begin{pmatrix} \dfrac{\sqrt{3}}{3} & -\dfrac{\sqrt{6}}{6}(\sin \delta + \sqrt{3}\cos \delta) & \dfrac{\sqrt{6}}{6}(\sqrt{3}\sin \delta - \cos \delta) \\ \dfrac{\sqrt{3}}{3} & -\dfrac{\sqrt{6}}{6}(\sin \delta - \sqrt{3}\cos \delta) & -\dfrac{\sqrt{6}}{6}(\sqrt{3}\sin \delta + \cos \delta \\ \dfrac{\sqrt{3}}{3} & \dfrac{\sqrt{6}}{3}\sin \varepsilon & \dfrac{\sqrt{6}}{3}\cos \varepsilon \end{pmatrix}, \qquad (6.15)$$

where the angle δ is between the axis Oz and the projection on the plane of the plate axis $\left[\bar{1}\bar{1}1\right]$.

In this case, the conversion of the magnetization vector using such a matrix occurs in accordance with the formula

$$\mathbf{M}' = \vec{A}_{111}(\delta)\mathbf{M}. \qquad (6.16)$$

Recall that the direction of the constant field $\mathbf{H}$ in the geometry under consideration (Fig. 6.1) is parallel to the Oz axis.

In general, the full transformation is rather cumbersome, so we confine ourselves to two special cases that are important for practice – when the field is parallel to the projection of the [111] axis on the plate plane, that is, δ = 0°, and also when the field is perpendicular to such a projection, that is, δ = 30°. Consider these cases sequentially.

6.1.3. Equations of motion at a field parallel to the projection axis [111]

We first consider the case when the field is parallel to the projection of the [111] axis on the plane of the plate, that is, δ = 0°. The transformation matrix takes the form:

$$
\overleftrightarrow{A}_{111}(0°) = \left(
\begin{array}{c|c|c}
\dfrac{\sqrt{3}}{3} & -\dfrac{\sqrt{2}}{2} & -\dfrac{\sqrt{6}}{6} \\
\hline
\dfrac{\sqrt{3}}{3} & -\dfrac{\sqrt{2}}{2} & \dfrac{\sqrt{6}}{6} \\
\hline
\dfrac{\sqrt{3}}{3} & 0 & \dfrac{\sqrt{6}}{3}
\end{array}
\right).
\tag{6.17}
$$

Performing the transformation in accordance with (6.16), we obtain

$$
M_{x'} = \frac{\sqrt{3}}{3} M_x - \frac{\sqrt{2}}{2} M_y - \frac{\sqrt{6}}{6} M_z;
\tag{6.18}
$$

$$
M_{y'} = \frac{\sqrt{3}}{3} M_x + \frac{\sqrt{2}}{2} M_y - \frac{\sqrt{6}}{6} M_z;
\tag{6.19}
$$

$$
M_{z'} = \frac{\sqrt{3}}{3} M_x + \frac{\sqrt{6}}{3} M_z.
\tag{6.20}
$$

Substituting these expressions into the energy density (6.4), we get

$$
U = \frac{2\pi M_0^2 - K}{M_0^2} M_x^2 - M_z H_0 - \frac{K_1}{12 M_0^4} \left(4 M_x^4 + 3 M_y^4 + 3 M_z^4 + \right.
$$

$$
\left. + 6 M_y^2 M_z^2 - 4\sqrt{2} M_x M_z^3 + 12\sqrt{2} M_x M_y^2 M_z \right).
\tag{6.21}
$$

We find the effective fields by differentiating U by the composition to magnets in accordance with (1.2):

$$H_{ex} = -\frac{2\left(2\pi M_0^2 - K\right)}{M_0^2} M_x + \frac{K_1}{3M_0^4}\left(4M_x^3 - \sqrt{2}M_z^3 + 3\sqrt{2}M_y^2 M_z\right);$$

(6.22)

$$H_{ey} = \frac{K_1}{M_0^4}\left(M_y^3 + M_y M_z^2 + 2\sqrt{2}M_x M_y M_z\right);$$

(6.23)

$$H_{ez} = H_0 + \frac{K_1}{M_0^4}\left(M_z^3 + M_y^2 M_z - \sqrt{2}M_x M_z^2 + \sqrt{2}M_x M_y^2\right);$$

(6.24)

Since further it is assumed linearization, that is, the components M_x and M_y will be small compared to M_z, and $M_z \rightarrow M_0$, it suffices to consider the equations of motion only for the components M_x and M_y. These equations, obtained from the full Landau–Lifshitz equation (2.20)–(2.22) without attenuation (i.e., at $\alpha = 0$) have the form

$$\frac{\partial M_x}{\partial t} = -\gamma\left(M_y H_{ez} - M_z H_{ey}\right);$$

(6.25)

$$\frac{\partial M_y}{\partial t} = -\gamma\left(M_z H_{ez} - M_x H_{ez}\right);$$

(6.26)

Substituting the effective fields (6.22)–(6.24) into these equations, we obtain:

$$\frac{\partial M_x}{\partial t} = -\gamma\left\{M_y\left[H_0 + \frac{K_1}{M_0^4}\left(M_z^3 + M_y^2 M_z - \sqrt{2}M_x M_z^2 + \sqrt{2}M_x M_y^2\right)\right] - \right.$$

$$\left. -M_z\frac{K_1}{M_0^4}\left(M_y^3 + M_y M_z^2 + 2\sqrt{2}M_x M_y M_z\right)\right\};$$

(6.27)

$$\frac{\partial M_y}{\partial t} = -\gamma \Bigg\{ M_z \Bigg[-\frac{2\left(2\pi M_0^2 - K\right)}{M_0^2} M_x +$$

$$+ \frac{K_1}{3 M_0^4} \left(4 M_x^3 - \sqrt{2} M_z^3 + 3\sqrt{2} M_y^2 M_z \right) \Bigg] -$$ (6.28)

$$- M_x \Bigg[H_0 + \frac{K_1}{M_0^4} \left(M_z^3 + M_y^2 M_z - \sqrt{2} M_x M_z^2 + \sqrt{2} M_x M_y^2 \right) \Bigg] \Bigg\}.$$

6.1.4. Two ways to linearize the equations

The linearization of the obtained equations can be performed in two ways. The classical method requires the preliminary determination of the equilibrium state of magnetization, the small amount of deviations from which allows one to neglect quadratic terms, as a result of which the equations become linear. Another way is to solve the equations in their original form, after which the resulting constant terms are assumed to be zero, which again leads to a linear form of the equations. Consider the general schemes of these methods in more detail.

Method one. With the usual linearization, that is, if we set $M_x = m_x$, $M_y = m_y$, $M_z = M_0$, where $m_x \ll M_0$, $m_y \ll M_0$, then in the second equation we get a constant term. It will work where M_z is multiplied by M_z^3, that is, in the internal bracket of the first addend, which, when linearized, gives M_z^4. In the first equation, such a product is absent, so a constant term is not formed here. Obviously, the derivative of the magnetization with respect to time cannot be constant, for this would mean an infinite increase in the magnetization time. It can be assumed that the presence of a constant term in the second equation indicates that the vector of magnetization does not lie in the equilibrium state along the Oz axis, that is, along the ferromagnetic resonance in the plate field, and out of the plane of the plate. This output is due to the asymmetry of the arrangement of [111] axes relative to the plate plane. The field is applied along the projection of one such monitor, and this axis is located on one side of the plane of the plate. On the other hand, two axes, symmetrically with respect to the field direction, are located, the projections of which on the plate plane are at angles of 60° relative to the projection of the first axis. The location of the axes can be seen in detail from Fig. 3.10. Thus, the magnetization vector in one direction from the plate surface pulls one axis of anisotropy, and in the other two, so that these three forces in the plane of the plate are not balanced. Therefore, never, even with a very large field, the vector of magnetization falls on the plane of the plate. From the symmetry of the arrangement of the axes relative to the plane Oyz, it follows that in the equilibrium state $M_y^{(0)} = 0$, however $M_x^{(0)} \neq 0$.

Generally speaking, in this case, that is, with $M_x^{(0)} \neq 0$, one must first find a stable equilibrium state of the magnetization vector, based on the minimum of the potential, that is, by differentiating the energy density (6.21) by the components of magnetization, in this case by M_x (the detailed procedure for the spherical coordinates θ and φ is discussed further in Section 7.3). Moreover, with respect to $M_x^{(0)}$, a third degree equation is obtained. After solving this equation, the coordinate system should be rotated so that the Oz axis is directed along the resulting equilibrium state of magnetization. Then one can carry out the linearization in the usual way, which will allow one to find the resonant frequency of the precession of the magnetization in the attached field.

The second way. But there is another way. Thus, it is possible to perform linearization right away, assuming that the equilibrium state of magnetization has already been found, and the resulting equation for M_x is set to zero. Again, one get a third-degree equation, but here one can do without first differentiating the energy density with respect to M_x. Therefore, this path is somewhat shorter than the traditional one, and the result is the same. Next this second path is selected, that is, one sets:

$$M_x = M_x^{(0)} + m_x; \qquad (6.29)$$
$$M_y = m_x; \qquad (6.30)$$
$$M_z = M_0. \qquad (6.31)$$

where $m_x \ll M_0$, $m_y \ll M_0$.

At the same time, in the resulting equations, the value $M_x^{(0)}$ is considered to have the same order as M_0, so: $m_x \ll M_x^{(0)}$ and $m_y \ll M_x^{(0)}$.

The powers of the M_x components are converted as follows:

$$M_x^2 \rightarrow \left(M_x^{(0)}\right)^2 + 2M_x^{(0)} m_x; \qquad (6.32)$$

$$M_x^3 \rightarrow \left(M_x^{(0)}\right)^3 + 3\left(M_x^{(0)}\right)^2 m_x; \qquad (6.33)$$

Comment. It should be borne in mind that here m_x and m_y denote small additions to the total length of the magnetization vector, whereas in Chapter 4, and also in Chapters 9 and 10, the normalized components of magnetization, that is, complete, divided by M_0 are indicated in the same letters. Such a different use of the same designations is largely a tribute to the historical tradition. So, in linear problems, normalization is usually not carried out, and small letters, because of their small size, they are used to designate small additives [6–8]. On the other hand, when considering nonlinearity, it is more convenient to use exactly the normalized components, since there (including in Chapters 4, 9, 10) the smallness of the change in magnetization is still not assumed, and the introduction of normalization reduces the bulkiness of the expressions obtained [264].

The physical essence of the constant term. Let's go back to linearization. Since the constant term is formed only in the second equation, we turn to its more detailed consideration. Substituting (6.29)–(6.31) into (6.28), taking into account (6.32) and (6.33) and leaving only the linear in m_x terms, we get:

$$\frac{\partial m_y}{\partial t} = -\gamma \left\{ -\frac{2(2\pi M_0^2 - K)}{M_0} M_x^{(0)} - \frac{2(2\pi M_0^2 - K)}{M_0} m_x + \right.$$
$$+ \frac{4K_1}{3M_0^4}(M_x^{(0)})^3 + \frac{4K_1}{M_0^4}(M_x^{(0)})^2 m_x - \frac{\sqrt{2}\,K_1}{3} -$$
$$- \underline{H_0 M_x^{(0)}} - \underline{\frac{K_1}{M_0} M_x^{(0)}} + \frac{\sqrt{2}\,K_1}{M_0^2}(M_x^{(0)})^2 + \frac{\sqrt{2}\,K_1}{M_0^2} M_x^{(0)} m_x -$$
$$\left. - H_0 m_x - \frac{K_1}{M_0} m_x + \frac{\sqrt{2}\,K_1}{M_0^2} M_x^{(0)} m_x. \right. \qquad (6.34)$$

This equation contains both variable terms dependent on m_x and constant terms (underlined) that do not depend on m_x. These constant terms give an equation to determine $M_x^{(0)}$ or $(M_x^{(0)}/M_0)$ in the form of:

$$\frac{4K_1}{3}\left(\frac{M_x^{(0)}}{M_0}\right) + \sqrt{2}K_1\left(\frac{M_x^{(0)}}{M_0}\right)^2 -$$
$$- \left[H_0 + \frac{2(2\pi M_0^2 - K)}{M_0} + \frac{K_1}{M_0}\right]\left(\frac{M_x^{(0)}}{M_0}\right) - \frac{\sqrt{2}K_1}{3} = 0. \qquad (6.35)$$

We introduce the notation:

$$H_a = \frac{2K}{M_0}; \qquad (6.36)$$

$$H_c = \frac{K_1}{M_0}. \qquad (6.37)$$

Comment. The first of these conventions H_a corresponds to the traditionally used concept of the field of uniaxial anisotropy [7, 8] (up to notation, defined by formulas (4.55), (4.69)). The quantity H_a is an effective field of the form (1.2) obtained by differentiating the variable component of the magnetization M_x of the energy density term (6.21) containing the constant K, in which, after differentiation, M_x is assumed to be M_0. Such the definition is justified by the fact that when a unlimited uniaxial medium is magnetized by a field perpendicular to the anisotropy axis, just at the field H_a, the magnetization vector falls exactly on the direction of the fields (section 5.1.1). The second of the notation H_c is the traditionally used field of cubic anisotropy (Section 4.3.2, formula (4.119)). Such a field is also obtained by differentiating the energy density according to one of the components of magnetization with the subsequent replacement of the remaining components by M_0, but the exact alignment of the magnetization vector along the field takes place only when some of the specific directions (for example, in Fig. 6.2, for $\delta = 30°$) are not perpendicular and one of the axes of cubic anisotropy, and the alignment field is also not equal to $2K_1/M_0$. On the other hand, when determining the constant K_1, the ratio K_1/M_0 is often used, for IYG it is 45 Oe [296] or 36 Oe [7] (Section 4.3.3), so the designation H_c introduced here should be understood in this way.

Separating equation (6.35) by M_0 and writing it down in notation (6.36)–(6.37) gives

$$\frac{4}{3}H_c\left(\frac{M_x^{(0)}}{M_0}\right)^3 + \sqrt{2}H_c\left(\frac{M_x^{(0)}}{M_0}\right)^2 -$$

$$-\left(H_0 + 4\pi M_0 - H_a + H_c\right)\left(\frac{M_x^{(0)}}{M_0}\right) - \frac{\sqrt{2}}{3}H_c = 0. \tag{6.38}$$

This is a cubic equation for the magnitude ($M_x^{(0)}/M_0$). The solution of such an equation for the case of a uniaxial magnetoelastic medium, performed by the Cardano method [261, 262], can be found, for example, in [297, 334,362]. A scheme of analytical solutions like equations is given in this monograph in Section 2.4. Here it will be assumed that the field H_0 is so large that the magnetization vector deviates from the direction of the field to such a small extent that $M_x^{(0)} \ll M_0$, so degrees above the first in this equation can be neglected. Solving the resulting equation of the first degree, we obtain

$$\frac{M_x^{(0)}}{M_0} = -\frac{\left(\sqrt{2}/3\right)H_c}{H_0 + 4\pi M_0 - H_a + H_c}. \tag{6.39}$$

This relation determines the equilibrium position of the magnetization vector for a given value of the field H_0. It can be seen that for a sufficiently large value of this field, the constant component of magnetization $M_x^{(0)}$ tends to zero, as it should be in accordance with the physical content of the phenomenon of magnetization turning in the direction of the field.

6.1.5. End result of linearization and solving linearized equations

We now turn to the dynamic side of the problem. Since the resulting value $M_x^{(0)}$ is determined from the condition of equality to zero of the constants terms in the equation for the component m_y (6.34), the remaining part of this equation determines the process oscillating in time, which is the precession of the magnetization. The equation for the m_x component is obtained from (6.27) by the same linearization, taking into account (6.29)–(6.33). The course of the calculations is similar to obtaining equation (6.34), and takes place without the formation of constant terms, so that we do not dwell on it in detail. In the end result, a system of linearized equations of the following form is obtained:

$$\frac{\partial m_x}{\partial t} = -\gamma\left(H_0 - 3\sqrt{2}H_c\frac{M_x^{(0)}}{M_0}\right)m_y; \tag{6.40}$$

$$\frac{\partial m_y}{\partial t} = \gamma\left[H_0 + 4\pi M_0 - H_a + H_c\left(1 - \sqrt{2}\frac{M_x^{(0)}}{M_0}\right)\right]m_x. \tag{6.41}$$

Assuming the dependence on time in the form of exp (iωt), we obtain the resonant frequency

$$\omega = \gamma\sqrt{\left(H_0 - 3\sqrt{2}H_c\frac{M_x^{(0)}}{M_0}\right) \times \\ \times\left[H_0 + 4\pi M_0 - H_a + H_c\left(1 - \sqrt{2}\frac{M_x^{(0)}}{M_0}\right)\right]} \tag{6.42}$$

It can be seen that in the absence of any anisotropy, that is, at $H_a = 0$ and $H_c = 0$, this expression transforms to the traditional Kittel formula for a tangently magnetized plate [6–8].

Definition of a practical parameter. If the frequency is known, for example, from the conditions of the experiment, then, provided that the field of cubic anisotropy is known, the expression (6.42) can be used to determine the quantity $4\pi M_0 - H_a$:

$$4\pi M_0 - H_a = \frac{\left(\dfrac{\omega}{\gamma}\right)^2}{H_0 - 3\sqrt{2}H\dfrac{M_x^{(0)}}{M_0}} - H_0 - H_c\left(1 - \sqrt{2}\frac{M_x^{(0)}}{M_0}\right). \tag{6.43}$$

This expression includes the quantity $M_x^{(0)}/M_0$, which is not known in advance, but can be determined from relation (6.39). However, this relationship includes the same desired value $4\pi M_0 - H_a$. Generally speaking, if in (6.43) we substitute $M_x^{(0)}/M_0$ in accordance with (6.39), then we get an equation for $4\pi M_0 - H_a$ of the third degree, a direct solution which will be quite cumbersome. If it is necessary to obtain a numerical solution, one can follow the path of successive approximations, for example, using the iteration method [261, 351, 352]. That is, in the first step we put the value of $M_x^{(0)}/M_0$ is equal to zero, substitute in equation (6.43), having decided to find $4\pi M_0 - H_a$ in the first approximation. Next, substitute the obtained value into (6.39), from where to find $M_x^{(0)}/M_0$ already in the second approximation. Then substitute this value in (6.43), find the next approximation of $4\pi M_0 - H_a$ and so on. In view of the smallness of $M_x^{(0)}/M_0$, it can be assumed that the convergence of such a process will be quite fast. Somewhat greater accuracy can be achieved if the term of the second order in $M_x^{(0)}/M_0$ is preserved in equation (6.38). In this case, the quantity $M_x^{(0)}/M_0$ will be determined by solving the corresponding quadratic equation. In the optimal case, the complete equation (6.38) can be solved, however, it must be taken into account that it is obtained by linearization, so it obviously contains some approximations, therefore the most accurate solution can be obtained only from consideration of a nonlinear problem.

Comment. The importance for the practice of the parameter $4\pi M_0 - H_a$ is determined by the fact that its knowledge makes it possible to estimate the uniaxial anisotropy field, especially in the case when it significantly exceeds the demagnetization field. Knowledge of the anisotropy field is important for the development of CMD devices [40]. On the other hand, the saturation magnetization M_0 included in this parameter can be determined by an independent method, for example, from the parameters of the domain structure [40], which allows one to determine the uniaxial anisotropy field in its pure form.

6.1.6. Solving a problem with a field perpendicular to the projection of the axis type [111]

We now consider the case when the field is perpendicular to the projection of the [111] axis on the plane of the plate, that is, $\delta = 30°$. The transformation matrix (6.15) takes the form

$$\ddot{A}_{111}(30°) = \begin{pmatrix} \dfrac{\sqrt{3}}{3} & \dfrac{\sqrt{3}}{3} & 0 \\ \dfrac{\sqrt{3}}{3} & \dfrac{\sqrt{6}}{6} & \dfrac{\sqrt{2}}{2} \\ \dfrac{\sqrt{3}}{3} & \dfrac{\sqrt{6}}{6} & \dfrac{\sqrt{2}}{2} \end{pmatrix} \tag{6.44}$$

Performing the transformation in accordance with (6.16), we obtain

$$M_{x'} = \frac{\sqrt{3}}{3} M_x - \frac{\sqrt{6}}{3} M_y; \tag{6.45}$$

$$M_{y'} = \frac{\sqrt{3}}{3} M_x - \frac{\sqrt{6}}{6} M_y - \frac{\sqrt{2}}{2} M_z; \tag{6.46}$$

$$M_{z'} = \frac{\sqrt{3}}{3} M_x - \frac{\sqrt{6}}{6} M_y + \frac{\sqrt{2}}{2} M_z; \tag{6.47}$$

Substituting these expressions into the energy density (6.4), we get

$$U = \frac{2\pi M_0^2 - K}{M_0^2} M_x^2 - M_z H - \frac{K_1}{12 M_0^4} \left(4M_x^4 + 3M_y^4 + 3M_z^4 + \right.$$

$$\left. + 6M_y^2 M_z^2 + 4\sqrt{2} M_x M_y^3 - 12\sqrt{2} M_x M_y M_z^2 \right). \tag{6.48}$$

We find the effective fields by differentiating U by the components of magnetization in accordance with (1.2):

$$H_{ex} = -\frac{2\left(2\pi M_0^2 - K\right)}{M_0^2}M_x + \frac{K_1}{3M_0^4}\left(4M_x^3 + \sqrt{2}M_y^3 - 3\sqrt{2}M_yM_z^2\right); \quad (6.49)$$

$$H_{ey} = \frac{K_1}{M_0^4}\left(M_y^3 + M_yM_z^2 + \sqrt{2}M_xM_y^2 - \sqrt{2}M_xM_z^2\right); \quad (6.50)$$

$$H_{ez} = H_0 + \frac{K_1}{M_0^4}\left(M_z^3 + M_y^2M_z - 2\sqrt{2}\,M_xM_yM_z\right). \quad (6.51)$$

Obtaining the equations of motion is completely analogous to the previous case. Here the arrangement of the axes of the [111] type with respect to the field direction is completely symmetrical, therefore $M_x^{(0)} = 0$ and one can use the usual linearization, that is, set $M_x = m_x$, $M_y = m_y$, $M_z = M_0$, where $m_x \ll M_0$, $m_y \ll M_0$. Substituting the effective fields (6.49)–(6.51) into the equations (6.25) and (6.26), after performing linearization, we obtain the equations of motion for the components of magnetization in the form(6.52)

$$\frac{\partial m_x}{\partial t} = -\gamma\left(H_0 m_y + \frac{\sqrt{2}K_1}{M_0}m_x\right); \quad (6.52)$$

$$\frac{\partial m_y}{\partial t} = \gamma\left\{\left[H_0 + \frac{2\left(2\pi M_0^2 - K\right)}{M_0} + \frac{K_1}{M_0}\right]m_x + \frac{\sqrt{2}K_1}{M_0}m_y\right\} \quad (6.53)$$

Substituting the notation for the fields H_a and H_c in accordance with (6.36) and (6.37), we obtain:

$$\frac{\partial m_x}{\partial t} = -\gamma\left(H_0 m_y + \sqrt{2}H_c m_x\right); \quad (6.54)$$

$$\frac{\partial m_y}{\partial t} = \gamma\left[\left(H_0 + 4\pi M_0 - H_a + H_c\right)m_x + \sqrt{2}H_c m_y\right]. \quad (6.55)$$

In these equations, the second terms in parentheses contain the same components of magnetization as the left side. When depending on the time of the form exp ($i\omega t$), they give the terms with different signs, so that when calculating the determinant we obtain the products of the form ($i\omega + A$) ($i\omega - A$), where $A = \gamma\sqrt{2}H_c$, which results in $-\omega^2 - A^2$, so that the imaginary is destroyed. As a result, we obtain the frequency in the form

$$\omega = \gamma\sqrt{H_0(H_0 + 4\pi M_0 - H_a + H_c) - 2H_c^2}. \quad (6.56)$$

Here, the last term under the root (taking into account the total coefficient γ) is exactly the value A^2.

It can be seen that at $H_c = 0$, which means the absence of cubic anisotropy, but the uniaxial preservation, formulas (6.56) and (6.42) coincide. When $H_a = 0$ and $H_c = 0$, that is, in the absence of anisotropy in general, the expression (6.56) goes into the traditional Kittel formula for a relatively magnetized plate [6–8].

Definition of a practical parameter. If the frequency is known, for example, from the conditions of the experiment, then this expression can be used to determine the $4\pi M_0$ value important for practice.

$$4\pi M_0 - H_a = \frac{1}{H_0}\left[\left(\frac{\omega}{\gamma}\right)^2 - H_0\left(H_0 + H_c\right) + 2H_c^2\right]. \quad (6.57)$$

6.1.7. Comparison of resonant frequencies for the examined orientations

From the comparison of the expressions for the resonant frequencies (6.42) and (6.56), corresponding to $\delta = 0°$ and $\delta = 30°$, it can be seen that in the first case both additives are proportional to $M_x^{(0)}/M_0$, are positive, since, in accordance with (6.39), the quantity $M_x^{(0)}/M_0$ is negative (in any case, for materials of the IYG type, where $4\pi M_0 \gg H_a$ and $K_1 > 0$). In the second case, the addition of $-2H_c^2$ is negative. At the same time, in both cases, the products of the subroots of the multipliers without these additives coincide. It follows from this that the resonance frequency at $\delta = 0°$ (6.42) exceeds the resonant frequency at $\delta = 30°$, that is, the relation is satisfied: $\omega|_{\delta = 0°} > \omega|_{\delta = 30°}$.

6.2. Normal magnetization

We now consider the case of normal magnetization, that is, when the external field is perpendicular to the plane of the plate. We assume that the uniaxial anisotropy field is larger than the demagnetization field, so that even in the absence of an external field, the equilibrium magnetization vector is oriented perpendicular to the plane of the plate. Since the [111] axis is also perpendicular to this plane, the cubic anisotropy does not deflect the magnetization vector from the normal. In this case, the application of an external field along the same normal does not change the orientation of the magnetization, that is, the magnetization vector for any field is always oriented

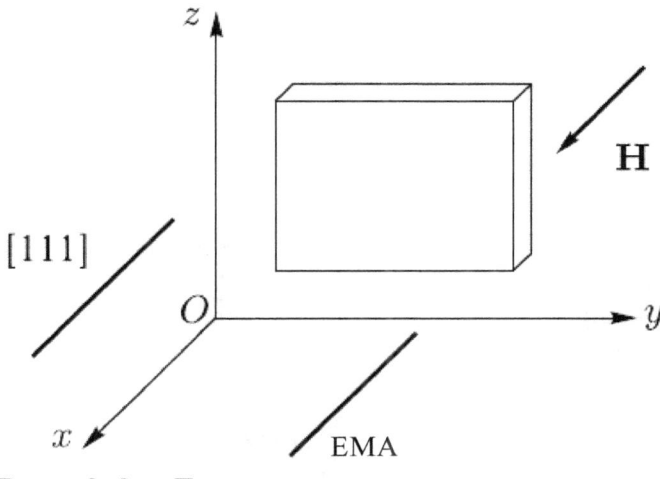

Fig. 6.3. Geometry of the problem in normal magnetization of a magnetic plate. EMA – easy magnetization axis.

perpendicular to the plane of the plate.

The geometry of the problem is shown in Fig. 6.3.

The coordinate system $Oxyz$ is oriented in such a way that the plane Oyz is parallel to the plane of the plate, and the axis Ox is perpendicular to it. The anisotropy axis is perpendicular to the plane of the plate, that is, directed along the axis Ox. The axis of cubic anisotropy of the type [111] will be assumed to be directed also perpendicular to the plane of the plate, that is, along the same axis Ox. That is, the orientation of the two coincides with that adopted in Section 6.1.1 and is due to technological practice [38, 40]. In the adopted geometry, the field vector has the form:

$$\mathbf{H} = \{H_0, 0, 0\}. \tag{6.58}$$

Since the field and the EMA are directed along the normal to the plane of the plate, by virtue of axial symmetry all directions in the plane of the plate are equivalent. Since in equilibrium the magnetization vector is also oriented along the same normal, when the cubic cell rotates around the normal, the energy density of the cubic anisotropy does not change. Therefore, the energy density can be taken in any convenient form, for example, in the form adopted in

Section 6.1.3 (formula (6.21)). In this case, the total energy density takes the form

$$U = \frac{2\pi M_0^2 - K}{M_0^2} M_x^2 - M_x M_0 - \frac{K_1}{12 M_0^4}(4M_x^4 + 3M_y^4 + 3M_z^4 +$$
$$+ 6M_y^2 M_z^2 - 4\sqrt{2}M_x M_z^3 + 12\sqrt{2}M_x M_y^2 M_z),$$

(6.59)

and the difference from (6.21) consists only in the second term, where instead of M_z there is M_x.

Differentiation of U according to the components of magnetization in accordance with (1.2) gives effective fields in the form:

$$H_{ex} = H_0 - \frac{2(2\pi M_0^2 - K)}{M_0^2} M_x + \frac{K_1}{3M_0^4}\left(4M_x^3 - \sqrt{2}M_z^3 + 3\sqrt{2}M_y^2 M_z\right);$$

(6.60)

$$H_{ey} = \frac{K_1}{M_0^4}\left(M_y^3 + M_y M_z^2 + 2\sqrt{2}M_x M_y M_z\right);$$

(6.61)

$$H_{ez} = \frac{K_1}{M_0^4}\left(M_z^3 + M_y^2 M_z + \sqrt{2}M_x M_y^2 - \sqrt{2}M_x M_z^2\right).$$

(6.62)

The difference from (6.22)–(6.24) consists only in moving H_0 from H_{ez} to H_{ex}. Since in the adopted geometry the equilibrium position of the magnetization vector is oriented along the axis Ox, then the precession of this vector also occurs around this axis, therefore it suffices to consider equations of motion only for the components M_y and M_z. In this case, instead of (6.25)–(6.26), from the full Landau–Lifshitz equation (2.20)–(2.22) without attenuation (that is, for $\alpha = 0$) we get:

$$\frac{\partial M_y}{\partial t} = -\gamma\left(M_z H_{ex} - M_x H_{ez}\right);$$

(6.63)

$$\frac{\partial M_z}{\partial t} = -\gamma\left(M_x H_{ey} - M_y H_{ex}\right);$$

(6.64)

Substituting the effective fields (6.60)–(6.62) into these equations, we obtain the equations of motion for M_y and M_z in the form:

$$\frac{\partial M_y}{\partial t} = -\gamma \left\{ M_z \left[H_0 - \frac{2(2\pi M_0^2 - K)}{M_0^2} M_x + \right. \right.$$

$$\left. + \frac{K_1}{3M_0^4} (4M_x^3 - \sqrt{2} M_z^3 + 3\sqrt{2} M_y^2 M_z) \right] - \qquad (6.65)$$

$$\left. - M_x \left[\frac{K_1}{M_0^4} (M_z^3 + M_y^2 M_z + \sqrt{2} M_x M_y^2 - \sqrt{2} M_x M_z^2) \right] \right\}.$$

$$\frac{\partial M_z}{\partial t} = -\gamma \left\{ M_x \left[\frac{K_1}{M_0^4} (M_y^3 + M_y M_z^2 + 2\sqrt{2} M_x M_y M_z) \right] - \right.$$

$$\left. - M_y \left[H_0 - \frac{2(2\pi M_0^2 - K)}{M_0^2} M_x + \frac{K_1}{3M_0^4} (4M_x^3 - \sqrt{2} M_z^3 + 3\sqrt{2} M_y^2 M_z) \right] \right\}. (6.66)$$

In the adopted geometry, the magnetization in the equilibrium state is aligned along the Ox axis; therefore, the constant terms in the equations for two or four equations can be obtained only proportional to M_x^4. It can be seen from the above equations that such terms are absent in both equations; therefore, we can perform the usual linearization, that is, put: $M_x = M_0$, $M_y = m_y$, $M_z = m_z$, where $m_y \ll M_0$, $m_z \ll M_0$.

In this case, the linearized equations of motion take the form:

$$\frac{\partial m_y}{\partial t} = -\gamma \left[H_0 - \frac{2(2\pi M_0^2 - K)}{M_0} + \frac{4K_1}{3M_0} \right] m_z; \qquad (6.67)$$

$$\frac{\partial m_z}{\partial t} = \gamma \left[H_0 - \frac{2(2\pi M_0^2 - K)}{M_0} + \frac{4K_1}{3M_0} \right] m_y; \qquad (6.68)$$

Introducing the notation (6.36), (6.37), we obtain:

$$\frac{\partial m_y}{\partial t} = -\gamma \left(H_0 - 4\pi M_0 + H_a + \frac{4}{3} H_c \right) m_z; \qquad (6.69)$$

$$\frac{\partial m_z}{\partial t} = \gamma \left(H_0 - 4\pi M_0 + H_a + \frac{4}{3} H_c \right) m_y; \qquad (6.70)$$

Assuming the dependence on time in the form of exp (iωt), we obtain the resonant frequency:

$$\omega = \gamma \left(H_0 - 4\pi M_0 + H_a + \frac{4}{3} H_c \right). \qquad (6.71)$$

It can be seen that for $H_a = 0$ and $H_c = 0$, this expression transforms into the traditional Kittel formula for a normally magnetized plate [6–8]. If the frequency is known, the formula (6.71) allows to obtain the value $4\pi M_0 - H_a$:

$$4\pi M_0 - H_a = H_0 + \frac{4}{3}H_a - \frac{\omega}{\gamma}. \tag{6.72}$$

6.3. Free ferromagnetic resonance with orientational transition

In the previous chapter, a quasistatic change in the orientation of the magnetization vector during the orientational transition was considered. We now turn to the dynamic properties of the precession of magnetization during such a transition. At the first stage, we consider free ferromagnetic resonance, which will allow us to find the change in the resonance frequency of free precession during the transition. Then we turn to stimulated ferromagnetic resonance, which will allow us to determine the dynamic magnetic susceptibility tensor. As an auxiliary task, we first consider the transformation of the magnetization vector during rotation of the coordinate axes corresponding to the orientational transition.

6.3.1. Magnetization vector transformation when rotating the coordinates

The orientational transition consists in changing the orientation of the magnetization vector. The problem of ferromagnetic resonance, as a rule, is solved in a coordinate system related to the equilibrium position of the magnetization vector, which makes it possible to use the linearization procedure, which greatly simplifies the calculations. In this case, since in the process of orientational transition the magnetization vector rotates, then the associated coordinate system in which the calculation is performed, also has to rotate. With this rotation, a certain transformation of the components of magnetization occurs, the consideration of which constitutes the content of this section. So, let us consider the transformation of the components of the magnetization vector when the coordinate system is rotated. We will follow the general presentation of a similar problem in the traditional forms of analytic geometry [292, 293]. The geometry for the particular case considered here is shown in Fig. 6.4.

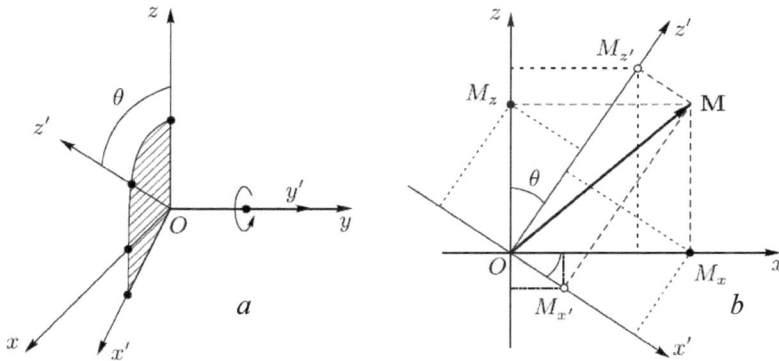

Fig. 6.4. Geometry of the problem on the rotation of the coordinate system associated with magnetization. a – scheme of rotation of the coordinate system at an angle θ around the axis Oy, the hatched plane contains the axes Ox, Oz, Ox', Oz'. b – diagram of the formation of the components of the magnetization vector, the view in the shaded plane along the positive direction of the axis Oy.

Figure 6.4 a shows a diagram of the rotation of the original coordinate system $Oxyz$ around the axis Oy at an angle θ. As a result of this rotation, the original $Oxyz$ system went into the $Ox'y'z'$ System, so that the angle between the Ox and Ox axes, as well as the Oz and Oz' is equal to θ. The Oy axis goes directly to the Oy' axis.

Figure 6.4 b shows a diagram of the transformation component of the magnetization vector with such rotation of the axes. Based on elementary geometric considerations, it can be seen that the direct transformation of the components is:

$$M_x = M_{x'}\cos\theta + M_{z'}\sin\theta, \tag{6.73}$$

$$M_z = -M_{x'}\sin\theta + M_{z'}\cos\theta. \tag{6.74}$$

The corresponding inverse transform is:

$$M_{x'} = M_x\cos\theta - M_z\sin\theta; \tag{6.75}$$

$$M_{z'} = M_x\sin\theta + M_z\cos\theta; \tag{6.76}$$

In these both cases, the y-component does not change, so that $M_y = M_{y'}$ or $M_{y'} = M_y$.

We write the transformation using the rotation matrices. Direct conversion has the form

$$\mathbf{M} = \ddot{A}_2 \mathbf{M}'.$$ (6.77)

where the index 2 corresponds to two dimensions of the problem, and the rotation matrix itself has the form

$$\ddot{A}_2 = \begin{pmatrix} \cos\theta & \sin\theta \\ -\sin\theta & \cos\theta \end{pmatrix}$$ (6.78)

The inverse transformation is performed by the formula

$$\mathbf{M}' = \ddot{A}_2^{-1} \mathbf{M}.$$ (6.79)

and the rotation matrix has the form

$$\ddot{A}_2^{-1} = \begin{pmatrix} \cos\theta & -\sin\theta \\ \sin\theta & \cos\theta \end{pmatrix}$$ (6.80)

The direct and inverse transformations in the three-dimensional case are determined by the formulas:

$$\mathbf{M} = \ddot{A}_3 \mathbf{M}';$$ (6.81)

$$\mathbf{M}' = \ddot{A}_3^{-1} \mathbf{M},$$ (6.82)

where the corresponding matrices are:

$$\ddot{A}_3 = \begin{pmatrix} \cos\theta & 0 & \sin\theta \\ 0 & 1 & 0 \\ \sin\theta & 0 & \cos\theta \end{pmatrix}$$ (6.83)

$$\ddot{A}_3 = \begin{pmatrix} \cos\theta & 0 & -\sin\theta \\ 0 & 1 & 0 \\ \sin\theta & 0 & \cos\theta \end{pmatrix}$$ (6.84)

Comment. It should be noted that in traditional lectures, for example [292, 293], in the formulas like (6.73)–(6.76), the sine has a sign opposite to that given. This happens for the reason that usually the direction of rotation of the axes of the coordinates is counted in positive direction, that is, in Fig. 6.4 *b* counterclockwise. Here, the rotation takes place clockwise, so that the angle θ is counted as if in a negative direction, which leads to a change in its sign.

The reason for this direction of rotation of the axes is that this task is preparatory for considering the orientational transition of the magnetization vector, which in the chosen geometry occurs from the axis Oz to the axis Ox (the magnetization vector rotates from the EMA in the field direction), that is, the angle θ changes in the opposite direction to the traditional one.

6.3.2. Features of ferromagnetic resonance conditions in the full field change interval

In the previous sections 6.1 and 6.2, we considered ferromagnetic resonance in an anisotropic medium with the orientation of the external field in some specific directions relative to the anisotropy axes. In this case, the applied field was assumed to be so much larger than the anisotropy field that in the stationary state the magnetization vector was oriented along this field or close to its direction.

The nature of the phenomenon in the case when the applied field is less than the anisotropy field is significantly different from that when such a field exceeds the anisotropy field. The difference is due to the fact that in the first case the magnetization vector is in a state of an orientational transition, and its direction changes with a change in the field, while in the second case the magnetization vector is oriented exactly along the field and its direction is constant. Consider further the full interval of the applied field from zero to large values in more detail.

6.3.3. The applied field is less than the anisotropy field

Let us first consider the case when the field is insufficient for a complete rotation of the magnetization vector to its direction. The sections 5.1.1, 5.1.2 show that in this case the magnetization vector undergoes an orientational transition from the direction determined by the anisotropy of the medium to the direction of the field. For simplicity, we limit ourselves to the case of uniaxial anisotropy in an unbounded environment, since the introduction of the cubic, due to the multiplicity of axes, leads to a significant complication of both the physical situation and the computational side of the problem. The field will be assumed to be directed perpendicular to the anisotropy axis.

The orientational transition occurs in the interval of increasing the field from zero to the anisotropy field. In a zero field, the

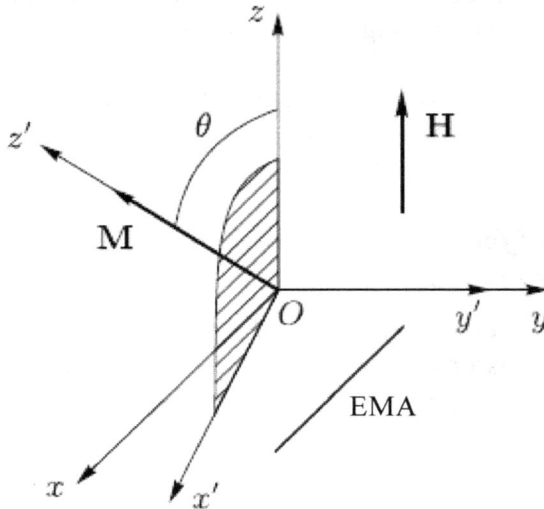

Fig. 6.5. Geometry of the problem on the rotation of the coordinate system associated with magnetization. The shading shows a single plane $Ox'xz'z$.

magnetization vector is oriented along the anisotropy axis. As the field increases, the magnetization vector rotates in the direction of the field. In a field equal to the anisotropy field, the magnetization vector is aligned along the field and then remains in this position, so that the orientational transition ends. Rotation of the magnetization vector occurs in one plane, passing through the axis of anisotropy and the direction of the field.

The geometry of the problem is shown in Fig. 6.5.

The basic coordinate system $Oxyz$ is oriented so that the axis Ox coincides with the axis of anisotropy, and the axis Oz coincides with the direction of the field. Auxiliary coordinate system $Ox'y'z'$ is oriented so that the Oz' axis is directed along the equilibrium direction of the magnetization vector, and the axis Ox' lies in the plane passing through the axes Oz and Oz'. With this the axes Oy and Oy' of both coordinate systems coincide with each other. So the axes are Ox, Oz, Ox' and Oz' lie in the same plane perpendicular to the Oy and Oy' axes, and when the field changes, leading to rotation of the magnetization vector in this plane, the $Ox'y'z'$ system, following the magnetization, rotates relative to the $Oxyz$ system around the Oy or Oy' axis. The angle of rotation between the axes Oz and Oz' is denoted by θ. Obviously, the angle between the axes Ox and Ox' also equal to θ.

For convenience of terminology, we will call the system $Oxyz$ – associated with the field, and the system $Ox'y'z'$ – related to magnetization. Ferromagnetic resonance is the precession of the magnetization vector around its equilibrium position. It means, that the precession problem (the Landau–Lifshitz equation) must be solved in a system related to magnetization, that is, $Ox'y'z'$. This approach with a small amplitude of oscillations allows linearization of the task, greatly simplifying the calculations. However, since the energy densities of anisotropy and the interaction of a field with magnetization are usually recorded in a system associated with the field, that is, $Oxyz$, then you first need to transfer all the energy density to the system $Ox'y'z'$. The total energy density in the coordinate system associated with the field is

$$U = -\frac{K}{M_0^2} M_x^2 - \mathbf{MH}, \tag{6.85}$$

where the first term corresponds to the anisotropy energy, and the second to the interaction energy of magnetization with the field.

In order to record the energy density in the system $Ox'y'z'$, it is necessary to convert the components of magnetization in (6.85) and field from the system $Oxyz$ into the system $Ox'y'z'$. The components along the axis Oy do not change, so it suffices to consider only the components along the axes Ox and Oz. To transform the component of magnetization M_x entering the first term (6.85), we use formula (6.77) with matrix (6.78), as a result of which we obtain

$$M_x = M_{x'} \cos\theta + M_{z'} \sin\theta. \tag{6.86}$$

The second term in (6.85) is a scalar product, independent of the choice of the coordinate system, so we write it down immediately in the system $Ox'y'z'$:

$$\mathbf{MH} = M_{x'} H_{x'} + M_{y'} H_{y'} + M_{z'} H_{z'}. \tag{6.87}$$

The field in the $Oxyz$ system is

$$\mathbf{H} = \{0, 0, H_0\} \tag{6.88}$$

that is, $H_x = 0$, $H_y = 0$, $H_z = H_0$. Expressing a field in the system $Oxy'z'$ in accordance with formula (6.79) using the matrix (6.80), gives

$$H_{x'} = H_x \cos\theta - H_z \sin\theta = -H_0 \sin\theta \qquad (6.89)$$
$$H_{y'} = H_y = 0 \qquad (6.90)$$
$$H_Z = H_z \sin\theta + H_z \cos\theta = H_0 \cos 0 \qquad (6.91)$$

Substituting in (6.87), we obtain

$$\mathbf{MH} = M_z H_0 \sin\theta + M_z H_0 \cos 0 \qquad (6.92)$$

Using (6.86) and (6.92), we obtain the total energy density in the form

$$U = \frac{K}{M_0^2}\left(M_{x'}^1 \cos^2\theta + M_{z'}^2 \sin^2\theta + 2M_{x'}M_{z'}\sin\theta\cos\theta\right) +$$
$$M_{x'}H_0 \sin\theta - M_{x'}H_0 \cos\theta, \qquad (6.93)$$

By opening the parentheses and ordering by variables and exponents, we get

$$U = -\frac{K}{M_0^2}\cos^2\theta M_x^2 - \frac{K}{M_0^2}\sin^2\theta M_{z'}^2 - \frac{2K}{M_0^2}\sin\theta\cos\theta M_{x'}M_{z'} +$$
$$+ H_0 \sin\theta M_{x'} - H_0 \cos\theta M_{z'}. \qquad (6.94)$$

We introduce auxiliary notation:

$$A_1 = -\frac{K}{M_0^2}\cos^2\theta; \qquad (6.95)$$
$$A_2 = -\frac{K}{M_0^2}\sin^2\theta; \qquad (6.96)$$
$$A_3 = -\frac{2K}{M_0^2}\sin\theta\cos\theta; \qquad (6.97)$$
$$A_4 = H_0 \sin\theta; \qquad (6.98)$$
$$A_5 = -H_0 \cos\theta. \qquad (6.99)$$

With these designations, the energy density (6.94) takes the form:

$$U = A_1 M_{x'}^2 + A_2 M_{z'}^2 + A_3 M_{x'}M_{z'} + A_4 M_{x'} + A_5 M_{z'}. \qquad (6.100)$$

In equilibrium, the effective field along the axis Ox' must be equal to zero that is

$$H_{ex'}^{(0)} = -\frac{\partial U}{\partial M_{x'}} = 0. \tag{6.101}$$

Differentiating (6.100) and equating the derivative to zero, we get

$$2A_1 M_{x'} + A_3 M_{z'} + A_4 = 0. \tag{6.102}$$

In equilibrium it should be: $M_{x'} = 0$, $M_{y'} = 0$, $M_{z'} = M_0$. Substituting these values into (6.102), we get

$$A_3 M_0 + A_4 = 0. \tag{6.103}$$

Substituting A_3 and A_4 in accordance with (6.97) and (6.98), we obtain the equation for determining the angle θ in the equilibrium state:

$$-\frac{2K}{M_0}\cos\theta + H_0 = 0, \tag{6.104}$$

whence, using in accordance with (6.36) the designation $H_a = 2K/M_0$, we obtain:

$$\cos\theta = \frac{H_0}{H_a}; \tag{6.105}$$

$$\sin\theta = \frac{\sqrt{H_a^2 - H_0^2}}{H_a}. \tag{6.106}$$

The formula (6.105), up to notation, coincides with the formula (5.23) obtained in Chapter 5, taking into account (5.14). The observed difference is due to the opposite choice of the orientation of the axes of coordinates relative to the field and the EMA, and also the reading of the angle θ in one case from the direction of the field, and in the other from the direction of the EMA.

Comment. From equation (6.104), only formula (6.105) is obtained, that is, for cos θ, and formula (6.106) is obtained from the equality of the unit of the sum of squares of sine and cosine. In this case, the sine can have two signs, that is, in addition to (6.106), the relation

$$\sin\theta = -\frac{\sqrt{H_a^2 - H_0^2}}{H_a}. \qquad (6.107)$$

In this case, the projection of the magnetization vector on the Ox axis is negative, and the vector itself is located in the Oxz plane symmetrically with respect to the Oz axis. This situation occurs if in the initial position in the absence of the field the magnetization vector is oriented in the negative direction of the Ox axis along the same EMA, the positive and negative directions along which are equal. As the field increases, the orientational transition occurs in the same Oxz plane, but the magnetization vector begins its movement from the negative direction of the Ox axis. This situation takes place in domains where initially there are two opposite directions of the magnetization vector. In the present review this question of principal importance, therefore here we confine ourselves only to the positive initial direction of the magnetization vector with respect to the axis Ox, that is, to determine θ, we will use the formula (6.106). Replacing the sine and cosine of the angle θ with the help of (6.105), (6.106), and also using the notation $H_a = 2K/M_0$ (6.36), we get the expressionsfor the parameters A_{1-5} (6.95)–(6.99) through the fields:

$$A_1 = -\frac{H_0^2}{2M_0 H_a}; \qquad (6.108)$$

$$A_2 = -\frac{H_a^2 - H_0^2}{2M_0 H_a}; \qquad (6.109)$$

$$A_3 = -\frac{H_0\sqrt{H_a^2 - H_0^2}}{M_0 H_a}; \qquad (6.110)$$

$$A_4 = -\frac{H_0\sqrt{H_a^2 - H_0^2}}{H_a}; \qquad (6.111)$$

$$A_5 = -\frac{H_0^2}{H_a}. \qquad (6.112)$$

Differentiating the expression for the energy density (6.100), we find the effective fields:

$$H_{ex'} = -\left(2A_1 M_{x'} + A_3 M_{z'} + A_4\right); \qquad (6.113)$$

$$H_{ey'} = 0; \tag{6.114}$$

$$H_{ez'} = -\left(A_3 M_{x'} + 2A_2 M_{z'} + A_5\right). \tag{6.115}$$

The equations of motion for magnetization obtained from the disclosure of the vector product in the Landau–Lifshitz equation (2.20)–(2.22) in the system $Ox'y'z'$ have the form:

$$\frac{\partial M_{x'}}{\partial t} = -\gamma\left(M_{y'} H_{ez'} - M_{z'} H_{ey'}\right); \tag{6.116}$$

$$\frac{\partial M_{y'}}{\partial t} = -\gamma\left(M_{z'} H_{ex'} - M_{x'} H_{ez'}\right); \tag{6.117}$$

$$\frac{\partial M_{z'}}{\partial t} = -\gamma\left(M_{x'} H_{ey'} - M_{y'} H_{ex'}\right); \tag{6.118}$$

Substituting the fields (6.113)–(6.115) and ordering by variables, we get:

$$\frac{\partial M_{x'}}{\partial t} = \gamma\left\{A_3 M_{x'} M_{y'} + 2A_2 M_{y'} M_{z'} + A_{5M_{y'}}\right\}; \tag{6.119}$$

$$\frac{\partial M_{y'}}{\partial t} = -\gamma\left\{A_3 M_{x'}^2 M_{y'} + A_3 M_{z'}^2 + 2\left(A_2 - A_1\right)M_{x'}M_{z'} - A_4 M_{z'}\right\}; \tag{6.120}$$

$$\frac{\partial M_{z'}}{\partial t} = -\gamma\left\{2A_1 M_{x'} M_{y'} + A_3 M_{y'} M_{z'} + A_4 M_{y'}\right\} \tag{6.121}$$

For linearization, we set: $M_{x'} = m_{x'}$, $M_{y'} = m_{y'}$, $M_{z'} = M_0$, where is $m_{x'} \ll M_0$, $m_{y'} \ll M_0$. Substituting these variables into (6.119) - (6.121) and leaving the terms of only zero and first orders in $m_{x'}$ and $m_{y'}$ we get:

$$\frac{\partial m_{x'}}{\partial t} = \gamma\left(2A_2 M_0 + A_5\right) m_{y'}; \tag{6.122}$$

$$\frac{\partial m_{y'}}{\partial t} = -\gamma\left\{\left[2\left(A_2 - A_1\right)M_0 + A_5\right]m_{x'} - \frac{\left(A_3 M_0 - A_4\right)M_0}{}\right\}; \tag{6.123}$$

$$\frac{\partial m_{z'}}{\partial t} = -\left(A_3 M_0 + A_4\right) m_{y'}. \tag{6.124}$$

It can be seen that here in the second equation we got a constant term (underlined). In Section 6.1.4, it is shown that the appearance of constants terms in such equations means that the equilibrium position of the magnetization vector is not taken into account. For determining equilibrium position these components must be set equal to zero. At the same time we get

$$A_3 M_0 + A_4 = 0. \tag{6.125}$$

It can be seen that this relation is identical (6.103), from which expressions (6.105), (6.106) are obtained, which determine the value of the angle θ of the equilibrium position of the magnetization. When taking into account the relation (6.125), the third equation (6.124) is satisfied identically (taking into account the fact that $M_{z'} = M_0$, that is, $\partial M_{z'}/t = 0$), and the first and second (6.122), (6.123) take the form:

$$\frac{\partial m_{x'}}{\partial t} = \gamma\left(2A_2 M_0 + A_5\right)m_{y'}; \tag{6.126}$$

$$\frac{\partial m_{y'}}{\partial t} = -\gamma\left[2\left(A_2 - A_1\right)M_0 + A_5\right]m_{x'}; \tag{6.127}$$

Using (6.108)–(6.112) we find the expression in brackets:

$$2A_2 M_0 + A_5 = -H_a; \tag{6.128}$$

$$2\left(A_2 - A_1\right)M_0 + A_5 = -\frac{H_a^2 - H_0^2}{H_a}. \tag{6.129}$$

Substituting these expressions into (6.126) and (6.127), we get:

$$\frac{\partial m_{x'}}{\partial t} = -\gamma H_a m_{y'}; \tag{6.130}$$

$$\frac{\partial m_{y'}}{\partial t} = \gamma \frac{H_a^2 - H_0^2}{H_a} m_{x'}; \tag{6.131}$$

When depending on the time of the form exp $(i\omega t)$, we obtain:

$$i\omega m_{x'}^{(0)} + \gamma H_0 m_{y'}^{(0)} = 0; \tag{6.132}$$

$$\gamma \frac{H_a^2 - H_0^2}{H_a} m_{x'}^{(0)} - i\omega m_{y'}^{(0)} = 0 \tag{6.133}$$

where $m_{x'}^{(0)}$ and $m_{y'}^{(0)}$ are the amplitudes of variable magnetization components. The equality to zero of the determinant of this system gives the resonant frequency in the form:

$$\omega_0 = \gamma\sqrt{H_a^2 - H_0^2}.$$

(6.134)

6.3.4. The applied field is larger than the anisotropy field

We now consider in the same geometry (Fig. 6.5) the case when the applied field is greater than the anisotropy field.

In equilibrium, the magnetization is directed along the field, that is:

$$\mathbf{M}^{(0)} = \{0, 0, M_0\}.$$

(6.135)

Thus, all calculations can be carried out in the coordinate system associated with the field, that is, $Oxyz$.

The total energy density in this system is

$$U = -\frac{K}{M_0^2}M_x^2 - M_z H_0,$$

(6.136)

where the first term corresponds to the anisotropy energy, and the second to the interaction energy of magnetization with the field.

Differentiating the expression for the energy density (6.136), we find the effective fields:

$$H_{ex} = \frac{2K}{M_0}M_x;$$

(6.137)

$$H_{ey} = 0;$$

(6.138)

$$H_{ez} = H_0.$$

(6.139)

The full equations of motion in accordance with (6.116)–(6.118) take the form:

$$\frac{\partial M_x}{\partial t} = -\gamma H_0 M_y;$$

(6.140)

$$\frac{\partial M_{y'}}{\partial t} = -\gamma\left(\frac{2K}{M_0^2}M_z - H_0\right)M_x;$$

(6.141)

$$\frac{\partial M_{z'}}{\partial t} = \gamma\frac{2K}{M_0^2}M_x M_y.$$

(6.142)

Performing linearization under the assumption: $M_x = mx$, $M_y = m_y$, $M_z = M_0$, we get:

$$\frac{\partial m_y}{\partial t} = -\gamma H_0 m_y;$$

(6.143)

$$\frac{\partial m_y}{\partial t} = \gamma \left(H_0 - H_a \right) m_x,$$

(6.144)

where, in accordance with (6.36), the notation is used: $H_a = 2K/M_0$. When depending on the time of the form $\exp(i\omega t)$, we obtain:

$$i\omega m_x^{(0)} + \gamma H_0 m_y^{(0)} = 0;$$

(6.145)

$$\gamma \left(H_0 - H_a \right) m_x^{(0)} - i\omega m_y^{(0)} = 0.$$

(6.146)

where $m_x^{(0)}$ and $m_y^{(0)}$ are the amplitudes of the variables of the magnetization. The equality to zero of the determinant of this system gives the resonant frequency in the form:

$$\omega_0 = \gamma \sqrt{H_0 \left(H_0 - H_a \right)}.$$

(6.147)

6.3.5. Arbitrary value of the attached field

Generalizing the formulas (6.134) and (6.147) to an arbitrary value of the applied field, and also performing the normalization of the frequency ω_0 to γ and the field H_0 to H_a, we obtain: with $H_0 \ll H_a$:

$$\frac{\omega_0}{\gamma} = \sqrt{1 - \left(\frac{H_0}{H_a} \right)^2};$$

(6.148)

with $H_0 \geq H_a$:

$$\frac{\omega_0}{\gamma} = \sqrt{\left(\frac{H_0}{H_a} \right) \left(\frac{H_0}{H_a} - 1 \right)};$$

(6.149)

The dependences corresponding to these formulas are illustrated in Fig. 6.6. The curves 1 and 2 are constructed according to the formulas

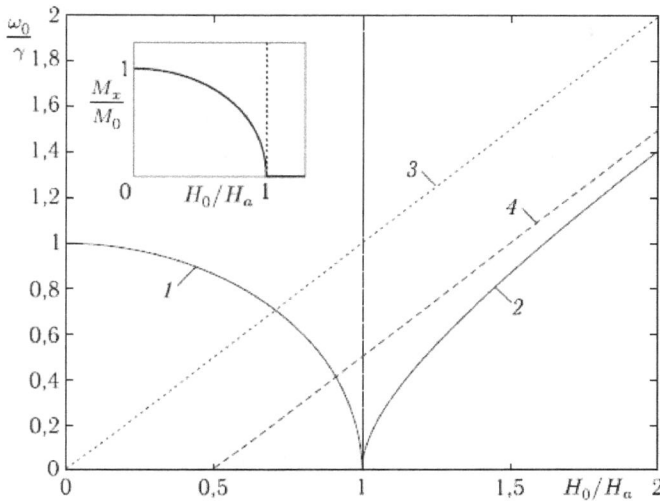

Fig. 6.6. The dependence of the resonance frequency ω_0 on the field H_0 at the orientational transition. The inset shows the dependence of the magnetization component M_x on the field H_0 at the same transition. All units are normalized. Curves correspond to the following dependencies: $1 - \omega_0/\gamma$ (H_0 / H_a) with $H_0 \leq H_a$; $2 - \omega_0/\gamma$ (H_0/H_a) with $H_0 \geq H_a$; $3 - \omega_0/\gamma = H_0 / H_a$; $4 - \omega_0/\gamma = H_0/H_a - 1/2$.

(6.134) and (6.147), curve 3 corresponds to a linear dependence of the frequency on the field in the absence of anisotropy, curve 4 id the same dependence shifted down by the normalized value of 0.5. The insert actually repeats the dependence of the transverse component of the magnetization on the field at the orientational transition, shown in Fig. 5.3 b.

It is seen from the figure that as the field increases from zero, the resonance frequency decreases, and when the field is equal to the anisotropy field, it drops to zero (curve 1). It is clear from the sidebar that this moment just corresponds to the end of the orientational transition, that is, the alignment magnetization exactly along the field. The vanishing of the frequency at this point corresponds to the 'softening' of the system, that is, the state of the 'soft mode' [5.133–136, 139]. When the field is larger than the anisotropy field, the magnetization of its position does not change, but the frequency increases due to an increase in the field H_0. The frequency dependence of the field (curve 2) as it increases tends to linear proportionality to the field itself (curve 4), however, due to the presence of the anisotropy field, H_a passes below that corresponding to $\gamma H_0/H_a$ (curve 3) by 0.5 normalized units.

Such a shift of asymptote 4 by curve 2 downwards relative to curve 3 is maintained up to the tendency of H_0 to infinity. The reason for the constancy of the magnitude of the shift can be explained if we find the asymptote of formula (6.149) as H_0 tends to infinity. We briefly illustrate the procedure for disclosing such uncertainty. Denoting the value of H_0/H_a by x, and the difference between the frequency $\gamma H_0/H_a$ (curve 3) and the frequency defined by the formula (6.149) (curve 2) by u, we find that this difference can be written in the form:

$$u = x - \sqrt{x(x-1)}. \qquad (6.150)$$

To determine the magnitude of the shift, it is necessary to find the $\lim\limits_{x \to \infty} u$, however, as x tends to infinity, this value goes into uncertainty of type $\infty - \infty$. To uncover such uncertainty, it must be converted to a form suitable for the application of the rule L'Hopital [268, 269]. Perform the following conversion

$$u = x - \sqrt{x(x-1)} = \frac{1 - \sqrt{1 - \dfrac{1}{x}}}{\dfrac{1}{x}}. \qquad (6.151)$$

Replace the variable:

$$1/x = y. \qquad (6.152)$$

Such a replacement at $x \to \infty$ corresponds to $y \to 0$. Thus, we obtain

$$u = \frac{1 - \sqrt{1 - y}}{y}. \qquad (6.153)$$

In this case, the required limit is replaced by $\lim\limits_{y \to \infty} y \to 0$ with uncertainty of type 0/0, which can now be found by the rule of L'Hôpital:

$$\lim_{x \to \infty} u = \lim_{y \to 0} u = \lim_{y \to 0} \frac{1 - \sqrt{1 - y}}{y} = \lim_{y \to 0} \frac{\dfrac{1}{2\sqrt{1 - u}}}{1} = \frac{1}{2}. \qquad (6.154)$$

Thus, the desired asymptote of curve 2 must lie below curve 3 by this value, which just gives curve 4.

6.4. Dynamic magnetic susceptibility under forced ferromagnetic resonance due to orientational transition

In previous Sections 6.1–6.3, free ferromagnetic resonance was considered under the conditions of the orientational transition. The resonance frequency was obtained and its dependence on the applied field was studied, the value of which can be both lower and higher than the anisotropy field. Let us now turn to the consideration of forced oscillations of magnetization in the same situation. The main purpose of this consideration will be to obtain the susceptibility tensor and its dependence on the field.

6.4.1. General geometry of the problem

The geometry of the problem is shown in Fig. 6.7. The difference from the geometry shown in Fig. 6.5, consists in adding a variable field **h**, whose orientation relative to a constant field is assumed arbitrary.

We will consider the medium as infinite, possessing uniaxial anisotropy, and the field is applied along the normal to this axis. We will consider two coordinate systems – associated with the field $Oxyz$ and associated with the magnetization $Ox'y'z'$. The axis Oz of the first system coincides with the direction of the field, the axis Oz' of the second system coincides with the equilibrium direction of

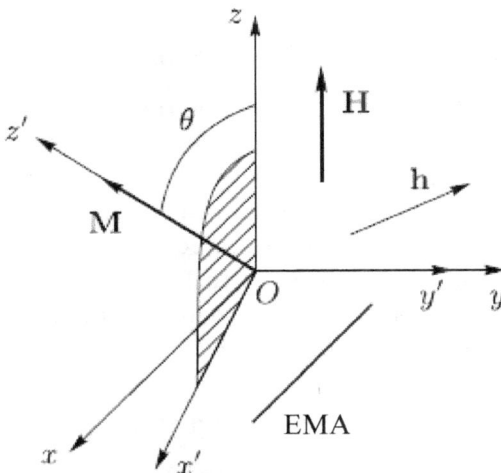

Fig. 6.7. Geometry of the problem of determining the dynamic magnetic susceptibility

the magnetization vector. The planes Oxz and $Ox'z'$ of both systems coincide with each other. The Oy and Oy' axes also match.

Since the susceptibility tensor can be assumed to be rigidly connected with the equilibrium vector of magnetization, since the precession will be considered precisely in this system, we define the variable field immediately as an expansion along the axes of the system $Ox'y'z'$

$$\mathbf{h} = \{h_{x'}, h_{y'}, h_{z'}\}. \tag{6.155}$$

On the other hand, the determination of the susceptibility tensor is a preparatory stage for solving the problem of the propagation of magnetostatic waves in a magnetized plate with normal uniaxial anisotropy. Since this task will be solved in the system associated with the field, then obtained in the system $Ox'y'z'$. The susceptibility tensor will then need to be converted to the $Oxyz$ system. Thus, the complete task will be solved in two stages:

1) consideration of forced oscillations in the system related to magnetization, linearization of the problem and obtaining the susceptibility tensor in this system;

2) the transformation of the resulting susceptibility tensor to the system associated with the field.

As before (Sections 6.3.3 and 6.3.4), we will carry out the consideration separately for the cases of an applied field of a smaller and larger field of anisotropy.

6.4.2. The applied field is less than the anisotropy field

Let us first consider the case when the field is insufficient for a complete rotation of the magnetization vector in its direction. Sections 5.1.1 and 5.1.2 show that in this case the magnetization vector undergoes an orientational transition from the direction determined by the anisotropy of the medium to the direction of the field. We first obtain the susceptibility tensor in the system associated with magnetization, and then transform this tensor to a system associated with the field.

6.4.3. Susceptibility in a magnetization system

Using the transformation (6.79) with the matrix (6.80), like (6.89)–(6.91) with account for (6.155), we find the field in the system $Ox'y'z$:

$$H_{x'} = -H_0 \sin\theta + h_{x'}; \qquad (6.156)$$

$$H_{y'} = h_{y'}; \qquad (6.157)$$

$$H_{z'} = H_0 \cos\theta + h_{z'}. \qquad (6.158)$$

The total energy density U is equal to the sum of the anisotropy energy densities and the interaction energy of magnetization with the field (6.85). In the $Ox'y'z$ system, like (6.93), the total energy density takes the form

$$U = -\frac{K}{M_0^2}\left(M_{x'}^2 \cos^2\theta + M_{z'}^2 \sin^2\theta + 2M_{x'}M_{z'}\sin\theta\cos\theta\right) +$$
$$+M_{z'}\left(-H_0\sin\theta + h_{x'}\right) - M_{y'}h_{y'} - M_{z'}\left(H_0\cos\theta + h_{z'}\right), \qquad (6.159)$$

Open the parentheses and order the items in terms of variables and exponents:

$$U = -\frac{K}{M_0^2}\cos^2\theta M_{x'}^2 - \frac{K}{M_0^2}\sin^2\theta M_{z'}^2 - \frac{2K}{M_0^2}\sin\theta\cos\theta M_{x'}M_{z'} +$$
$$+\left(H_0\sin\theta - h_{x'}\right)M_{x'} - h_{y'}M_{y'} + (-H_0\cos\theta - h_{z'})M_{z'}. \qquad (6.160)$$

Introducing the notation analogous to (6.95)–(6.99), we bring the total energy density to the form

$$U = A_1 M_{x'}^2 + A_2 M_{z'}^2 + A_3 M_{x'}M_{z'} +$$
$$+\left(A_4 - h_{x'}\right)M_{x'} - h_{y'}M_{y'} + \left(A_5 - h_{z'}\right)M_{z'}. \qquad (6.161)$$

Differentiating this expression by the components of magnetization (with the opposite sign), we find the effective fields:

$$H_{ex'} = -\left[2A_1 M_{x'} + A_3 M_{z'} + \left(A_4 - h_{x'}\right)\right]; \qquad (6.162)$$

$$H_{ey'} = h_{y'}; \qquad (6.163)$$

$$H_{ez'} = -\left[A_3 M_{x'} + 2A_2 M_{z'} + \left(A_5 - h_{z'}\right)\right]. \qquad (6.164)$$

The equations of motion of magnetization obtained from the disclosure of the vector product in the Landau–Lifshitz equation (2.20)–(2.22), in the $Ox'y'z$ system have the form (6.116)–(6.118).

Substituting the fields (6.162)–(6.164) into these equations and ordering by variables, we get:

$$\frac{\partial M_{x'}}{\partial t} = \gamma\left\{A_3 M_{x'} M_{y'} + 2A_2 M_{y'} M_{z'} + (A^5 - h_{z'}) M_{y'} + h_{y'} M_{z'}\right\};$$

$$(6.165)$$

$$\frac{\partial M_{y'}}{\partial t} = -\gamma\left\{A_3 M_{x'}^2 - A_3 M_{z'}^2 + 2(A_2 - A_1) M_{x'} M_{z'} + \right.$$
$$\left. + (A_5 - h_{z'}) M_{x'} - (A_4 - h_{x'}) M_{z'}\right\};$$

$$(6.166)$$

$$\frac{\partial M_{z'}}{\partial t} = -\gamma\left\{2A_1 M_{x'} + A_3 M_{y'} M_{z'} + h_{y'} M_{x'} + (A_4 - h_{x'}) M_{y'}\right\}.$$

$$(6.167)$$

For linearization, we set: $M_{x'} = m_{x'}$, $M_{y'} = m_{y'}$, $M_{z'} = M_0$, where is $m_{x'} \ll M_0$, $m_{y'} \ll M_0$. We will also assume that the excitation is small, that is, $h_{x'} \sim h_{y'} \sim h_{z'} \ll H_0$, with $H_0 \sim M_0$, that is, the terms of the form mimk, and also hjml have the second order of smallness, so that with linearization discarded. Substituting such components of magnetization into (6.165)–(6.167), taking into account that, according to (6.125), $A_3 M_0 + A_4 = 0$ is in equilibrium, and leaving only terms of zero and first orders in $m_{x'}$ and $m_{y'}$, we get:

$$\frac{\partial m_{x'}}{\partial t} = \gamma(2A_2 M_0 + A_5) m_{y'} + \gamma M_0 h_{y'};$$

$$(6.168)$$

$$\frac{\partial m_{y'}}{\partial t} = -\left\{\left[2(A_2 - A_1) M_0 + A_5\right] m_{x'} - \gamma M_0 h_{x'}\right\};$$

$$(6.169)$$

$$\frac{\partial m_{z'}}{\partial t} = 0,$$

$$(6.170)$$

The last of these equations corresponds to the condition $M_{z'} = M_0$, that is, is satisfied identically, so that is not considered further. Using (6.95)–(6.99), taking into account (6.105), (6.106), we convert the expressions in (6.168)–(6.169) in brackets:

$$2A_2 M_0 + A_5 = -H_0;$$

$$(6.171)$$

$$2(A_2 - A_1)M_0 + A_5 = -\frac{H_a^2 - H_0^2}{H_a}.$$
(6.172)

Substituting these expressions into (6.168) and (6.169), we obtain the system of equations of motion for small components of magnetization in the form

$$\frac{\partial m_{x'}}{\partial t} = -\gamma H_a m_{y'} + \gamma M_0 h_{y'};$$
(6.173)

$$\frac{\partial m_{y'}}{\partial t} = \gamma \frac{H_a^2 - H_0^2}{H_a} m_{x'} - \gamma M_0 h_{x'}.$$
(6.174)

When depending on the time of the form exp($i\omega t$), we obtain:

$$i\omega m_{x'}^{(0)} + \gamma H_a m_{y'}^{(0)} = \gamma M_0 h_{y'}^{(0)},$$
(6.175)

$$\gamma \frac{H_a^2 = H_0^2}{H_a} m_{x'}^{(0)} - i\omega m_{y'}^{(0)} = \gamma M_0 h_{x'}^{(0)},$$
(6.176)

where $m_{x'}^{(0)}$ and $m_{y'}^{(0)}$ are the amplitudes of variable magnetization components.

As for equations (6.132) and (6.133), the equality to zero of the determinant of this system gives the resonant frequency in the form

$$\omega_0 = \gamma \sqrt{H_a^2 - H_0^2}.$$
(6.177)

To simplify further writing of the formulas, we introduce the auxiliary notation:

$$B_1 = H_a;$$
(6.178)

$$B_2 = \frac{H_a^2 - H_0^2}{H_a}.$$
(6.179)

We also omit further the superscript '0', not forgetting, however, that the values denoted further by $m_{x'}$, $m_{y'}$, $h_{x'}$, $h_{y'}$ are the amplitudes of the sinusoidal variables and do not depend on time. As a result, we obtain the equations of motion in the form:

$$i\omega m_{x'} + \gamma B_1 m_{y'} = \gamma M_0 h_{y'};$$
(6.180)

$$\gamma B_2 m_{x'} - i\omega m_{y'} = \gamma M_0 h_{x'}; \tag{6.181}$$

Solving this system, we find the expression $m_{x'}$, $m_{y'}$ through $h_{x'}$ and $h_{y'}$:

$$m_{x'} = G_1 h_{x'} + H h_{y'}; \tag{6.182}$$

$$m_{y'} = -H h_{x'} + G_2 H_{y'}, \tag{6.183}$$

where the auxiliary notation is entered:

$$G_1 = \frac{\gamma^2 B_1 M_0}{D_0} = \frac{\gamma^2 H_a M_0}{\gamma^2 \left(H_a^2 - H_0^2 \right) - \omega^2}; \tag{6.184}$$

$$G_2 = \frac{\gamma^2 B_2 M_0}{D_0} = \frac{\gamma^2 (H_a^2 - H_0^2) M_0}{H_a \left[\gamma^2 \left(H_a^2 - H_0^2 \right) - \omega^2 \right]}; \tag{6.185}$$

$$H = \frac{i\omega\gamma M_0}{D_0} = \frac{i\omega\gamma M_0}{\gamma^2 \left(H_a^2 - H_0^2 \right) - \omega^2}; \tag{6.186}$$

including:

$$D_0 = \gamma^2 B_1 B_2 - \omega^2 = \gamma^2 \left(H_a^2 - H_0^2 \right) - \omega^2. \tag{6.187}$$

Thus, the solution of the problem of forced oscillations on magnetization in a system associated with magnetization is obtained. We write this solution using matrices:

$$\mathbf{m}' = \ddot{\mathbf{X}}' \mathbf{h}', \tag{6.188}$$

where is χ' is the susceptibility tensor, having the form

$$\bar{X}' = \begin{pmatrix} G_1 & & 0 \\ \hline -H & G_2 & 0 \\ \hline 0 & - & 0 \end{pmatrix} \tag{6.189}$$

where G_1, G_2, H are defined by the formulas (6.184)–(6.187).

6.4.4. Susceptibility in the system associated with the field

So, the susceptibility tensor was obtained in the system related to magnetization (6.189), that is, the first part of the problem is solved. To resolve the second part, this tensor must be transformed into a system associated with the field. This problem can be solved in two ways:

1) based on equations (6.182) and (6.183), to successively convert the components of the magnetization vectors **m'** and field **h'** in the system associated with the field, then from the obtained equations to find the susceptibility tensor;

2) convert the tensor obtained in the system related to the magnetization to the system associated with the field using the appropriate rotation matrix of the coordinate systems.

Both of these methods lead to the same results, but the second one is somewhat simpler and less cumbersome, therefore, we consider the transformation in the second way.

So, according to (6.188), there is a ratio

$$\mathbf{m'} = \ddot{\mathbf{X}}'\mathbf{h'}, \tag{6.190}$$

which should be brought to mind

$$\mathbf{m} = \ddot{X}\mathbf{h}. \tag{6.191}$$

In Section 3.2.5, for the case of transition from one coordinate system to another, the rules for the transformation of vectors are considered. We first consider how these rules can be applied to the transformation of tensors in general.

Let for transformation of vectors there are matrices $\ddot{A}$ and $\ddot{A}^{-1}$ such that the direct transformation has the form (3.10):

$$\mathbf{m} = \ddot{A}\,\mathbf{m'}; \tag{6.192}$$

$$\mathbf{h} = \ddot{A}\mathbf{h'}, \tag{6.193}$$

and the corresponding inverse transformation has the form (3.14):

$$\mathbf{m'} = \ddot{A}^{-1}\mathbf{m}; \tag{6.194}$$

$$\mathbf{h'} = \ddot{A}^{-1}\mathbf{h}. \tag{6.195}$$

Using these formulas in conjunction with (6.190), we perform a chain of successive transformations:

$$\mathbf{m} = \ddot{A}\mathbf{m}' = \ddot{A}(\mathbf{h}') = \ddot{A}\ddot{X}'\mathbf{h}' = \ddot{A}\ddot{X}'(\ddot{A}^{-1}\mathbf{h}) =$$

$$= \ddot{A}\ddot{X}'\mathbf{A}^{-1}\mathbf{h} = \left(\ddot{A}\ddot{X}'\ddot{A}^{-1}\right)\mathbf{h}. \tag{6.196}$$

Comparing the first and last members of this chain with (6.191) we get

$$\ddot{X} = \mathbf{A}\ddot{X}'A^{-1} \tag{6.197}$$

From the comparison of Fig. 6.7 with Fig. 6.4 one can see that the *Oxyz* system present in the field is connected with the field, and the system *Ox'y'z'* associated with magnetization. Therefore, the matrices present in expression (6.197) $\ddot{A}$ and $\ddot{A}^{-1}$ are identical with matrices $\ddot{A}_3$ and $\ddot{A}_3^{-1}$ described by the formulas (6.83) and (6.84). Thus, for the matrices $\ddot{A}$ and $\ddot{A}^{-1}$ entering into expression (6.197) we get:

$$\ddot{A} = \begin{pmatrix} \cos\theta & 0 & \sin\theta \\ 0 & 1 & 0 \\ -\sin\theta & 0 & \cos\theta \end{pmatrix} \tag{6.198}$$

$$\ddot{A}^{-1} = \begin{pmatrix} \cos\theta & 0 & -\sin\theta \\ 0 & 1 & 0 \\ \sin\theta & 0 & \cos\theta \end{pmatrix} \tag{6.199}$$

Here the mutual orientation of the systems *Oxyz* and *Ox'y'z'* is determined by the equilibrium position of the magnetization vector, that is, according to (6.105) and (6.106), the angle θ is related to the fields H_0 and H_a by the relations:

$$\cos\theta = \frac{H_0}{H_a}; \tag{6.200}$$

$$\sin\theta = \frac{\sqrt{H_a^2 - H_0^2}}{H_a}, \tag{6.201}$$

Substituting matrices (6.198) and (6.199), as well as the susceptibility tensor (6.189) into expression (6.197), and then performing the matrix multiplication from right to left, we obtain the susceptibility tensor in the system associated with the field in the form

(6.202)

where the values of G_1, G_2, H are determined by the relations (6.184) to (6.187).

So, the susceptibility tensor is obtained in the system associated with the field, so that the magnetization vector **m** is determined through the field **h** by the relation (6.191).

Let us make some simplifications, for which we write the susceptibility tensor (6.202) in the shortened notation:

(6.203)

where in accordance with (6.184) - (6.187) taking into account (6.178) - (6.179), as well as (6.200) - (6.201), the following relations take place:

$$A_{11} = G_1 \cos^2 \theta = \frac{\gamma^2 M_0 H_0^2}{H_a \left[\gamma^2 \left(H_a^2 - H_0^2 \right) - \omega^2 \right]};$$
(6.204)

$$A_{12} = H \cos \theta = \frac{i\omega\gamma M_0 H_0}{H_a \left[\gamma^2 \left(H_a^2 - H_0^2 \right) - \omega^2 \right]};$$
(6.205)

$$A_{12} = -G_1 \sin\theta\cos\theta = \frac{\gamma^2 M_0 H_0 \sqrt{H_a^2 - H_0^2}}{H_a \left[\gamma^2 \left(H_a^2 - H_0^2 \right) - \omega^2 \right]};$$
(6.206)

$$A_{22} = G_2 = \frac{\gamma^2 M_0 \left(H_a^2 - H_0^2 \right)}{H_a \left[\gamma^2 \left(H_a^2 - H_0^2 \right) - \omega^2 \right]};$$
(6.207)

$$A_{23} = H \sin \theta = \frac{i\omega\gamma M_0 \sqrt{H_a^2 - H_0^2}}{H_a \left[\gamma^2 \left(H_a^2 - H_0^2 \right) - \omega^2 \right]};$$
(6.208)

$$A_{23} = G_1 \sin^2 \theta = \frac{\gamma^2 M_0 \left(H_a^2 - H_0^2 \right)}{H_a \left[\gamma^2 \left(H_a^2 - H_0^2 \right) - \omega^2 \right]}.$$
(6.209)

The components of the magnetization vector are expressed in terms of the components of the field using the tensor (6.203) as follows:

$$m_x = A_{11}h_x + A_{12}h_y + A_{13}h_z; \tag{6.210}$$

$$m_y = A_{12}h_x + A_{22}h_y + A_{23}h_z; \tag{6.211}$$

$$m_z = A_{13}h_x + A_{23}h_y + A_{33}h_z; \tag{6.212}$$

Like (2.56), (2.57), we introduce the normalized frequencies:

$$\Omega = \frac{\omega}{4\pi\gamma M_0}; \tag{6.213}$$

$$\Omega_H = \frac{H_0}{4\pi M_0}; \tag{6.214}$$

$$\Omega_A = \frac{H_a}{4\pi M_0}. \tag{6.215}$$

Sharing numerators and denominators of fractions included in (6.184)–(6.186), on $(4\pi\gamma M_0)^2$ and introducing the frequencies (6.213)–(6.215), we get:

$$G_1 = \frac{1}{4\pi} \frac{\Omega_A}{\left(\Omega_A^2 - \Omega_H^2\right) - \Omega^2}; \tag{6.216}$$

$$G_2 = \frac{1}{4\pi} \frac{\Omega_A^2 - \Omega_H^2}{\Omega_A\left[\left(\Omega_A^2 - \Omega_H^2\right) - \Omega^2\right]}; \tag{6.217}$$

$$H = \frac{1}{4\pi} \frac{\Omega}{\left(\Omega_A^2 - \Omega_H^2\right) - \Omega^2}. \tag{6.218}$$

Using these sets, as well as taking into account (6.200) and (6.201), we express the components of the tensor (6.204)–(6.209) in terms of the normalized frequencies:

$$A_{11} = \frac{1}{4\pi} \frac{\Omega_H^2}{\Omega_A \left[\left(\Omega_A^2 - \Omega_H^2 \right) - \Omega^2 \right]};$$

(6.219)

$$A_{12} = \frac{i}{4\pi} \frac{\Omega\Omega_H}{\Omega_A \left[\left(\Omega_A^2 - \Omega_H^2 \right) - \Omega^2 \right]};$$

(6.220)

$$A_{13} = \frac{1}{4\pi} \frac{\Omega_H \sqrt{\Omega_A^2 - \Omega_H^2}}{\Omega_A \left[\left(\Omega_A^2 - \Omega_H^2 \right) - \Omega^2 \right]};$$

(6.221)

$$A_{22} = \frac{1}{4\pi} \frac{\Omega_A^2 - \Omega_H^2}{\Omega_A \left[\left(\Omega_A^2 - \Omega_H^2 \right) - \Omega^2 \right]};$$

(6.222)

$$A_{33} = \frac{1}{4\pi} \frac{\Omega_A^2 - \Omega_H^2}{\Omega_A \left[\left(\Omega_A^2 - \Omega_H^2 \right) - \Omega^2 \right]};$$

(6.223)

$$A_{33} = \frac{1}{4\pi} \frac{\Omega_A^2 - \Omega_H^2}{\Omega_A \left[\left(\Omega_A^2 - \Omega_H^2 \right) - \Omega^2 \right]};$$

(6.224)

From such a record it can be seen that

$$A_{22} = A_{33},$$

(6.225)

that is, not five, but only five components are independent in the tensor (6.203), so it can be written in a somewhat simpler form:

$$\ddot{X} = \begin{pmatrix} A_{11} & A_{12} & A_{13} \\ -A_{12} & A_{22} & A_{23} \\ A_{13} & -A_{23} & A_{22} \end{pmatrix}$$

(6.226)

Comment. It is important to note that the obtained tensor, although it possesses certain symmetry properties, is neither completely symmetric nor fully antisymmetric. So the χ_{12} and χ_{21} components, as well as χ_{23} and χ_{32}, which are symmetrical with respect to the main diagonal, have opposite signs, whereas for the

χ_{13} and χ_{31} members that are symmetric from the same diagonal, the signs are the same.

By this property, the obtained tensor differs significantly from the classical tensor for a magnet in a saturated state, which is antisymmetric [6–8]. It can be assumed that such properties of symmetry of the tensor, which determines the susceptibility of a magnetic material under the conditions of the transition, should have a noticeable effect on the dispersion and properties of the reciprocal for various wave processes in ferrites.

6.4.5. The applied field is larger than the anisotropy field

We now consider the case when the applied field is greater than the anisotropy field. As shown in Section 5.1.1, due to the perpendicularity between the directions of the field and the anisotropy axis (Fig. 6.7), the magnetization vector is oriented exactly along the field (as shown in Fig. 5.3 up to the notation), so that the problem can be completely solved in the system associated with the field. For consideration of forced events, we will assume that the applied field has the form

$$\mathbf{H} = \{h_x, \, h_y, \, H_0\}, \tag{6.227}$$

where is $h_x \sim h_y \ll H_0$. Energy density has the form

$$U = -\frac{K}{M_0^2} M_x^2 - h_x M_x - h_y M_y - H_0 M_z, \tag{6.228}$$

Differentiating this expression, we obtain the effective fields in the form:

$$H_{ex} = \frac{2K}{M_0^2} M_x + h_x; \tag{6.229}$$

$$H_{ey} = h_y; \tag{6.230}$$

$$H_{ez} = H_0. \tag{6.231}$$

The equations of motion of magnetization obtained from the disclosure of the vector product in the Landau–Lifshitz equation (2.20)–(2.22) in the $Oxyz$ system have the form:

$$\frac{\partial M_x}{\partial t} = -\gamma \left(M_y H_{ez} - M_z H_{ey} \right); \tag{6.232}$$

$$\frac{\partial M_y}{\partial t} = -\gamma \left(M_z H_{ez} - M_x H_{ez} \right);$$

(6.233)

$$\frac{\partial M_x}{\partial t} = -\gamma H_0 M_y + \gamma h_y M_z;$$

(6.234)

Substituting the effective fields (6.229)–(6.231), we obtain the equations of motion in the form:

$$\frac{\partial M_x}{\partial t} = -\gamma H_0 M_y + \gamma h_y M_z;$$

(6.235)

$$\frac{\partial M_y}{\partial t} = -\gamma \left(\frac{2K}{M_0^2} M_z - H_0 \right) M_x + \gamma h_x M_z;$$

(6.236)

$$\frac{\partial M_z}{\partial t} = -\gamma \left\{ h_y M_x - \left(\frac{2K}{M_0^2} M_x + h_x \right) M_y \right\}.$$

(6.237)

Performing linearization under the assumption: $M_x = m_x$, $M_y = m_y$, $M_z = M_0$, and also taking into account that $h_x \sim h_y \ll H_0$, that is, leaving only terms of the first order in m_x, m_y, h_x, h_y, we get:

$$\frac{\partial m_x}{\partial t} = -\gamma H_0 m_y + \gamma M_0 h_y;$$

(6.238)

$$\frac{\partial m_y}{\partial t} = \gamma \left(H_0 - H_a \right) m_x + \gamma M_0 h_x;$$

(6.239)

where in accordance with (6.36) the following notation is used: $H_a = 2K/M_0$. The third equation is not written out here, since it is satisfied identically. When depending on the time of the form $\exp(i\omega t)$, we obtain:

$$i\omega m_x^{(0)} + \gamma H_0 m_y^{(0)} = \gamma M_0 h_y^{(0)};$$

(6.240)

$$\gamma \left(H_0 - H_a \right) m_x^{(0)} - i\omega m_y^{(0)} = \gamma M_0 h_x^{(0)},$$

(6.241)

where $m_x^{(0)}$ and $m_y^{(0)}$ are the amplitudes of the variables of the magnetization. Solving these equations, we get:

$$m_x^{(0)} = P_1 h_x^{(0)} + Q h_y^{(0)};$$

(6.242)

$$m_x^{(0)} = Qh_x^{(0)} + P_2 h_y^{(0)};$$

(6.243)

where the notation is entered:

$$P_1 = \frac{\gamma^2 M_0 H_0}{\gamma^2 H_0 (H_0 - H_a) - \omega^2};$$

(6.244)

$$P_2 = \frac{\gamma^2 M_0 (H_0 - H_a)}{\gamma^2 H_0 (H_0 - H_a) - \omega^2};$$

(6.245)

$$Q = \frac{i\omega\gamma M_0}{\gamma^2 H_0 (H_0 - H_a) - \omega^2};$$

(6.246)

Thus, the susceptibility tensor has the form:

$$\ddot{X} = \begin{pmatrix} P_1 & Q & 0 \\ -Q & P_2 & 0 \\ 0 & 0 & 0 \end{pmatrix}$$

(6.247)

Writing the components of the tensor through the normalized frequencies (6.213)–(6.215), we obtain:

$$P_1 = \frac{1}{4\pi} \frac{\Omega_H}{\Omega_H (\Omega_H - \Omega_A) - \Omega^2};$$

(6.248)

$$P_2 = \frac{1}{4\pi} \frac{\Omega_H - \Omega_A}{\Omega_H (\Omega_H - \Omega_A) - \Omega^2};$$

(6.249)

$$Q = \frac{1}{4\pi} \frac{\Omega}{\Omega_H (\Omega_H - \Omega_A) - \Omega^2};$$

(6.250)

Note that, in contrast to (6.226), this susceptibility tensor has a purely antisymmetric form, which agrees with the classical tensor for the precession of magnetization in a saturated state [6–8].

Conclusions for chapter 6

This chapter is devoted to the description of the phenomenon of ferromagnetic resonance in an anisotropic medium, including the conditions of the orientational transition.

The main results obtained in this chapter are as follows.

1. The ferromagnetic resonance in a plate with uniaxial and cubic anisotropy is considered in the case when the EMA and the [111] axis are perpendicular to the plane of the plate. The principal difference between the cases of tangential and normal magnetization, due to the asymmetric arrangement of the [111] axes relative to the plane of the plate, is noted. For the case of tangential magnetization, the consideration is performed in the coordinate system associated with the field. In this case, the energy density of cubic anisotropy, originally recorded in the system associated with the edges of the cubic cell, is also converted to a system associated with the field, for which the corresponding rotation matrix is obtained. There is a variety of possible types of rotation matrix entries, due to different choices of the orientation of the Euler angles relative to the original system coordinates. The equivalence of matrices is demonstrated for some practical cases of such an orientation.

2. Two important cases of the practice of special cases of tangential magnetization are considered: when the field is parallel to the projection of the [111] axis on the film plane and when the field is perpendicular to such a projection. For both of the rays, the energy density, effective fields, and the equations of motion are written. Two equivalent options for performing the task linearization are given. According to the first variant, for a given direction and magnitude of the field, an equilibrium state of magnetization is first found, the deviation of which vector from this direction is obviously assumed to be small, and in the resulting equations of motion all terms, the order of which in magnetization exceeds the first, are completely omitted. According to the second variant, the equilibrium orientation of the magnetization is not determined in advance, and the equations of motion are written through the full components of the magnetization vector, after which the constant terms appearing in these equations are assumed to be zero.

The complete equivalence of both variants is shown, and it is noted that the equality to zero of the constant terms in the second variant corresponds exactly to the definition of the equilibrium state in the first.

3. A method is given for determining the material parameter of practical importance – the difference between the demagnetization fields and uniaxial anisotropy, taking into account the cubic anisotropy field. The possibility of a numerical solution of the resulting self-consistent third degree equation is noted. The case of normal magnetization of the plate is considered. The energy density,

effective fields and linearized equations of motion are obtained. The resonant frequency is found and the possibility of determining the difference parameter between the demagnetization fields and the uniaxial anisotropy is shown.

4. The ferromagnetic resonance is considered at a simple orientational transition. The equilibrium orientation of the magnetization vector is determined with a change in the field perpendicular to the EMA, in the range of fields less than the anisotropy field. A transformation matrix was obtained that corresponds to the transition from a laboratory coordinate system to a system associated with magnetization. In such a coordinate system, the equations of motion of the magnetization are obtained and their linearization carried out. The resonant frequency of the free precession is obtained and it is shown that it is determined by the square root of the difference between the squares of the anisotropy field and the external field. The resonance conditions are considered when the external field is oriented perpendicular to the EMA, and when the external field is greater than the anisotropy field. The linearized equations of motion are obtained and the resonant frequency of free precession is found. It is shown that in this case it is determined the square root of the product of the external field by the difference between the external field and the anisotropy field. The obtained two solutions are generalized to the full range of the field change from zero to infinity. It is shown that as the field increases, when its value is less than the anisotropy field, the resonance frequency decreases to zero at the phase transition point, after which, in a field larger than the anisotropy field, it increases according to a law asymptotically tending to increase linearly.

A constant downward shift of the asymptote downward as compared with the linear dependence corresponding to the direct proportionality of the resonant frequency to the magnitude of the field is noted, which is independent of the magnitude of the field. The final magnitude of the shift is determined by disclosing the uncertainty between the asymptote and the linear dependence that occurs as the field tends to infinity.

5. The forced precession of magnetization under conditions of an experimental transition is considered. The formulation of the problem of obtaining the dynamic susceptibility tensor for ferromagnetic resonance in a plate with normal uniaxial anisotropy magnetized by a field perpendicular to the anisotropy axis is presented. It is shown that the solution of such a problem for the case of an external field,

a smaller anisotropy field, requires a consistent implementation of two sets.

The first stage is to consider the forced oscillations in the system associated with the magnetization, the linearization of the equations of motion and the obtaining of the susceptibility tensor in this system. The second stage consists in converting the resulting susceptibility tensor to the laboratory coordinate system associated with the field. In the framework of the first stage, linearized equations of the forced precession of magnetization are obtained, from which the found tensor of the dynamic susceptibility is antisymmetric. In the framework of the second stage, the transformation of the obtained tensor to the laboratory coordinate system was performed. Two possible ways of performing such a transformation are given, their equivalence is shown, and the matrices of direct and inverse transformations are found. It is shown that the obtained susceptibility tensor in the system associated with the field is not completely symmetric or antisymmetric. Thus, the components located in the immediate vicinity of the main diagonal of the tensor matrix are antisymmetric, and those more distant from the same diagonal are symmetric. It is suggested that the specific tensor is of particular symmetry in the formation of the dispersion and the non-reciprocal characteristics of the wave processes in ferrites.

6. The solution of the same problem is considered for the case of an external field with a larger anisotropy field. It is shown that here, due to the orientation of the magnetization vector exactly along the field direction, the problem can be completely solved in one stage in the system associated with the field. The linearized equations of motion are obtained for the forced precession of magnetization. From the solution of the equations, the dynamic susceptibility tensor is obtained, it is shown that it is completely antisymmetric. The resonant nature of the components of the tensor is noted, differing from the classical one only in the presence of the difference between the anisotropy field and the external field.

7

Ferromagnetic resonance in films with tips of axis of easy magnetization

This chapter is devoted to the description of the phenomenon of ferromagnetic resonance in films with the deviation of the magnetic anisotropy axis from the normal to the film plane. The conditions for ferromagnetic resonance with magnetization of films with inclined uniaxial anisotropy by fields of different directions are considered. The possibility is noted and the conditions for the separate measurement of the anisotropy field and the saturation magnetization in the case of a deviation of the anisotropy axis from the normal to the film plane are given. The results of experiments on ferromagnetic resonance in real fibres of garnet ferrites are briefly considered. When considering the surveys listed in the survey, we will mainly follow the works [6–8, 20,22–25], the remaining references are indicated in the text.

7.1. General provisions for the calculation of ferromagnetic resonance in films with an inclination of the axis of easy magnetization

7.1.1. General calculation method

In the preceding section, the ferromagnetic resonance in uniaxial bands, the anisotropy axis of which is oriented perpendicular to their surface, is considered. However, in practice films with an anisotropy axis, which is normal to the film plane, are often encountered significant angle [25, 30, 40, 42, 275, 278, 355-358]. Let us consider some features of FMR in such lines.

When calculating the FMR frequency in such fibers, the Smit–Suhl method is often used, which is a solution of the linearized Landau–Lifshitz equation in a spherical coordinate system [6–8, 31, 267]. In a number of cases, this method allows one to measure the FMR frequency under the conditions of the orientational transition [31], as well as to determine the anisotropy constant and orientation of the EAM with respect to the film plane [23–25]. It is also possible to determine the orientation of the axes and the value of the cubic anisotropy constant [22]. Another advantage of the method is the possibility of separate measurement of the uniaxial anisotropy constant and saturation magnetization of the film material [23, 25]. However, in spite of a certain convenience, which allows one to do without calculating cumbersome distributors, this method has a significant drawback, which consists in the divergence of the obtained value of the resonant frequency when the constant field is oriented near the polar axis spherical coordinate system. Therefore, for successful use of the method, the polar axis should be oriented in such a way as to avoid its proximity to the direction of the constant field. If in the experiment the field is always oriented in one plane, then the normal to this plane should be chosen as the polar axis. The experiments described in [22–25], where the field rotated in a circle in one plane, belong to such a case. Consider this situation in more detail.

7.1.2. General geometry of the problem

The overall geometry of the problem is shown in Fig. 7.1. It is assumed that the external field **H** always remains in the same plane during its change, and the field direction within this plane can be arbitrary. The Cartesian coordinate system $Oxyz$ is chosen so that the coordinate plane Oxy coincides with the plane of change of the magnetic field. In this case, the Oz axis is obtained perpendicular to the plane of the field change. Such an orientation of the axes of the Cartesian system is necessary so that when moving to a spherical coordinate system, the polar axis which will be oriented along the Oz axis, the angle between the direction of the field and the polar axis has always been sufficiently far from zero. In this case, this angle is 90°, which is quite enough to eliminate the divergence noted above.

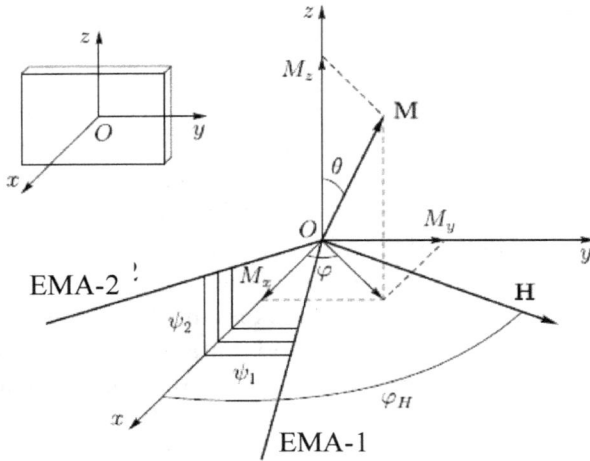

Fig. 7.1. General geometry of the problem of ferromagnetic resonance in films with an inclined axis of easy magnetization. In the upper left corner is a diagram of film orientation.

Since the field specifies only the orientation of the Oz axis, the orientation of the Oxz and Oyz planes can be chosen arbitrary. It is convenient to combine one of these planes with the film plane, for example, the Oyz plane. In this case, the Oxz plane will pass through the normal to the film plane perpendicular to this plane itself. By the condition of the problem, the film has uniaxial anisotropy, the axis which is deflected from the normal to the film plane, that is, from the axis Ox. As noted above, the orientation of the field is assumed only in the Oxy plane. Further, for simplicity, two particulars will be considered considering the orientation of the easy magnetization axis (EMA):

1) The EMA is in the Oxy plane and makes an angle ψ_1 with the Ox axis, that is, the EMA is in the plane of the field change;

2) The EMA is in the plane Oxz and makes an angle ψ_2 with the axis Ox, that is, the EMA is in a plane perpendicular to the plane of the field change.

7.1.3. Introduction of spherical coordinates

We introduce a spherical coordinate system whose polar axis coincides with the axis Oz, and the azimuthal axis with the axis Ox. The components of the magnetization vector will have the form (the index m for simplicity is omitted):

$$M_x = M_0 \sin\theta\cos\varphi; \tag{7.1}$$

$$M_y = M_0 \sin\theta\sin\varphi; \tag{7.2}$$

$$M_z = M_0 \cos\theta. \tag{7.3}$$

In the form normalized to M_0, the same components take the form:

$$m_x = \sin\theta\cos\varphi; \tag{7.4}$$
$$m_y = \sin\theta\sin\varphi; \tag{7.5}$$
$$m_z \cos\theta. \tag{7.6}$$

The field has components:
$$H_x = H_0 \cos\varphi_H; \tag{7.7}$$
$$H_y = H_0 \sin\varphi_H; \tag{7.8}$$
$$H_z = 0. \tag{7.9}$$

In the form normalized to H_0, the field components take the form:

$$h_x = \cos\varphi_H; \tag{7.10}$$
$$h_y = \sin\varphi_H; \tag{7.11}$$
$$h_z = 0. \tag{7.12}$$

7.1.4. Energy density of various kinds

In this formulation of the problem, the total energy density has three components: the demagnetizating energy U_p, the interaction energy of the magnetization with the external field U_m, and the anisotropy energy U_a. Consider these types of energy sequentially.

The energy density of demagnetization. Demagnetizing energy is determined by the shape of the sample, that is, in this case a thin film, the plane of which is perpendicular to the axis Ox. In accordance with (2.5), this energy has the form

$$U_p = -\frac{1}{2}\mathbf{M}\mathbf{H}_m, \tag{7.13}$$

where $\mathbf{H}_m$ is the magnetic field occurring inside the sample and due to its shape, usually called the 'demagnetization field'. In the particular case of an ellipsoid, this field is uniform and can be defined by the formula

$$\mathbf{H}_m = -\ddot{N}\mathbf{M}, \tag{7.14}$$

where $\ddot{N}$ is the ellipsoid demagnetizing factor tensor

$$\ddot{N} = \begin{pmatrix} N_x & 0 & 0 \\ 0 & N_y & 0 \\ 0 & 0 & N_z \end{pmatrix} \tag{7.15}$$

A thin film (or infinite plate) can be considered as a special case of an ellipsoid. When the plane of such a film is oriented perpendicular to the Ox axis, the demagnetizing factor tensor takes the form

$$\ddot{N} = \begin{pmatrix} 4\pi & 0 & 0 \\ 0 & 0 & 0 \\ 0 & 0 & 0 \end{pmatrix} \tag{7.16}$$

Substituting (7.16) into (7.14), performing multiplication and substituting the resulting expression in (7.13), taking into account (7.1)–(7.3), we obtain the demagnetization energy density in the form similar to (4.204):

$$U_m^{(1)} = 2\pi M_0^2 m_x^2 = 2\pi M_0^2 \sin^2\theta \cos^2\varphi. \tag{7.17}$$

Taking into account the condition of preserving the length of the magnetization vector (4.205), we obtain another, analogous to (4.206), equivalent to (7.17), the expression for the demagnetization energy density:

$$U_m^{(2)} = -2\pi M_0^2 \left(m_y^2 + m_z^2\right) = -2\pi M_0^2 \left(\sin^2\theta \sin^2\varphi + \cos^2\theta\right). \tag{7.18}$$

Comment. It should be noted that the reduced form of the demagnetizing factor (7.15) tensor takes place only for an ellipsoid in the Cartesian coordinate system, the axes of which are oriented along the main axis of the ellipsoid. The same applies to the form of the tensor (7.16), which is a special case (7.15) to the extent that the thin film boundless in the plane can be considered a special case of an ellipsoid, the two axes of which are infinitely large. In the general case, when the sample shape does not correspond to the ellipsoid shape, the expression for energy density (7.13) retains its shape, but the demagnetization field is non-uniform, so that all (or almost all) components of the demagnetizing $\ddot{N}$ tensor are non-zero.

The density of the interaction energy of magnetization with the field. The energy density of the interaction of magnetization with the external field U_h, according to (4.207), has the form

$$U_h = -\mathbf{MH}. \tag{7.19}$$

or in the record through the normalized components

$$U_h = -M_0 H_0 \left(m_x h_x + m_y h_y + m_z h_z \right). \tag{7.20}$$

Substituting (7.4)–(7.6) and (7.10)–(7.12), after a small number of transformations, we get:

$$U_h = -M_0 H_0 \sin \theta \cos \left(\varphi - \varphi_H \right). \tag{7.21}$$

Anisotropy energy density. As follows from Section 4.2, the anisotropy energy density with an arbitrary orientation of the EMA with respect to the initial coordinate system can have a rather complicated form. Therefore, we will consider this issue gradually, moving from a simpler situation to a more complex one.

7.2. Parallel-perpendicular geometry

From the comparison of Fig. 7.1 with Fig. 4.5 it can be seen that in this problem (in both its variants) the orientation of the EMA relative to the original Cartesian coordinate system is, as it were, a special case of the more general one shown in Fig. 4.5, therefore possible calculations are somewhat simplified. We first consider the simplest case of orientation of the EMA exactly along the normal to the plane of the film, that is, when $\psi = 0$. Since in this case any geometric elements of the structure are only in a parallel or perpendicular relation, we will conventionally call this case 'parallel–perpendicular geometry'. As shown in Section 4.5, the energy density of the uniaxial anisotropy, as well as the demagnetization energy, allow two different recordings, passing one into another while maintaining the length of the magnetization vector. Consider both options here.

7.2.1. Equivalence of two expressions for energy density

In this case, in the geometry of Fig. 7.1, taking into account the physical content of the various entries (Sections 4.5.1, 4.5.2), the anisotropy energy density is

$$U_a^{(1)} = K\left(m_y^2 + m_z^2\right), \tag{7.22}$$

and the demagnetization energy density has the form

$$U_m^{(1)} = 2\pi M_0^2 m_x^2. \tag{7.23}$$

The total energy density is equal to the sum of these expressions:

$$U = U_a^{(1)} + U_m^{(1)} = K\left(m_y^2 + m_z^2\right) + 2\pi M_0^2 m_x^2. \tag{7.24}$$

From the condition of preservation of the length of the magnetization vector

$$m_x^2 + m_y^2 + m_z^2 = 1, \tag{7.25}$$

we get

$$m_y^2 + m_z^2 = 1 - m_x^2. \tag{7.26}$$

Substituting this expression into the first term (7.24), we get

$$U = K - \left(K - 2\pi M_0^2\right)m_x^2 \rightarrow -K_a m_x^2. \tag{7.27}$$

Such a record with $K > 2\pi M_0^2$, up to a constant term, is equivalent to uniaxial anisotropy with a constant

$$K_a = K - 2\pi M_0^2. \tag{7.28}$$

where the axis of such anisotropy is oriented along the coordinate axis Ox. When $K < 2\pi M_0^2$, the same expression describes easy-plane anisotropy, the plane of which is parallel to the coordinate plane Oyz.

On the other hand, from (7.25) one can get

$$m_x^2 = 1 - \left(m_y^2 + m_z^2 \right). \tag{7.29}$$

Substituting this expression into the second term (7.24), we obtain

$$U = 2\pi M_0^2 + \left(K - 2\pi M_0^2 \right)\left(m_y^2 + m_z^2 \right) \to K_a \left(m_y^2 + m_z^2 \right), \tag{7.30}$$

which also for $K > 2\pi M_0^2$ gives uniaxial or for $K < 2\pi M_0^2$ easy-plane anisotropy with the same constant K.

With opposite records of the same uniaxial anisotropy and the same demagnetization (sections 4.5.1, 4.5.2), we get

$$U = U_a^{(2)} + U_m^{(2)} = -Km_x^2 - 2\pi M_0^2 \left(m_y^2 + m_z^2 \right), \tag{7.31}$$

whence, substituting (7.26) or (7.29), we obtain the same types of easy-axis or easy-plane anisotropy.

Thus, in the case of uniaxial anisotropy, whose axis is perpendicular to the film plane, the total anisotropy is equivalent to uniaxial or easy-plane anisotropy, the contributions of both types of anisotropy being indistinguishable, and the character is determined by the difference between the corresponding constants.

Comment. The mentioned circumstance (equivalence of contributions to the energy density from both anisotropies) leads to the impossibility in the experiment on the classical ferromagnetic resonance to distinguish one anisotropy from another. If such a difference were possible, then this would allow experimentally to measure separately the uniaxial anisotropy constant and the saturation magnetization of the film material.

In the well-known second monograph of experiments, the difference between the contributions of the two anisotropies is manifested in two cases:

1) the anisotropy axis is deviated from the normal to the film plane;

2) the demagnetization of the film is inhomogeneous, for example, a small field is applied in its plane, which is periodic in space (for example, the field of a magnetostatic wave [1]). Separate measurement of the anisotropy and magnetization constants in accordance with the first case was implemented in [22–25], in accordance with the second, in [66]. Somewhat further (in Sections

7.5 and in part 7.6) it is the first case that will be considered in more detail.

7.2.2. Two typical inclined geometry options

In real films the anisotropy axis often deviates from the normal to the film plane [25, 30, 40, 42, 275, 278, 355–358]. According to the historically established tradition, such a deviation is referred to as the 'inclination of the anisotropy axis', therefore we will conditionally call this case 'inclined geometry'. Since the experiments on FMR, as a rule, are carried out when the film is rotated in some sufficiently strong constant field [6–8, 22–25], with respect to the coordinate system associated with the film the field experiences a turn at certain angles, remaining all the time in one plane. If we assume that the plane of rotation of the field passes through the normal to the plane of the film, then we can consider two cases as fundamentally different. The first one is when the EMA lies in the same plane, the second one is when the plane passing through the EMA and the normal to the film plane is perpendicular to the plane of field rotation. These cases just correspond to two different orientations of the EMA with respect to the plane of the field change, which were noted when considering Fig. 7.1. For the sake of clarity, we give here their wording again:

1) The EMA is in the Oxy plane and makes an angle ψ_1 with the Ox axis, that is, the EMA is in the plane of the field change;

2) The EMA is in the Oxz plane and makes an angle ψ_2 with the axis Ox, that is, the EMA is in a plane perpendicular to the plane of the field change.

Consider these cases sequentially.

7.3. Option one: the easy magnetization axis lies in the plane of the field change

In this case, in the geometry of Fig. 7.1 the polar and azimuth angles of the EMA (more precisely, the EMA-1) are equal to:

$$\theta_a = 90°;$$

(7.32)

$$\varphi_a = \psi_1.$$

(7.33)

Within the scope of the present case (until the end of section 7.3) in order to simplify further recording, the index 1 with ψ is further omitted, that is, we assume:

$$\varphi_a = \psi. \tag{7.34}$$

From the general geometry (Fig. 7.1) one can see that when the field **H** is rotated in the Oxy plane, it passes a position that exactly corresponds to the orientation of the EMA, that is, $\varphi_H = \varphi_a$.

In the sections 4.2.2 and 4.2.6, two variants of a general expression for the energy density of uniaxial anisotropy with an arbitrary direction of the axis are obtained: $U_a^{(1)}$ (4.68) and $U_a^{(2)}$ (4.86). First we turn to the expression for $U_a^{(1)}$:

$$U_a^{(1)} = K\left\{ m_x^2\left(\cos^2\theta_a\cos^2\varphi_a + \sin^2\varphi_a\right) + m_y^2\left(\cos^2\theta_a\sin^2\varphi_a + \cos^2\varphi_a\right) + \right.$$
$$+ m_z^2\sin^2\theta_a - 2m_x m_y \sin^2\theta_a \sin\varphi_a \cos\varphi_a -$$
$$\left. -2m_y m_z \sin\theta_a\cos\theta_a\sin\varphi_a - 2m_z m_x \sin\theta_a\cos\theta_a\cos\varphi_a.\right\} \tag{7.35}$$

Substituting θ_a and φ_a in accordance with (7.32), (7.34), and also m_x, m_y, m_z in accordance with (7.4)–(7.6), we get

$$U_a^{(1)} = K\left(\sin^2\psi\sin^2\theta\cos^2\varphi + \cos^2\psi\sin^2\theta\sin^2\varphi + \right.$$
$$+ \cos^2\theta - 2\sin\psi\cos\psi\sin^2\theta\sin\varphi\cos\varphi) =$$
$$= K\left[\sin^2\theta\sin^2(\varphi - \psi) + \cos^2\theta\right]. \tag{7.36}$$

Consider now the expression U_a^2:

$$U_a^{(2)} = -K\left\{ m_x^2\sin^2\theta_a\cos^2\varphi_a + m_y^2\sin^2\theta_a\sin^2\varphi_n + \right.$$
$$+ m_z^2\cos^2\theta_a + 2m_z m_y \sin^2\theta_a\sin\varphi_0\cos\varphi_a +$$
$$\left. +2m_x m_z\sin\theta_a\cos\theta_a\cos\varphi_a + 2m_y m_z\sin\theta_a\cos\theta_a\sin\varphi_a \right\}. \tag{7.37}$$

Substituting θ_a and φ_a in accordance with (7.32), (7.34), and also m_x, m_y, m_z in accordance with (7.4)–(7.6), we get

$$U_a^{(2)} = K\left(\cos^2\psi\sin^2\theta\cos^2\varphi + \sin^2\psi\sin^2\theta\sin^2\varphi + \right.$$
$$\left. 2\sin\psi\cos\psi\sin^2\theta\sin\varphi\cos\varphi\right) =$$
$$= -K\left[\sin^2\theta\cos^2\left(\varphi - \psi\right)\right]. \tag{7.38}$$

Expressing the second factor in parentheses through a sine, we get

$$U_a^{(2)} = -K + \left[\sin^2\theta\sin^2\left(\varphi - \psi\right) + \cos^2\theta\right]. \tag{7.39}$$

It can be seen that this expression coincides with (7.36) up to a constant term, which disappears in the subsequent differentiation. Therefore, for further consideration, we use expression (7.38) as the energy density of anisotropy as having a simpler spelling. Thus, the total energy density, up to a constant term, equal to the sum (7.38), (7.17) and (7.21), takes the form

$$U = -K\sin^2\theta\cos^2\left(\varphi - \psi\right) +$$
$$+ 2\pi M_0^2\sin^2\theta\cos^2\varphi - M_0 H_0\sin\theta\cos\left(\varphi - \varphi_H\right). \tag{7.40}$$

In this expression, θ and φ are the spherical coordinates of the magnetization vector **M** (Fig. 7.1). According to the general procedure for finding the extremum of a two-variable function [261, 262], the equilibrium position of the magnetization vector (φ_0, θ_0) corresponding to the minimum energy density U, is found from the solution of the system of equations:

$$\frac{\partial U}{\partial\theta} = 0; \tag{7.41}$$

$$\frac{\partial U}{\partial\varphi} = 0; \tag{7.42}$$

under the following sustainability terms:

$$\frac{\partial^2 U}{\partial\theta^2} > 0; \tag{7.43}$$

$$\frac{\partial^2 U}{\partial \theta^2} \frac{\partial^2 U}{\partial \varphi^2} - \left(\frac{\partial^2 U}{\partial \theta \partial \varphi} \right)^2 > 0.$$
(7.44)

From condition (7.41), we immediately find $\theta_0 = 90°$. It is not necessary to check condition (7.43), since it is immediately apparent from the symmetry of the geometry of the problem that with the accepted orientation of the EMA, the equilibrium position of the magnetization vector always lies in the Oxy plane.

Substituting the obtained value $\theta_0 = 90°$ into (7.40), up to a constant term, we get

$$U|_{\theta=90°} = K \sin^2 (\varphi - \psi) + \\ + 2\pi M_0^2 \cos^2 \varphi - M_0 H_0 \cos(\varphi - \varphi_H).$$
(7.45)

Differentiating this expression by φ, substituting in (7.42) and performing simplifying trigonometric transformations, we obtain an equation for φ of the following form:

$$K \sin\left[2(\varphi - \psi)\right] - 2\pi M_0^2 \sin(2\varphi) + M_0 H_0 \sin(\varphi - \varphi_H) = 0.$$
(7.46)

7.3.1. The equilibrium position of magnetization in the absence of an external field

The analytical solution of equation (7.46), apparently, is quite simple only with $H = 0$. In this case, we obtain

$$\varphi_0(H = 0) = \frac{1}{2} \operatorname{arctg}\left(\frac{K \sin 2\psi}{K \cos 2\psi - 2\pi M_9^2} \right) + 90°n,$$
(7.47)

where n is an integer, so that the last term reflects the cyclicality of the arctangent through $180°$.

The verification of conditions (7.43)–(7.44) is also not required in this case, since it can be seen that as $M_0 \to 0$, the angle φ tends to ψ (since the demagnetizing factor of the film (plate) is out of the game, and the equilibrium orientation of the magnetization vector controls the anisotropy field), that is, the solution (7.47) describes a notorious minimum. We introduce the effective fields: anisotropy field

$$H_a = \frac{2K}{M_0};$$

(7.48)

demagnetizing field

$$H_m = 4\pi M_0,$$

(7.49)

as well as their relationship

$$Q = \frac{H_a}{H_m} = \frac{K}{2\pi M_0^2}.$$

(7.50)

This parameter is widely used in the technique of cylindrical magnetic domains (CMD) and is called there the 'quality factor' of a magnetic film [40] (Section 1.3.1, formula (1.8)). The practical sense of the quality factor is especially evident for the situation when the anisotropy axis is exactly perpendicular to the film plane (that is, the angles $\psi_{1,2}$ in Fig. 7.1 are equal to zero). In the case of $Q > 1$, the equilibrium position of the magnetization vector in the absence of an external field is normal to the film plane, which ensures the possibility of the existence of a labyrinth domain structure, as well as the CMD. Typical materials of this type are some mixed garnet ferrites, orthoferrites or hexaferrites. When $Q < 1$ in the equilibrium position, the magnetization vector is oriented in the film plane, so that the domain structure has a linear band-like character, and CMDs are not formed. Typical materials of this type are permalloy, as well as some species of mixed breeds of pomegranate [40]. A special case is the iron–yttrium garnet (IYG) widely used in engineering, where, along with uniaxial anisotropy, cubic and also magnetoelastic interaction plays a significant role. Thus, depending on the ratio of the constants, the domain structure, as well as the resonant properties in the IYG\ have a great variety [333, 363–367]. In more detail this question, in detail, as applied to the properties of magnetostatic waves, is supposed to be considered in subsequent monographs of this series.

Let us return to the formula (7.47) and using the quality factor (7.50) we write it in the form

$$\varphi_0 \left(H = 0 \right) = \frac{1}{2} \text{arctg} \left(\frac{Q \sin 2\psi}{Q \cos 2\psi - 1} \right) + 90°n.$$

(7.51)

In practical use of formula (7.51), the choice of n should be guided by the physical meaning of the problem, that is, at low anisotropy, the magnetization vector is oriented near the film plane, and at large anisotropy, near the anisotropy axis. That is, when $Q \to 0$, the angle φ_0 tends to 90°, so one should take $n = 1$, and as $Q \to \infty$, the angle φ_0 tends to ψ, so one should take $n = 0$. When $\psi = 0°$ and $Q \to 1$, the formula (7.51) has a divergence. However, if we set $Q = 1$, then as ψ tends to zero in brackets, it indicates an uncertainty of type 0/0, the disclosure of which, according to the L'Hôpital's rule [261, 268, 269], gives infinity, which corresponds to $\varphi_0 = 45°$ (more precisely, minus infinity, that is, $\varphi_0 = -45°$, which for $n = 1$ gives the same). Such an angle corresponds to the equality of forces tending to turn the magnetization vector to the film plane or to the anisotropy axis. As an illustration, we give an example of the dependence of the orientation of the magnetization vector in the Oxy plane (Fig. 7.1) on the value of the quality factor Q, for which we turn to Fig. 7.2, where similar dependences are shown for several angle values ψ. Due to the mirror symmetry of the geometry of the problem with respect to the Oxz plane, as well as the bi-directionality of the uniaxial anisotropy (that is, the anisotropy energy does not change when the direction of the magnetization vector is reversed), it is sufficient to consider only positive interval of angle ψ from zero

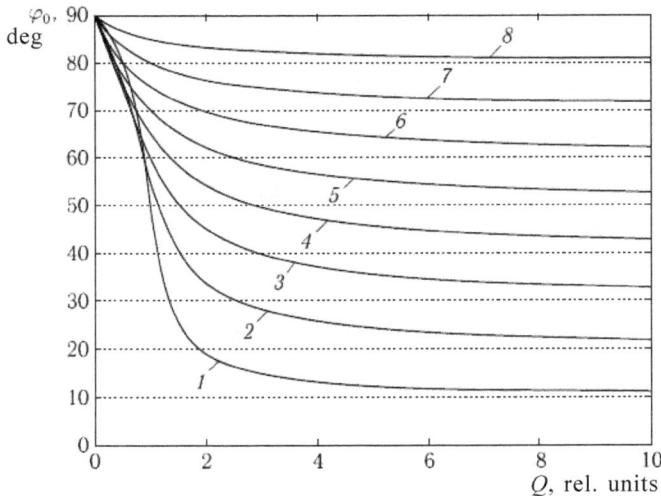

Fig. 7.2. The dependence of the azimuthal angle of the equilibrium position of the magnetization vector on the quality factor of the magnetic film for various angles ψ of deviations of the EMA from the normal to the film plane: 1 – 10°; 2 – 20°; 3 – 30°; 4 – 40°; 5 – 50°; 6 – 60°; 7 – 70°; 8 – 80°.

to 90°. In this case, the equilibrium orientation of the magnetization vector will always correspond to the smaller of the two angles between the anisotropy axis and the film plane. Curves in Fig. 7.2 are constructed according to the following formulas: when $\psi < 45°$ (curves 1–4) in the interval $0 < Q < 1 / \cos(2\psi)$

$$\varphi_0 = \frac{1}{2}\text{arctg}\left(\frac{Q\sin 2\psi}{Q\cos 2\psi - 1}\right) + 90°;$$

(7.52)

when $\psi < 45°$ (curves 1–4) in the interval $1/\cos(2\psi) < Q < +\infty$

$$\varphi_0 = \frac{1}{2}\text{arctg}\left(\frac{Q\sin 2\psi}{Q\cos 2\psi - 1}\right)$$

(7.53)

when $\psi > 45°$ (curves 5–8) in the whole interval $0 < Q < +\infty$

$$\varphi_0 = \frac{1}{2}\text{arctg}\left(\frac{Q\sin 2\psi}{Q\cos 2\psi - 1}\right) + 90°;$$

(7.54)

It can be seen from the figure that all curves in the absence of an anisotropy field, that is, at $Q = 0$, begin at $\varphi_0 = 90°$, which corresponds to the orientation of the magnetization vector exactly in the film plane. As Q increases, each curve gradually approaches its horizontal dotted line corresponding to a given angle, that is, from the orientation in the film plane, the magnetization vector rotates in the direction of the anisotropy axis.

From the figure one can see that already with $Q \sim 2$ rel. units all curves are close enough to fit in their horizontal asymptotes. The largest difference is about two times, it occurs for curve 1, corresponding to $\psi = 10°$, whereas for curve 8, corresponding to $\psi = 80°$, this difference does not exceed 10%. When $Q \sim 10$, the difference is significantly reduced and for technical curves it is 10% and less than 2%, respectively. Thus, for practical purposes, it can be assumed that, within units of percent, the equilibrium direction of the magnetization vector coincides with the direction of the axis anisotropy, starting from Q on the order of several dinits.

7.3.2. The equilibrium position of magnetization during the extrusion of the external field

In the presence of an external field, the magnetization vector, in addition to the double field, the anisotropy field and the demagnetization field, is acted upon by a third force due to the external field. This force, as well as the force of anisotropy, tends to turn the vector in its direction, so that its position is determined by the equilibrium three forces. In contrast to the anisotropy field, the external field is unidirectional, that is, the interaction energy of the magnetization with this field, when the direction of the magnetization vector is reversed, also changes. Therefore, the interval of angles from zero to 90° is not enough here, and all angles from zero to 180° must be considered. When analyzing the equilibrium orientation of the magnetization vector in the presence of a field, we will continue to rely on the expression for the energy density (7.45), the minimum of which will also be obtained by differentiation with respect to the angle φ, that is, we require the fulfillment of condition (7.42). Analytical solution of this equation with respect to the angle φ, apparently, is quite difficult, if at all possible, therefore, we turn to a numerical solution based on the zero search algorithm. We will look for the function zero:

$$F(\varphi) = K \sin\left[2(\varphi - \psi)\right] - 2\pi M_0^2 \sin(2\varphi) + M_0 H_0 \sin(\varphi - \varphi_H). \quad (7.55)$$

representing the left side of equation (7.46). Here, the field H_0, the angle ψ and the magnetization M_0 act as parameters. Changing the value of the argument φ in sufficiently small steps, we will follow the change in the sign of the function $F(\varphi)$, which will give the desired value of φ. If necessary, to obtain higher accuracy, after registering the change of the sign of $F(\varphi)$, one can go back φ one step, then reduce the step, for example, by an order of magnitude, and again pass the interval of angles φ corresponding to the original step length. Such a step reduction can be repeated several times until the technical measures until the required accuracy is achieved.

7.3.3. Orientational dependence of minimum energy

Generally speaking, in the interval of changing the angle φ from zero to 360°, the sign of $F(\varphi)$ can change more than once, so for a start it is useful to consider the general form of this function

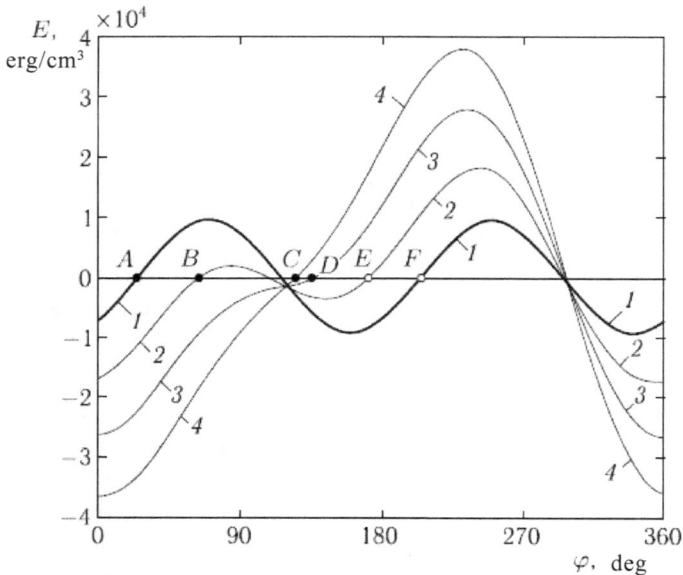

Fig. 7.3. Dependences of the function $F(\varphi)$ on the angle of orientation of the magnetization φ at $\varphi_H = 120°$ different values of the field H_0: 1 – 0 Oe; 2 – 500 Oe; 3 – 1000 Oe; 4 – 1500 Oe. Parameters: $K = 1.15 \cdot 10^4$ erg×cm^{-3}, $M_0 = 22$ G, $\psi = 20°$

depending on the angle φ for several typical field values. An example of such consideration is shown in Fig. 7.3 with the parameters of the film material: $K = 1.15 \times 10^4$ erg $\cdot$ cm^{-3}, $M_0 = 22$ G, and also with the angle ψ equal to 20°. For mixed garnet films used in the CMD technique, such parameters are quite typical [40] and adopted, for example, in the experiments described in [22–25]. For clarity of consideration, the angle φ_H is set equal to 120. By virtue of the periodicity of the function $F(\varphi)$ equal to 360°, it suffices to consider only the interval of values of $0° \leq \varphi \leq 360°$. It can be seen from the figure that in the absence of a field (curve 1, indicated by thickening), the function $F(\varphi)$ has four zeros, corresponding to the values of the angle φ equal to 26°, 116°, 206°, 296°. In this case, the first and the third of these beginnings fall on the increasing portions of the function $F(\varphi)$, and the second and fourth fall on the decreasing ones. Since the function $F(\varphi)$ is the first derivative of the energy density (7.45), the tangents at each point of this function are the second derivatives. The minimum condition is the positivity of the second derivative, so we can conclude that the points $\varphi = 26°$ (point A) and 206° (point F) correspond to the minima of the energy density, that is, the equilibrium positions of the magnetization vector in the absence of a field. In this case, the angles $\varphi = 116°$ and $296°$ correspond to maxima $F(\varphi)$ and for this review is of no interest.

The number of minima is two due to the bi-directionality of the anisotropy axis. The difference in angle between the minima is $180°$, and the difference between the smaller value of $26°$ and the angle $\psi = 20°$, which determines the direction of the anisotropy axis, is caused by the form demagnetization field film $4\pi M_0$, seeking to arrange the magnetization in the plane of the film. The same applies to the difference of the angle $\varphi = 206°$ from the opposite direction of the anisotropy axis, for which the angle ψ, measured from the axis Ox, is $200°$.

As the field H increases, the $F(\varphi)$ curve is deformed. The left of its maxima in the figure decreases, and the right increases. This leads to the fact that instead of quadruls in the interval from zero to $360°$ (curves 1, 2), only two remain (curves 3, 4). In this case, zero, in the absence of a field corresponding to point A, sequentially moves to point B, then D and after all C. The null, corresponding to point F in the absence of field, moves to point E, then D and C.

These two different paths correspond to the two initial positions of magnetization along the positive and negative directions of the anisotropy axis. The movement along the F-E-D-C chain is continuous, and the A-B-D-C chain undergoes a gap between points B and D. This gap corresponds to a sharp reorientation of the magnetization vector from one direction of the anisotropy axis to the opposite. Since the end result of the displacement of the magnetization vector is its alignment as close as possible to the field direction $\varphi_H = 120°$, from the initial position $26°$ to $120°$ the magnetization vector should turn to $94°$, and from the position $206°$ only $86°$. At the same time, the second of these heads ($86°$) is less than $90°$, so the projection of the magnetization vector on the field direction always remains positive, while the first angle ($94°$) is more than $90°$, so the projection of the magnetization vector on the field direction is initially negative and only then becomes positive. That is, in this case, the magnetization vector is forced to reorient from one direction of the anisotropy axis to the opposite, which happens in a jump-like manner. This jump can be traced in more detail with the help of Fig. 7.4, where the dependences of the function $F(\varphi)$ on the angle of orientation of the magnetization φ are shown under the same conditions as in Fig. 7.3 for two different quantities of the field H: 1 – 650 Oe; 2 – 700 Oe, located at equal distances on both sides of the 675 Oe field, corresponding to the jump. It can be seen

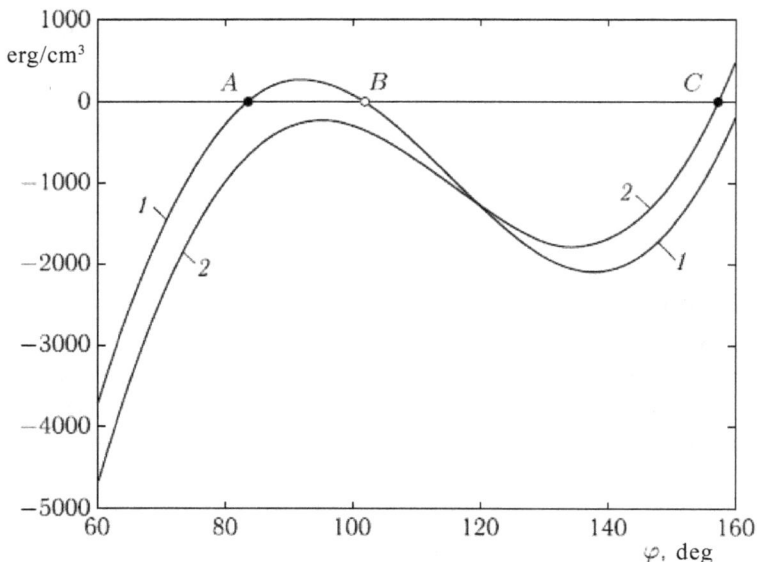

Fig. 7.4. Dependences of the function $F(\varphi)$ on the angle of orientation of the magnetization φ at $\varphi_H = 120°$ and various values of the field H: 1 – 650 Oe; 2 – 700 Oe. Parameters – the same as taken in the construction of Fig. 7.3

from the figure that, at a field of 650 Oe (curve 1), there is a zero corresponding to a minimum at 82 (point A). In this case, the sign +of the projection of the magnetization vector on the anisotropy axis is opposite to the sign of the projection of the field vector on the same axis. Zero at point B at 103° corresponds to the maximum and is not important for the present consideration.

With a field of 700 Oe (curve 2), the $F(\phi)$ dependence drops below the zero line, and the zero disappears. However, the magnetization is forced to search for a minimum, which it finds at 157° (point C). At this point, the signs of the projections of the magnetization vector and the field on the anisotropy axis coincide. Due to the significant difference between the previous and the new minima (from 82° to 157°), and these minima are on different branches of the curve, the transition from one minimum (point A) to another (point C) occurs abruptly. Change field size. As an illustration, consider

The typical dependences of the orientation of the magnetization vector ϕ on the field H_0 are presented in Fig. 7.5. For clarity, the field orientations are presented in the interval φH from zero to 360° with a step of 20°.

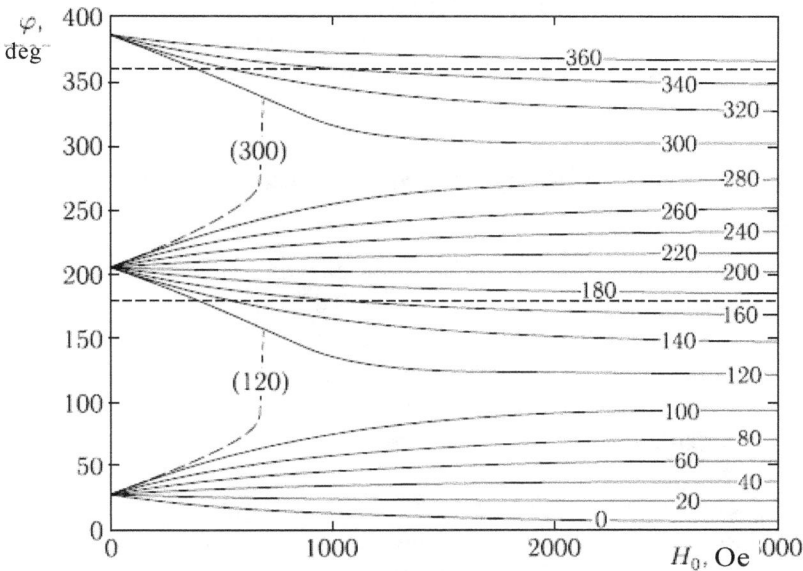

Fig. 7.5. Dependences of the azimuthal angle of the equilibrium position of the magnetization vector on the field strength for different orientations in the *Oxy* plane (Fig. 7.1). The numbers on the curves correspond to the values of the angle φ_H between the field vector and the axis *Ox* in degrees. Dotted curves marked with numbers in brackets correspond to magnetization jumps from one direction of the anisotropy axis to another. The horizontal dotted lines correspond to the angles $\varphi = 180°$ and $360°$. Parameters are the same as taken in the construction of Fig. 3.

It can be seen from the figure that all the curves, when passing through 180°, periodically repeat, which reflects the geometric symmetry of the problem in combination with the bi-directionality of the anisotropy axis. In the interval from zero to 180°, all the curves at $H_0 = 0$ begin at one point corresponding to $\varphi = 26°$, which is determined by the equilibrium between the anisotropy and demagnetization fields. In the interval from 180° to 360°, the curves begin, respectively, at 206° (180° + 26°). As the field H is increased, it tends to drag the magnetization vector to its own direction, with the result that the curves diverge like a fan, tending each to an angle corresponding to its own field orientation. It can be seen that, already at a field of about 1500 Oe, the deviation of the magnetization vector from the field direction in any case does not exceed fifteen degrees, and with further increase in the field decreases even more. For example, for a curve corresponding to

$\varphi_H = 100°$ (that is, the field deviation from the EMA direction by an angle close to the right angle), the angle φ with a field of 1500 Oe is about 85°, with a field of 2000 Oe – about 90°, and with a field 3000 Oe reaches already 95°, that is, the magnetization deviates from the field direction by only 5°.

7.3.4. Intermittent change in the orientation of the magnetization vector

With an initial orientation of 26°, the curves corresponding to fields with angles φ_H above 26° + 90° = 116°, jump to curves starting at an angle of 206°. One of these, corresponding to $\varphi_H = 120°$, is shown by a dotted line. The jump occurs at a field of about 675 Oe in accordance with the mechanism illustrated in Fig. 7.4. A similar jump experiences a curve starting at 206° and corresponding to $\varphi_H = 300°$.

The curves starting at 26° and corresponding to $\varphi_H = 140°$ and 160° also experience a jump to the curves corresponding to the same angles, but starting at 206°. For the purpose of simplification, these curves are not shown in the figure. Similar jumps have the curves starting at 206° and corresponding to $\varphi_H = 320°$ and 340° (also not shown).

The field at which the jump occurs is strongly dependent on the orientation of the external field. The approximate dependence of the hopping field H_c on the angle φ_H is illustrated in Fig. 7.6, which is built with the technical parameters as Fig. 7.5 for the initial position of the magnetization $\varphi = 26°$. It can be seen that the nature of the dependence is rather complicated: there are two minima of unequal depth and a maximum between them. The sidebar shows a diagram of characteristic orientations of the external field using a unit circle. The circle is depicted in the Oxy plane (view from the Oz axis). The small rectangle inside the circle is the projection of the film on the Oxy plane (a view of the film from the end). the oblique thickened line is a projection of the EMA on the same plane. First of all, it should be noted that in the range of external field orientations $-10 \leq \varphi_H \leq 100°$ a jump is absent. On the unit circle, the ends of this interval are shown by the points A and B. The minima correspond to angles of 160° (point C) and 270° (point E), the maximum is 200° (point D). From the sidebar it can be seen that the position of the maximum (point D) exactly corresponds to the orientation of the field along the EMA, but in the opposite direction to the original

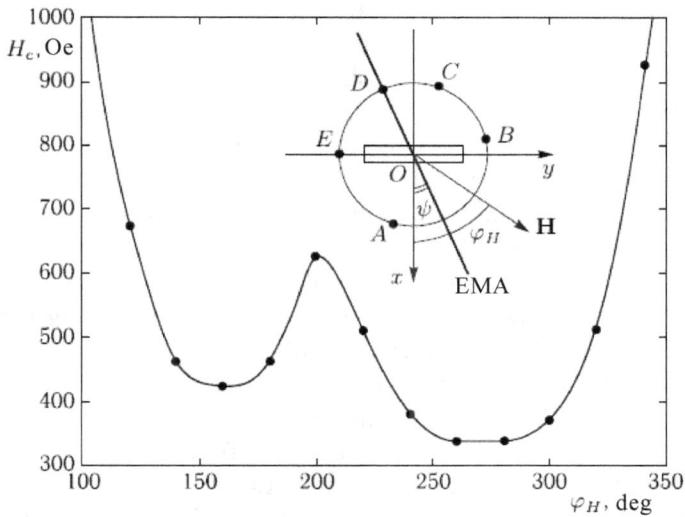

Fig. 7.6. The dependence of the jump field on the angle of orientation of the external field at the initial position of the magnetization φ = 26°. Points are the results of a machine experiment. Line – the result of the approximation. The sidebar shows the basic angle diagram on the unit circle. Parameters – the same as taken in the construction of Fig. 7.3

(26°). In these words, the field tends to reorient the magnetization vector from the original direction to the opposite direction in the near opposite direction of both directions (magnetization and field). At the same time, as can be seen from (7.40), although the energy of interaction of magnetization with the field is large, but the shoulder of the force turning the magnetization in the direction of the field, proportional to the sine of the angle between the vectors (close to 180°), is small, which makes this hopping threshold difficult H_c rises. At the same time, the shoulder of the same force at points C and E, where the angle between the field and the magnetization approaches the straight line, is relatively large. In this case, the magnetization in the direction of the field rotates easily, which ensures minima H_c at angles of 160 °and 270°.

The minimum at φ_H = 270° is deeper than at 160°, apparently because this angle just corresponds to the orientation of the field in the film plane, whereas at φ_H = 160°, the field to the film plane is almost perpendicular.

Note that the reasons for the nature given here on the curve in Fig. 7.6, are of rather qualitative character, corresponding to the accepted values of the parameters. More precisely, the nature of the jump can be traced by analyzing the energy density (7.40) and also character of the curves similar shown in Figs. 7.3 and 7.4.

Summarizing the above consideration, we note that the equilibrium direction of the magnetization vector approaches the direction of the fields starting from fields of the order of 600–700 Oe. It can be seen that this field size is quite close to the magnitude of the demagnetization field ($H_m = 4\pi M_0 = 276$ Oe), so that it exceeds it no more than two or three times. This proximity becomes clear if we consider that the jump is determined by the struggle between the total action of the external field and the anisotropy field on the one hand and the action of the demagnetization field on the other. Thus, the first of these magnets, when magnetized near the normal to the film plane, tends to turn off the magnetization from the film plane, while the second one turns the magnetization into this plane. The action of the second is manifested in fields that are only comparable with the magnitude of the demagnetization field, whereas the action of the first has no upper limit. Therefore, a jump is possible only when the first field is less than the second or the same order with it. When the external field reaches sufficiently large quantities, the demagnetization field becomes negligible, and for the magnetization of the choice between the orientations of the pulling fields no longer remains, so that it asymptotically tends to a resultant external field and anisotropy field. It should be noted that in practical calculations it is necessary to take into account the possibility of the magnetization vector hopping, which can be avoided by choosing the initial direction of this vector that differs from the field direction by no more than 90°.

7.3.5. Ferromagnetic resonance conditions

The main objective of this study is to consider the resonant vibrations of magnetization in films with uniaxial anisotropy, the axis of which is deviated from the normal to the film plane. Since the conditions of the equilibrium position of the magnetization for this case (field changes in the plane containing the EMA) are now identified, consider the natural frequency of ferromagnetic resonance, for which we use the Smit–Sul method (Section 2.3), according to which the FMR frequency is (2.149):

$$\omega = \frac{\gamma}{M_0 \sin \theta_0} \sqrt{U_{\theta\theta} U_{\varphi\varphi} - U_{\theta\varphi}^2},$$

$$(7.56)$$

where the derivatives of the energy density are taken in the equilibrium position of the magnetization vector.

Let us proceed from the expression for the energy density (7.40):

$$U = -K\sin^2\theta\cos^2(\varphi-\psi) +$$
$$+2\pi M_0^2 \sin^2\theta\cos^2\varphi - M_0 H_0 \sin\theta\cos(\varphi-\varphi_H). \tag{7.57}$$

Differentiating with respect to θ and φ, we obtain the derivatives:

$$U_{\theta\theta} = -2K\cos2\theta\cos^2(\varphi-\psi) +$$
$$+4\pi M_0^2 \cos2\theta\cos^2\varphi + M_0 H_0 \sin\theta\cos(\varphi-\varphi_H), \tag{7.58}$$

$$U_{\varphi\varphi} = 2K\sin^2\theta\cos[2(\varphi-\psi)] -$$
$$-4\pi M_0^2 \sin^2\theta\cos 2\varphi + M_0 H_0 \sin\theta\cos(\varphi-\varphi_H), \tag{7.59}$$

$$U_{\theta\varphi} = K\sin2\theta\sin[2(\varphi-\psi)] -$$
$$-2\pi M_0^2 \sin2\theta\sin 2\varphi + M_0 H_0 \cos\theta\sin(\varphi-\varphi_H), \tag{7.60}$$

Comment. Here, the anisotropy energy density is used in the form (7.38). By direct verification, we can verify that the use of the formula (7.36) leads to the same expressions for the derivatives (7.58)–(7.60). The same applies to the demagnetization energy density (7.17), which allows the replacement of m_x^2 by $1 - m_y^2 - m_z^2$. According to (7.32), in the equilibrium position, $\theta_0 = 90°$. Substituting this value into (7.58)–(7.60), we get:

$$U_{\theta q} = 2K\cos^2(\varphi-\psi)4\pi M_0^2\cos^2\varphi + M_0 H_0 \cos(\varphi-\varphi_H); \tag{7.61}$$

$$U_{\varphi\varphi} = 2K\cos[2(\varphi-\psi)] - 4\pi M_0^2 \cos2\varphi + M_0 H_0 \cos(\varphi-\varphi_H); \tag{7.62}$$

$$U_{\theta\varphi} = 0. \tag{7.63}$$

To use the Smit–Sul method in these expressions, the angle φ must correspond to the equilibrium position of the magnetization. We introduce auxiliary notation:

$$A = H_a\cos^2(\varphi-\psi) - H_m\cos^2\varphi; \tag{7.64}$$

$$B = H_a \cos[2(\varphi - \psi)] - H_m \cos 2\varphi; \qquad (7.65)$$

where the standard notation is used – anisotropy fields (7.48)

$$H_a = 2K/M_0 \qquad (7.66)$$

and demagnetization fields (7.49)

$$H_m = 4\pi\, M \qquad (7.67)$$

With the notation (7.64) and (7.65) from (7.61) and (7.62) we get:

$$U_{\theta\theta} = M_0 \left[A + H_0 \cos(\varphi - \varphi_H) \right] \qquad (7.68)$$

$$U_{\varphi\varphi} = M_0 \left[B + H_0 \cos(\varphi - \varphi_H) \right] \qquad (7.69)$$

Substituting these derivatives into the Smit–Sul formula (7.56) and taking into account that $\theta_0 = 90°$, we obtain the frequency of ferromagnetic resonance:

$$\omega = \gamma \sqrt{\left[A + H_0 \cos(\varphi - \varphi_H) \right]\left[B + H_0 \cos(\varphi - \varphi_H) \right]} \qquad (7.70)$$

Change of the field orientation. We now consider how the frequency of the ferromagnetic resonance ω depends on the field H_0 with different orientations in the *Oxy* plane (Fig. 7.1). The corresponding dependences with a step in the field orientation of 20° are shown in Fig. 7.7. The curves are plotted in the interval φ_H from zero to 180°, and the curve for 180° coincides with the curve for $\varphi_H = 0°$, and with a further increase in the angle, all the curves in frequency repeat from time to time.

It can be seen from the figure that all the curves emanate from one point corresponding to the angle $\varphi = 26°$, which is determined by the equilibrium between the anisotropy and demagnetization fields. Further, with sufficiently large fields (here, more than 1000 Oe), the curves have an increasing character, which corresponds to a linear increase in the resonance frequency from a field of type $\omega \sim \gamma_H$, when this field dominates in magnitude over all others, in this case over anisotropy and demagnetization . In cases when the field orientation is close to the initial orientation of the magnetization

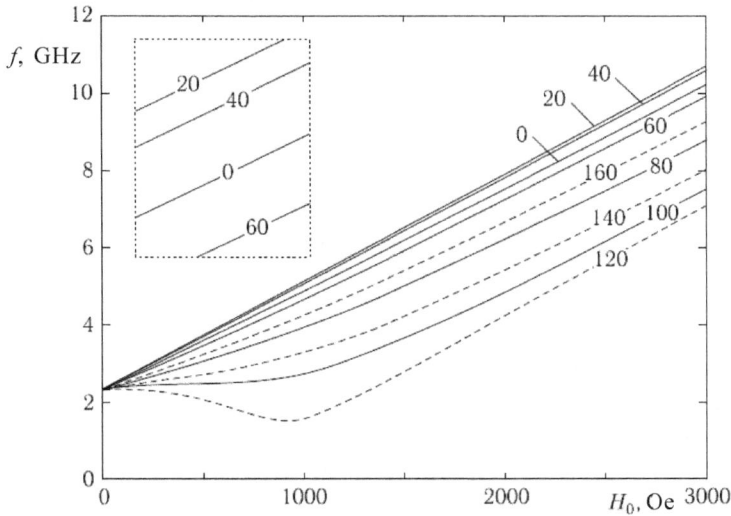

Fig. 7.7. Dependences of the frequency of ferromagnetic resonance on the field H_0 for different orientations in the Oxy plane (Fig. 7.1). The numbers on the curves correspond to the values of the angle φ_H between the field vector and the axis Ox in degrees. At the top left of the inset is a diagram of the location of curves at low angles. The vertical axis represents the linear frequency associated with the circular relationship: $f = \omega/2\pi$, with $\gamma = 2.8$ MHz/Oe. Parameters – the same as taken in the construction of Fig. 7.3: $K = 1.15 \cdot 10^4$ erg cm^{-3}, $M_0 = 22$ G, $\psi = 20°$

vector ($0° \leq \varphi_H \leq 80°$ and $140° \leq \varphi_H \leq 180°$), the curves all increase only. If the field orientation is different from the initial one for magnetization, which is the case for curves with $\varphi_H = 100°$ and $120°$ (i.e. the difference close to $90°$), the growth of the curves on the initial section slows down and can generally become negative, as for the curve $\varphi_H = 0°$, which corresponds to a decrease in the frequency in the 'soft mode' (section 6.3.5, Fig. 6.6) [6–8, 31, 133, 134].

The mutual arrangement of curves is illustrated in addition to Fig. 7.8, where the dependences of the frequency of the ferromagnetic resonance on the angle φ_H are shown for some typical values of the field H_0. It can be seen from the figure that with fixed fields in the region of average angles (over $30°$), as the angle φ_H increases, the frequency to $\varphi_H = 120°$ first decreases, after which, passing through a minimum, up to $\varphi_H = 180°$ increases, tending to the value at $\varphi_H = 0°$. Such a minimum of frequency corresponds to its decrease in the process of approaching the conditions of the 'soft mode'. In the region of small angles (less than $30°$), the frequency first increases to $\varphi_H = 20°$, after which it decreases (as shown in the inset in Fig. 7.7).

This local increase corresponds to the orientation of the field near the initial position ($\varphi = 26°$), which is determined by the equilibrium between the anisotropy and demagnetization fields.

At a field of about 1000 Oe, the minimum of the dependence of the frequency f on the angle φ_H is the deepest, since here the deviation of the magnetization vector from the field is still noticeably manifested, and at large fields (2000–3000 Oe) the minimum depth decreases and the curves become similar to each other.

7.3.6. Resonance field

The curves shown in Figs. 7.7 and 7.8 as well as the formula (7.70) reflect the change in the resonant frequency with different field variations. However, in the experiment using standard cut-off spectrometers, the frequency, as a rule, remains constant, and the resonance conditions are achieved by changing the field. For example, in the EPA-2M spectrometer, experiments with which are described in [22–25], the frequency is 10 GHz, and the field changes from 0 to 5000 Oe. To find the value of the resonant field at a constant frequency, it is necessary to resolve the formula (7.70) relative to the field. After squaring, this formula takes the form of a quadratic equation, the solution of which has the form (only the positive root is preserved)

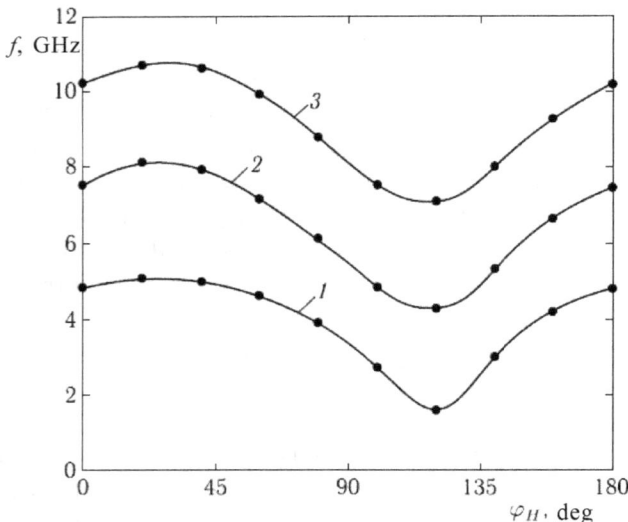

Fig. 7.8. Dependences of the frequency of ferromagnetic resonance on the angle φ_H for various amounts of $\underline{H}_0$: 1 – 1000 Oe; 2 – 2000 Oe; 3 – 3000 Oe. Parameters - the same as taken in the construction of Fig. 7.7

$$H_0 = \frac{1}{\cos(\varphi - \varphi_H)}\left[-\frac{A+B}{2} + \sqrt{\left(\frac{\omega}{\gamma}\right)^2 + \left(\frac{A-B}{2}\right)^2}\,\right] \cdot \qquad (7.71)$$

where A and B are defined by formulas (7.64) and (7.65), in which the angle φ corresponds to the equilibrium orientation of the magnetization (and here also).

In this form, formula (7.71) is 'self-consistent', as to use it, one must first find the equilibrium position of the magnetization, which itself is determined by the value of the applied field through the angle φ entering into A and B. That is, (7.71) is essentially an equation for H_0, which contains the solution of equation (7.46). As noted above, the solution of equation (7.46) in an analytical form is hardly possible, therefore, equation (7.71) cannot be resolved either. In a numerical solution, it is apparently possible to use the iteration method [261,351,352], that is, in the first step, specify some field value that is more or less close to reality, and then find the equilibrium position of the magnetization from equation (7.46), which can be used to find fields according to the formula (7.71). Further, this obtained field value should be again substituted in (7.46), find the new value of the angle φ, find H_0 through it and so on until the desired accuracy is achieved. On the other hand, for practice, you can use the formula (7.70), with the help of which, by setting the field with a certain step, one can get the dependence of frequency on the field. In this case, the equilibrium value φ will already be included in the resulting dependence as a necessary previous step of using formula (7.70).

According to the obtained dependence of frequency on the field, one can determine the value of the field that corresponds to the desired frequency. This procedure should be done for all required values of the field orientation (angle φ_H), which will result in the desired dependence of the resonance field on the angle φ_H at a given frequency.

The dependences of the resonant field on the angle φ_H obtained as a result of this procedure for given values of the stack are shown in Fig. 7.9. Points are the results of a computer experiment, the curves are constructed according to the following empirical formulas: curve 1 – frequency 10 GHz,

$$H = 630\sin\left[2(\varphi_H - 70°)\right] + 3400; \qquad (7.72)$$

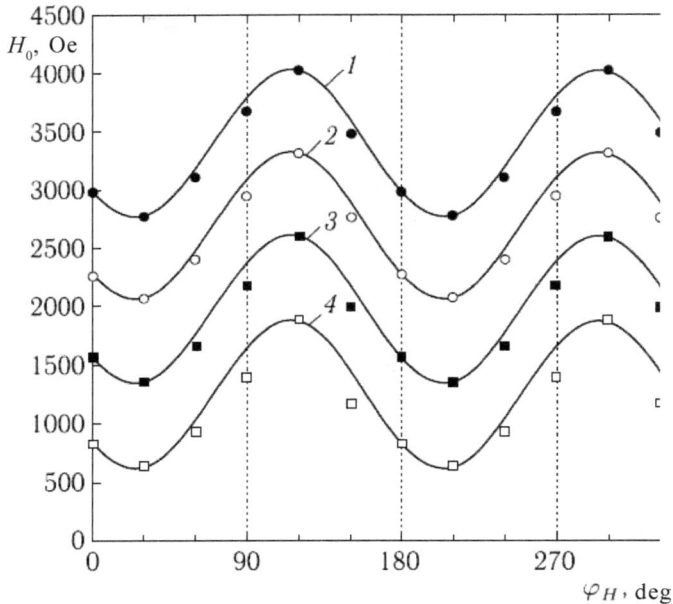

Fig. 7.9. Dependence of the resonant field on the angle of its orientation for given values of the frequency: 1 - 10 GHz; 2 - 8 GHz; 3 - 6 GHz; 4 - 4 GHz. Points are the results of a machine experiment. Curves are constructed according to the formulas (7.72) - (7.75) given in the text. Parameters - the same as taken in the construction of Fig. 7.7

curve 2 – frequency 8 GHz,

$$H = 630\sin\left[2\left(\varphi_H - 70°\right)\right]2700;$$ (7.73)

curve 3 – frequency 6 GHz,

$$H = 630\sin\left[2\left(\varphi_H - 70°\right)\right]+1980;$$ (7.74)

curve 4 – frequency 4 GHz,

$$H = 630\sin\left[2\left(\varphi_H - 70°\right)\right]+1260;$$ (7.75)

where the angle φ_H is taken in degrees, and the field is obtained in Oersteds.

It can be seen from the figure that all curves have a periodicity of 180°, which requires the coefficient 2 under the sine sign in formulas (7.72)–(7.75). All curves have the first minimum at $\varphi_H = 26°$, which corresponds to the equilibrium orientation of the magnetization in the absence of a field. The maxima of the curves are at $\varphi_H = 116°$, which corresponds to the perpendicular to the equilibrium orientation. With increasing frequency, the curves rise up the field fairly evenly in increments of about 350 Oe by 1 GHz.

It can be seen that the machine curves are rather close to sinusoids, however there is some difference on the slopes on both sides of the maximum, where the points lie slightly below the curves constructed by formulas (7.72)–(7.75). This mismatch with decreasing frequency increases. So at the midpoint of the clones, the difference at a frequency of 10 GHz does not exceed 6%, and at a frequency of 4 GHz it reaches 36%. It can be assumed that such an increase in the mismatch is due to an increase in the deviation of the magnetization vector from the field direction with decreasing frequency, since the field required for the resonance decreases.

7.3.7. Strong field approximation

The above consideration does not contain any assumptions about the mutual magnitude of the components included in formula (7.71), so this formula is suitable for any relationship between the anisotropy field and the magnetization of the film. However, very important for practice, especially in the technique of CMD, are films of mixed ferrite garnet, typical parameters for which the following [40]:

$$M_0 \sim 10. \ . \ . \ 100 \ \text{Gs}; \ K \sim 10^3...10^5 \ \text{erg} \times \text{cm}^{-3}; \ \psi \sim 5°...30°.$$

With such frequencies on the order of 10 GHz, the second term under the sign of the radical, as a rule, is two orders of magnitude less than the first, so that it can be neglected. Further, with a field net of not more than 1000 Oe, we can assume that the external field corresponding to the resonance at a frequency of about 10 GHz is 3000–4000 Oe, so that it is strong enough to rotate the magnetization in its direction. That is, we can put $\varphi = \varphi_H$. In these cases, the formula (7.71) is greatly simplified and takes the form

$$H_0 \frac{\omega}{2} + \frac{A+B}{2}.$$

$$(7.76)$$

where A and B are defined by the formulas (7.64) and (7.65), which under the assumption made (that is, $\varphi = \varphi_H$) have the form:

$$A = H_0 \cos^2\left(\varphi_H - \psi\right) - H_m \cos^2 \varphi_H; \qquad (7.77)$$

$$AB = H_0 \cos^2\left[\left(\varphi_H - \psi\right)\right] - H_m \cos 2\, \varphi_H; \qquad (7.78)$$

Substituting (7.77) and (7.78) into (7.76), we get:

$$H_0 - H\,\frac{\omega}{\gamma} + \frac{1}{2}\left(H_a - H_m\right) - \frac{3}{2}\left[H_a \cos^2\left(\varphi_H - \varphi\right) - H_m \cos^2 \varphi_H\right]. \,(7.79)$$

The dependences of the resonance field on the angle φ_H in the approximation of a sufficiently strong field is illustrated by Fig. 7.10. Here the lines are constructed according to the formula (7.79).

From the figure we can see that the dependence is similar to that shown in Fig. 7.9, has a sinusoidal character with a frequency of 180°. The amplitude of dependence is also close to that in Fig. 7.9. The magnitude of the resonant field with decreasing frequency also decreases linearly in the field, which is provided by the first term in formula (7.79). A more detailed examination shows that the lines

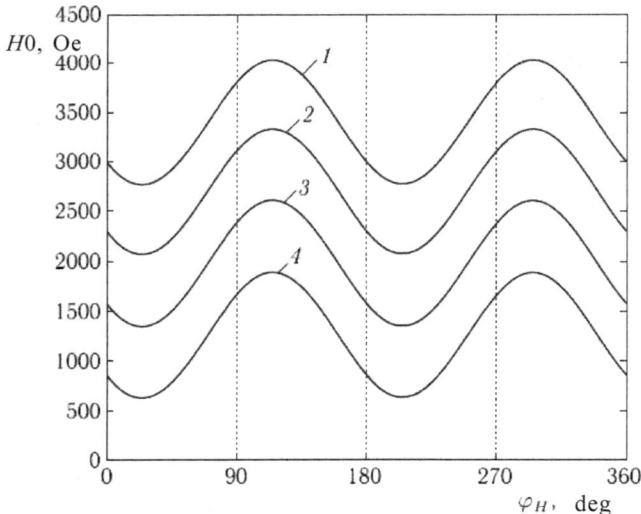

Fig. 7.10. Dependence of the resonant field on the angle of its orientation for given values of the frequency: 1 – 10 GHz; 2 – 8 GHz; 3 – 6 GHz; 4 – 4 GHz. The curves are constructed according to the formula (7.79). Parameters – the same as taken in the construction of Fig. 7.7

in this figure with a high degree of accuracy coincide with those in Fig. 7.9 (the difference is no more than a few percent, especially in areas in the growing curves), therefore, the approximation accuracy is about the same. Thus, the accuracy of the approximation of a strong field is the better, the higher the frequency and coincides with the accuracy of the curves constructed in Fig. 7.9 according to empirical formulas (7.72)–(7.75).

Comment. The strong-field approximation given in this section is focused precisely on films of mixed ferrite garnets with the parameters mentioned above. However, in other materials, the parameters may differ significantly. So orthoferrites, widely studied at the beginning of the development of CMD techniques [40], have anisotropy fields of up to tens of thousands of oersteds, so that in practically achievable fields up to 20 thousand oersteds the equilibrium direction of the magnetization vector may differ very strongly from the field direction. The barium hexaferrite, studied in [31] by the magnitude of the anisotropy field, occupies an intermediate position between orthoferrites and mixed garnets (H_a about 19 000 Oe), but even there the authors had to use a magnet that gives up to 25 000 Oe. On the other hand, for mixed garnet-ferrites in [22–25], the strong field approximation gave results that agree quite well with the experiment (with an accuracy of 10%). Therefore, for such materials it can be considered quite suitable, then as for other materials, the applicability of the method requires a separate check, taking into account the correctness of the made approximations.

7.4. Option two: the axis of easy magnetization lies in the plane perpendicular to the plane of the field change

In this case, in the notation of Section 7.3, the polar and azimuth angles of the EMA are equal to:

$$\theta_a = 90° - \psi_2 \qquad (7.80)$$
$$\varphi_a = 0° \qquad (7.81)$$

From the general geometry (Fig. 7.1), it can be seen that when the field **H** is rotated in the Oxy plane, it passes a position that exactly corresponds to the projection of the EMA on this plane, that is, such that $\varphi_H = \varphi_a = 0°$.

Given the dual nature of the expressions for the energy density of anisotropy and demagnetization (Sections 4.5.1, 4.5.2), we will use those expressions that are more consistent with the following the qualitative physical meaning of the problem under consideration, namely: for the energy density of uniaxial anisotropy with an arbitrary direction of the axis (4.86):

$$U_a^{(2)} = -K\left\{ m_x^2 \sin^2\theta_a \cos^2\varphi_a + m_y^2 \sin^2\theta_a \sin^2\varphi_a + \right.$$

$$m_z^2 \cos^2\theta_a + 2m_x m_y \sin^2\theta_a \sin\varphi_a \cos\varphi_a +$$

$$\left. +2m_x m_z \sin\theta_a \cos\theta_a \cos\varphi_a + 2m_y m_z \sin\theta_a \cos\theta_a \sin\varphi_a \right\}, \tag{7.82}$$

and for the demagnetization energy density (4.206):

$$U_m^{(2)} = -2\pi M_0^2 \left(m_y^2 + m_z^2 \right). \tag{7.83}$$

Substituting into these formulas (7.80) and (7.81) and writing the components of magnetization through spherical variables (7.4)–(7.6), and also considering that the total energy, in addition to U_a, also contains the demagnetization energies (7.83) and interactions with the external field (4.208), we obtain the total energy in the form (as in the previous section, here the index 2 at the angle ψ is omitted, not forgetting, however, that the angles ψ_1 and ψ_2 differ from each other according to the geometry shown in Fig. 7.1)

$$U = -K\left(\cos\psi\sin\theta\cos\varphi + \sin\psi\cos\theta\right)^2 -$$

$$-2\pi M_0^2\left(\sin^2\theta\sin^2\varphi + \cos^2\theta\right) - M_0 H_0 \sin\theta\cos\left(\varphi - \varphi_H\right). \tag{7.84}$$

7.4.1. The equilibrium position of magnetization is the absence of an external field

We first consider the equilibrium position of the magnetization in the absence of an external field ($H_0 = 0$). We will assume that the anisotropy field exceeds the demagnetization field. In this case, from the geometric symmetry of the problem (Fig. 7.1), it can be seen that the magnetization vector is always in the Oxz plane, so that $\varphi = 0$. Suppose further that the EMA is perpendicular to the film plane, that is, oriented along the Ox axis, so that $\psi = 0$. In this case, the energy density (7.84), up to a constant term, takes the form

$$U = -\left(K - 2\pi M_0^2\right)\sin^2\theta. \tag{7.85}$$

In accordance with (7.48) and (7.49), passing from the constants K and M_0 to the anisotropy H_a and the demagnetization fields H_m, we can write (7.85) as

$$U = -\frac{M_0}{2}\left(H_a - H_m\right)\sin^2\theta. \tag{7.86}$$

It can be seen that at $H_a > H_m$ the minimum of energy, that is, the equilibrium position of the magnetization, corresponds to $\theta = 90°$, so that the magnetization is perpendicular to the film plane and coincides in direction with the EMA. When $H_a < H_m$, the energy minimum corresponds to $\theta = 0°$, so that the magnetization vector lies in the film plane.

When $\psi \neq 0$, by analogy with (7.85), up to a constant term, we get

$$U = -\left(K\frac{\sin^2(\psi+\theta)}{\sin^2\theta} - 2\pi M_0^2\right)\sin^2\theta, \tag{7.87}$$

or in the record through the fields

$$U = -\frac{M_0}{2}\left(H_a\frac{\sin^2(\psi+\theta)}{\sin^2\theta} - H_m\right)\sin^2\theta. \tag{7.88}$$

It can be seen that when $\psi = 0$, these expressions go to (7.85) and (7.86), respectively.

Comment. It is important to note that when $\psi = 0$ (that is, the EMA is exactly perpendicular to the film plane), the dependences of the demagnetization energy density and uniaxial anisotropy on the angle θ are completely equivalent, since both terms in brackets go to expressions (7.85) and (7.86) equally. However, when $\psi = 0$ (that is, when the EMA is deviated from the normal to the film plane), this equivalence is violated, since in the first expressions (7.87) and (7.88), the dependence on the angle θ appears. The equivalence of the K and M_0 contributions to the energy density (7.85), which takes place in films with normal EMA, prevents the experiment from distinguishing these parameters separately, that is, measuring the magnetization separately and the anisotropy constant. However, in films with deviations of the EMA from the normal, that is, when $\psi \neq 0$, due to the marked nonequivalence of the K and M_0 contributions, such a separation is possible, and experiments with separate determination of the magnetization and anisotropy constant are more successful than the angle is greater.

We now find the equilibrium position of the magnetization vector in the absence of a field for an arbitrary angle. As noted above, when $H_a > H_m$ should be $\varphi = 0$. The energy density (7.84) in this case takes the form

$$U = -K\sin^2(\psi + \theta) - 2\pi M_0^2 \cos^2\theta. \qquad (7.89)$$

Differentiating with respect to θ and equating the derivative to zero (in accordance with condition (7.41)), we obtain

$$-K\sin\left[2(\psi + \theta)\right] + 2\pi M_0^2 \sin^2\theta = 0. \qquad (7.90)$$

Transforming the first term by the sum sine formula, giving similar terms, solving the equation for θ and considering that, according to the geometry of the spherical coordinate system, it should always be $\theta \geq 0$, we get

$$\theta_0 - \frac{1}{2}\arctan\left(\frac{K\sin 2\psi}{K\cos 2\psi - 2\pi M_0^2}\right) + 90°\cdot n. \qquad (7.91)$$

where $n = 0$ with $K\cos 2\psi \leq 2\pi M_0^2$ and $n = 1$ for $K\cos 2\psi > 2\pi M_0^2$. Writing the same through the field, we get

$$\theta_0 = -\frac{1}{2}\arctan\left(\frac{H_a \sin 2\psi}{H_2 \cos 2\psi - H_m}\right) + 90°\cdot n. \qquad (7.92)$$

where $n = 0$ with $H_a \cos 2\psi \leq H_m$ and $n = 1$ for $H_a \cos 2\psi > H_m$.
Like (7.50), introducing the 'quality factor' of the film

$$Q = H_a/H_m, \qquad (7.93)$$

we get

$$\theta_0 - \frac{1}{2}\arctan\left(\frac{Q\sin 2\psi}{Q\cos 2\psi - 1}\right) + 90°\cdot n. \qquad (7.94)$$

where $n = 0$ with $Q \leq 1/\cos 2\psi$ and $n = 1$ for $Q > 1 / \cos 2\psi$.

Comment. The definition of the polar angle of the equilibrium position of the magnetization given here is based on an expression for the anisotropy energy density of the form (4.206) (or here (7.83)). At the same time, as shown in Section 4.5.2, the equal right to existence has an expression for the energy density of the form (4.204). However, performing actions similar to those shown in this section shows that the expression (4.204) leads to the same value of the equilibrium angle θ_0, as the formulas (7.91), (7.92) and (7.94).

The dependence of the polar angle of the equilibrium position of the magnetization θ_0 on the magnitude of the deviation of the EMA from the normal to the film plane ψ is shown in Fig. 7.11. Different curves correspond to different values of the quality factor. One can see from the figure that the curves corresponding to $Q > 1$ (1, 2, 3) begin at $\theta_0 = 90°$ (the magnetization is perpendicular to the film plane), curve 4, corresponding to $Q = 1$, starts at $\theta_0 = 45°$ (the magnetization is inclined midway between the EMA and the film plane), and all curves corresponding to $Q < 1$ (5, 6, 7) begin at $\theta_0 = 0°$ (the magnetization lies in the film plane).

Fig. 7.11. The dependence of the equilibrium angle θ_0 on the angle ψ in the absence of a field with different values of the film quality factor Q: 1– 20; 2 – 2; 3 – 1.2; 4 – 1.0; 5 – 0.8; 6 – 0.4; 7 – 0.1

In this case, all curves terminate at $\theta_0 = 0°$ (the magnetization is in the film plane). With a large quality factor (curve 1, $Q = 20$), the dependence is almost a straight line, the magnetization vector is almost always oriented along the EMA, the demagnetization field has little effect. With a decrease in the quality factor, a dip appears in the dependence of θ_0 on ψ in the region of average values (curve 2, $Q = 2$), which is deeper than the quality factor less (curve 3, $Q = 1.2$). However, all the curves still begin at $\theta_0 = 90°$, since at small angles ψ the shoulder of the force acting on the magnetization vector from the demagnetizing field is not enough to turn the vector in its direction.

The critical value of the quality factor, in which the forces from the anisotropy field and the demagnetizing field of the film are compared to each other, corresponds to $Q = 1$ (curve 4, shown by a thickened line). In this case, when $\psi = 0°$, the magnetization vector is established exactly in the middle between the direction of the EMA and the film plane under 45° to each of them. Further, as ψ increases, the equilibrium position of the magnetization shifts closer to the film plane, and at $\psi = 90°$ it exactly fits into this plane. When the value of the quality factor is less than unity (curves 5, 6, 7), both ends of the dependences fall on $\theta_0 = 0°$. The position of the lower end here is determined by the smallness of the force shoulder on the side of the demagnetizing field, and the position of the upper end is determined by the EMA orientation exactly in the film plane. However, between these positions, the anisotropy field still manages to extend the magnetization vector from the zero orientation, with the result that a maximum appears on curves 5, 6, 7, the less pronounced the smaller the quality factor. At $Q = 0.1$ (curve 7), the magnetization vector already almost always remains near the film plane for any σ values, since here the demagnetization field exceeds the anisotropy field by an order of magnitude. The curves shown in Fig. 7.11 reflect the equilibrium orientation of the magnetization as a function of the normalized parameter Q. However, in practice it is often useful to know this orientation, expressed in terms of the absolute parameters of the film, such as the saturation magnetization M_0 and the anisotropy constant K. First, find the normalized parameter Q, then by the specified angle знач using formula (7.94) find the value θ_0. So, with the values of film parameters taken in previous cases: $K = 1.15 \times 10^4$ erg cm^{-3}, $M_0 = 22$ G, we obtain an anisotropy field $H_a = 1046$ Oe, a demagnetization field $H_m = 276$ Oe, from which $Q = 3.78$.

In the case of $\psi = 20°$, from (7.94) we obtain $\theta_0 = 64°$, that is, the value that just complements the value obtained in the previous sections from $26°$ to $90°$. This coincidence is due to the fact that the geometry of the problem (Fig. 7.1) with respect to the Ox axis in the absence of an external field has cylindrical symmetry. In this case, the rotation of the film around this axis by $90°$ (from position 1 to position 2) does not change the orientation of the equilibrium position of the magnetization relative to the anisotropy axis.

Comment. The formulas (7.91)–(7.92), (7.94) given here, as well as Fig. 7.11, were obtained under the assumption that $\varphi = 0$, which is unconditionally performed for $H_a > H_m$, that is, $Q > 1$. The generalization of these textures to the case $H_a < H_m$, that is, $Q < 1$ is not quite legitimate, since this can be $\varphi \neq 0$, so here is mainly to illustrate the behaviour of the angle θ in this case. A more correct solution, that is, an appeal to the complete system of equations for θ and φ, obtained by differentiating the expression for energy density (7.84), in an analytical form, if it is possible, then, apparently, it is extremely cumbersome. Therefore, it is more appropriate to use a

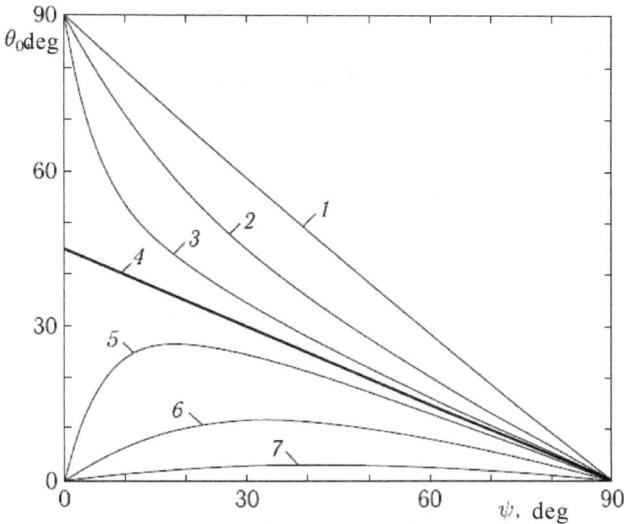

Fig 7.11. Dependence of the equilibrium angle θ0 on the angle ψ in the absence of a field at various values of the film quality factor Q: $1 - 20$; $2 - 2$; $3 - 1.2$; $4 - 1.0$; $5 - 0.8$; $6 - 0.4$; $7 - 0.1$

numerical solution by the method of determination (Section 5.1.4). An example of such a solution, including in the presence of an external field, is given in the following sections.

7.4.2. The equilibrium position of the magnetization during the extrusion of the external field

As in the previous case, in the presence of an external field on the magnetization vector, besides the double field, the anisotropy field and the demagnetizing field, a third force acts due to the external field. This force, like the first two, seeks to turn the vector to its direction, so that its position is determined by the balance of the three forces. In contrast to the anisotropy field, the external field is unidirectional, that is, the interaction energy of the magnetization with this field, when the direction of the magnetization vector is reversed, also changes. Therefore, the interval of angles from zero to 90° is not enough here, and all angles from zero to 180° must be considered. When analyzing the equilibrium orientation of the magnetization vector in the presence of a field, we will continue to rely on the expression for the energy density (7.84), the minimum of which, unlike the previous case, requires the fulfillment of two conditions (7.41) and (7.42) at once. That is, the energy density (7.84) must now be differentiated by two variables, θ and φ, after which the resulting derivatives equate to zero.

Performing differentiation by θ and equating the result to zero, we get the equation

$$-K\left(\cos^2\psi\sin 2\theta\cos^2\varphi - \sin^2\psi\sin 2\theta + \sin 2\psi\cos 2\theta\cos\varphi\right) +$$
$$+2\pi M_0^2\sin 2\theta\cos^2\varphi - M_0 H_0\cos\theta\cos\left(\varphi - \varphi_H\right) = \dot{0} \qquad (7.95)$$

Similar differentiation with respect to φ gives the equation

$$-K\left(-\cos^2\psi\sin^2\theta\sin 2\varphi - \sin\psi\cos\psi\sin 2\theta\sin\varphi\right) -$$
$$-2\pi M_0^2\sin^2\theta\sin 2\varphi + M_0 H_0\sin\theta\sin\left(\varphi - \varphi_H\right) = 0. \qquad (7.96)$$

These two equations form a system for θ and φ, which must be solved together. The expression from the first equation of θ through φ leads to a complete fourth-degree equation for cos θ, the coefficients of which contain the sines and cosines of φ also to the fourth degree. The analytical resolution of the resulting equation for cos θ, although possible (for example, by the Ferrari method or other equivalent [261, 262]), is extremely cumbersome. Moreover, the substitution of the obtained solution into the second equation will lead to its so complex form that the possibility of an analytical solution in general becomes a big question. An alternative to an analytical solution here may be a numerical solution by the establish method, discussed in the next section.

Comment. In this section, to calculate the equilibrium position of the magnetization, we used the expression for the energy density of the form (7.84), obtained in turn from the basic equations $U_a^{(2)}$ (4.182) or (7.72), as well as $U_m^{(2)}$ (4.206) or (7.83). But, as shown in Sections 4.5.1 and 4.5.2, along with (4.182) and (4.206) for the energy density, there are equivalent expressions $U_a^{(1)}$ (4.184) and $U_m^{(1)}$ (4.204). The use of these expressions to minimize energy leads to a system, although somewhat different from (7.95) and (7.96), but also analytically difficult to solve. Therefore, here it makes sense to use a numerical solution by the establishment method.

7.4.3. Solution by the establishment method

If the system (7.95)–(7.96) is undecidable analytically, the equilibrium position of the magnetization can be found numerically using the establishment method (Section 5.1.4). Recall that the establishment method is a series of consecutive steps, as a result of which the calculated position of the magnetization vector approaches its equilibrium position.

In this case, at the first step, the direction of magnetization is arbitrary, based on which the anisotropy field is determined. This resulting field is added to the external field and the direction of their vector sum is taken as the new direction of magnetization.

On this basis, the directions of magnetization, the anisotropy field is again determined, it is again added to the external field, and all procedures are repeated until the technical measures until the calculated direction of magnetization ceases to change (with a given degree of accuracy).

The direction of magnetization obtained in this way is taken as the equilibrium one. Let us show how the setting method can be applied to the problem considered here. Since in the process of implementing the establishment method, it is necessary to perform vector summation, it is more convenient to carry out all calculations in the Cartesian coordinate system (Fig. 7.1). So, the anisotropy energy density, according to (4.182) (or here (7.82)), taking into account (7.80)–(7.81), takes the form

$$U_a^{(2)} = -K\left(m_x^2 \cos^2 \psi + m_z^2 \sin^2 \psi + 2m_x m_z \sin \psi \cos \psi\right). \qquad (7.97)$$

The energy density of the demagnetization, in accordance with (4.206) (or here (7.83)), has the form

$$U_m^{(2)} = -2\pi M_0^2 \left(m_y^2 + m_z^2\right). \qquad (7.98)$$

The density of the interaction energy of magnetization with the external field (7.20) with (7.10)–(7.12) has the form

$$U_h = -M_0 H_0 \left(m_x \cos\varphi_H + m_y \sin\varphi_H\right). \qquad (7.99)$$

Since here the anisotropy and demagnetization fields work simultaneously against the external field, it is necessary to find equilibria between, on the one hand, the external field energy U_h and, on the other hand, the sum of the anisotropy and demagnetization energies $U_a^{(2)} + U_m^{(2)}$.

The effective fields in this case are determined by the classical formula (1.2), for the normalized magnetization taking the form

$$\mathbf{H}_e = -\frac{1}{M_0}\frac{\partial U}{\partial \mathbf{m}}. \tag{7.100}$$

Performing the differentiation of expressions (7.97)–(7.99) and introducing anisotropy (7.48) $H_a = 2K/M_0$ and demagnetization fields (7.49) $H_m = 4\pi M_0$, we find the components of the total anisotropy and demagnetization field:

$$H_{px} = H_a \cos^2 \psi m_x + H_a \sin \psi \cos \psi m_z; \tag{7.101}$$

$$H_{py} = H_m m_y; \tag{7.102}$$

$$H_{pz} = H_a \sin^2 \psi m_z + H_a \sin \psi \cos \psi m_x + H_m m_z. \tag{7.103}$$

In this case, the components of the external field, in accordance with the geometry of the task (Fig. 7.1), are equal to:

$$H_x = H_0 \cos \varphi_H; \tag{7.104}$$

$$H_y = H_0 \sin \varphi_H; \tag{7.105}$$

$$H_z = 0. \tag{7.106}$$

Finding the components of the vector sum and normalizing them to the full length of the vector constituting this sum, we get:

$$h_{Sx} = \frac{H_{px} + H_x}{\sqrt{H_{Sx}^2 + H_{Sy}^2 + H_{Sz}^2}}; \tag{7.107}$$

$$h_{Sy} = \frac{H_{py} + H_y}{\sqrt{H_{Sx}^2 + H_{Sy}^2 + H_{Sz}^2}}; \tag{7.108}$$

$$h_{Sz} = \frac{H_{pz} + H_z}{\sqrt{H_{Sx}^2 + H_{Sy}^2 + H_{Sz}^2}}; \tag{7.109}$$

These field components are calculated under the assumption that at the time of obtaining the formulas (7.101)–(7.103), the components of magnetization $m_{x,y,z}$ were known. Thus, at the first stage of calculation, these components can be specified arbitrarily, without forgetting, however, that the total length of the normalized magnetization vector **m** should be equal to one. Now suppose that the new normalized components of magnetization are just equal to these values (7.107)–(7.109), that is:

$$m_x = h_{Sx};$$ (7.110)
$$m_y = h_{Sy};$$ (7.111)
$$m_z = h_{Sz}$$ (7.112)

Obviously, the normalization is not violated in this case, so that the total length of the vector **m** remains equal to unity. Thus, the first step of the establishment method is completed, that is, refined values are obtained from arbitrary charges of the components of the magnetization. The next step of the establishment method is to substitute these refined entries into the formulas (7.101)–(7.103), after which the whole procedure is repeated, so that new, even more refined values are found, which can again be used to substitute into the formulas (7.101)–(7.103), and so on. It makes sense to repeat the procedure until the technology until the new obtained values of the components of the magnetization will no longer continue to change within the limits of the accuracy specified. Thus, the result will be the required equilibrium position of the magnetization vector for a given value and orientation of the external field.

7.4.4. Choosing expressions for energy density

Here, the energy densities in the form of $U_a^{(2)}$ (4.184) or (7.97) and $U_m^{(2)}$ (4.206) or (7.98) were used to calculate the effective anisotropy and demagnetization parameters. At the same time, as shown in sections 4.5.1 and 4.5.2, for technical energies there are equivalent expressions $U_a^{(1)}$ (4.182) and $U_m^{(1)}$ (4.204). However, as shown in the sections 4.2.4 and 4.2.6, the use of equivalent expressions for energy leads to nonequivalent field expressions. So, for the anisotropy field $U_a^{(1)}$ of the form (4.182) (or in the expanded form – (4.70)), the fields (4.72)–(4.74) are obtained, and for the anisotropy field of $U_a^{(2)}$ of the form (4.184) (or in the expanded as – (4.87)) – fields (4.88) – (4.90). From the comparison of such variants of expressions for the fields, one can see the obvious difference. A similar pattern holds for demagnetization fields. In the present work, the choice between those and other expressions for the energy density was made on the basis of the qualitative physical meaning of the results obtained. So, for example, with the adopted material parameters $K = 1.15 \cdot 10^4$ erg×cm^{-3}, $M_0 = 22$ G, and also at $\psi = 0°$ and the field $H_0 = 1000$ Oe, it can be seen that the equilibrium magnetization vector must lie along the Ox axis (since $H_a > H_m$), that is, should be $\theta_0 = 0°$. This value is obtained by using the expressions $U_a^{(2)}$ and $U_m^{(2)}$. However, the use of the expressions $U_a^{(1)}$ and $U_m^{(1)}$ instead of $U_a^{(2)}$ and $U_m^{(2)}$ for technical values, leads to the value $\theta_0 = 49°$, which clearly contradicts the qualitative physical meaning. Therefore, the choice stops precisely at $U_a^{(2)}$ and $U_m^{(2)}$. The authors of this monograph at the moment find it difficult to give a correct physical conclusion of expediency in the calculations given here in choosing $U_a^{(2)}$ and $U^{(2)}$, however, they can indicate a simple rule of thumb. So, in the case of an easy-axis type anisotropy (that is, for an easy magnetization axis), use the expression for energy containing only one component of magnetization along this axis (like (4.184)), and in the case of an easy-plane type anisotropy (i.e., for anisotropy of the film shape), an expression for the energy should be used that contains both components of magnetization lying in this plane (of the type (4.182)). However, in order to prevent possible physical errors when using the establishment method, it is recommended to always follow the qualitative physical sense of the solution obtained in all the cases.

7.4.5. Some features of the method of establishing

The method of establishing implemented here is suitable for any orientation of the anisotropy axis, any value of its constant, for any saturation magnetization, as well as for any given value and orientation of the external field (of course, all this is within reasonable limits from the physical side). The method is not critical to the specification of the initial values of the components of the magnetization vector. However, if the problem in question is algorithmicized in the event of a multiple launch of the establishment process, such values should be updated (i.e, restore the original ones) with each new launch (since otherwise unrealizable hysteretic solutions can be obtained).

Only the sign of the initial component of magnetization m_z is critical with respect to the obtained solutions in the considered geometry (Fig. 7.1), that is, specifying one or another sign results in two solutions corresponding to the positive or negative orientation of the magnetization relative to the anisotropy axis. Thus, in the considered calculations, the initial values were assumed as follows:

the first task was: $m_{x0} = 0.1$, $m_{y0} = 0.1$, $m_{z0} = +1.0$; the second task: $m_{x0} = 0.1$, $m_{y0} = 0.1$, $m_{z0} = -1.0$. Examples of such solutions will be given below.

Practical experience shows that the method converges fairly quickly. So in the real cases considered in the framework of this work, as a rule, no more than ten steps were enough to obtain an equilibrium orientation of the magnetization with an accuracy of several radii. However, to get the smoother dependencies, you should still take the number of steps a bit more. Practice has shown that the number of steps of the order of 1000 is optimal from the point of view of the ratio of sufficient accuracy and smallness of the counting time. In the problem considered here, the accuracy was tenths of a degree when the counting time is about ten seconds.

7.4.6. Equilibrium orientation of magnetization

We now turn to the study by the method of establishing the equilibrium orientation of the magnetization vector. As before, we will assume that the film plane coincides with the Oyz coordinate plane, the anisotropy axis makes an angle ψ with the normal to the film plane and lies in the Oxz plane, and the field changes in the Oxy plane, making an angle φ_H with the Ox axis (Fig. 7.1, case 2, that is, $\psi = \psi_2$).

Figure 7.12 shows the dependences of the equilibrium azimuthal angle φ_0 and the equilibrium component m_{z0} of the magnetization vector from the azimuthal angle of the external field φ_H for various values of this field. The choice of the component m_{z0} instead of the angle θ_0 was made in favour of clarity of consideration, since it is this component that characterizes the deviation of the magnetization vector from the plane of the field change.

If the reader wants to work with the polar angle of magnetization θ_0, it can be easily obtained from m_{z0} using the formula

$$\theta_0 = \arccos(m_{z0}). \qquad (7.113)$$

It can be seen from the figure that as the field orientation changes, the angle φ_0 mainly follows φ_H, and, the larger the field, the more accurately the dependence of φ_0 on φ_H approaches the straight line corresponding to $\varphi = \varphi_H$. The magnetization component mz0 deviates from the film plane in the same direction as the EMA, that is, when

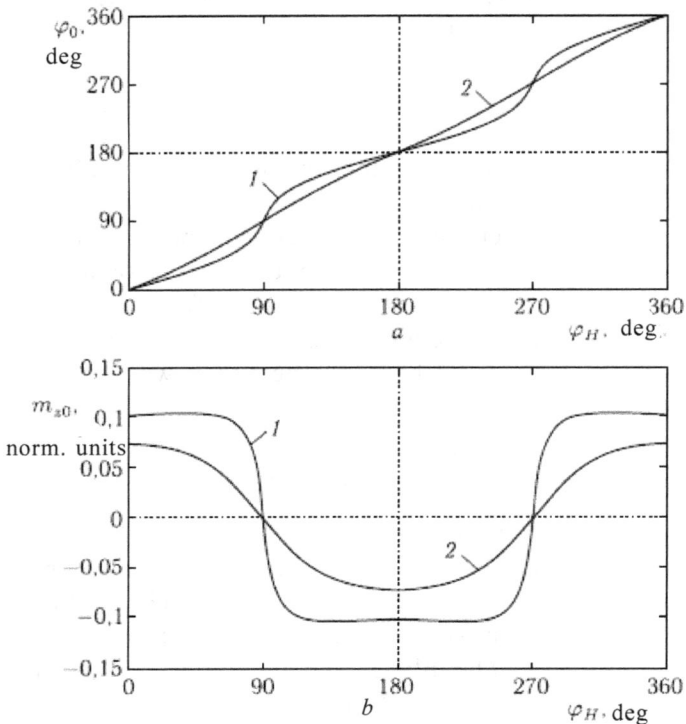

Fig. 7.12. The dependences of the azimuthal angle (*a*) and *z*-components (*b*) of the magnetization vector on the azimuthal angle of the external field for different values: 1 – 1000 Oe; 2 – 4000 Oe. Parameters: $K = 1.15 \cdot 10^4$ erg cm^{-3}, $M_0 = 22$ G, $\psi = 20°$

$x > 0$ up, and when $x < 0$, down, so that the deviation changes sign when passing φ_H through 90° and 270° (as well as EMA).

The deviation is the smaller, the larger the field, and at a field of 4000 Oe (that is, just in the FMR region at a frequency of 10 GHz), the normalized value of the *z*-component does not exceed 0.07, which corresponds to the deviation of the magnetization from the film plane by only 4°. From the latter position, it can be concluded that, at a field of about 4000 Oe, the Smit–Sul method can be used to find the FMR frequency using the strong field approximation. If one countd the value of the deviation of the *z*-component as a measure of accuracy, in this case it is 0.07, that is, 7%.

7.4.7. Features of the orientation of magnetization in a small field

The above properties of the equilibrium orientation of the

magnetization relate to the case when the field is large enough to force the magnetization, mainly to follow. However, as shown in section 7.3.3, with fields less than 1000 Oe, the magnetization it does not always follow the field unconditionally, and in some cases it can deviate significantly, including at small field changes, jump over large angles (Fig. 7.5).

Although the situation shown in Fig. 7.5, corresponds to the location of the EMA in the Oxy plane, but also in the geometry considered here, that is, for an EMA lying in the Oxz plane, an abrupt change in the orientation of the magnetization also takes place, and is more pronounced than the угол angle.

At $\psi = 20°$, the critical value of the field below which magnetization jumps take place, and absent above, is about 700–900 Oe. Therefore, we further take the field H_0 to be 500 Oe and consider how the magnetization changes as its orientation changes in the Oxy plane.

Turn to Fig. 7.13, where the dependences of the equilibrium azimuthal angle φ_0 and the equilibrium component mz_0 of the

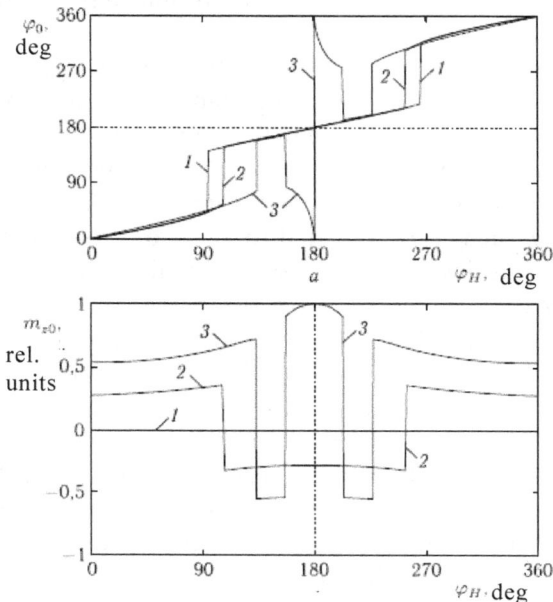

Fig. 7.13. The dependences of the azimuthal angle (a) and zcomponents (b) of the magnetization vector on the azimuthal angle of the external field at its value of 500 Oe, and different values of the angle: 1 – 0°; 2 – 20°; 3 – 40°. Parameters: $K = 1.15 \cdot 10^4$ erg cm^{-3}, $M_0 = 22$ G, $H_0 = 500$ Oe.

magnetization vector on the azimuthal angle of the external field φ_H are shown for various values of angle ψ.

It can be seen from the figure that, as the angle φ_H changes, the magnetization not only monotonically follows the field, but also, for some value, triggers a jumplike change to significant angles.

So, within the positive EMA value (that is, where it is located above the Oxy plane), the magnetization first monotonously follows the field with some φ_H, the larger, the more the field orientation moves away from the projection of the EMA onto the Oxy plane, and then experiences a jump, moving it forward with respect to the field, and with a further increase in φ_H this advance decreases, turning to zero at $\varphi_H = 180°$. With a further increase in φ_H, the magnetization again lags behind the field, then experiences a jump, leading it to advance the field, after which the advance further decreases, falling to zero at $\varphi_H = 360°$. Thus, the dependence of the equilibrium magnetization angle φ_0 on the angle of the field φ_H with respect to the value $\varphi_H = 180°$ is symmetric. The jump of magnetization to $\varphi_H = 180°$ occurs at a larger value of φ_H, the greater the angle ψ, and the reverse jump occurs symmetrically with respect to the value $\varphi_H = 180°$.

If the magnitude of the angle ψ is at $0°$ and $20°$ (curves 1 and 2), there is only one magnetization jump, but when the angle ψ is $40°$ (curve 3) there are two jumps: near the angle $170°$, the reverse, and at $180°$ straight side (i.e. 'two-step'). That is, near $\varphi_H = 180°$, the external field 'compensates' the anisotropy field. The component of magnetization m_{z0} at $\psi = 0°$ is equal to zero for any value of φ_H, and with increasing ψ it deviates from zero more than this angle is larger.

With increasing φ_H before the jump, the magnetization m_{z0} slowly increases, and when it jumps, it changes sign sharply — it is reoriented relative to the Oxy plane in the opposite direction. After jumping to $\varphi_H = 180°$, it decreases slightly, and after $\varphi_H = 180°$ varies mirror with respect to $\varphi_H = 180°$. With a two-step hopping, the magnetization extends almost perpendicular to the film plane. In general, for all angles ψ, the deviation of the magnetization from the plane of the field change (Oxy) is significantly greater than in the first case (Fig. 7.12), when the field is strong enough to almost lay the magnetization in its plane.

7.4.8. The influence of the initial orientation of magnetization

When the fields are above the critical, the behaviour of the equilibrium position of the magnetization with a change in the field orientation does not depend on the initial orientation of the magnetization corresponding to $\varphi_H = 0°$. However, in the field below the critical one, such a dependence appears, all the more pronounced than the value of the angle ψ is greater.

A certain idea of this dependence is given in Figs. 7.14 and 7.15, the first of which corresponds to $\psi = 20°$, and the second $\psi = 40°$. In both cases, two initial orientations are considered: at $m_{z0} = +1.0$ (curves 1) and $m_{z0} = -1.0$ (curves 2). It can be seen that in the first case ($\psi = 20°$, Fig. 7.14), when the initial orientation changes from positive to negative jump to negative direction of the EMA (down relative to the Oxy plane), it occurs at smaller angles φ_H than before, since now for the field there is no need to turn the magnetization from a positive direction to a negative one, because it initially is in a negative direction.

It is important to note that when the initial deflection changes, the jump occurs asymmetrically with respect to the plane of the field change, since at a small angle the field 'overpowers' the anisotropy and the magnetization at small φ_H always orientates in the positive direction given by the sum of the anisotropy and external fields.

When changing the value of φ to negative, the whole picture is reflected in the plane of the field change. That is, we can say that there is a certain 'hysteresis'. The reverse jump occurs symmetrically with respect to $\varphi_H = 180°$.

In the second case ($\psi = 40°$, Fig. 7.15), where the jump is two-step, the magnetization vector at the initial negative position is immediately set in the negative direction relative to the Oxy plane. Further, as φ_H increases, the magnetization vector approaches this plane, abruptly jumps to the other side, after which it soon jumps back and then all the time remains on the negative side of the Oxy plane up to $\varphi_H = 180°$. After the end of these shortcuts, that is, when φ_H passes through the value of $180°$, the whole picture repeats symmetrically. Thus, most motion of magnetization occurs in a negative half-space relative to the plane of variation of the field.

7.4.9. Note on the potential nature of the observed dependences

The above examination shows that, above and below the critical value of the external field, the behaviour of the magnetization has a fundamentally different character. It can be assumed that a convenient

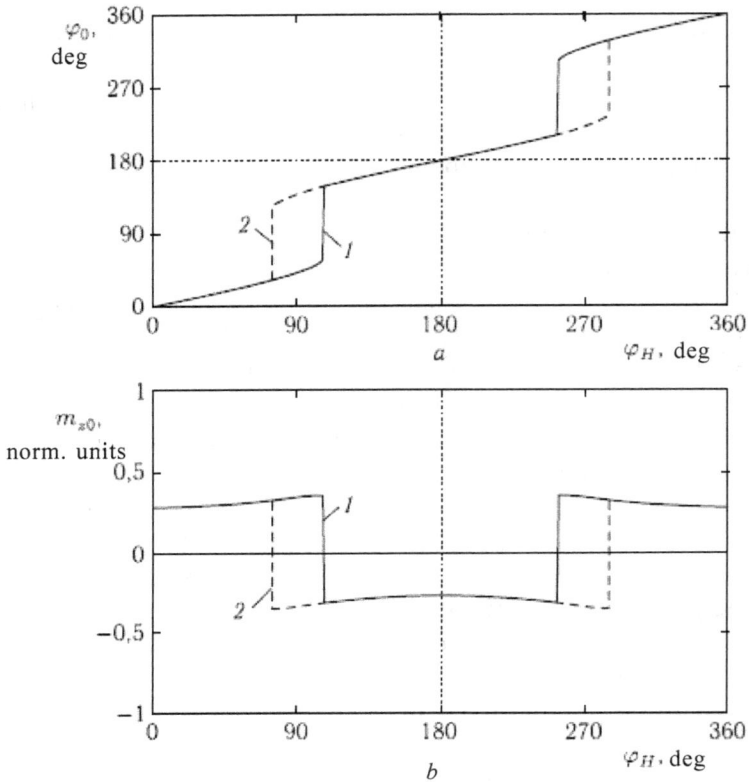

Fig. 7.14. Dependencies of the azimuth angle (*a*) and *z*-components (*b*) of the magnetization vector from the azimuth angle of the external field with its value of 500 Oe and different initial orientation of the magnetization: 1 – initial orientation – in the positive direction of the axis *Oz*: $m_{x0} = 0.1$, $m_{y0} = 0.1$, $m_{z0} = +1.0$; 2 – initial orientation – in the negative direction of the *Oz* axis: $m_{x0} = 0.1$, $m_{y0} = 0.1$, $m_{z0} = -1.0$. Parameters: $K = 1.15 \cdot 10^4$ erg cm^{-3}, $M_0 = 22$ G, $H_0 = 500$ Oe, $\psi = 20°$. Curves 1 here coincide with curves 2 in Fig. 7.13

interpretation of such a diversity of behavior would be an analysis of the dependencies obtained on the basis of the configuration of potential energy, similar to that carried out in Section 7.3.3. However, it should be noted that there the equilibrium orientation of the magnetization vector for any orientation of the external field did not go out of the plane of variation of this field, that is, the polar angle of magnetization was always exactly 90° (Fig. 7.1, Case No. 1). Therefore, in the expression for the energy density (7.40) it was possible to immediately put the polar angle equal to 90°, which led to the expression (7.45), which depends only on the azimuth angle.

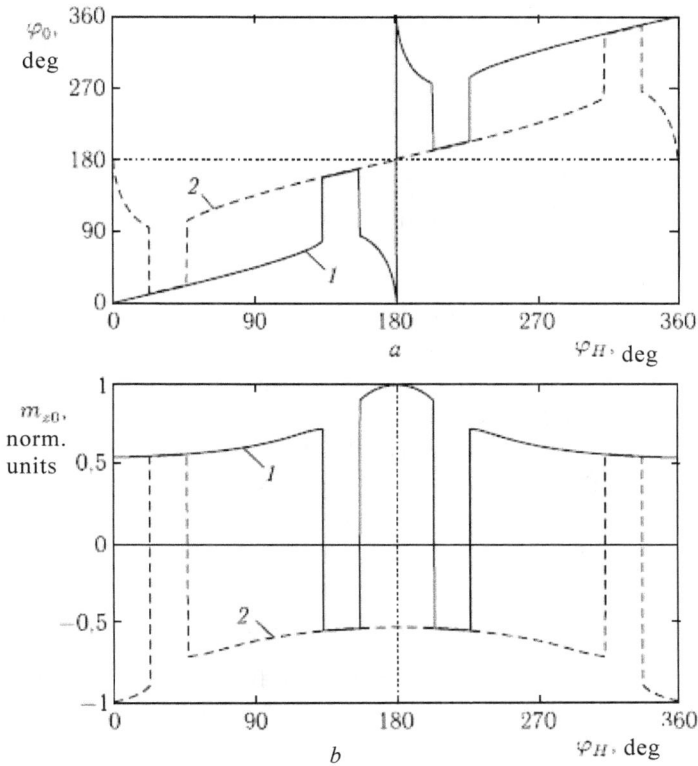

Fig. 7.15. Dependences of the azimuth angle (*a*) and *z*-components (*b*) of the magnetization vector from the azimuth angle of the external field with its value of 500 Oe and different initial orientation of the magnetization: 1 – initial orientation – in the positive direction of the axis *Oz*: $m_{x0} = 0.1$, $m_{y0} = 0.1$, $m_{z0} = +1.0$; 2 – initial orientation – in the negative direction of the *Oz* axis: $m_{x0} = 0.1$, $m_{y0} = 0.1$, $m_{z0} = -1.0$. Parameters: $K = 1.15 \cdot 10^4$ erg cm^{-3}, $M_0 = 22$ G, $H_0 = 500$ Oe, $\psi = 40°$. Curves 1 here coincide with curves 3 in Fig. 7.13

So minimization could be carried out only on this corner, which gave expression (7.46), the analysis of which was carried out further.

Such an analysis was carried out on the basis of the one-dimensional potential curves shown in Figs. 7.3 and 7.4.

In the case considered here (Fig. 7.1, Case No. 2), the polar angle of the magnetization vector as the orientation of the external field changes may change, which is caused by the EMA coming out of the plane of field change. Moreover, in the expression for the energy density (7.84) the polar angle of magnetization cannot be set equal to 90°, that is, a simplification similar to that carried out in deriving the expression (7.45) cannot be obtained here. The need to perform a minimization at both angles leads to a system of two equations (7.95), (7.96), the joint solution of which in an analytical form, if possible, is extremely difficult.

Thus, here are potential curves similar to those shown in Figs. 7.3 and 7.4, should be built depending not on one, but on two variables, that is, the potential must be represented as a specific three-dimensional surface. Depending on the construction parameters, more or less localized minima should appear on such a potential surface, so that the transition of the operating point from one minimum to another, caused by a change in the field, will just reflect the jump-like nature of the curves in Figs. 7.13–7.15. In the present work, such an analysis is not carried out, but the authors believe that it should be the basis for independent work. As an example of such an analysis of jump dependences based on the surface of a three-dimensional potential, for example, the work [335,368–374].

7.4.10. Ferromagnetic resonance conditions

We now turn to the consideration of the conditions of ferromagnetic resonance in this geometry (Fig. 7.1, case 2). We will look for the resonant frequency and the resonant field, for which, as before, we use the Smit–Sul method (Section 2.3).

Section 4.5.1 shows that, depending on the recording form (determined by the condition that the magnetization vector is constant), the anisotropy energy density can be represented as two equivalent expressions (4.182) and (4.184). The same applies to the representation of the demagnetization energy density in the form of (4.204) and (4.206) (Section 4.5.2). However, at the beginning of Section 7.4, the choice was made in favour of the expressions $U_a^{(2)}$ (7.82) and $U_m^{(2)}$ (7.83), as being more appropriate to the physical content of the problem. We give here these expressions in the form written through magnetization components:

$$
\begin{aligned}
U_a^{(2)} = -K\big\{ & m_x^2 \sin^2\theta_a \cos^2\varphi_a + m_y^2 \sin^2\theta_a \sin^2\varphi_a + \\
& m_z^2 \cos^2\theta_a + 2m_x m_y \sin^2\theta_a \sin\varphi_a \cos\varphi_a + \\
& +2m_x m_z \sin\theta_a \cos\theta_a \cos\varphi_a + 2m_y m_z \sin\theta_a \cos\theta_a \sin\varphi_a \big\};
\end{aligned}
$$

$$(7.114)$$

$$U_m^{(2)} = -2\pi M_0^2 \left(m_y^2 + m_z^2 \right).$$

(7.115)

The density of the interaction energy of magnetization with an external field has a single form (7.20)

$$U_h = -M_0 H_0 \left(m_x h_x + m_y h_y + m_z h_z \right).$$

(7.116)

According to (7.80), (7.81), in the geometry under consideration one can put:

$$\theta_a = 90° - \psi;$$

(7.117)

$$\varphi_a = 0°.$$

(7.118)

In this case, the field components have the form (7.10)–(7.12):

$$h_x = \cos \varphi_H;$$

(7.119)

$$h_y = \sin \varphi_H;$$

(7.120)

$$h_z = 0.$$

(7.121)

Taking into account (7.117) and (7.118), the anisotropy energy density (7.114) and the interaction of magnetization with the field (7.116) take the form:

$$U_a^{(2)} = -K \left(m_x^2 \cos^2 \psi + m_z^2 \sin^2 \psi + 2 m_x m_z \sin \psi \cos \psi \right),$$

(7.122)

$$U_h = -M_0 H_0 \left(m_x \cos \varphi_H + m_y \sin \varphi_H \right).$$

(7.123)

Adding (7.122), (7.115), (7.123) and recording the components of the magnetization through the polar and azimuth angles, we obtain the total energy density in the form

$$U = U_a^{(2)} + U_m^{(2)} + U_h = -K(\cos^2 \psi \sin^2 \cos^2 \varphi^2 +$$

$$+ \sin^2 \psi \cos^2 \theta + 2 \sin \psi \cos \psi \sin \theta \cos \theta \cos \varphi) -$$

$$-2\pi M_0^2 \left(\sin^2 \theta \sin^2 \varphi + \cos^2 \theta \right) = M_0 H_0 \sin \theta \cos \left(\varphi - \varphi_H \right).$$

(7.124)

Performing trigonometric transformations, and also taking into account the constancy of the angle ψ and the saturation magnetization M0, we bring (7.124) to a simpler form:

$$U = -K \left[\sin^2 \theta \left(\cos^2 \psi \cos^2 \varphi - \sin^2 \psi \right) + \sin \psi \cos \psi \sin 2\theta \cos \varphi \right] +$$

$$2\pi M_0^2 \sin^2 \theta \cos^2 \varphi - M_0 H_0 \sin \theta \cos \left(\varphi - \varphi_H \right).$$

(7.125)

Differentiating with respect to θ and φ we obtain the derivatives:

$$U_{\theta\theta} = -2\left(K\cos^2\psi - 2\pi M_0^2\right)\cos 2\theta\cos^2\varphi +$$
$$+2K\left(\sin^2\psi\cos 2\theta + \sin 2\psi\sin 2\theta\cos\varphi\right) +$$
$$+ M_0 H_0 \sin\theta\cos\left(\varphi - \varphi_H\right), \quad (7.126)$$

$$U_{\varphi\varphi} = 2\left(K\cos^2\psi - 2\pi M_0^2\right)\sin^2\theta\cos 2\varphi +$$
$$+ K\sin\psi\cos\psi\sin 2\theta\cos\varphi + M_0 H_0 \sin\theta\cos\left(\varphi - \varphi_H\right). \quad (7.127)$$

$$U_{\theta\varphi} = \left(K\cos^2\psi - 2\pi M_0^2\right)\sin 2\theta\sin 2\varphi + K\sin 2\psi\cos 2\theta\sin\varphi +$$
$$M_0 H_0 \cos\theta\sin\left(\varphi - \varphi_H\right). \quad (7.128)$$

We introduce auxiliary notation:

$$A = \frac{1}{M_0}\left[-2\left(K\cos^2\psi - 2\pi M_0^2\right)\cos 2\theta\cos^2\varphi +\right.$$
$$\left. +2K\left(\sin^2\psi\cos 2\theta + \sin 2\psi\sin 2\theta\cos\varphi\right)\right]; \quad (7.129)$$

$$B = \frac{1}{M_0}\left[2\left(K\cos^2\psi - 2\pi M_0^2\right)\sin^2\theta\cos 2\varphi +\right.$$
$$\left. + K\sin\psi\cos\psi\sin 2\theta\cos\varphi\right]; \quad (7.130)$$

$$C = \frac{1}{M_0}\left[\left(K\cos^2\psi - 2\pi M_0^2\right)\sin 2\theta\sin 2\varphi + K\sin 2\psi\cos 2\theta\sin\varphi\right]. \quad (7.131)$$

Being written using the standard notation for the anisotropy field H_a (7.48) and demagnetization H_m (7.49), these expressions have the form:

$$A = \left[-\left(H_a\cos^2\psi - H_m\right)\cos 2\theta\cos^2\varphi +\right.$$
$$\left. +2H_a\left(\sin^2\psi\cos 2\theta + \sin 2\psi\sin 2\theta\cos\varphi\right)\right]; \quad (7.132)$$

$$B = \left[\left(H_a\cos^2\psi - H_m\right)\sin^2\theta\cos 2\varphi +\right.$$
$$\left. +\frac{1}{2}H_a\sin\psi\cos\psi\sin 2\theta\cos\varphi\right]; \quad (7.133)$$

$$C = \frac{1}{2}\left[\left(H_a\cos^2\psi - H_m\right)\sin 2\theta\sin 2\varphi + H_a\sin 2\psi\cos 2\theta\sin\varphi\right]. \quad (7.134)$$

With the notation A, B, and C introduced, the derivatives (7.126) –(7.128) take the form:

$$U_{\theta\theta} = M_0 \left[A + H_0 \sin\theta\cos(\varphi - \varphi_H) \right];$$

(7.135)

$$U_{\varphi\varphi} = M_0 \left[B + H_0 \sin\theta\cos(\varphi - \varphi_H) \right];$$

(7.136)

$$U_{\theta\varphi} = M_0 \left[C + H_0 \cos\theta\sin(\varphi - \varphi_H) \right];$$

(7.137)

Substituting these derivatives into the Smit–Sul formula (2.148), we obtain the frequency in the form:

$$\omega = \frac{\gamma}{\sin\theta} \left\{ H_0^2 \left[\sin^2\theta\cos^2(\varphi - \varphi_H) + \cos^2\theta\sin^2(\varphi - \varphi_H) \right] + \right.$$
$$+ H_0 \left[(A+B)\sin\theta\cos(\varphi - \varphi_H) - 2C\cos\theta\sin(\varphi - \varphi_H) \right] +$$
$$\left. + (AB) - C^2 \right\}^{1/2}.$$

(7.138)

Here, the values of θ and φ must correspond to the equilibrium position of the magnetization vector in a given field H_0, the azimuthal orientation of which is determined by the angle φ_H. That is, the first step in solving the problem should be the determination of such an equilibrium position along both coordinates, approximately as shown in Section 7.3.

7.4.11. Resonance field

In real experiments on FMR, in connection with the design of commonly used spectrometers, the frequency is set constant, and the resonance conditions are achieved by changing the field [22–25]. In this case, expression (7.138) should be resolved with respect to H_0 by solving the corresponding quadratic equation, in much the same way as was done in Section 7.3. For such permission we introduce auxiliary notation:

$$a = \sin^2\theta\cos^2(\varphi - \varphi_H) + \cos^2\theta\sin^2(\varphi - \varphi_H);$$

(7.139)

$$b = (A+B)\sin\theta\cos(\varphi - \varphi_H) - 2C\cos\theta\sin(\varphi - \varphi_H);$$

(7.140)

$$c = (AB) - C^2.$$

(7.141)

At the same time, from (7.138) we obtain with respect to H_0 a quadratic equation

$$aH_0^2 + bH_0 + \left[c - \left(\frac{\omega}{\gamma} \right)^2 \sin^2\theta \right] = 0,$$

(7.142)

whose solution has the form

$$H_0 = -\frac{b}{2a} + \sqrt{\left(\frac{\omega}{\gamma}\right)^2 \sin^2\theta + b^2 - 4ac}. \qquad (7.143)$$

As in the previous case, the formula obtained here (7.143) is self-consistent, that is, the right-hand side must be calculated at the equilibrium values of the polar and azimuthal angles, which are determined by the field value, therefore, the same remarks related to section 7.3.6 concerning formula (7.71) apply to it.

7.4.12. Strong field approximation

Since in case 1 (Fig. 7.1) with accepted frequencies and parametres of the film, the strong field approximation gave quite acceptable results (Section 7.3.7), we can expect that here (that is, in case 2) with the technical parameters of the approximation strong field will have about the same magnitude. Therefore we will not here, it is difficult to determine the problem of determining the resonance conditions by first finding the equilibrium position of the magnetization by the establishment method, and immediately restrict ourselves to the approximation of a strong field. So, we will assume that the magnetization vector in the equilibrium the state is built exactly along the field, that is, in formulas (7.139), (7.140), as well as in the incoming formulas (7.129)–(7.131), you can put $\theta = 90°$ and $\varphi = \varphi_H$. Conveniently also, like (7.48) and (7.49), introduce effective anisotropy fields

$$H_a = 2K/M_0 \qquad (7.144)$$

and degaussing:

$$H_m = 4\pi M_0 \qquad (7.145)$$

Further, assuming that the second and third terms under the root in (7.143) are significantly less than the first, we can bring (7.143) to the form

$$H_0 = \frac{\omega}{\gamma} - \frac{A+B}{2}, \qquad (7.146)$$

where A and B take the form:

$$A = \left(H_a \cos^2 \psi - H_m\right)\cos^2 \varphi_H - H_a \sin^2 \psi; \tag{7.147}$$

$$B = \left(H_0 \cos^2 \psi - H_m\right)\cos\left(2\varphi_H\right). \tag{7.148}$$

Note that expression (7.146) is similar to (7.76), but the meanings of A and B, unlike (7.77) and (7.78), are different.

Substituting (7.147) and (7.148) into (7.146), we obtain the resonant field in the form

$$H_0 = \frac{\omega}{\gamma} + \frac{H_0}{2}\sin^2 \psi - \frac{1}{2}\left(H_a \cos^2 \psi - H_m\right)\left(3\cos^2 \varphi_H - 1\right). \tag{7.149}$$

Figure 7.16 shows the dependence of the resonant field on the angle of its orientation, constructed by the formula (7.149) with the technical parameters of the film and the specified values, as in case 1 (Fig. 7.9). From the figure we can see that the dependence is similar to that shown in Fig. 7.10, has a sinusoidal character with a frequency of 180°. The amplitude of dependence is also close to that in Fig. 7.10. The magnitude of the resonant field with decreasing

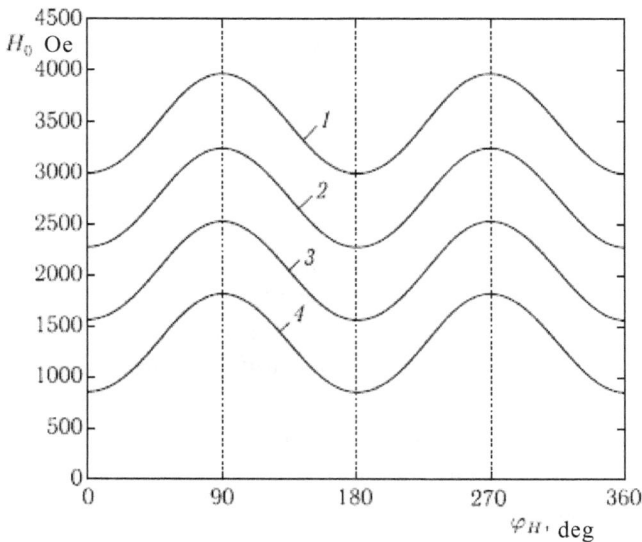

Fig. 7.16. Dependence of the resonance field on the angle of its orientation frequency values: 1 – 10 GHz; 2 – 8 GHz; 3 – 6 GHz; 4 – 4 GHz. Curves are structured by formula (7.149). Parameters: $K = 1.15 \times 10^4$ erg $\times$ cm^{-3}, $M_0 = 22$ G, $\psi = 20°$.

frequency also decreases linearly, which is ensured by the first term in formula (7.149). However, here, unlike in Fig. 7.10, the shift of the curve as a whole along the horizontal axis is missing. The curve minima are exactly at the corners: 0°, 180°, 360°, and the maxima are at 90° and 270°. This lack of shear can be understood from consideration of Fig. 7.1, whence it is seen that the projection of the EMA on the film plane, that is, on the plane in which the field vector **H** is rotated, is attached just to the axis Ox, from which the angle φ_H is measured. That is, the dependence of the resonant field on this angle with respect to the Ox axis is completely symmetric.

Comment. The review here is carried out under the technical conditions, as is the similar consideration of case 1 in Section 7.3. Therefore, regarding the correctness of such a review should be guided by the same considerations as set out in the final remark of Section 7.3. That is, for mixed garnets at a frequency of about 10 GHz and higher, the approximation error of a strong field is no more than a few percent, whereas for orthoferrites and even for barium hexaferrite in fields less than 20 kOe, a complete solution should be used. That is, the use of the strong field approximation should be approached with some caution, taking into account the operating frequency and specific material parameters.

7.5. On the possibility of separate experimental determination of the constant uniaxial anisotropy and magnetization

Above, we considered two cases of fundamentally different orientational dependences of the FMR frequency on the field in films with a deviation of the anisotropy axis from the normal to the film plane. Let us consider these cases from the point of view of the possibility of experimentally determining the anisotropy and magnetization constants. We keep the numbering of cases, adopted in their initial review.

Case number 1. According to (7.57), the total energy density is

$$U = -K\sin^2\theta\cos^2(\varphi-\psi)+$$
$$+2\pi M_0^2\sin^2\theta\cos^2\varphi - M_0 H_0\sin\theta\cos(\varphi-\varphi_H). \tag{7.150}$$

We introduce an auxiliary parameter

$$K_{a1} = K\cos^2(\varphi-\psi) - 2\pi M_0^2\cos^2\varphi. \tag{7.151}$$

With this parameter (7.150) takes the form

$$U = -K_{a1} \sin^2 \theta - M_0 H_0 \sin \theta \cos (\varphi - \varphi_H).$$

(7.152)

It can be seen that here the orientational dependences of anisotropy and demagnetization (the first two terms in (7.150), grouped together in expression (7.151), minus the total coefficient $\sin^2\theta$), are the same. The difference in phase between the first and second terms in (7.151) does not matter, since both dependences are proportional to the square of the cosine of φ, so that the total dependence will also be proportional to the square of the cosine of the same angle.

Case number 2. According to (7.125), the total energy density is:

$$U = -K\left[\sin^2 \theta \left(\cos^2 \psi \cos^2 \varphi - \sin^2 \varphi \right) + \sin \psi \cos \psi \sin 2\theta \cos \varphi \right] +$$
$$+2\pi M_0^2 \sin^2 \theta \cos^2 \varphi - M_0 H_0 \sin \theta \cos \left(\varphi = \varphi_H \right).$$

(7.153)

We introduce an auxiliary parameter:

$$K_{a2} = K \cos^2 \psi - 2\pi M_0^2.$$

(7.154)

This parameter takes into account the total contribution of the demagnetization energy and that part of the anisotropy energy, which has the same orientation dependence as the demagnetization energy. In this case, the total energy density (7.125) takes the form:

$$U = -K_{a2} \sin^2 \theta \cos^2 \varphi + K \sin \psi \left(\sin \psi \sin^2 \theta - \cos \psi \sin 2\theta \cos \varphi \right) -$$
$$- MH_0 \sin \theta \cos \left(\varphi - \varphi_H \right).$$

(7.155)

Here, the first term also describes the same orientation dependence for anisotropy and demagnetization, however, the second term corresponds to the part of the anisotropy energy that has an orientation dependence that differs from that for magnetization. Thus, with respect to the angle φ, the first term is proportional to the square of the cosine, that is, corresponds to periodicity in π, while the second term is proportional to only the cosine in the first degree, that is, corresponds to the periodicity in 2π. Thus, in the first case, due to the identity of orientational dependences of anisotropy and demagnetization, experimental separation of the contributions of both energies and, therefore, separate measurement of anisotropy and magnetization constants is not possible. In the second case, in addition to the identical part of the orientational dependences, due to the anisotropy energy there is an additive that has a different

372

periodicity, which makes it possible to determine the experimentally determined anisotropy and magnetization constants separately. The easiest way to perform such a measurement is to use the calculation formulas for resonance oleys in the strong field approximation, such as (7.79) and (7.149). The angle ψ between the EMA and the normal to the film plane must be known in advance from independent measurements, for example, by the disappearance of domains in a sufficiently strong field, as described in Section 5.3.

To determine the anisotropy and magnetization, the field of ferromagnetic resonance should be measured experimentally at a fixed frequency with two different orientations of the external field in the film plane: when this field is parallel and when it is perpendicular to the EMA projection on the film plane. Next, using the two derivations, of the two expressions for the resonance oles (7.79) and (7.149) for a given angle ψ, a system of two equations should be constructed, containing anisotropy and demagnetization as the unknowns. The resolution of such a system will give the values of these variables, whence the desired values of the constant

The anisotropy and saturation magnetization of the film are obtained by reversing relations (7.48) and (7.49).

A description of the implementation of such a measurement is contained in [23]. According to the data presented there, for films of mixed garnet films with an inclination of the axis from 3 to 30°, the accuracy of separate measurements of anisotropy and magnetization was about 20%. If one because of these parameters was known from independent measurements, the accuracy of determining the other increased to a few percent.

7.6. Experimental study of ferromagnetic resonance in ferrite–garnets

In this section, we briefly list the results of experiments on the study of ferromagnetic resonance in films of mixed garnet ferrites with a significant deviation of the axis of uniaxial anisotropy from the normal to the film plane. The experiments were performed in 1972–1980 with the participation of one of the authors of this monograph and described in detail in [22–25]. Here we present ihod results, without going into unnecessary details. An experimental study of ferromagnetic resonance, described in the above-mentioned studies, was carried out on the same mixed garnet–garnet films as described in Section 5.3, the study of orientational permeation properties. As samples, small pieces of films cut on a substrate with a size of 2–3 by 3–4 mm were used, cut with a disk diamond saw. Such a small sample size was due to the need for their placement inside a two-half-wave pass-through resonator of the 3 cm range, which was a section of a standard waveguide with a cross section of 18 × 10 mm, enclosed by two diaphragms placed about 35 mm apart from each other. The field from the controlled electromagnet was oriented perpendicular to the wide wall of the waveguide. The sample was fixed on a thin dielectric rod, which was introduced into the resonator through a window with a diameter of 6 mm in its narrow wall in the middle between the diaphragms.

The sample could be fixed (with glue) on the rod in two positions, so that the film plane was parallel or perpendicular to the axis of the rod. The rotation of the rod around its axis corresponded to the rotation of the field around the sample in two reciprocal perpendicular planes. An additional rotation of the sample around an axis perpendicular to the axis of the rod was carried out by glueing it to a new position. Thus, it was possible to change the orientation of the sample in its plane with an accuracy of 10° and perpendicular to its plane with an accuracy of 5°. With such a change in the orientation of the sample, the direction of the field relative to the laboratory coordinate system remained unchanged, that is, the rotation of the sample was, as it were, equivalent to the rotation of the field in the entire possible range of polar and azimuthal angles. The main results of the experiments are given in [22, 23, 25]. The orientational dependences of the resonance fields were measured when the film was rotated around an axis lying in its plane (by turning the dielectric rod). Two options were investigated: when the inclined EMA lay in the plane of rotation of the field (case 1 in Fig. 7.1) and when the projection of the EMA into the plane of the film lay in this plane (Case 2 in Fig. 7.1). The theoretical interpretation was carried out in the strong field approximation. The values of resonant fields given in [22, 23, 25] with characteristic film orientations differ somewhat in form from the sections given here in Sections 7.3–7.5, which is associated with a different notation adopted there. However, the final results of the calculations completely coincide. Comparison of calculated results with experimental values, performed in [22, 23, 25], revealed good agreement. The possibility of separate measurement of the uniaxial anisotropy constant and saturation magnetization, performed in accordance with the technique of magnetization of a film along and across the projection of the EMA on the film plane, as described in section 7.5. In [23, 25], the measurement of the constant of magnetoelastic interaction was performed. The constant was determined by registering the change in the resonant field under mechanical pressure on the sample. The dosed pressure was carried out using a vertical dielectric rod introduced into the resonator surface, at the upper end of which was placed the cargo area with a water tank placed on it, the level of which determined the degree of mechanical load.

It was found that, in the studied samples, the EMA from the normal to the film plane exceeds the [111] axis of the substrate from the same normal by 3–5 times (up to 20 compared to 4° for the substrate). To interpret such an increase in the deviation, the role of the magnetostriction mechanism was evaluated (Section 5.2.1). It was shown that the measured value of the magnetoelasticity constant can explain a significant part of the observed deviation. As additional results obtained in [22,23,25], one can note the observation of the fine structure of the resonance line due to the inhomogeneity of the films, as well as the measurement of the decay constant of the ferromagnetic resonance. Separately, note should be made of work [24] (and also the corresponding part of work [25]) performed on films of the remaining $(YGdYb)_3(FeGa)_5O_{12}$, which do not have a pronounced deviation of the EMA from the normal to the film plane (Section 5.2.1). In these films resonance measurements revealed strongly manifested cubic anisotropy. Some uniaxial anisotropy also occurred, but there was no increase in the deviation of its axis from the normal to the film plane. Moreover, measurements by the magneto-optical method for the disappearance of the domain structure during magnetization of the film by the field in its plane did not reveal cubic anisotropy, while the uniaxial value was almost twice the value measured by the resonance method. That is, it was concluded that the sensitivity and objectivity of the resonance method are higher than the optical one. The orientational dependence of the resonant field on the angle in the plane passing through the [111] axis, perpendicular to the film plane, and the projection of another [111] axis to the plane of the same film was measured. The theoretical interpretation was carried out using the Smit–Sul method in the strong field approximation. Since the constants of cubic and uniaxial anisotropy were not known beforehand, first the saturation magnetization of the film was determined by an independent method (using the collapse fields and elliptical instability of the CMD), as described in [25, 42, 375]. Further, on the basis of the theoretical interpretation of the orientational dependence of the resonant field, three field values were noted, corresponding to the characteristic crystallographic orientations. Using these three field values as parameters, taking into account the previously measured magnetization value, a system of equations was constructed containing the anisotropy constants and the gyromagnetic constant, which is part of the Landau–Lifshitz equation.

The resolution of this system allowed us to determine the uniaxial and cubic anisotropy constants separately, as well as the gyromagnetic constant, which turned out to be somewhat (approximately 3.7%) higher than the traditional value (2.91 MHz/Oe instead of 2.80 MHz /Oe). The theoretical dependence of the resonance field on the orientation of the film constructed using the parameters obtained in this way showed excellent agreement with the experiment in the entire range of the studied angles (from zero to 360°). As an additional result, we can also note the measurement of the resonance line width (100–150 Oe), as well as an estimate of the attenuation parameter of the magnetization precession.

Conclusions for chapter 7

This chapter is devoted to the description of the phenomenon of ferromagnetic resonance in films with the deviation of the magnetic anisotropy axis from the normal to the film plane.

The main results obtained in this chapter are as follows.

1. The conditions for the excitation of ferromagnetic resonance in a film of a uniaxial magnet, the axis of easy magnetization of which is inclined from the normal to the film plane, are considered. The general geometry of the problem is presented; from the entire variety of possible field orientations with respect to the film and the EMA, a change in the field in a single plane passing through the normal to the film plane is selected. As the most characteristic orientations of an EMA relative to the field, two cases are highlighted: the first is when the EMA lies in the plane of the field change, the second is when the EMA is in a plane perpendicular to the plane of the field change. As a rather convenient method for calculating the resonant frequency, the Smit–Sul method in a spherical coordinate system was noted and the conditions for its applicability were revealed. In the selected geometry, expressions for the energy densities of anisotropy, demagnetization, and the interaction of magnetization with the external field, written in terms of spherical variables – the polar and azimuth angles of the magnetization vector, are obtained.

Two equivalent forms of recording the energy of uniaxial anisotropy are given, their relationship with the demagnetization energy density is revealed, and a fundamental difference of this ratio is noted for cases of magnetization parallel and perpendicular to the projection of the EMA on the film plane.

2. The first of the above cases is considered, namely, the magnetization of a film by a field, the plane of variation of which passes through the normal to the plane of the film, and the film is oriented in such a way that the EMA does lie in the plane of variation of the field. By minimizing the expression for the total energy density, the equilibrium position of the magnetization vector is determined both in the absence of a field and in fields of various orientations. It is shown that, in the absence of a field, the quality parameter of the film, which is the ratio of the anisotropy field to the demagnetization field, is a convenient parameter to characterize the equilibrium position of the magnetization. It is noted that in the case when the quality factor is less than unity, the magnetization vector in a free state is oriented near the film plane, and when the quality factor is one, the preferred direction of magnetization is normal to the film plane. Examples of real materials of both types are given. Thus, the quality factor less than a unit is characteristic of permalloy and some mixed ferrite garnet, and a larger unit is characteristic of orthoferrites, hexaferrites, and the overwhelming majority of mixed granite. On the basis of a numerical analysis of the expression for the first derivative of the total energy density, a discontinuous character of the change in the orientation of the magnetization vector, which occurs when the field value changes, is revealed. It is noted that such a character of orientation appears noticeably only when the fields are smaller or of the order of the demagnetization field (no more than two or three times greater than this field). A model was constructed to explain the jumps in the orientation of the magnetization by the presence of two or more miniatures of the film energy potential, the depth of which, as the magnitude or orientation of the external field changes, also changes, which leads to a jump in the operating point of the system from one minimum to another. A qualitative interpretation of the magnetization jump is given, which clearly explains the limitation of the field interval, within which this jump is possible, by a value close to the demagnetization field.

3. For the first of the above-mentioned cases, that is, when the EMA is oriented in the plane of the field change, the frequency of ferromagnetic resonance is calculated based on the obtained

equilibrium orientation of the magnetization using the Smit–
Sul method. The dependence of the resonance frequency on the
external field is considered for various orientations in the plane
passing through the EMA and the normal to the film plane. It
is shown that with a strong difference in the orientation of the
field from the direction of the EMA, the resonance frequency
first decreases, after which, after passing through a minimum, it
increases, asymptotically aiming for a linear dependence passing
through the point corresponding to the resonance in zero field.
The minimum of dependence falls on the field, which exceeds the
demagnetization field by two or three times. With a small difference
in the orientation of the field from the direction of the EMA, the
dependence frequency from the field has a monotonically increasing
character, asymptotically striving for the same straight line as with
a strong difference. The minimum of the observed dependence is
interpreted on the basis of the physics of a simple and partially
tightened orientational transition, which occurs when the film is
magnetized near the direction perpendicular to the EMA. For the
case of a fixed frequency, the dependence of the magnitude of the
ferromagnetic resonance field on the orientation of the external field
is considered. It is shown that the nature of this dependence is close
to sinusoidal with a period of 180° and some phase shift relative
to the direction corresponding to the normal to the film plane. It is
noted that such a shift is determined by the angle of deviation of
the EMA from the normal to the film plane with allowance for the
additive introduced by the demagnetization field. In order to simplify
the obtained expressions, the same dependence was considered in the
strong field approximation. Based on a comparison with a complete
solution, the correctness of such an approximation is estimated,
and it is shown that its accuracy increases as the magnitude of the
external field increases, and the frequency at which the resonance
field is determined.

4. The second of the above cases is considered, namely, the
magnetization of a film by a field, the plane of variation of which
passes through the normal to the plane of the film, with the EMA
being in a plane perpendicular to the plane of variation of the field.
By minimizing the expression for the total energy density, the
equilibrium position of the magnetization vector is determined both
in the absence of a field and in fields of various orientations. It is
noted that in the absence of a field, the quality factor of the film is
a convenient parameter for characterizing the equilibrium position

of the magnetization. The dependence of the equilibrium value of the polar angle of the magnetization vector is considered from the magnitude of the deviation from the normal to the film plane for various factors of quality. The critical value of the quality factor has been revealed, in which the forces acting on the magnetization from the anisotropy field and the demagnetizing field of the film are compared to each other. It is shown that at this value the dependence of the equilibrium polar angle on the magnitude of the quality factor is a straight line corresponding to a change in the polar angle from 45° to zero. With values of the quality factor, larger and smaller than the critical, the dependence of the equilibrium polar angle on the quality factor, starting at 0° and 90, respectively, aim for their critical values. The equilibrium orientation of the magnetization vector in the presence of an external field is considered. It was noted that in contrast to the previous case, when it was enough to minimize the energy density only by the azimuth angle, here it is necessary to carry out the minimization at both angles polar and azimuth simultaneously. Such a difference is due to the ejection from the plane of variation of the external field, which causes a deviation of the equilibrium magnetization from this plane. Two possible expressions for the energy density, related through the constancy of the length of the magnetization vector, are discussed, practical recommendations for choosing the optimal expression are given. A system of two equations for the azimuthal and polar angles was obtained, its equivalence to a very complex fourth-degree full equation was noted, as a result of which it was concluded that its numerical solution was more expedient by the method of establishment. Analytical expressions for the fields required for the implementation of the establishment method are given. The convergence of the iterative establishment process is estimated. It is shown that for the problem considered here, the number of iterations of the order of 1000 is optimal from the point of view of the relation of sufficient accuracy with smallness of the counting time. The accuracy is tenths of a degree when the counting time is about ten seconds. The method of establishing the dependences of the equilibrium azimuthal angle and the equilibrium normal to the film plane of the component of magnetization on the azimuthal angle of the field change in its plane. The critical value of the field is revealed, above and below which the obtained dependences have a fundamentally different character. The magnitude of the critical field is defined as exceeding the demagnetization field by two to three times. It is shown that in a field larger than the critical one, as the field orientation changes, the magnetization vector basically follows the field direction, and the normal component of the magnetization deviates from the film plane in the same direction as the EMA. In a field less critical, the dependence of the cacazimuth angle and the normal component of magnetization on the orientation of the field have a strongly pronounced jump-like character. It has been established that in the formation of the nature of the jumps, an important role is played by both the magnitude of the field and the initial orientation of the magnetization vector. The possibility of analyzing magnetization jumps on the basis of a model of a three-dimensional surface of the energy potential is noted, references are made to works where such an analysis is performed to solve similar problems.

5. For the second of the above cases, that is, when the EMA is oriented in a plane perpendicular to the plane of the field change, based on the obtained equilibrium orientation of the magnetization, the frequency of ferromagnetic resonance is calculated. Two equivalent energy density records are considered, related to the constancy of the length of the magnetization vector, the choice of the one that most corresponds to the qualitative physical content of the problem is made. A formula is obtained for the dependence of the resonance frequency on the magnitude and orientation of the external field, expressed in terms of the polar and azimuthal angles of equilibrium position of the magnetization vector. It is shown that the formula for frequency can be represented as a quadratic equation with respect to the external field. From the solution of this equation, the dependence of the ferromagnetic resonance field on the polar and azimuth angles of the magnetization vector is obtained under the condition of a given frequency value. In the strong field approximation, the orientational dependence of the resonance field on the azimuth angle of the external field is obtained. It is shown that the nature of this dependence is close to sinusoidal with a period of 180°, without any phase shift relative to the direction corresponding to the normal to the film plane. It is noted that the absence of a shift is determined by the coincidence of the projection of the EMA on the plane of the field change with the direction normal to the film plane, so that the orientation of the equilibrium position of the magnetization relative to the plane of the field change is completely symmetrical.

6. The considered cases of a change in the external field in the plane passing through the normal to the film plane with two different orientations of the EMA with respect to this normal are analyzed from the point of view of the possibility of separately measuring the saturation magnetization and the uniaxial anisotropy constant. A method for implementing such a measurement in the experiment is described, which includes determining the two extensions of the resonant field corresponding to the magnetization of a film in its plane by a field parallel to and perpendicular to the projection of the EMA onto the film plane, followed by using these expressions to compile a system of equations whose resolution gives the desired magnetization values and anisotropy constants.

7. A brief review of the results of previously performed experiments on the study of ferromagnetic resonance in films of mixed garnets, with a significant deviation of the axis of uniaxial anisotropy from the normal to the film plane, was performed. A measurement technique with brief characteristics of the equipment used, as well as methods of fixing and orientation of the sample, is described. The compositions and basic parameters of the samples studied, including those measured by independent methods, are given. The results of a study of the orientation dependences of the resonant fields for the two external orientation orientations relative to the EMA and normal to the film plane, the theoretical interpretation of which was performed in the strong field approximation, are presented. A good agreement between the experimental cuts of the ultats and the orientational dependences obtained theoretically on the basis of the Smit–Sul method and the solution was noted. A separate measurement of the uniaxial anisotropy constant and saturation magnetization, performed in accordance with the method described above, is demonstrated. The technique is described and the results of measuring the constant of the magnetoelastic interaction of the films are presented, including the role of the magnetostriction mechanism in the formation of the EMA deviation from the normal to the film plane. The results of the observation of the fine structure of the line and the measurement of the damping parameter of the ferromagnetic resonance are briefly listed. The results of a study of the orientation dependences of the ferromagnetic resonance field on the orientation of the external field in films possessing, along with uniaxial, also cubic anisotropy, are presented. The technique is presented and the results of measuring the uniaxial and cubic anisotropy constants are described under the condition that the saturation magnetization from

independent sources is known. The higher sensitivity of measuring the parameters of cubic anisotropy by the resonance method is noted compared with similar measurements performed by the magneto-optical method.

8

Ferromagnetic resonance in a composition environment consisting of anisotropic ferrite particles

This chapter is devoted to the description of the phenomenon of ferromagnetic resonance in a composite medium consisting of spherical spherical particles with uniaxial magnetic anisotropy. The general case is considered when the anisotropy axes in individual particles are oriented in a chaotic manner, so that when such a medium is magnetized by a uniform constant field, the magnetization vectors in the individual particles rotate in the direction of the field. Using the averaging method over the polar and azimuthal angles, we obtained the tensor of the dynamic magnetic susceptibility of the medium as a whole. The possibility of the formation of specified values of tensor properties by partial orientation of the anisotropy axes in individual particles is considered. When considering the surveys listed in the survey, we will mainly follow the works [6–8, 22–25, 78–117, 376–395], the remaining references are indicated in the text.

8.1. General formulation of the problem

Consider a composite medium, which is a combination of ferrite grains interspersed in the form of a lattice into a rigid non-magnetic matrix. We assume that the matrix material is isotropic and has dielectric properties, and the dielectric constant is set equal to unity. For the sake of simplicity, we assume that the particles have the

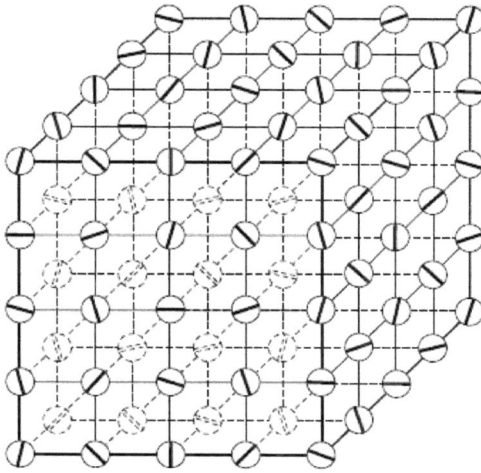

Fig. 8.1. Possible geometry of the compositional environment structure.

same spherical shape and their genetic parameters coincide. Let all particles have the same uniaxial anisotropy, the field of which is of the same order as the saturation magnetization of their material. The orientations of the anisotropy axes in individual particles are distributed in space in an arbitrary chaotic manner, all directions of the anisotropy axes being equally probable. Let the distances between the particles be so large that there is no interaction between them (similar to the 'independent green' model for a polycrystal [6–8, 376]). The medium is magnetized by a constant uniform field, the magnitude of which is sufficient to destroy the domains inside the particles.

One of the possible variants of the geometry of the structure of the proposed medium is shown in Fig. 8.1. As an example, a cubic lattice was chosen, however, since there is no interaction between the particles, the lattice type does not matter and in the further calculation is not counted. Spherical particles are schematically depicted in the form of circles located in its nodes. The thickened lines inside the circles indicate the location of the anisotropy axes of the individual spheres, oriented randomly in a random way. Since the ferrite spheres occupy only a part of the space, in order to obtain the electrodynamic parameters of the medium as a whole, it is necessary to multiply the corresponding particle parameters by a filling factor equal to the ratio of the volumes of the filled and empty parts of the space. We estimate the fill factor for the structure shown in Fig. 8.1.

Let the radius of the sphere be r, and the distance between adjacent lattice sites be d. Obviously, one cell corresponds to one sphere. Then the fill factor, equal to the ratio of the volume of the sphere to the volume of the cell is equal to:

$$N_p = \frac{4}{3}\pi\left(\frac{r}{d}\right)^3. \tag{8.1}$$

The total number of particles per unit volume is equal to:

$$N_0 = d^{-3} \tag{8.2}$$

8.1.1. Coordinate systems

The main objective of this work is the calculation of the dynamic susceptibility, that is, the tensor connecting the dynamic magnetization with a variable field. Since the medium consists of a set of identical interacting particles oriented in different ways, the dynamic susceptibility of the medium as a whole is determined by the sum of the susceptibilities of its constituent particles. Since the particles are distributed uniformly in orientation, the sum can be replaced by the product of the susceptibility of one particle, averaged over all possible orientations, by the total number of particles per unit volume of the medium. Thus, as the first task, we can single out the determination of susceptibility for a single particle with a

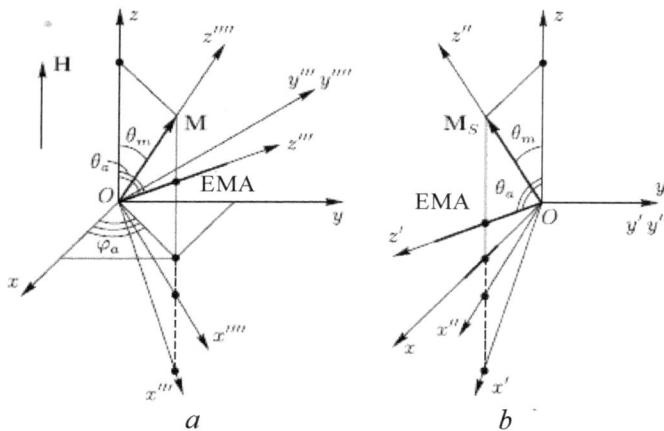

Fig. 8.2. Coordinate systems used to obtain the susceptibility tensor of a single particle with an EMA anisotropy axis oriented at angles θ_a and φ_a.

given orientation of the anisotropy axis. To solve this problem, we use the coordinate systems shown in Fig. 8.2.

Figure 8.2 *a* shows the general location of the EMA anisotropy axis (EMA is an abbreviation of the expression 'easy magnetization axis') given by the angles θ_a and φ_a in the laboratory coordinate system associated with a constant field, as well as the orientation of the magnetization vector **M** defined by the angles θ_m and φ_a in that same system. Note that here, for the azimuthal angle of the magnetization vector, no eigenvalue φ_m, different from φ_a, is introduced, since the symmetry of the problem implies that the equilibrium magnetization is oriented in the plane passing through the EMA and the external field (that is, in the zOz''' plane).

Figure 8.2 *b* shows the orientations of technical quantities at $\varphi_a = 0$. The enlarged points, which denote the intersection of the corresponding lines, are shown in the figure to show the relative position of the axes lying in the plane passing through the Oz axis and the magnetization vector **M** coordinate systems.

The following coordinate systems are used in solving the problems:

System No. 1 – $Oxyz$ – laboratory system, associated with the direction of the external field. The system has axes Ox, Oy, Oz. The axis Oz is directed along the external field **H**, that is, it coincides with the direction of the axis Or of the spherical system associated with the field. The axes Ox and Oy are perpendicular to the external field. The plane Oxz corresponds to $\varphi = 0$.

The system is designed to obtain a susceptibility tensor in a form convenient for its use in problems of the propagation of electromagnetic waves. In this system, the susceptibility tensor must be obtained in its final form.

System number 2 – $Ox'y'z'$ – associated with the axis of anisotropy – EMA in the case when this axis lies in the Oxz plane of the system No. 1.

The system has axes Ox', Oy', Oz'. Oz' is directed along the EMA. The axis Oy' coincides with the axis Oy of the system No. 1. The angle between the axes Oz' and Oz (or Oz' and Or) is eqjual to θ_a. The angle between the axes Ox' and Ox is also equal to θ_a. Plane $Ox'z'$ coincides with the Oxz plane of system No. 1, corresponding to $\varphi = 0$. The system corresponds to the case of zero value of the azimuthal angle of the EMA. This system is used when finding the equilibrium position of the magnetization vector.

System number 3 – $Ox''y''z''$ – associated with the equilibrium orientation of the magnetization vector $\mathbf{M}_S$.

The system has axes Ox'', Oy'', Oz''. The Oz'' axis is directed along the equilibrium direction of the magnetization vector. The Oy'' axis coincides with the axis Oy of the system number 1. Plane $Ox''z'$ coincides with the Oxz plane of system No. 1. The angle between the axes Oz'' and Oz (or Oz'' and Or) is θ_m. The angle between the axes Ox'' and Ox is also equal to θ_m. The spherical coordinates of the vector M_S correspond to θ_m and φ_m. By virtue of the axial symmetry of the problem with respect to the Oz axis (that is, with respect to the direction of the field $\mathbf{H}$), we can set $\varphi_m = \varphi_a$.

The system corresponds to the case of a zero value of the azimuthal angle of the EMA (that is, $\varphi_m = 0$).

In this system, the equation of motion of the magnetization vector is written, linearized, and solved for the case $\varphi_a = 0$.

System number 4 – $Ox'''y'''z'''$ – associated with the axis of anisotropy, when the EMA is oriented relative to the Oxz plane of system No. 1 in an arbitrary manner, that is, rotated around the axis Oz of system No. 1 by an angle φ_a.

The system has the axes Ox''', Oy''', Oz'''. Oz''' axis is directed along the EMA. Oz''' axis projection (i.e., also the EMA) on the Oxy plane makes an angle φ_a with the Ox axis of the system No. 1. Ox'' axis projection on the Oxy plane, the angle φ_a is with the Ox axis of system No. 1. Oy''' axis projection on the Oxy plane, the angle φ_a is with the Oy axis of system No. 1 Since the equilibrium vector of magnetization $\mathbf{M}$ lies in the plane passing through the EMA and the direction of the constant field, that is, through the axis Oz''' and Oz, then the projection of the vector $\mathbf{M}$ on the Oxy plane forms with the axis Ox the angle φ_a (that is, $\varphi_m = \varphi_a$, as noted when considering system No. 3).

The system corresponds to the case of an arbitrary value of the azimuthal angle of the EMA.

In this system, the energy density of uniaxial anisotropy is recorded in the simplest form.

System number 5 – $Ox''''y''''z''''$ – associated with equilibrium magnetization, when the EMA is oriented relative to the Oxz plane of the system number 1 in an arbitrary way, that is, rotated around the axis Oz of the system number 1 at an angle φ_a.

The system has the axis Ox'''', Oy'''', Oz''''. Oz'''' axis is directed along the equilibrium magnetization vector $\mathbf{M}_S$, that is, the angle θ_m is made with the Oz axis of system No. 1. Oz'''' axis projection (that is, the vector $\mathbf{M}_s$) on the plane Oxy is the angle φ_a with the axis Ox of the system No. 1. Ox'''' axis projection on the Oxy plane forms the angle φ_a with the Ox axis of system No. 1. Oy'''' axis projection on the Oxy plane forms the angle φ_a with the Oy axis of system No. 1.

The system corresponds to the case of an arbitrary value of the azimuthal angle of the EMA, denoted as φ_a.

In this system, the equation of motion of the magnetization vector is written, linearized, and solved for the case of an arbitrary angle φ_a.

From the above description it is clear that the orientation of the axes of the systems $Ox'y'z'$ (No. 2) and $Ox'''y'''z'''$ (No. 4) with respect to the laboratory system $Oxyz$ (No. 1) is completely determined by specifying the polar and azimuthal angles of the anisotropy axis (EMA) θ_a and φ_a. To determine the orientation of the axes of the systems $Ox''y''z''$ (No. 3) and $Ox''''y''''z''''$ (No. 5), in addition to the azimuthal angle φ_a, knowledge of the polar angle θ_m, which determines the equilibrium orientation of the magnetization vector $\mathbf{M}_s$, is necessary.

Thus, the first task is to find the equilibrium position of the vector $\mathbf{M}_s$, which is determined by the interaction of the magnetization with the anisotropy field and the constant field.

8.1.2. Equilibrium orientation of magnetization

Let us find the equilibrium orientation of the magnetization vector in a medium with uniaxial anisotropy (EMA) placed in a constant field $\mathbf{H}$ uniform in space. The geometry of the problem is shown in Fig. 8.3. The coordinate system $Oxyz$ is chosen in such a way that the axis Oz is parallel to the vector of the field $\mathbf{H}$, and the plane Oxz passes through the field vector and the axis of anisotropy. In this case, the Oy axis is oriented perpendicular to the field vector and the anisotropy axis (in Fig. 8.3 from the drawing plane to the reader). Since the magnetization vector $\mathbf{M}$ is acted upon only by the field and anisotropy, it is located in the equilibrium state in the same Oxz plane as the field vector and the anisotropy axis, that is, the problem reduces to two-dimensional in this plane.

To clarify the essence of the problem, it is useful to consider in the plane Oxz two auxiliary coordinate systems: $Ox'z'$, the axis Oz' of which is parallel to the axis of anisotropy, and $Ox''z''$, the axis Oz'' of which is parallel to the magnetization vector. The angle between the axes of Oz and Oz' will be equal to the angle θ_a, and the angle between the axes Oz and Oz'' to the angle θ_m. In this formulation of

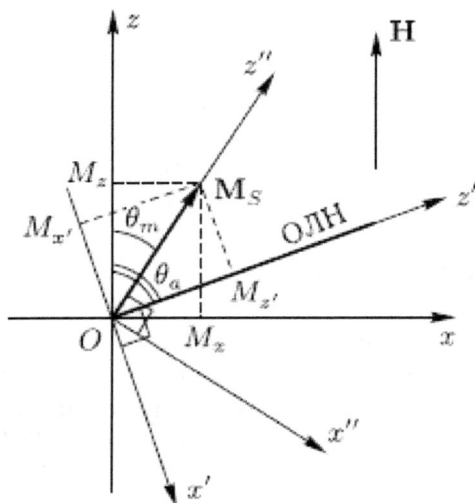

Fig. 8.3. Auxiliary coordinate systems used to find the equilibrium position of the magnetization vector

the system Oxz, $Ox'z'$ and $Ox''z''$ are special cases of systems $Oxyz$ (No. 1), $Ox'y'z'$ (No. 2) and $Ox''y''z''$ (No. 3), introduced in the previous section, with $\varphi_a = 0$. In this case, the angle θ_a, the field $\mathbf{H}$ and the anisotropy field are assumed to be known, and the angle θ_m is to be determined. The constant field vector in the $Oxyz$ system is

$$\mathbf{H} = \{0, 0, H_0\}, \tag{8.3}$$

the magnetization vector $\mathbf{M}$ in the $Oxyz$ system is

$$\mathbf{M}_S = \{M_0 \sin\theta_m, 0, M_0 \cos\theta_m\}; \tag{8.4}$$

the same vector in the system $Ox'y'z'$ has the appearance

$$\mathbf{M}_S = \{M_0 \sin(\theta_a - \theta_m), 0 M_0 \cos(\theta a - \theta_m)\}. \tag{8.5}$$

The energy density of uniaxial anisotropy in the system $Ox'y'z'$ has the form

$$U_a = -\frac{K}{M_0^2} M_{z'}^2, \tag{8.6}$$

where K is the anisotropy constant ($K \geq 0$). The energy density of the interaction of magnetization with a constant field in the $Oxyz$ system is:

$$U_H = -\mathbf{MH} = -H_0 M_z.$$
(8.7)

The overall energy density of the system as a whole is

$$U = -\frac{K}{M_0^2} M_{z'}^2 - H_0 M_z.$$
(8.8)

To find the equilibrium state of magnetization, it is necessary to express M_z and $M_{z'}$ through θ_m, substitute in (8.8), differentiate the resulting expression for the energy density U by θ_m and equate the first derivative is zero, which will result in an equation to determine θ_m. So, from (8.4) and (8.5) we get:

$$M_z = M_0 \cos\theta_m;$$
(8.9)

$$M_{z'} = M_0 \cos(\theta_a - \theta_m).$$
(8.10)

Substituting (8.9) and (8.10) into (8.8), we obtain the energy density in the form

$$U = -K\cos^2(\theta_a - \theta_m) - H_0 M_0 \cos\theta_m.$$
(8.11)

Differentiating (8.11) with respect to θ_m and equating the derivative to zero, we obtain the equation for determining θ_m:

$$K\sin\left[2(\theta_a - \theta_m)\right] - H_0 M_0 \sin\theta_m = 0.$$
(8.12)

According to the definition of the anisotropy field (4.55)

$$H_a = \frac{2K}{M_0}.$$
(8.13)

Usi|ng (8.13), we can write equation (8.12) in the form:

$$H_a \sin\left[2(\theta_a - \theta_m)\right] - 2H_0 \sin\theta_m = 0.$$
(8.14)

Expanding the brackets in the first term, expressing $\cos\theta_m$ in the resulting expression through $\sin\theta_m$, as well as introducing the notation

$$h = \frac{H_0}{H_a},$$

(8.15)

This equation can be reduced to the following form:

$$4\sin^4\theta_m + 8h\sin 2\theta_a \sin^3\theta_m + 4(h^2 - 1)\sin^2\theta_m -$$
$$- 4h\sin 2\theta_a \sin\theta_m + \sin^2 2\theta_a = 0.$$

(8.16)

It can be seen that the resulting equation (8.16) is very similar in structure to equation (5.34), which reflects the equilibrium orientation of the magnetization vector when the orientation of the external field changes. Some difference is due only to the fact that the field there rotates relative to the laboratory coordinate system, and here the anisotropy axis. For the rest, the character of equations (8.16) and (5.34) is completely identical, so that we can expect a similar behavior of the equilibrium position of magnetization.

Regarding the solvability of this equation, we can say the same thing that was said regarding equation (5.34). That is, in analytical form, a solution, apparently, perhaps, for example, by the Ferrari method (or other similar method), for which the Cardano equation will be the resolving equation [261, 262]. The scheme of the analytical solution of such equations is given in this monograph in section 2.4. However, as can be seen from [297, 334, 362], where such equations are solved by analytical methods, the general form of the solution is somewhat cumbersome. Therefore, for a single particle, that is, for a single solution of equation (8.16) or (8.14), the numerical solution seems to be more convenient. In this case, as with the solution of equation (5.34), you can use the method of finding zero, applying it not to equation (8.16), but directly to equation (8.14). Such a solution of equation (8.14) with respect to the polar angle of the equilibrium magnetization vector θm is illustrated in Fig. 8.4, where the dependences of θ_m on the constant field are shown, constructed for different values of the anisotropy ratio and the orientation of the axis.

By virtue of the parity (bidirectionality) of uniaxial anisotropy with respect to the z-component of magnetization, the real physical meaning is only considering the polar angle of the anisotropy axis in the interval $0° \le \theta_a \le 90°$. The value of the polar angle of the magnetization vector in the equilibrium state θ_m also falls on

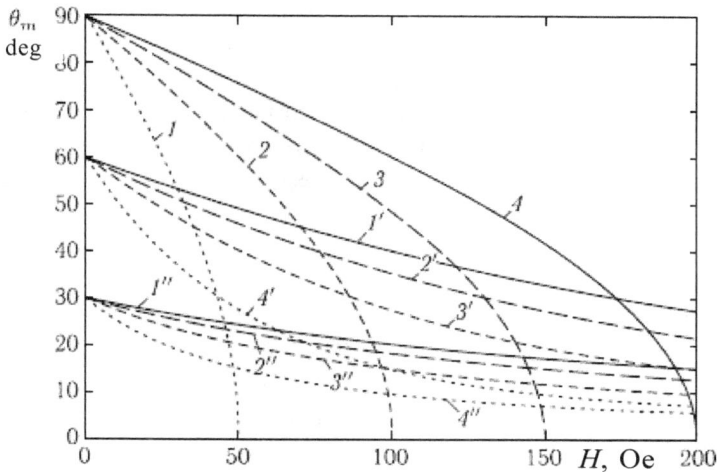

Fig. 8.4. Dependences of the polar angle of the equilibrium magnetization vector θ_m on the constant field for different values of the fields H_a and the anisotropy axis θ_a. Curves 1, 1', 1'' correspond to the field H_a = 50 Oe, 2, 2', 2'' – field H_a = 100 Oe, 3, 3', 3'' – field H_a = 150 Oe, 4, 4', 4'' – to the field H_a = 200 Oe. Curves without strokes correspond to θ_a = 90°, curves with one stroke – θ_a = 60°, with two strokes – θ_a = 30°.

this interval. Orientation of the anisotropy axis in the range of $90° \leq \theta_a \leq 180°$ is equivalent to an orientation in the range of $0° \leq \theta_a \leq 90°$ and leads to the same equilibrium values of the polar angle of the magnetization vector. It can be seen from the figure that the character of the curves is completely similar to that which occurs with a tightened orientation transition (Section 5.1.2). Similar curves are shown in Figs. 5.5 and 5.6. The only difference is that the field direction changed there, while the orientation of the anisotropy axis remained unchanged, here the position is reversed – the orientation of the anisotropy axis changes, and the field direction remains unchanged. That is, the difference is reduced only to the choice of a coordinate system, and the physics of the process is completely preserved. So here, as the constant field increases, the magnetization vector rotates in the direction of the field, and this rotation occurs the faster, the smaller the anisotropy field. If the anisotropy axis is oriented perpendicular to the constant field, then the rotation of the magnetization vector ends at a constant field equal to the anisotropy field, after which the magnetization vector orientates exactly along the field direction (curves 1–4). If the anisotropy axis is deflected from the normal to a constant field, then the magnetization vector tends to turn in the direction of the field, the sooner the closer the direction of the anisotropy axis to the field direction. In this case, the magnetization vector exactly along the field never aligns, and even when the constant field exceeds the anisotropy field several times (for example, for curves 1' and 1''), the deviation of the anisotropy vector from the field direction can be quite significant and amount to 10 degrees.

8.2. The sequential order of solving the problem

To calculate the dynamic susceptibility of the compositional medium, we will use the coordinate systems shown in Fig. 8.2. When moving from one coordinate system to another, we use the method of transition matrices [261], described in detail in Chapter 3.

To obtain the susceptibility tensor in the system $Oxyz$, associated with a constant field, with an arbitrary orientation of the EMA with respect to this field, that is, with arbitrary values of the angles θ_a and φ_a, you first need to find the susceptibility tensor in the system associated with the equilibrium position of the magnetization vector $Ox''''y''''z''''$, then convert the resulting tensor to the system $Oxyz$.

Since the magnetization vector $\mathbf{M}$ in the steady state is aff3ected by only two forces − on the side of the constant field and on the side of uniaxial anisotropy, then in equilibrium it lies in a plane passing through the direction vectors of the field and the axis of anisotropy, that is, in the plane Ozz'''. The direction of the magnetization vector $\mathbf{M}$ is uniquely determined by the orientation of the axis Oz'''', which makes the angle θ_m with the axis Oz. Moreover, the equality $\varphi_m =$ holds.

Axis orientation Ox'''' and Oy'''' relative to the rotation around the axis Oz'''' is not determined. Therefore, based on the convenience of calculation, we assume that the axis Ox'''' lies in the same plane Ozz''''. The angle between the axis Ox'''' and the projection of the magnetization vector $\mathbf{M}$ on the plane Oxy is obtained equal to θ_m, and the axis Oy'''' lies in the Oxy plane and makes an angle φ_a with the Oy axis.

Thus, if we assume that the angles θ_a and φ_a are given, and the angles θ_m and φ_m are known, then the first step in the calculation is to find the susceptibility tensor in the system $Ox''''y''''z''''$. For this, all forces acting on the magnetization vector $\mathbf{M}$ must be transferred to this system. In the steady state, there are two such forces: the strength of a constant field and the strength of the anisotropy field.

But the constant field is set in the $Oxyz$ system, and the anisotropy field is set in the system $Ox'''y'''z'''$. Therefore, we need two transition matrices: from the system $Oxyz$ to the system $Ox''''y''''z''''$ and from the system $Ox'''y'''z'''$ to the system $Ox''''y''''z''''$.

Since the susceptibility tensor is determined by the dynamic response of a system that is in a stationary state to an alternating field specified in the $Oxyz$ system, the vector of the alternating field **h** must also be transferred from the $Oxyz$ system to the $Ox''''y''''z''''$ system, for which we can use the first of the mentioned transitions atrits.

Further in the system $Ox''''y''''z''''$ it is necessary to write the equation of motion for magnetization, whose solution will give the susceptibility tensor in the system $Ox''''y''''z''''$. After finding the tensor of susceptibility in the system $Ox''''y''''z''''$, the next calculation step will be to transfer it to the $Oxyz$ system. For this we need matrices of forward and reverse transition from the system $Ox''''y''''z''''$ to the $Oxyz$ system.

8.2.1. Transition matrices

Thus, we need three transition matrices:

1) $\ddot{A}_{04}$ – from system $Oxyz$ to the system $Ox''''y''''z''''$

2) $\ddot{A}_{40}$ – from the system $Ox''''y''''z''''$ to the Oxyz system;

3) $\ddot{A}_{34}$ – from the $Ox'''y'''z'''$ system to the $Ox''''y''''z''''$ system.

These matrices are expressed in terms of the cosines of the angles between the coordinate axes of the corresponding system and have the form:

$$\ddot{A}_{04} = \begin{pmatrix} \cos(x,x''') & \cos(x,y''') & \cos(x,z''') \\ \cos(y,x''') & \cos(y,y''') & \cos(y,z''') \\ \cos(z,x''') & \cos(z,y''') & \cos(z,z''') \end{pmatrix} \tag{8.17}$$

$$\ddot{A}_{40} = \begin{pmatrix} \cos(x''',x) & \cos(x''',y) & \cos(x''',z) \\ \cos(y''',x) & \cos(y''',y) & \cos(y''',z) \\ \cos(z''',x) & \cos(z''',y) & \cos(z''',z) \end{pmatrix} \tag{8.18}$$

$$\ddot{A}_{34} = \begin{pmatrix} \cos(x'',x''') & \cos(x'',y''') & \cos(x'',z''') \\ \cos(y'',x''') & \cos(y'',y''') & \cos(y'',z''') \\ \cos(z'',x''') & \cos(z'',y''') & \cos(z'',z''') \end{pmatrix}$$

(8.19)

where the indices in the notation of matrices correspond to the number of strokes in the notation of the coordinates.

To obtain the cosine of the angle between any two axes, it is necessary to define a unit vector along the initial axis and find its projection on the final axis. If the initial and final axes are located in different coordinate planes, the preliminary step is to find the projection of the original unit vector onto the intersection line of these points, after which the resulting projection is once again projected onto the final axis. The projection of the initial unit vector on the final axis found as a result of this procedure is equal to the cosine of the angle between the axes.

The transition matrices obtained in this way are:

$$\ddot{A}_{34} = \begin{pmatrix} \cos(x''',x'''') & \cos(x''',y'''') & \cos(x''',z'''') \\ \cos(y''',x'''') & \cos(y''',y'''') & \cos(y''',z'''') \\ \cos(z''',x'''') & \cos(z''',y'''') & \cos(z''',z'''') \end{pmatrix}$$

(8.20)

$$\ddot{A}_{40} = \begin{pmatrix} \cos\theta_m \cos\varphi_m & \cos\theta_m \sin\varphi_a & -\sin\theta_m \\ -\sin\varphi_a & \cos\varphi_a & 0 \\ \sin\theta_m \cos\varphi_a & \sin\theta_m \sin\varphi_a & \cos\theta_m \end{pmatrix}$$

(8.21)

$$\ddot{A}_{34} = \begin{pmatrix} \cos\theta_m(\theta_a - \theta_m) & 0 & -\sin(\theta_a - \theta_m) \\ 0 & 1 & 0 \\ \sin(\theta_a - \theta_m) & 0 & \cos(\theta_a - \theta_m) \end{pmatrix}$$

(8.22)

In this case, the magnetization vectors are converted by the formulas:

$$\mathbf{m} = \ddot{A}_{04}\vec{m}'''';$$

(8.23)

$$\mathbf{m}'''' = \ddot{A}_{40}\vec{m};$$

(8.24)

$$\mathbf{m}''' = \ddot{A}_{34}\vec{m}'''';$$
(8.25)

The susceptibility tensor is transformed by a formula similar to (6.197):

$$\chi = \ddot{A}_{04}\ddot{\chi}''''\ddot{A}_{40}.$$
(8.26)

8.2.2. The implementation of the sequential steps of solving the problem

STEP # 1.

The vector of a constant field (the sum of the vectors of the external field and the demagnetization field) in the $Oxyz$ system is

$$\mathbf{H} = \begin{pmatrix} 0 \\ 0 \\ \hline H_0 \end{pmatrix}$$
(8.27)

Transition formula:

$$\mathbf{H}'''' = \ddot{A}_{40}\mathbf{H}.$$
(8.28)

Transition matrix (8.21):

$$\ddot{A}_{40} = \begin{pmatrix} \cos\theta_m\cos\varphi_a & \cos\theta_m\cos\varphi_a & -\sin\theta_m \\ -\sin\varphi_a & \cos\varphi_a & 0 \\ \sin\theta_m\cos\varphi_a & \sin\theta_m\sin\varphi_a & \cos\theta_m \end{pmatrix}$$
(8.29)

As a result of the transition, we obtain the vector of the constant field in the form

$$\mathbf{H}'''' = \begin{pmatrix} -H_0\sin\theta_m \\ 0 \\ H_0\cos\theta_m \end{pmatrix}$$
(8.30)

STEP #2.

Write the variable field in the $Ox''''y''''z''''$ system.

The variable field $\mathbf{h}$ is included in the expression for the energy density as a factor in the scalar product by the magnetization of the form $\mathbf{Mh}$. Since the magnitude of the scalar product does not depend on the choice of the coordinate system, and the variable field is not included in the final expression for susceptibility, the field $\mathbf{h}$ can be immediately set in the system $Ox''''y''''z''''$.

In this system, the vector of the variable field has the form

$$\mathbf{h} = \left(\begin{array}{c} h_{x'''} \\ \hline h_{y'''} \\ \hline h_{z'''} \end{array} \right) \tag{8.31}$$

STEP #3

Transfer the vector of magnetization from the $Ox'''y'''z'''$ system to the $Ox''''y''''z''''$ system

***

This transfer is required to ensure that the energy density of uniaxial anisotropy in the $Ox'''y'''z'''$ system is transferred to the $Ox''''y''''z''''$ system. The transition formula from the $Ox'''y'''z'''$ to the $Ox''''y''''z''''$ for the magnetization vector $\mathbf{M}$ has the form

$$\mathbf{M}''' = \ddot{A}_{34} \mathbf{M}'''' \tag{8.32}$$

$$\ddot{A}_{34} = \left(\begin{array}{c|c|c} \cos(\theta_a - \theta_m) & 0 & -\sin(\theta_a - \theta_m) \\ \hline 0 & 1 & 0 \\ \hline \sin(\theta_a - \theta_m) & 0 & \cos(\theta_a - \theta_m) \end{array} \right) \tag{8.33}$$

$$M''' = \left(\begin{array}{c} M_{x'''} \cos(\theta_a - \theta_m) - M_{z'''} \sin(\theta_a - \theta_m) \\ \hline M_{y'''} \\ \hline M_{x'''} \sin(\theta_a - \theta_m) + M_{z'''} \cos(\theta_a - \theta_m) \end{array} \right) \tag{8.34}$$

STEP #4.

The total energy density includes the uniaxial anisotropy energy density U_a, the interaction energy density of the magnetization with a constant field U_H and the interaction energy density of the magnetization with a variable field U_h:

$$U = U_a + U_H + U_h. \tag{8.35}$$

The terms in the system (8.35) in the system $Ox''''y''''z''''$ have the form:

uniaxial anisotropy energy density

$$U_a = -\frac{K}{M_0^2} M_z^2 m = -\frac{K}{M_0^2} \left[M_{x''''} \sin(\theta_a - \theta_m) + M_{x''''} \cos(\theta_a - \theta_m) \right]^2 =$$

$$= -\frac{K}{M_0^2} [M_{x''''}^2 \sin(\theta_a - \theta_m) + M_{z''''}^2 \cos^2(\theta_a - \theta_m) +$$

$$+ 2M_{x''''} M_z \sin(\theta_a - \theta_m) \cos(\theta_a - \theta_m)]; \tag{8.36}$$

– density of interaction energy of magnetization with constant field

$$U_H = -\mathbf{MH} = -\mathbf{M}'''\mathbf{H}''' = -M_{x''''} H_0 \sin\theta_m + M_{z''''} H_0 \cos\theta_m; \tag{8.37}$$

- density of interaction energy of magnetization with a variable field

$$U_h = -\mathbf{Mh} = -\mathbf{M}''''\mathbf{h}'''' =$$
$$= -M_{x''''} h_{x''''} - M_{y''''} h_{y''''} - M_{z''''} h_{z''''}. \tag{8.38}$$

Substituting (8.36)–(8.38) into (8.35), we obtain the total energy density in the form

$$U = -\frac{K}{M_0^2} [M_{x''''}^2 \sin^2(\theta_a - \theta_m) + M_{z''''}^2 \cos^2(\theta_a - \theta_m) +$$
$$+ 2M_{x''''} M_{z''''} \sin(\theta_a - \theta_m) \cos(\theta_a - \theta_m)] +$$
$$+ M_{x''''} H_0 \sin\theta_m - M_{z''''} H_0 \cos\theta_m -$$
$$- M_{x''''} h_{x''''} - M_{y''''} h_{y''''} = M_{z''''} h_{z''''}. \tag{8.39}$$

To shorten further entries, we introduce the following notation:

$$A_1 = -\frac{K}{M_0^2} \sin^2(\theta_a - \theta_m), \tag{8.40}$$

$$A_2 = -\frac{K}{M_0^2} \cos^2(\theta_a - \theta_m), \tag{8.41}$$

$$A_3 = -\frac{2K}{M_0^2} \sin(\theta_a - \theta_m) \cos(\theta_a - \theta_m), \tag{8.42}$$

$$A_4 = H_0 \sin\theta_m, \tag{8.43}$$

$$A_5 = -H_0 \cos\theta_m, \tag{8.44}$$

In these terms, the expression (8.39) for the total energy density takes the form

$$U = A_1 M^2_{x''''} + A_2 M^2_{z''''} + A_3 M_{x''''} +$$
$$+(A_4 - h_{x''''})M_{z''''} - h_{y''''}M_{y''''} + (A_5 - h_{z''''})M_{z''''}. \qquad (8.45)$$

STEP #5

To find effective fields acting on the magnetization vector in the system $Ox''''y''''z''''$.

***

To simplify the further writing it should be remembered that all calcu;ations are conducted on the $Ox''''y''''z''''$ system, temporatily abandon the index $''''$. The expression (8.45) for the energy density takes the form

$$U = A_1 M^2_x + A_2 M^2_z + A_3 M_x M_z +$$
$$A(A_4 - H_x)M_x - h_y M_y = (A_5 - h_z)M_z. \qquad (8.46)$$

We find effective fields using formula (1.2):

$$H_i = -\frac{\partial U}{\partial M_i}, \qquad (8.47)$$

where $i = x, y, z$.

In this case, we obtain:

$$H_z = -\left[2A_1 M_x + A_3 M_z + \left(A_4 - h_x\right)\right]; \qquad (8.48)$$

$$H_y = h_y; \qquad (8.49)$$

$$H_z = -\left[A_3 M_x + 2A_2 M_z + \left(A_5 - h_z\right)\right]. \qquad (8.50)$$

STEP #6.

Write the system of three equations of motion for components of the magnetization vector $M_{y''''}, M_{y''''}, M_{z''''}$ *under the effective fields in the system* $Ox''''y''''z''''$.

***

The equations of motion for magnetization have the form (2.20)–(2.22):

$$\frac{\partial M_x}{\partial t} = -\gamma\left(M_y H_z - M_z H_y\right) + \frac{\alpha}{M_0}\left(M_y \frac{\partial M_z}{\partial t} - M_z \frac{\partial M_y}{\partial t}\right); \qquad (8.51)$$

$$\frac{\partial M_y}{\partial t} = -\gamma \left(M_z H_x - M_x H_z \right) + \frac{\alpha}{M_0} \left(M_z \frac{\partial M_x}{\partial t} - M_x \frac{\partial M_x}{\partial t} \right); \quad (8.52)$$

$$\frac{\partial M_z}{\partial t} = -\gamma \left(M_x H_y - M_y H_x \right) + \frac{\alpha}{M_0} \left(M_x \frac{\partial M_y}{\partial t} - M_y \frac{\partial M_x}{\partial t} \right). \quad (8.53)$$

Substituting the effective fields (8.48)–(8.50) into (8.51)–(8.53), we obtain the equations of motion in the form:

$$\frac{\partial M_x}{\partial t} = \gamma \left[A_3 M_x M_y + 2 A_2 M_y M_z + \left(A_5 - h_z \right) M_y + h_y M_z \right] +$$
$$\frac{\alpha}{M_0} \left(M_y \frac{\partial M_z}{\partial t} - M_z \frac{\partial M_y}{\partial t} \right);$$

$$(8.54)$$

$$\frac{\partial M_x}{\partial t} = -\gamma [A_3 M_x^2 - A_3 M_z^2 + 2 \left(A_2 - A_1 \right) M_z +$$
$$+ \left(A_5 - h_z \right) M_y + \left(A_4 - h_x \right) M_z] +$$
$$+ \frac{\alpha}{M_0} \left(M_z \frac{\partial M_x}{\partial t} - M_x \frac{\partial M_z}{\partial t} \right);$$

$$8.55)$$

$$\frac{\partial M_z}{\partial t} = -\gamma \left[2 A_1 M_x M_y + A_3 M_y M_z + h_y M_x + (A_4 - h_x) M_y \right] +$$
$$+ \frac{\alpha}{M_0} \left(M_x \frac{\partial M_y}{\partial t} - M_y \frac{\partial M_x}{\partial t} \right). \quad (8.56)$$

STEP #7
Carry out linearization of the resultant system of equations assuming: $M_{z''''} \sim M_{y''''} \ll M_{z'''} \approx M_0$, as a result of which we obtain a system of two linear equations for the components of magnetization $M_{x''''}$ and $M_{y''''}$.
The equations of motion (8.540–(8.56) are non-linear and valid for any cases of amplitudes of a precession of the magnetization vector. Consider further the case of small amplitudes, allowing linearization. To do this, assume that:

$$M_x = m_x; \ M_y = m_y; \ M_z = M_0 + m_z; \quad (8.57)$$

moreover: $m_x \sim m_y \sim m_z \ll M_0$ and also: $h_x \sim h_y \ll M_0$.

Substituting (8.57) into (8.54)–(8.56) and leaving the terms not higher than the first order in magnetization and a variable field, we obtain:

$$\frac{\partial m_x}{\partial t} = \gamma\left(2A_2 M_0 + A_5\right)m_y + \gamma M_0 h_y - \alpha\frac{\partial M_y}{\partial t};$$

(8.58)

$$\frac{\partial M_z}{\partial t} = -\gamma\left(A_3 M_0 + A_4\right)m_y.$$

(8.59)

$$\frac{\partial M_z}{\partial t} = -\gamma\left(A_3 M_0 + A_4\right)m_y.$$

(8.60)

The equations (8.59) and (8.60) include the expression $A_3 M_0 + A_4$. Substituting A_3 and A_4 in accordance with the formulas (8.42) and (8.43), and also considering that in an equilibrium state, the orientation of the magnetization vector obeys equation (8.14)

$$H_a \sin\left[2\left(\theta_a - \theta_m\right)\right] - 2H_0 \cdot \sin\theta_m = 0,$$

(8.61)

we get

$$A_3 M_0 + A_4 = -\frac{2K}{M_0}\sin\left(\theta_a - \theta_m\right)\cos\left(\theta_a - \theta_m\right) + H_0 \sin\theta_m =$$

$$= \frac{1}{2}\left\{H_a \sin\left[2\left(\theta_a - \theta_m\right)\right] - 2H_0 \sin\theta_m\right\} = 0.$$

(8.62)

In view of (8.62), the linearized equations of motion (8.58)–(8.60) take the form

$$\frac{\partial m_x}{\partial t} - \gamma\left(2A_2 M_0 + A_5\right)m_y + \alpha\frac{\partial M_y}{\partial t} = \gamma M_0 h_y;$$

(8.63)

$$\frac{\partial m_y}{\partial t} + \gamma\left[2\left(A_2 - A_1\right)M_0 = A_5\right]m_x - \alpha\frac{\partial M_x}{\partial t} = -\gamma M_0 h_x.$$

(8.64)

We introduce the notation:

$$B_1 = -\left(2A_2 M_0 + A_5\right);$$

(8.65)

$$B_2 = -[2\left(A_2 - A_1\right)M_0 + A_5];$$

(8.66)

Substituting (8.65) and (8.66) into (8.63) and (8.64), we obtain the equations of motion in the form

$$\frac{\partial m_x}{\partial t} + \gamma B_1 m_y + \alpha \frac{\partial M_y}{\partial t} = \gamma M_0 h_y;$$

(8.67)

$$\frac{\partial m_y}{\partial t} - \gamma B_2 m_x - \alpha \frac{\partial M_x}{\partial t} = -\gamma M_0 h_x;$$

(8.68)

Next, we set the time dependence in the form of $e^{i\omega t}$, that is:

$$m_x = m_{x0} e^{i\omega t};$$

(8.69)

$$m_y = m_{y0} e^{i\omega t};$$

(8.70)

$$h_x = h_{x0} e^{i\omega t};$$

(8.71)

$$h_y = h_{y0} e^{i\omega t};$$

(8.72)

Substituting (8.690–(8.72) into (8.67)–(8.68) and dividing by $e^{i\omega t}$, we get:

$$i\omega m_x + (\gamma B_1 + i\omega\alpha) m_y = \gamma M_0 h_y;$$

(8.73)

$$(\gamma B_2 + i\omega\alpha) m_x - i\omega m_y = \gamma M_0 h_x;$$

(8.74)

Divide everything by $4\pi\gamma M_0$ and enter the notation:

$$\Omega = \frac{\omega}{4\pi\gamma M_0};$$

(8.75)

$$\Omega_1 = \frac{B_1}{4\pi\gamma M_0};$$

(8.76)

$$\Omega_2 = \frac{B_2}{4\pi\gamma M_0};$$

(8.77)

In this case, from (8.73) and (8.74) we obtain the system of equations of motion in the final form:

$$i\Omega m_x + (\Omega_1 + i\Omega\alpha) m_y = \frac{1}{4\pi} h_y;$$

(8.78)

$$(\Omega_2 + i\Omega\alpha) m_x - i\Omega m_y = \frac{1}{4\pi} h_x;$$

(8.79)

STEP No. 8.

Solve the linearized systemof equations as a result of which the components of the susceptibility tensor $\tilde{\chi}''''$ in the $Ox''''y''''z''''$ system.

402

The determinant of the system of equations (8.78)–(8.79) is

$$D = \left(1+\alpha^2\right)\Omega^2 - \Omega_1\Omega_2 - i\alpha\Omega\left(\Omega_1 + \Omega_2\right). \qquad (8.80)$$

The solution of the system (8.780–(8.79) is:

$$m_x = -\frac{\Omega_1 + i\Omega\alpha}{4\pi D}h_x - \frac{i\Omega}{4\pi D}h_y; \qquad (8.81)$$

$$m_y = \frac{i\Omega}{4\pi D}h_x - \frac{\Omega_2 + i\Omega_a}{4\pi D}h_y; \qquad (8.82)$$

We introduce the notation:

$$G_1 = -\frac{\Omega_1 + i\Omega_\alpha}{4\pi D}; \qquad (8.83)$$

$$G_2 = -\frac{\Omega_2 + i\Omega_\alpha}{4\pi D}; \qquad (8.84)$$

$$H = -\frac{i\Omega}{4\pi D}; \qquad (8.85)$$

In these expressions (8.81) and (8.82) take the form:

$$m_x = G_1 h_x + H h_y; \qquad (8.86)$$

$$m_y = -H h_x + G_2 h_x; \qquad (8.87)$$

Returning in the system $Qx''''y''''z''''$ and writing (8.86)–(8.87) in the vector form gives:

$$\mathbf{m}'''' = \vec{\vec{\chi}}\,'''' \mathbf{h}\,'''', \qquad (8.88)$$

$$\begin{pmatrix} m_{x'''} \\ m_{y'''} \\ m_{z'''} \end{pmatrix} = \begin{pmatrix} G_1 & H & 0 \\ -H & G_2 & 0 \\ 0 & 0 & 0 \end{pmatrix}\begin{pmatrix} h_{x'''} \\ h_{y'''} \\ h_{z'''} \end{pmatrix} \qquad (8.89)$$

Thus, the susceptibility tensor in the $Ox''''y''''z''''$ system has the form

$$\vec{X}'''' = \begin{pmatrix} G_1 & H & 0 \\ -H & G_2 & 0 \\ 0 & 0 & 0 \end{pmatrix} \qquad (8.90)$$

where G_1, G_2, H are determined by the expressions (8.83)–(8.85).

STEP #9

To convert the susceptibility tensor $\ddot{\chi}''''$, produced in the system $Ox''''y''''z''''$ to the system $Oxyz$.

***

The susceptibility tensor is transferred from the $Ox''''y''''z''''$ system to the $Oxyz$ system using the formula (8.26)

$$\ddot{\chi} = \ddot{A}_{04}\ddot{\chi}''''\ddot{A}_{40}. \tag{8.91}$$

where the transition matrices are defined by the formulas (8.20) and:(8.21)

$$\ddot{A}_{04} = \begin{pmatrix} \cos\theta_m\cos\varphi_a & -\sin\varphi_a & \sin\theta_m\cos\varphi_a \\ \cos\theta_m\cos\varphi_a & \cos\varphi_a & \sin\theta_m\sin\varphi_a \\ -\sin\theta_m & 0 & \cos\theta_m \end{pmatrix} \tag{8.92}$$

$$\ddot{A}_{04} = \begin{pmatrix} \cos\theta_m\cos\varphi_a & \cos\theta_m\sin\varphi_a & -\sin\theta_m \\ -\sin\varphi_a & \cos\varphi_a & 0 \\ \sin\theta_m\cos\varphi_a & \sin\theta_m\sin\varphi_a & \cos\theta_m \end{pmatrix} \tag{8.93}$$

Equation (8.91) is calculated in two stages: initially the product of the matrices $\ddot{\chi}''''\ddot{A}_{40}$ is determined and this is followed by obtaining the multiplication to the right by $\ddot{A}_{04}$.

Susceptibility tensor. As a result of performing the above steps, we obtain the susceptibility tensor in the $Oxyz$ system in the form

$$\ddot{X} = \begin{pmatrix} X_{xx} & X_{xy} & X_{xz} \\ X_{xz} & X_{yy} & X_{yz} \\ X_{zx} & X_{zy} & X_{zz} \end{pmatrix} \tag{8.94}$$

where

$$X_{xx} = G_1\cos^2\varphi_a\cos^2\theta_m + G_2\sin^2\varphi_a; \tag{8.95}$$

$$X_{xy} = G_1\sin\varphi_a\cos\varphi_a\cos^2\theta_m - G_2\sin\varphi_a\cos\varphi_a + H\cos\theta_m; \tag{8.96}$$

$$X_{xz} = -G_1\cos\varphi_a\sin\theta_m\cos\theta_m - H\sin\varphi_a\sin\theta_m; \tag{8.97}$$

$$X_{yz} = G_1 \sin\varphi_a \cos\varphi_a \cos^2\theta_m - G_2 \sin\varphi_a \cos\varphi_a - H\cos\theta_m; \quad (8.98)$$

$$X_{yy} = G_1 \sin^2\varphi_a \cos^2\theta_m + G_2 \cos^2\varphi_a; \quad (8.99)$$

$$X_{yz} = -G_1 \sin\varphi_a \sin\theta_m \cos\theta_m + H\cos\varphi_a \sin\theta_m; \quad (8.100)$$

$$X_{zx} = -G_1 \cos\varphi_a \sin\theta_m \cos\theta_m + H\sin\varphi_a \sin\theta_m; \quad (8.101)$$

$$X_{zy} - G_1 \sin\varphi_a \sin\theta_m \cos\theta_m - H\cos\varphi_a \sin\theta_m; \quad (8.102)$$

$$X_{zz} = G_1 \sin^2\theta_m. \quad (8.103)$$

This concludes the basic calculation of the components of the susceptibility tensor in steps.

8.2.4. Components of the susceptibility tensor in a complex form

We write the obtained components of the tensor in a complex form, that is, as the sum of the real and imaginary parts. To do this, we select the real and imaginary parts in the determinant (8.80) of system (8.78)and (8.79), and then we write down the values of G_1, G_2, H, defined by the expressions (8.83)–(8.85), as a sum of such parts.

As a result, for the determinant (8.80) we get

$$D = A + iB, \quad (8.104)$$

Where:

$$A = (1+\alpha^2)\,\Omega^2 - \Omega_1\Omega_2 \quad (8.105)$$

$$B = \alpha\Omega\,(\Omega_1 + \Omega_2) \quad (8.106)$$

In this case, we obtain:

$$G_1 = G_1' + iG_1'' \quad (8.107)$$

$$G_1 = G_1' + iG_1'' \quad (8.108)$$

$$H = H' + iH'' \quad (8.109)$$

where:

$$G_1' = -\frac{1}{4\pi}\frac{\Omega_1 A + \alpha\Omega B}{A^2 + B^2}; \quad (8.110)$$

$$G_2'' = \frac{1}{4\pi}\frac{\Omega_1 B - \alpha\Omega A}{A^2 + B^2}; \quad (8.111)$$

$$G_2' = -\frac{1}{4\pi}\frac{\Omega_2 A + \alpha\Omega B}{A^2 + B^2}; \quad (8.112)$$

$$H_{ex'} = -\left(2A_1 M_{x'} + A_3 M_{z'} + A_4\right);$$

(8.113)

$$H' = -\frac{1}{4\pi} \frac{\Omega B}{A^2 + B^2};$$

(8.114)

$$H'' = -\frac{1}{4\pi} \frac{\Omega A}{A^2 + B^2};$$

(8.115)

As a result, we obtain the components of the susceptibility tensor in a complex form:

$$X_{xx} = \left(G_1' \cos^2 \varphi_a \cos^1 \theta_m + G_2' \sin^2 \varphi_a\right) +$$
$$+ i\left(G_1'' \cos^2 \varphi_a \cos^2 \theta_m + G_2'' \sin^2 \varphi_a\right);$$

(8.116)

$$X_{xy} = \left(G_1' \sin \varphi_a \cos \varphi_a \cos^2 \theta_m - G_2' \sin \varphi_a \cos \varphi_a + H' \cos \theta_m\right) +$$
$$+ i\left(G_1'' \sin \varphi_a \cos \varphi_a \cos^2 \theta_m - G_2'' \sin \varphi_a \cos \varphi_a + H'' \cos \varphi_m\right);$$

(8.117)

$$X_{xz} = \left(-G_1' \cos \varphi_a \sin \theta_m \cos \theta_m - H' \sin \varphi_a \sin \theta_m\right) +$$
$$+ i\left(-G_1'' \cos \varphi_a \sin \theta_m \cos \theta_m - H'' \sin \varphi_a \sin \theta_m\right);$$

(8.118)

$$X_{yz} = \left(G_1' \sin \varphi_a \cos \varphi_a \cos^2 \theta_m - G_2' \sin \varphi_a \cos \varphi_a - H' \cos \theta_m\right) +$$
$$+ i\left(G_1'' \sin \varphi_a \cos \varphi_a \cos^2 \theta_m - G_2'' \sin \varphi_a \cos \varphi_a - H' \cos \theta_m\right);$$

(8.119)

$$X_{yy} = \left(G_1' \sin^2 \varphi_a \cos^2 \theta_m + G_2' \cos^2 \varphi_a\right) +$$
$$+ i\left(G_1'' \sin^2 \varphi_a \cos^2 \theta_m + G_2'' \cos^2 \varphi_a\right);$$

(8.120)

$$X_{yz} = \left(-G_1' \sin \varphi_a \sin \theta_m \cos \theta_m + H' \cos \varphi_a \sin \theta_m\right) +$$
$$+ i\left(-G_1'' \sin \varphi_a \cos \theta_a \cos \theta_m + H'' \cos \varphi_a \sin \theta_m\right);$$

(8.121)

$$X_{zz} = \left(-G_1' \cos \varphi_a \sin \theta_m \cos \theta_m + H' \sin \varphi_a \sin \theta_m\right) +$$
$$+ i\left(-G_1'' \cos \varphi_a \sin \theta_m \cos \theta_m + H'' \sin \varphi_a \sin \theta_m\right);$$

(8.122)

$$X_{zy} = \left(-G_1' \sin \varphi_a \sin \theta_m \cos \theta_m - H' \cos \varphi_a \sin \theta_m\right) +$$
$$+ i\left(-G_1'' \sin \varphi_a \sin \theta_m \cos \theta_m - H'' \cos \varphi_a \sin \theta_m\right);$$

(8.123)

$$X_{yz} = \left(G_1' \sin^2 \theta_m\right) + i\left(G_1'' \sin^2 \theta_m\right).$$

(8.124)

8.3. Representation of the components of the susceptibility tensor through the parameters of the material and the constant field

Imagine the components of the susceptibility tensor defined by formulas (8.116)–(8.124), through the parameters of the material and the constant field. To do this, we substitute in (8.116)–(8.124) the expressions for G_1, G_2 and H, given by formulas (8.107)–(8.109) with regard to (8.110)–(8.115), in which, in our turn, the value D will be expressed in accordance with the formula (8.104) in view of (8.105), (8.106). In this case, we obtain the components of the susceptibility tensor in the form:

$$\chi_{xx} = \frac{(\Omega_1 + i\Omega\alpha)\cos^2\varphi_a\cos^2{}+\theta_m+(\Omega_2+i\Omega\alpha)\sin^2\varphi_a}{4\pi\left[\Omega_1\Omega_2-(1+\alpha^2)\Omega^2+i\alpha\Omega(\Omega_1+\Omega_2)\right]};$$
(8.125)

$$\chi_{xy} = \frac{(\Omega_1+i\Omega\alpha)\sin\varphi_a\cos\varphi_2\cos^2{}+\theta_m+(\Omega_2+i\Omega\alpha)\sin\varphi_a\cos\varphi_2+i\Omega\cos\theta_m}{4\pi\left[\Omega_1\Omega_2-(1+\alpha^2)\Omega^2+i\alpha\Omega(\Omega_1+\Omega_2)\right]};$$
(8.126)

$$\chi_{xz} = \frac{-(\Omega_1+i\Omega\alpha)\cos\varphi_a\sin\theta_2\cos_m-i\Omega\sin\varphi_a\sin\theta_m}{4\pi\left[\Omega_1\Omega_2-(1+\alpha^2)\Omega^2+i\alpha\Omega(\Omega_1+\Omega_2)\right]};$$
(8.127)

$$\chi_{yz} = \frac{(\Omega_1+i\Omega\alpha)\sin\varphi_a\cos\varphi_2\cos^2\theta_m+(\Omega_2+i\Omega\alpha)\sin\varphi_a\cos\varphi_a-i\Omega\cos\theta_m}{4\pi\left[\Omega_1\Omega_2-(1+\alpha^2)\Omega^2+i\alpha\Omega(\Omega_1+\Omega_2)\right]};$$
(8.128)

$$\chi_{yy} = \frac{(\Omega_1+i\Omega\alpha)\sin^2\varphi_a\cos^2\theta_m+(\Omega_2+i\Omega\alpha)\cos^2\varphi_a}{4\pi\left[\Omega_1\Omega_2-(1+\alpha^2)\Omega^2+i\alpha\Omega(\Omega_1+\Omega_2)\right]};$$
(8.129)

$$\chi_{yz} = \frac{-(\Omega_1+i\Omega\alpha)\sin\varphi_a\sin\theta_m\cos\theta_m+i\Omega\cos\varphi_a\sin\theta_m}{4\pi\left[\Omega_1\Omega_2-(1+\alpha^2)\Omega^2+i\alpha\Omega(\Omega_1+\Omega_2)\right]};$$
(8.130)

$$\chi_{zx} = \frac{-(\Omega_1+i\Omega\alpha)\cos\varphi_a\cos\theta_m\cos\theta_m+i\Omega\sin\varphi_a\sin\theta_m}{4\pi\left[\Omega_1\Omega_2-(1+\alpha^2)\Omega^2+i\alpha\Omega(\Omega_1+\Omega_2)\right]};$$
(8.131)

$$\chi_{zy} = \frac{-(\Omega_1+i\Omega\alpha)\sin\varphi_a\sin\theta_m\cos\theta_m+i\Omega\cos\varphi_a\sin\theta_m}{4\pi\left[\Omega_1\Omega_2-(1+\alpha^2)\Omega^2+i\alpha\Omega(\Omega_1+\Omega_2)\right]};$$
(8.132)

$$\chi_{zz} = \frac{\left(\Omega_1 + i\Omega\alpha\right)\sin^2\theta_m}{4\pi\left[\Omega_1\Omega_2 - \left(1+\alpha^2\right)\Omega^2 + i\alpha\Omega\left(\Omega_1+\Omega_2\right)\right]}. \tag{8.133}$$

The quantities Ω_1, Ω_2 and Ω entering into these expressions will be expressed in accordance with formulas (8.75)–(8.77), where B_1 and B_2 are given by the formulas (8.65) and (8.66), and the expressions A_1 (8.66), (8.66), A_2 and A_5 are given by the formulas (8.40), (8.41) and (8.44). We also introduce the notation for the anisotropy field (4.55), (8.13):

$$H_a = \frac{2K}{M_0}, \tag{8.134}$$

and the field corresponding to the resonance of uniform precession at the frequency ω [6–8]:

$$H_r = \frac{\omega}{\gamma}, \tag{8.135}$$

As a result, we obtain the following formulas for Ω_1, Ω_2, and Ω:

$$\Omega_1 = \frac{1}{4\pi M_0}\left[H_a\sin^2\left(\theta_a - \theta_m\right) + H_0\cos\theta_m\right], \tag{8.136}$$

$$\Omega_2 = \frac{1}{4\pi M_0}\left\{H_a\cos 2\left[2\left(\theta_a - \theta_m\right)\right] + H_0\cos\theta_m\right\}; \tag{8.137}$$

$$\Omega = \frac{H_r}{4\pi M_0}, \tag{8.138}$$

which, being substituted into the formulas (8.1250–(8.133), together with the equilibrium condition (8.12) or (8.14), completely determine the components of the susceptibility tensor for a given orientation of the anisotropy axis relative to the direction of the constant field. It should be noted that in the particular case of orientation of the axes of anisotropy of all the particles along the polar axis, as well as the vector of the constant field along the same axis, that is, for $\theta_a = 0$, $\varphi_a = 0$, and also $\theta_m = 0$, the formulas (8.125)(8.133) go into the components of the susceptibility tensor in an isotropic medium of the form (2.38) or (2.49), taking into account (2.58)–(2.61), and also (2.56)–(2.57)

Comment. From the general geometry of the problem (Fig. 8.2, a), it can be seen that turning the EMA around the Oz axis (around the constant field direction) while keeping the angle θ_a constant, that is, changing only the angle φ_a, does not change the conditions for the excitation of ferromagnetic resonance. In this case, the mutual arrangement of the equilibrium vector of magnetization, the axis of easy magnetization, and the vector of the alternating field does not change. Therefore, one should expect that the susceptibility tensor should not change with such a rotation, that is, its dependence on the angle φ_a should be absent. However, it can be seen from the formulas (8.125)–(8.132) that such dependence takes place everywhere except for the formula (8.133). This is due to the fact that the orientation of the EMA is written in the spherical coordinate system, whereas the components of the magnetization vector and the variable field included in the equations of motion (8.51)–(8.53), and the following expressions (8.81)–(8.82), directly determining the components of susceptibility (8.83)–(8.85), are written in the Cartesian system coordinates. Therefore, the components of the susceptibility tensor (8.125)–(8.132) are obtained recorded in the same Cartesian system, whereas the orientation of the EMA with respect to the same system remains recorded through the spherical coordinates. That is, if a fixed value of the angle φ_a is specified in the Cartesian system, the Cartesian components of the susceptibility tensor written for the EMA, turned with respect to the same Cartesian system exactly by this angle φ_a, are obtained. From this, we obtain the dependence of the components of the tensor (8.125)–(8.132) on φ_a. For the χ_{zz} component (8.133), there is no such dependence for the reason that the direction of the polar axis of the spherical system just coincides with the Oz axis of the Cartesian system, so that when the EMA rotates around this axis, the equilibrium direction of magnetization with respect to either coordinate system does not change.

8.4. Averaging procedure

In the previous section, the susceptibility tensor was found for a single particle, the direction of the anisotropy axis of which with respect to a constant field is determined by the fixed values of the polar and azimuthal angles θ_a and φ_a. The real medium contains a large number of particles, the anisotropy axes of which are oriented in arbitrary directions. Therefore, to obtain the susceptibility tensor of the medium as a whole, the tensor components obtained for a

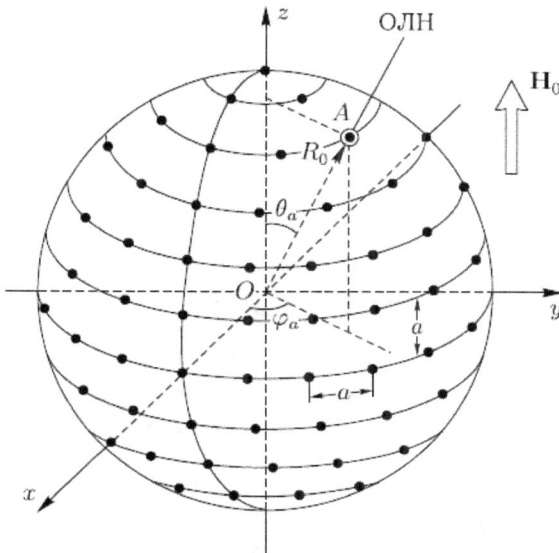

Fig. 8.5. Illustration of the general scheme of averaging procedure. EMA – axis of easy magnetization passing through the surface of the sphere at point *A*.

single particle must be averaged over all possible orientations of the anisotropy axis, that is, over all possible values of the angles θ_a and φ_a.

8.4.1. General scheme of averaging procedure

The general scheme of the averaging procedure is illustrated in Fig. 8.5, which is based on the coordinate system $Oxyz$ ($Oz\|\mathbf{H}_0$) associated with the external field. The figure shows the auxiliary sphere of an arbitrary given radius R_0, on the surface of which the enlarged points show the intersections with the possible directions of the anisotropy axes of individual particles.

The figure shows, as an example, the reading of the azimuth and polar angles for the axis passing through one of the enlarged points, denoted by the letter *A*. The points of intersection of the anisotropy axes with the surface of the sphere are distributed so that the distances by the angular coordinates between each two adjacent points were equal to the same given value *a*. Thus, the problem is considered on a discrete set of orientations of the anisotropy axes distributed uniformly over the surface of a sphere with the centre at point *O*.

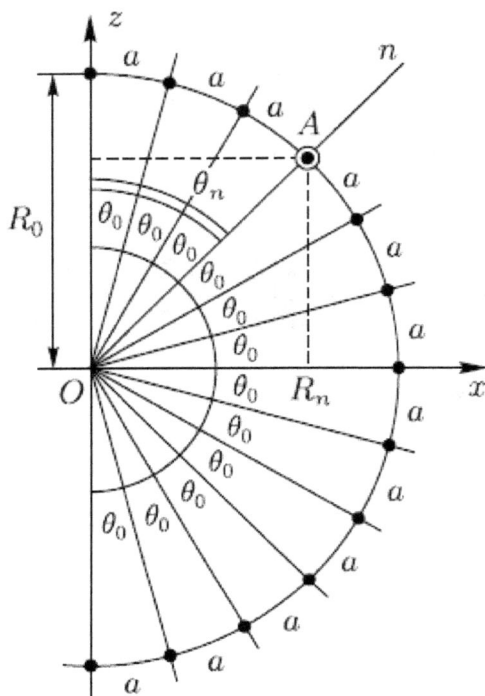

Fig. 8.6. Averaging procedure over the polar angle.

8.4.2. *Averaging over the polar angle*

The first part of the task is to find the totality of such orientations. We first consider the averaging procedure over the angle θ, which is illustrated in Fig. 8.6. Due to the cylindrical symmetry of the problem it suffices to consider a change in the angle θ in the interval from 0 to π, and set the angle φ equal to zero. In the figure in the Oxz plane with a radius R_0, an auxiliary semicircle is constructed, which is a section of the surface of the sphere with this plane in Fig. 8.5. Radially spaced solid straight lines indicate the possible directions of the anisotropy axes, spaced from each other by equal angles θ_0. Larger points show the intersection of these directions with a circle of radius R_0. It is seen that along the arc of a circle, these points are separated by equal distances

$$a = R_0\,\theta_0 \tag{8.139}$$

that is, the condition of uniform distribution of the anisotropy axes over the angle θ is satisfied.

Let us number the directions of the anisotropy axes by integers n. Obviously, in the range of $0 \leq \theta \leq \pi$ value n takes values from 0 to π/θ_0. In the figure, as an example, one of the possible directions of the anisotropy axis is shown, passing through point A and making an angle with the axis Oz

$$\theta_n = n\theta_0 \tag{8.140}$$

It can be seen that the projection of point A on the axis Ox is separated from point O by a distance

$$R_n = R_0 \sin(n\theta_0) \tag{8.141}$$

8.4.3. Averaging over azimuthal angle

We now turn to the distribution of intersection points by angle φ, the scheme of which is illustrated in Fig. 8.7. It shows the coordinate plane Oxy, on which sections of the surface of the sphere are projected by planes parallel to Oxy, passing through the intersection points shown in Fig. 8.6 for various values n.

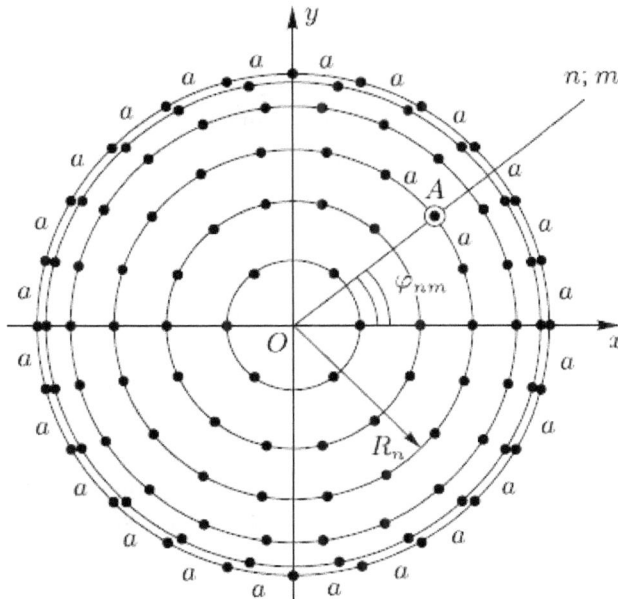

Fig. 8.7. Illustration of the azimuthal angle averaging procedure

The outer circumference in Fig. 8.7, having a radius R_0, corresponds to the equator of the sphere lying in the Oxy plane. Here we set the distances between the points of intersection along the arc of a circle equal to a. The remaining circles correspond to different values of n and have radii R_n. On each of such circles, we set the distances between the points of intersection along the arc of a circle also equal to a. At the same time, on each circle corresponding to the number n, m points will fit, where the maximum value is:

$$m_{n\max} = \frac{2\pi R_n}{a} = \frac{2\pi R_0 \sin(n\theta_0)}{a},$$

$$(8.142)$$

The step in the azimuthal angle for the circle with the number n is equal to

$$\varphi_{n0} = \frac{2\pi}{m_{n\max}} = \frac{a}{R_0 \sin(n\theta_0)}.$$

$$(8.143)$$

Given that $a = R_0\theta_0$, we get the azimuthal angles of the intersection points:

$$\varphi_{mn} = \frac{m\theta_0}{\sin(n\theta_0)}.$$

$$(8.144)$$

Thus, the angular coordinates of possible directions of anisotropy axes are determined in the polar plane n by the values of the polar angle

$$\theta_n = n\theta_0,$$

$$(8.145)$$

changing with step θ_0, and:

$$0 \le n \le \frac{\pi}{\theta_0}$$

$$(8.146)$$

In the azimuthal plane with the number n, the coordinates are determined by the m values of the azimuthal angle

$$\varphi_{mn} = \frac{m\theta_0}{\sin(n\theta_0)},$$

$$(8.147)$$

changing with the step

$$\varphi_{n0} = \frac{\theta_0}{\sin(n\theta_0)},$$

(8.148)

where

$$0 \le m_n \le \frac{2\pi \sin(n\theta_0)}{\theta_0}.$$

(8.149)

Thus, setting the pitch of the polar angle θ_0 allows one to uniquely determine all possible values of both the polar and azimuthal angles of the anisotropy axes. The maximum values of n and m_n correspond to the right-hand sides of inequalities (8.146) and (8.149):

$$n_{max} = \frac{\pi}{\theta_0};$$

(8.150)

$$m_{n\,max} = \frac{2\pi \sin(n\theta_0)}{\theta_0}.$$

(8.151)

8.4.4. Summation over all possible directions

The further course of determining the total susceptibility of a system of particles with arbitrarily oriented anisotropy axes consists in calculating the susceptibility for all possible orientations of these axes (that is, for all the n and m permissible values), followed by summation and division by the total number of possible orientations.

The summation over all possible anisotropy orientations for the susceptibility tensor component χ_{ik} (where $i, k = x, y, z$) gives:

$$S_{ik} = \sum_{\substack{n=0 \\ m=0}}^{\substack{n_{max} \\ m_{n\,max}}} \chi_{ik}^{(n,m)} = \sum_{n=0}^{n_{max}} \left(\sum_{m=0}^{m_{n\,max}} \chi_{ik}^{(n,m)} \right).$$

(8.152)

The total number of orientations is

$$N = 2\pi \sum_{n=0}^{n_{max}} \frac{\sin(n\theta_0)}{\theta_0}.$$

(8.153)

From here we obtain the average value of the susceptibility component χ_{ik} in the form

$$\chi_{ik} = \frac{S_{ik}}{N} = \frac{\sum\limits_{n=0}^{n_{max}}\left(\sum\limits_{m=0}^{m_{n\,max}} \chi_{ik}^{(n,m)}\right)}{2\pi \pm \sum\limits_{n=0}^{n_{max}} \dfrac{\sin(n\theta_0)}{\theta_0}}. \tag{8.154}$$

We note that the susceptibility tensor thus obtained does not take into account the fraction of the total volume of the particle material in the entire volume of the medium. Here it is assumed that the volume of one cell of the medium is equal to the volume of one particle. To obtain the real value of susceptibility, it is necessary to multiply all the obtained χ_{ik} components by the filling factor N_p given by formula (8.1). As a result of (8.154) we get:

$$\chi_{ik} = N_p\frac{S_{ik}}{N} = \frac{4}{3}\pi\left(\frac{r}{d}\right)^3 \frac{\sum\limits_{n=0}^{n_{max}}\left(\sum\limits_{m=0}^{m_{n\,max}} \chi_{ik}^{(n,m)}\right)}{2\pi\sum\limits_{n-0}^{n_{max}} \dfrac{\sin\left(n\theta_0\right)}{\theta_0}}. \tag{8.155}$$

8.4.5. On the order of numerical calculation of the susceptibility tensor

From the above consideration, it can be seen that the general procedure for calculating the susceptibility tensor includes many more or less identical actions. Apparently, the analytical execution of such a calculation would be extremely cumbersome, so we briefly discuss the possible algorithm for its numerical implementation.

So, first of all, one should set the environment parameters that are common to all components. These are the saturation magnetization, the field or anisotropy constant, the size of the particles and the distances between them. One should also specify the applied constant field and the frequency at which the components of the tensor will be determined. In addition, it is necessary to determine the intervals of change of the polar and azimuthal angles of orientation of the axes of anisotropy of the particles, as well as the value of the step of the polar angle θ_0. It should be taken into account that the limiting values of the azimuthal angle depend on the magnitude of the polar angle in accordance with formula (8.147), that is, these values (and with them the required number of steps m in azimuthal angle φ_a) should be determined during the calculation for each given polar angle θ_a. The full number n as the number of admissible values of the polar angle θ_a can be found at the initial stage of the calculation, since it depends on the step θ_0 and when this angle changes from zero to 180°, it is determined by the right equality in the formula (8.146). If the conditions of the problem require a smaller than 180° interval of change of the polar angle, then the number n is determined by dividing the specified interval by the step size θ_0.

The first step in calculating the component of the susceptibility tensor itself should be to specify some initial value of the polar angle θ_a, and first on the edge of changing its range. At the same time, the initial value of the azimuthal angle φ_a should be found by the formula (8.147) with $n = 1$ and $m = 1$ (from Figs. 8.5 and 8.7, we can see that for $n = 0$, the value of m is not defined, so there is nothing to average at to complete the picture, it can be taken into account once, without performing the summation over the azimuthal angle).

Further, when these parameters are found, find the equilibrium orientation of the magnetization vector in one particle, that is, determine the angle θ_m using equation (8.12) or (8.14). For the numerical solution, apparently, the equation (8.16) is more convenient, which can be solved, for example, by the zero search method.

The obtained value of θ_m should be substituted into formulas (8.136)–(8.138), from which the frequencies Ω_1, Ω_2, Ω will be obtained (the latter frequency does not depend on the parameters of the medium, so it can be determined at the stage of setting the initial parameters). With the help of these frequencies, the components of the susceptibility tensor χ_{ik} are determined by the formulas (8.125)–(8.133).

These components will be obtained for one given value of the polar and azimuthal angles of one particle at a given frequency. The next step should be setting a new value of the azimuthal angle φ_a at the same value of the polar angle θ_a, separated from the initial value φ_a by step φ_{n0}, defined by formula (8.143). At this value of the azimuthal angle and the former value of the polar angle, all components of the susceptibility tensor are again found by formulas (8.125)–(8.133).

Next, the value of the azimuthal angle is again increased by the same step and the components of the susceptibility tensor are calculated again. Such a cyclic calculation is made until the technical conditions, until the entire interval of permissible values of the azimuthal angle φ_a, corresponding to the specified value of the polar angle θ_a, that is, divided by the formula (8.151), is passed. All obtained values of the components of the susceptibility tensor are summarized in the m_{mmax} content, in agreement with the internal sum of the formula (8.152).

The next step should be to increase the value of the polar angle θ_a by the value of the same step, after which the interval of permissible values of the azimuthal angle φ_a is determined from this value in accordance with formula (8.149). From this interval is selected the first value at which the equilibrium value of the polar angle of the magnetization vector θ_m is found, followed by the frequencies Ω_1 and Ω_2, which then determine the components of the susceptibility tensor by the formulas (8.125) and (8.133). That is, the whole procedure of calculating these components is repeated. Next, we perform the next step in the azimuthal angle with the calculation of the components of the susceptibility tensor, and so on, until the entire interval of the azimuthal angle change is exhausted. The obtained values of the components of the tensor are again summed in accordance with the internal sum of the formula (8.152).

The next step is to increase the value of the polar angle by one more step, after which everything repeats until tech, until the entire interval of values of the polar angle is passed. At each stage the obtained internal values of the formula (8.152) are summed in accordance with the external sum of this formula.

The final step is the calculation of the total number of orientations using formula (8.153) (note that such a calculation can be performed at the initial stage of solving the problem, once the value θ_0 is specified).

The sum obtained by formula (8.152) is divided by this number of orientations in accordance with formula (8.154), from which, taking into account the particle size and the distance between them, the components of the resulting susceptibility tensor are obtained for a unit of the medium as a whole in accordance with formula (8.155).

The calculation algorithm given here implies that the susceptibility is determined at one given frequency, which is included in (8.125) –(8.133) through the normalized frequency Ω, corresponding to formula (8.138). If necessary, to obtain the dependence of the susceptibility tensor on the frequency, the required frequency interval with a given step, at each step doing all the calculations given here again.

Performing a numerical calculation in accordance with the above algorithm makes it possible to determine both the real and imaginary parts of the tensor components, for which the complex nature of the variables should be taken into account when performing calculations using the formulas (8.125)–(8.133). Further sections of this chapter are devoted to the description of calculations of the susceptibility tensor of the composition medium, performed in accordance with the described algorithm under various conditions.

8.5. The dependence of the components of the tensor on the frequency

We now consider some properties of the resulting susceptibility tensor. First of all, it is clear from the formulas (8.125)–(8.133) that all nine of its components are non-zero. The tensor is not diagonal, symmetric or antisymmetric. All of its components are complex, that is, they have non-zero real and imaginary parts. Recall that in a similar tensor for a homogeneous isotropic medium (formulas (2.49), (2.58)–(2.61) in Chapter 2), only χ_{xx}, χ_{xy}, χ_{yx}, χ_{yy} and in some cases χ_{zz} are non-zero, and all other components are zero, moreover, $\chi_{xx} = \chi_{yy}$, $\chi_{xy} = -\chi_{yx}$. In this case, in the absence of damping, χ_{xx} and χ_{yy} are purely real, and χ_{xy} and χ_{yx} are purely imaginary [6–8]. All this in the resulting tensor is not observed. It must be assumed that the listed differences of the obtained tensor from the known one are caused by the anisotropic character particle compositional medium.

Despite what has been said, comparing the obtained expressions with those given in Chapter 2, one can see the well-known similarity of the structure of those other expressions corresponding to the components of tensors that are the same in the columns and rows. This correspondence is most pronounced when $\varphi_a = 0$ and $\theta_m = 0$, so that, up to notation, the following relations hold: $\chi_{xx} \to \chi$, $\chi_{xy} \to i\chi_a$, $\chi_{zz} \to \chi_\parallel$.

We further note that the necessary step for calculating the components of the tensor for a composite medium is averaging over all possible directions of orientation of the anisotropy axes. In this case, for each direction, its own resonant frequency is obtained, at which the components of the tensor in the absence of attenuation tend to infinity. Such a circumstance can lead to the tendency of the components of the resulting tensor to infinity over the entire range of changes in the resonant frequencies of individual particles, which causes some uncertainty, especially when the machine is in the engine count. The easiest way out of the situation, apparently, is to take direct account of the attenuation that is present in all the expressions obtained.

We now turn to the dependence of the components of the tensor on the frequency. For simplicity and preserve the generality of the solution, the fill factor will be considered equal to one. For consideration, we choose the following parameters: the saturation magnetization of the particle material: $4\pi M_0$ = 1750 G; uniaxial anisotropy constant:K = 7000 erg·cm^{-3}, which corresponds to the anisotropy field H_a = 100 Oe; gyromagnetic constant: γ = 2.8 MHz Oe^{-1}; attenuation constant: α = 0.01. We take the constant field equal to H = 500 Oe. Note that this field corresponds to the internal field in the particle, that is, to go to an external field, one must take into account the demagnetizing factor, equal to $4\pi/3$ for a sphere, or the demagnetization field equal to $H_d = 4\pi M_0/3$ = 583 Oe, that is, to create a field of 500 Oe inside the sphere, a field equal to $H_s = H + H_d$ = 1083 Oe should be applied to it from the outside. Averaging over the angle θ_a will be carried out with a step $\theta0$ equal to 0.1°.

8.5.1. Extreme frequencies of susceptibility resonance

In accordance with what has been said regarding the dependence of the components of the tensor (8.125)–(8.133) on the angle φ_a, we first consider the particular case with φ_a = 0, that is, at this stage we will not average this angle. This case corresponds to the fact that the misorientation of the axes of easy magnetization of individual particles occurs only in one plane and this plane coincides with the coordinate plane Oxz in Fig. 8.2, a.

Under these conditions, the extreme frequencies of resonances, corresponding to the orientation of the anisotropy axis across and along the constant field, are respectively:

$$f_{1p} = \gamma\sqrt{H(H - H_0)} = 1252 \text{ MHz},$$

(8.156)

$$f_{2p} = \gamma(H + H_0) = 1680 \text{ MHz},$$

(8.157)

and the full width of the resonance line of a single particle at half height, defined by the formula $\Delta f_r = \alpha f_{1,2}$, is 80–100 MHz. As an illustration of what is said, Fig. 8.8 shows the dependences of the real (*a*) and imaginary (*b*) parts of the χ_{xx} susceptibility component on frequency with parallel (1) and perpendicular (2) field orientations relative to the anisotropy axis. In the construction of this figure, no averaging over the angles was carried out, so that the curves obtained here correspond to just the extreme cases of parallel and perpendicular orientations of the anisotropy axis with respect to the field.

From the figure one can see a pronounced resonance character of all the dependences, similar to that shown in Fig. 2.1 (Chapter 2), up to the values of the resonant frequencies that are here determined by the presence of anisotropy. The resonance frequencies coincide with those obtained by the formulas (8.156) and (8.157) (the figure shows vertical dotted lines).

Since there are two extreme possible orientations of the anisotropy axis relative to the field direction, that is, two extreme values of the angle θ_a, namely, 0° and 90°, it should be expected that at all intermediate angles the resonant frequencies will be located between these extreme ones. That is, the full spectrum of susceptibility will be concluded between the frequencies f_{1p} and f_{2p}, defined by the formulas (8.156) and (8.157).

Changing the angle θ_a from 90° to 180° does not bring anything new, because, due to the bi-directionality of the anisotropy axis, the magnetization vector in the equilibrium position will always be oriented as close as possible to the field direction, that is, from 0 to 90°.

Comment. Note that a change in the orientation of the field in the full range of angles θ_a from 0° to 180° in the case of sufficient magnetic rigidity of the material of the spheres, especially if their size allows domain structure, can lead to phenomena of hysteresis nature. That is, the orientation of the magnetization vector in the positive and negative directions relative to the anisotropy axis will depend on the prehistory of the magnetization reversal process, so that the resonance frequencies can go beyond the marked interval. However, this issue requires separate consideration and will not be touched upon in this monograph, that is, we will assume that the material of the spheres is completely soft in magnetic terms, and the field is completely sufficient to saturate it.

420

8.5.2. Features of the averaging procedure when changing step size

Let us now consider how the total frequency dependence of the susceptibility changes, taking into account the different orientations of the anisotropy axes in the constituent spherical particles. As an illustration, we turn to Fig. 8.9, where the dependences of the real (a) and imaginary (b) parts of the susceptibility component χ_{xx} on the frequency are shown at different averaging steps over the angle θ_a. In this case, the angle φ_a, as before, is set equal to zero. The left column of the picture (a) corresponds to the real part χ'_{xx}, right column (b) – imaginary part χ''_{xx}. From top to bottom of both columns of the figure, the averaging step is reduced. It can be seen from the figure that with a sufficiently large step equal to 100° (fragments of Fig. 1 and 5), that is, in the absence of averaging, the dependences of both the real and imaginary parts practically coincide with those shown in Fig. 8.8. The difference in amplitude is two times due to

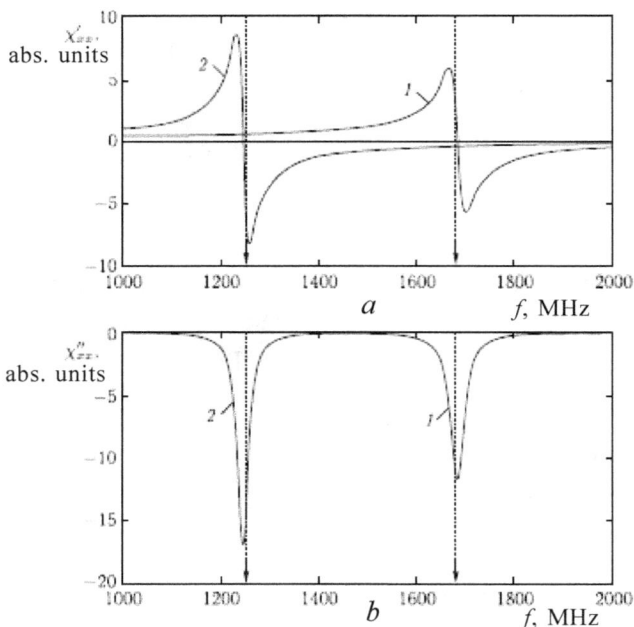

Fig. 8.8. Dependences of the real (a) and imaginary (b) parts of the susceptibility component χ_{xx} on the frequency with parallel (1) and perpendicular (2) orientations of the axis relative to the anisotropy axis. Vertical dotted lines (shown by arrows) are the limiting frequencies f_{1p} and f_{2p}, corresponding to the formulas (8.156) and (8.157). Parameters: $4\pi M_0 = 1750$ G, $H = 500$ Oe, $H_a = 100$ Oe, $\alpha = 0.01$, $\gamma = 2.8$ MHz Oe^{-1}.

the fact that when plotting the curves in Fig. 8.8, the initial and final values of the angle θ_a were assumed to be equal, in one case 0° (curves 2 in Fig. 8.8), in the other case 90° (curves 1 in Fig. 8.8). In this case, the averaging procedure was carried out in full, so that both of these values, the initial and final, were taken into account twice (as the beginning and end of the account). When building curves in Fig. 8.9 the initial and final values of the angle θ_a were assumed to be 0° and 180°, so that all values of the angles, including 0° and 90°, were counted only once, therefore here the amplitude was twice as small as in Fig. 8.8. It must be assumed that the same reason is also responsible for a small (at ~10 MHz) change in the resonant frequency observed at low-frequency resonance.

The main feature of the curves shown in Fig. 8.9, it is possible to consider a discrete, sharply cut up sawtooth-like character, especially pronounced with an average step size of 10° and 20° (fragments of Figs. 3, 4, as well as 6, 7). Such a discrete character is due to the superposition of resonant curves of the individual pieces, axes of anisotropy corresponding to different directions. That is, due to the discrete orientation of the axes and the location of the resonant frequencies also turned out to be discrete.

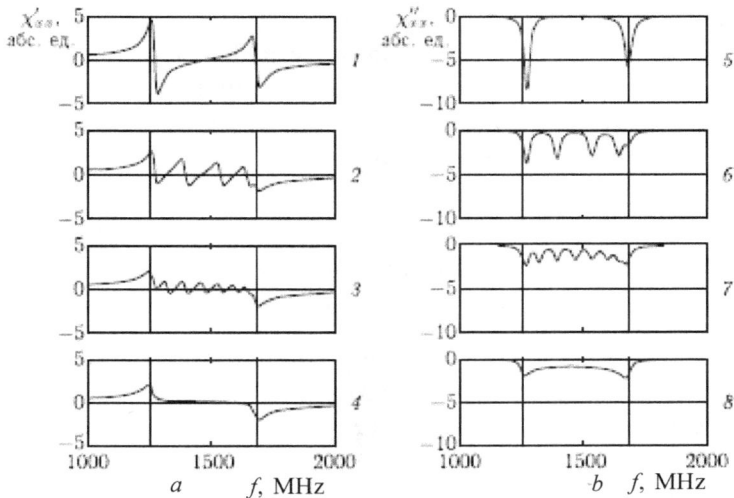

Fig. 8.9. Dependences of the real (a) and imaginary (b) parts of the susceptibility component χ_{xx} on the frequency at different averaging steps θ^0 over the angle θ_a: 1, 5 – 100°; 2, 6 – 20°; 3, 7 – 10°; 4, 8 – 1°. Vertical dotted lines – limiting frequencies f_{1p} and f_{2p}, corresponding to the formulas (8.156) and (8.157). Parameters – the same as taken in the construction of Fig. 8.8

It can be seen that as the pitch decreases, the teeth of the saw both in height and length decrease. At the same time, their number of increases, and with a sufficiently small step equal to 1° (fragments of Figs. 4 and 8), all the teeth merge so that the curves as a whole acquire a completely smooth character.

Thus, since the teeth are formed from fairly narrow resonant ejections, similar to those on fragments of Figs. 1 and 5, the discrete nature of the curves consisting of them can manifest itself only up to the time, while the width of these outliers will be less than the distance between them. That is, the reason for the discreteness is the small, compared to the step size in frequency, the width of the ferromagnetic resonance line. When the curve shift at one averaging step is less than the width of the ferromagnetic resonance line, all curves from individual particles should merge into a single whole, as is observed with a step of 1° (fragments of Figs.4 and 8).

Note that the amplitude of the resultant curves also decreases as the pitch decreases (that is, from the top to the bottom of the figure). The reason for this decrease is the uniform distribution of the resonant frequencies in frequency, provided that the total number of particles per unit volume is maintained (formula (8.2)). That is, with a large step of orientation, the anisotropy axes of the particles are concentrated at angles corresponding to the ends of this step, as a result of which their amplitudes are added. At the same small step, these orientations are distributed more or less evenly over the entire interval of variation of the angle θ_a, as a result of which the amplitudes are smeared over a considerable interval of angles, so that the total amplitude falls.

8.5.3. The dependence of the components of the tensor on the frequency when averaged over the polar angle

We now consider the complete dependence of all susceptibility components on frequency, for which we turn to Fig. 8.10. The dependences shown in this figure are pairs (vertically), geometrically located in the same way as the components of the susceptibility tensor in the matrix record (8.94):

$$\ddot{X} = \begin{pmatrix} X_{xx} & X_{xy} & X_{xz} \\ X_{yx} & X_{yy} & X_{yz} \\ X_{zx} & X_{zy} & X_{zz} \end{pmatrix} \tag{8.158}$$

moreover, the upper dependence of each pair corresponds to the real part of the tensor component, and the lower to the imaginary part of the same component.

Figure 8.10 shows that all the components of the tensor are non-zero, mainly within the limits of f_{1p} and f_{2p}, and there are extremes near the extreme frequencies f_{1p} and f_{2p}. Between the extremes the magnitudes of the components of the tensor take relatively small final values. Such a course of curves (two extrema at extreme frequencies with a flat area between them) is due to the fact that the resulting curve is obtained by summing the resonant modes of individual particles shifted about the frequency relative to each other. At the same time, due to the fact that the resonance curve of each particle at the resonant frequency changes sign, within the interval of the extreme frequencies, the positive and negative contributions cancel each other, resulting in relatively small values of the total susceptibility. At the edges of the screw, there was no such compensation, therefore, the resonance curves give edge extrema of the general curve.

Consider, for example, the χ_{xx} component as the result of susceptibility calculation using averaging. From Fig. 8.10 it is clear that the frequency dependence of the real part of the susceptibility $\chi'_{xx}(f)$ has two extremes of different directions – a maximum height of 2.1126 at a frequency of 1244 MHz and a minimum of -1.9206 at a frequency of 1690 MHz, between which the curve passes through zero at a frequency of 1614 MHz. The frequency dependence of the imaginary part of the susceptibility $\chi''_{xx}(f)$ also has two extremes, but of the same sign – minima at a frequency of 1259 MHz, a value of 2.0188 and at a frequency of 1670 MHz, a value of -2.1323. Between these minima, the curve goes smoothly, but has another weakly pronounced extremum – a maximum of -1.1000 at a frequency of 1389 MHz. It can be seen from the above figures that the frequencies of extremes of the dependences $\chi'_{xx}(f)$ and $\chi''_{xx}(f)$ coincide with the values of frequencies f_{1p} = 1252 MHz and f_{2p} = 1680 MHz, corresponding to the average orientations of the anisotropy axes, with an accuracy of no worse than 1%.

We now turn our attention to the comparative values of the individual components of the obtained tensor. Fig 8.10 shows that the diagonal components χ_{xx} (Fig. 8.10, a), χ_{yy} (Fig. 8.10, e), as well as the components χ_{xy} symmetrically located relative to the main diagonal (Fig. 8.10, b), χ_{yx} (Fig. 8.10 , d). All other components χ_{xz}, χ_{zx}, χ_{yz}, χ_{zy} (Fig. 8.10, c, e, g, h) as well as χ_{zz} (Fig. 8.10 i) have a

424

value significantly smaller (approximately two orders of magnitude). On the other hand, the curves for the components χ_{xx} (Fig. 8.10, a) and χ_{yy} (Fig. 8.10, e) are very similar to each other and differ by no more than 10–15%. The curves for the χ_{xy} components (Fig. 8.10, b), χ_{yx} (Fig. 8.10, d) are also very similar and differ from each other in quite a few with accuracy to the sign. Thus, in general, the comparative values of the components of the averaged tensor of the

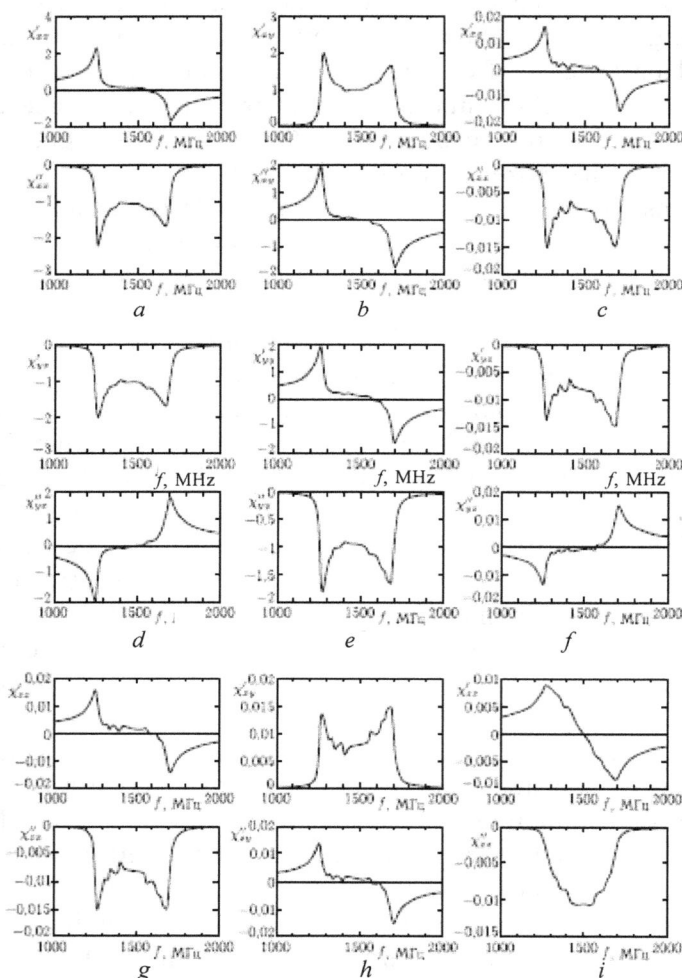

Fig. 8.10. The dependence of the components of the susceptibility tensor on frequency. Averaging is performed only over the angle θ_a with a step of 1°, the angle φ_a is set to be constant, equal to zero. Parameters – the same as taken in the construction of Fig. 8.8

composite medium behave quite similarly to the components of the tensor for an isotropic medium (formulas (2.58)–(2.61)), and the observed differences do not exceed 10–15%. This means that the composite medium mainly retains the gyrotropic properties close to those of a homogeneous isotropic medium.

However, the most important differences are the non-zero nature of all the ten tensor components, as well as the significantly wider range of overlapping frequencies (in the example given, 400 MHz compared to 80 MHz for an isotropic medium). A significant difference is also the presence of significant imaginary components of the χ_{xx} and χ_{yy} components of the tensor, as well as the actual component χ_{xy} and χ_{yx}.

A few words can be said about the properties of tensor immetry. From the comparison of curves in Fig. 8.10, b and Fig. 8.10, d that the components χ_{xy} and χ_{yx} are antisymmetric with an accuracy of 10–15%. The same can be said about the components χ_{yz} (Fig. 8.10, e), χ_{zy} (fig. 8.10, h). From the comparison of curves in Fig. 8.10 c, and 8.10, g it is clear that the components χ_{xz} and χ_{zx} are symmetrical with the same accuracy. Thus, the susceptibility tensor of the composite medium partially exhibits the properties of both symmetry and antisymmetry.

8.5.4. Dependences of tensor components on environmental parameters

We now note some features of the change in the components of the tensor with a change in the parameters of the medium.

Additional studies have shown that with an increase in the constant field by a factor of 10, the value of χ_{xx} decreases by a factor of 3, and the frequency increases approximately in proportion to the field. Extreme extremes converge together, as for a single resonance.

With the anisotropy increases 10 times the frequency increases, and the value χ_{xx} decreases, both by about 10 times. When the attenuation decreases, the position of the extrema in frequency does not change, but their magnitudes increase in absolute value, tending to infinity. When the attenuation is reduced 10 times, the curves exhibit a discrete character, which disappears when the averaging step is reduced 10 times. In this case, the χ_{zz} value decreases by 2 times. With an increase in attenuation by 10 times, everything smoothes, the curves for χ_{zz} take the form of a single resonance, the value of χ_{zz} decreases by 5 times.

When the magnetization decreases by 10 times, the frequency increases: the frequencies of the extreme extremes become 2400 and 4200 MHz. The curve for the real part of χ_{zz} becomes sharply asymmetric: from the low side it is a sharp peak with an amplitude of 0.02, from the high side it is a dull peak with an amplitude of -0.01. The curve for the imaginary part χ_{zz} also becomes sharply asymmetrical: from the low frequency side, a sharp drop down to amplitude -0.015, then a smooth rise to zero at the edge of the range from high frequencies, while the peak is absent.

With an increase in magnetization 10 times the frequency χ_{zz} decreases and becomes equal to 1400 MHz, the extreme peaks converge, the curves become similar to a single resonance. The height of the peaks decreases slightly and becomes 0.008 for the real part, and 0.010 for the imaginary part. The curve for the imaginary part has only one peak (down). The distance between the peaks in the real part is about 80 MHz.

With an increase in magnetization by 10 times, the real part of χ_{xx} increases to amplitude 55, that is, about 25–30 times, the real part of χ_{xz} increases to 0.5, that is, more than 20 times. The extreme peaks of all fish converge, and the curves take the form of a single resonance. The distance between the extreme peaks of the real part is about 40 MHz.

8.5.5. Averaging over azimuthal angle

Up to this point, the uniform distribution of the anisotropy axes was considered only over the polar angle θ_a from 0° to 180°, and the azimuthal angle φ_a was assumed to be constant, equal to zero. That is, the anisotropy axes of all the particles are located in an arbitrary direction, but only in one plane. Let us now consider how the components of the susceptibility tensor vary with an arbitrary orientation of the anisotropy axes of particles throughout the three-dimensional space, that is, besides changing θ_a from 0 to 180°, also changes in φ_a from 0 to 360°.

In order to clarify in more detail the role of the angle φ_a in the formation of the components of susceptibility, we will assume that the angle θ_a still varies in the full range from 0° to 180°, and for the angle φ_a we will set the interval from zero to a certain fixed value, whose value will be denoted through φ_s.

First of all, we note that a change in the azimuthal angle from zero to φ_s for any values of φ_s for the principal components of the susceptibility tensor χ_{xx}, χ_{xy}, χ_{yz}, χ_{yy} does not lead to any change in them. At the same time, the components χ_{xz}, χ_{yz}, χ_{zx}, χ_{zy}, as well as χ_{zz}, change significantly. That is, those components that are nonzero in the classical notation of the susceptibility tensor matrix (formulas (2.38), (2.49)) remain unchanged, while those that are zero in the same record change. In this case, as noted when considering Fig. 8.10, the values of unchanged components exceed those for the components varying, approximately by two orders of magnitude. That is, in all cases, the sensitivity of susceptibility is the largest in terms of its value, and the component retains properties similar to the tensor for an isotropic medium.

Comment. Note that this similarity occurs only for the values of the components, but not for their cost properties. So when considering the same Fig. 8.10, it was shown that, in all rays, the frequency range of the susceptibility components remains confined between the characteristic frequencies f_{1p} and f_{2p} defined by the formulas (8.156) and (8.157), which can significantly exceed the frequency range of the components corresponding to a single fixed orientation of the anisotropy axis.

Let us now consider what happens when the azimuthal angle changes with the other components, namely: χ_{xz}, χ_{yz}, χ_{zx}, χ_{zy}, χ_{zz}. A detailed study shows that the nature of changes in all these components is about the same, so we limit ourselves to considering only one component χ_{xz}. Turn to Fig. 8.11, where the dependences of the real (1, 3) and imaginary (2, 4) components of the susceptibility tensor component χ_{xz} on the averaging interval over the azimuthal angle φ_s at different frequencies: 1, 2 – 1400 MHz, 3, 4 – 1600 MHz are shown. Averaging over θ_a was carried out in the interval from $0°$ to $180°$ with a step of $1°$. Averaging over φ_a is in the interval from $0°$ to φ_c with a step of $5°$.

It can be seen from the figure that as the averaging interval over the angle φ_a changes, all dependences of the tensor components on the limiting angle φ_s change smoothly, and the character of the change is close to sinusoidal. In the entire range of changes in φ_a from zero to $360°$, the amplitudes of the curves become sufficiently small, approximately of the same order as in the absence of averaging over φ_a (that is, at $\varphi_s = 0°$).

However, it can be clearly seen that all curves, starting at $\varphi_s = 0°$ at different levels, at $\varphi_s = 360°$ converge to one point corresponding to the zero level (in the figure this level is shown by a horizontal dotted curve).

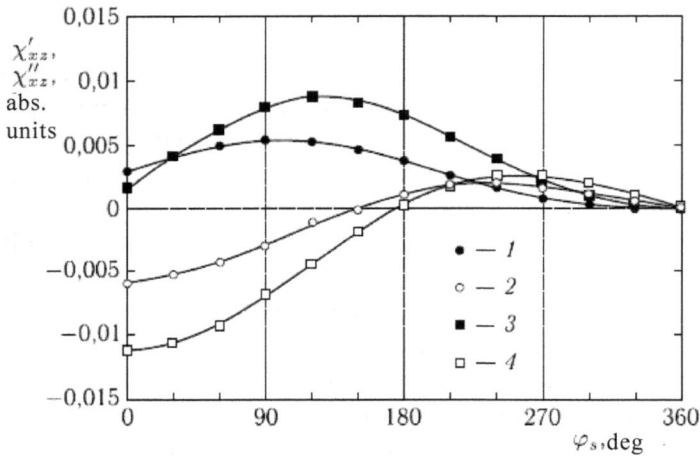

Fig. 8. 11. Dependences of the real (1, 3) and imaginary (2, 4) parts of the susceptibility tensor component χ_{xz} on the value of the averaging interval in azimuth angle ϕ_s at different frequencies: 1, 2 – 1400 MHz, 3, 4 – 1600 MHz. Parameters: $4\pi M_0 = 1750$G, $H = 500$ Oe, $H_a = 100$ Oe, $\alpha = 0.01$, $\gamma = 2.8$ MHz Oe^{-1}.

That is, it can be argued that when the orientation distributions of the anisotropy axes of individual particles are uniform throughout the three-dimensional space, the components of the susceptibility tensor χ_{xz}, χ_{yz}, χ_{zx}, χ_{zy}, χ_{zz} tend to zero, as is the case for an isotropic medium.

We note that the vanishing of these components from a physical point of view seems quite natural, since with an even distribution of the orientation of the anisotropy axes over the entire three-dimensional space, the selection of any directions in the medium as a whole disappears, which makes it related to the isotropic case.

It should not be forgotten, however, that the difference between such an environment and an isotropic one remains in a significant range of frequencies overlapped by the main components, that is, the susceptibility medium becomes broadband, which can be very useful in an applied sense.

8.6. Possible case of analytical averaging

In the particular case of a uniform distribution of the axes of anisotropy of particles over the azimuthal angle, the formulas (8.125)–(8.133) allow convenient analytical simplification.

So from their lines of the structure one can see that the angle ϕa is present in all forms of rmulas under the sine or cosine signs in the same terms, acting as coefficients in other expressions. Given the uniform distribution of anisotropy axes along this angle, one can average the trigonometric functions entering into these formulas over the angle φ_a from 0 to 2π using the following formula:

$$\overline{f(\varphi_a)\big|_0^{2\pi}} = \frac{1}{2\pi}\int_0^{2\pi} f(\varphi_a)\,d\varphi_a,$$

(8.159)

In this case, we obtain the averaged values:

$$\overline{\sin\varphi_a} = \overline{\cos\varphi_a} = \overline{\sin\varphi_a \cos\varphi_a} = 0,$$

(8.160)

$$\overline{\sin\varphi_a} = \overline{\cos^2\varphi_a} = \frac{1}{2},$$

(8.161)

Substituting these values into the formulas (8.125)–(8.133), resulted in the components of the susceptibility tensor in the form:

$$\chi_{xx} = \chi_{yy} = \frac{(\Omega_1 + i\Omega\alpha)\cos^2\theta_m + (\Omega_2 + i\Omega\alpha)}{4\pi\left[\Omega_1\Omega_2 - (1+\alpha^2)\Omega^2 + i\alpha\Omega(\Omega_1 + \Omega_2)\right]},$$

(8.162)

$$\chi_{xy} = -\chi_{yx} = \frac{i\Omega\cos\theta_m}{4\pi\left[\Omega_1\Omega_2 - (1+\alpha^2)\Omega^2 + i\alpha\Omega(\Omega_1 + \Omega_2)\right]};$$

(8.163)

$$\chi_{xz} = \chi_{yz} = \chi_{zx} = \chi_{zy} = 0;$$

(8.164)

$$\chi_{zz} = \frac{(\Omega_1 + i\Omega\alpha)\sin^2\theta_m}{4\pi\left[\Omega_1\Omega_2 - (1+\alpha^2)\Omega^2 + i\alpha\Omega(\Omega_1 + \Omega_2)\right]},$$

(8.165)

In this case, the symmetry properties of the susceptibility tensor are the same as for a homogeneous isotropic medium:

$$\chi_{xx} = \chi_{yy},$$

(8.166)

$$\chi_{xy} = -\chi_{yz},$$

(8.167)

$$\chi_{xz} = \chi_{yz} = \chi_{zx} = \chi_{zy} = 0. \tag{8.168}$$

Note that the zero components of the tensor χ_{xz}, χ_{yz}, χ_{zx}, and χ_{zy}, taking place according to formula (8.164), coincide with the vanishing of these components at $\varphi_s = 360°$, as shown in Fig. 8.11.

It can be seen that for $\theta_a = \theta_m = 0$, as well as $H_a = 0$, the formulas (8.162)–(8.165) go to the known expressions for the components of the susceptibility tensor of the isotropic medium (2.58)–(2.59), and also (2.60)–(2.61) subject to (2.56)–(2.57).

8.7. Formation of specified properties magnetic susceptibility of the environment

In the framework of the model of arbitrary orientation of the anisotropy axes of individual particles considered above, the main frequency parameters of the medium are not independent. Thus, for a given external field, the value of the anisotropy field of the particles simultaneously determines both the maximum (or minimum) working frequency and the width of the overlapped frequency range, which is associated with the chaotic nature of the orientation of the anisotropy axes. The interdependence of environmental parameters reduces the possibilities of its technical application, therefore, it is of interest to identify the possibility of independent control of its properties.

Let us now consider the possibility, important for practice, of forming specified magnetic susceptibility properties of a medium by partially ordering the orientation of the anisotropy axes of individual particles.

8.7.1. Streamlining particle orientation

Consider an environment in which the orientation of the anisotropy axes of individual particles is close to some particular direction. Possible variants of the medium structure geometry are shown in Fig. 8.12. As an example, a cubic lattice was chosen for the figure, however, since there is no interaction between the particles, the lattice type does not matter and is not taken into account in the further calculation. Spherical particles are schematically depicted in the form of circles located in its nodes. The thickened lines inside the circles indicate the location of the individual anisotropy axes with spheres. Figure. 8.12 a corresponds to an arbitrary orientation of the axes of anisotropy of particles, Fig. 8.12, b to the partially

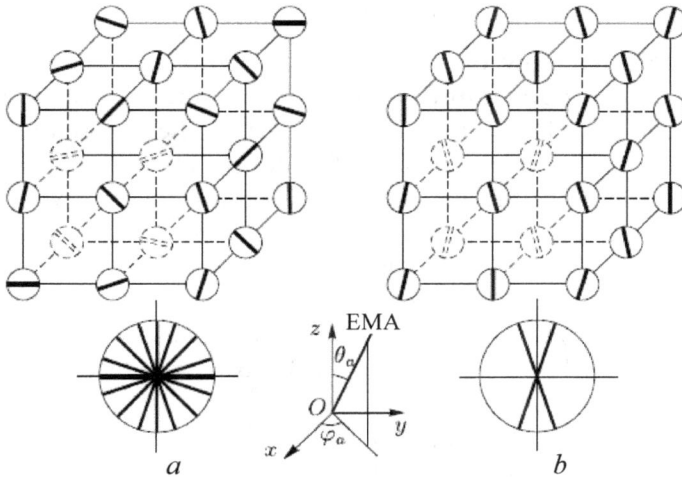

Fig. 8.12. Geometry of the structure of the compositional medium of anisotropic ferrite particles. *a* is an arbitrary orientation of the anisotropy axes; *b* is the partially ordered orientation of the anisotropy axes.

ordered orientation of the anisotropy axes along the vertical along the Figure. Below are shown the schemes of orientation of the axes of anisotropy of various particles for the same cases. Below in the middle is a diagram of the reference angles of orientation of the axes of easy magnetization (EMA) in the Cartesian and spherical systems of coordinates.

Technologically, an environment with a preferred direction of orientation of the anisotropy axes can be made by placing spherical particles in an initially liquid gradually thickening matrix in the presence of a constant magnetic field orienting particles during solidification. In this case, the field of one direction will lead to a partial ordering of the orientation of the axes of anisotropy of particles around this direction. Smooth quasistatic rotation of the field along a cone or swinging it in a plane will lead to the separation of preferential orientation within defined intervals along the polar and azimuthal angles.

The boundaries of the range of ordering in angles: polar – through θ_{1s} and θ_{2s}, azimuthal – through φ_{1s} and φ_{2s}. Because of the parity of the uniaxial anisotropy of the disordered state $0° \leq \theta_a \leq 90°$, i.e. $\theta_{1s}=0°$, $\theta_{2s}=0°$ and the completely ordered along the polar axis $\theta_a=0°$, i.e. $\theta_{1s} = \theta_{2s}=0°$.

8.7.2. Ordering by the polar angle

We first consider how the susceptibility properties depend on the ordering of the polar angle θ_a in the absence of ordering over the azimuthal angle φ_a (that is, at $\varphi_a = 0$). The behaviour of the susceptibility at different values of ordering in θ_a is illustrated in Fig. 8.13, where are the dependences on the frequency $\chi'_{xx} = \text{Re}\chi_{xx}$ (solid lines 1', 2', 3') and imaginary $\chi''_{xx} = \text{Im}\chi_{xx}$ (dotted line 1", 2", 3") of the components of the susceptibility tensor component χ_{xx} at different intervals of the ordering angle θ_a. Curves 1', 1" correspond to an unordered state, that is, $0° \leq \theta_a \leq 90°$, curves 2', 2" - partial ordering at $0° \leq \theta_a \leq 45-$, curves 3', 3" to complete ordering, that is, $\theta_a = 0°$.

The parameters of construction in Fig. 8.13 are given in its signature. The frequencies of the resonances corresponding to the orientation of the anisotropy axis along and across the constant field, defined by the formulas (8.156) and (8.157), are respectively: $f_{1p} =$ 1252 MHz; $f_{2p} = 1680$ MHz. The width of the resonance line of a single particle is $\Delta f = 80-100$ MHz.

It can be seen from the figure that, in an unordered state (curves 1', 1"), the boundary ejections of both components of the dependence $\chi_{xx}(f)$ fall at the limiting resonance frequencies f_{1p} and f_{2p}. As the ordering increases (curves 2', 2"), the frequency of the lower burst increases, while the upper one remains close to f_{2p}. In a fully ordered state (curves 3', 3"), the frequencies of both ejections approach f_{2p}, differing from each other by the width of the resonance line Δf. Emission amplitude χ'_{xx} as the ordering increases in module from 1.5 to 3.0, and emissions χ''_{xx} from 1.5 to 5.7.

Comment. The curves 1', 1", presented in Fig. 8.13, were constructed with the technical parameters like both curves in Fig. 8.10, a. However, it can be seen that there the amplitude of the curves is two more than here. Such a difference is due to the fact that there averaging over θa was carried out in the interval from 0° to 180°, and here from 0° to 90°. That is, in the first case, twice as many particles were taken into account, which led to an increase in the amplitude of the susceptibility by half. It should be noted that here the averaging was carried out in a half interval in the expectation that the anisotropy axis is bidirectional, that is, the equilibrium position of the magnetization is always oriented in the positive direction. That is, even if at the initial moment the magnetization vector was oriented in the negative direction, then under the action of the field it is still set in the positive direction of the same axis, although perhaps with some hysteresis, for the present consideration is not significant. Thus, if we assume that, in the initial state, the angle θ_a has all possible values from 0° to 180°, then under the action of the field it immediately reorients to the interval of angles from 0° to 90°. However, it should not be forgotten that in this case the number of particles, where the magnetization vector is oriented in the positive direction of the anisotropy axis, doubles, that is, the amplitudes of the curves shown in Fig. 8.13 must be doubled. In this case, they will be similar in amplitude to those shown in Fig. 8.10. When plotting the curves in Fig. 8.13 and the following doubling was not carried out in order to more clearly reveal the role of the value of the interval in θ_a, when it constitutes from a full angle of 90°.

The behaviour of the susceptibility during the ordering within the same interval in angle θ_a, but starting at different initial angles, in the absence of ordering in φ_a, is illustrated in Fig. 8.14, built with the technical parameters as Fig. 8.13. This figure shows the dependence on the frequency of the actual $\chi'_{xx} = \mathrm{Re}\chi_{xx}$ (solid lines – 1', 2', 3') and imaginary $\chi''_{xx} = \mathrm{Im}\chi_{xx}$ (dotted – 1", 2", 3") of the components of the susceptibility tensor component χ_{xx} for different intervals of ordering the angle θ_a (indicated in the figure caption).

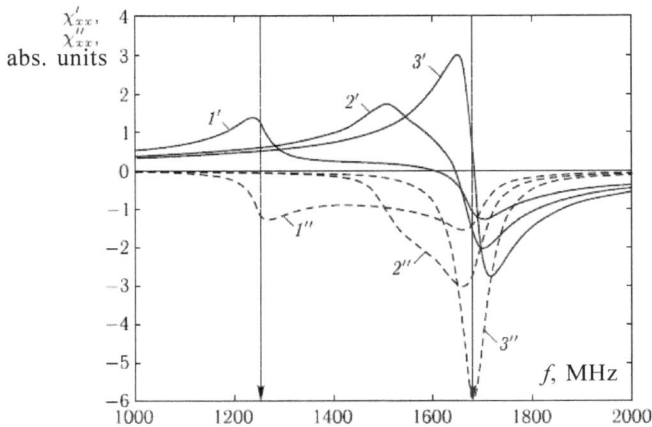

Fig. 8.13. Dependences of real χ'_{xx} and imaginary χ''_{xx} parts of the susceptibility tensor component χ_{xx} versus frequency at different intervals of ordering in the angle θ_a: 1 – 0°$\leq \theta_a \leq$ 90°; 2 – 0° $\leq \theta_a \leq$ 45°; 3 – θ_a = 0°. Parameters: $4\pi M_0$ = 1750 G, H = 500 Oe, H_a = 100 Oe, α = 0.01, γ = 2.8 MHz Oe^{-1}.

Figue 8.14 that the change in the position of the ordering interval in the polar angle changes the general form of the frequency dependence of the susceptibility is quite small. When the interval θ_a is shifted up from 0° to 90°, the susceptibility curve, basically remaining self-similar, shifts down in frequency from 1700 to 1300 MHz. Such a strong change is caused by a change in the predominant position of the anisotropy axis of the main mass of particles from a parallel to an external field to a perpendicular one.

8.7.3. Azimuthal angle alignment

In both of the considered rays, the ordering along the azimuthal angle φ_a was absent, that is, only the components of the susceptibility tensor χ_{xx}, χ_{xy}, χ_{yx}, χ_{yy}, χ_{zz} were non-zero, and the components $\chi_{xz} = \chi_{yz} = \chi_{zx} = \chi_{zy}$ were zero. We now consider the case of partial ordering with respect to the angle φ_a, when all nine components of the tensor are non-zero. For example, we restrict ourselves to the χ_{xz} component, the frequency dependences of real $\chi'_{xz} = \mathrm{Re}\chi_{xz}$ (solid lines 1'–5') and imaginary $\chi''_{xz} = \mathrm{Im}\chi_{xz}$ (dotted line 1''–5'') of the parts of which at different intervals of the angle φ_a are shown in Fig. 8.15. All curves are plotted with the same parameters as in the previous

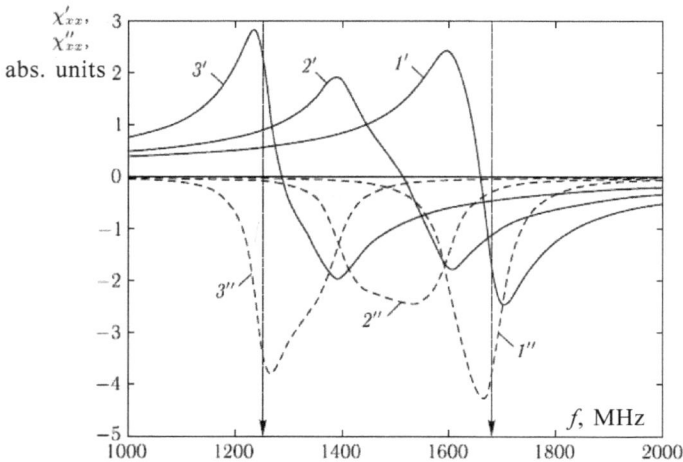

Fig. 8.14. The dependence of the real χ'_+ and imaginary χ''_{xx} parts of the component of the susceptibility tensor χ_{xx} on the frequency at different intervals of ordering in the angle θ_a: $1 - 0° \le \theta_a \le 30°$; $2 - 30° \le \theta_a \le 60°$; $3 - 60° \le \theta a \le 90°$.
The parameters are the same as those taken when constructing Fig. 8.13

figures, and there is no ordering along the polar angle – $0°\leq \theta_a \leq 90°$. Curves 1', 1" correspond to the absence of a change in the azimuthal angle, that is, the state of this angle is completely ordered. Curves 5', 5" – change in azimuthal angle in full possible interval from $0°$ to $360°$, that is, the state is completely disordered. The remaining ordering intervals for φ_a, which are intermediate between these two, are indicated in the figure caption.

From Fig. 8.15 it can be seen that the boundaries of the frequency interval, in which the susceptibility component χ_{xz} is noticeably different from zero, do not depend on the degree of ordering in angle φ_a. The real component of χ'_{xz} takes the maximum values ± 0.080 near the edges of the frequency interval at maximum ordering $\varphi_a = 0°$ (curve 1'). As ordering decreases, the maximum values of χ'_{xz} decrease with $0°\leq \varphi_a \leq 270°$ are 0.010 and -0.025 (curve 4'), And with complete disordering, $0° \leq \varphi_a \leq 360°$ everywhere $\chi'_{xz} = 0$ (curves 5', 5"). Imaginary component χ''_{xz} behaves in the same way: as the ordering decreases in amplitude, decreases, and with complete disordering it vanishes, which corresponds to the zero value in the formula (8.164).

8.7.4. General possibilities of forming environmental parameters

From the review we can see that the partial ordering of the orientation of the anisotropy axes of individual particles provides the possibility of independent formation of the specified parameters of the composite medium, such as the operating frequency and the overlapped frequency interval. So, from Fig. 8.13 it can be seen that changing the magnitude of the ordering interval of the particle orientation at the polar angle allows changing the frequency range without changing the upper operating frequency, and from Fig. 8.14, it can be seen that changing the angular position of the ordering interval by the polar angle allows changing the working frequency without changing the width of the interval of the overlapping frequencies. From Fig. 8.15 it can be seen that a change in the ordering of the orientation of the particles at the azimuthal angle leads to the appearance of an additional susceptibility tensor, which may be of interest for expanding the functional capabilities of microwave devices.

436

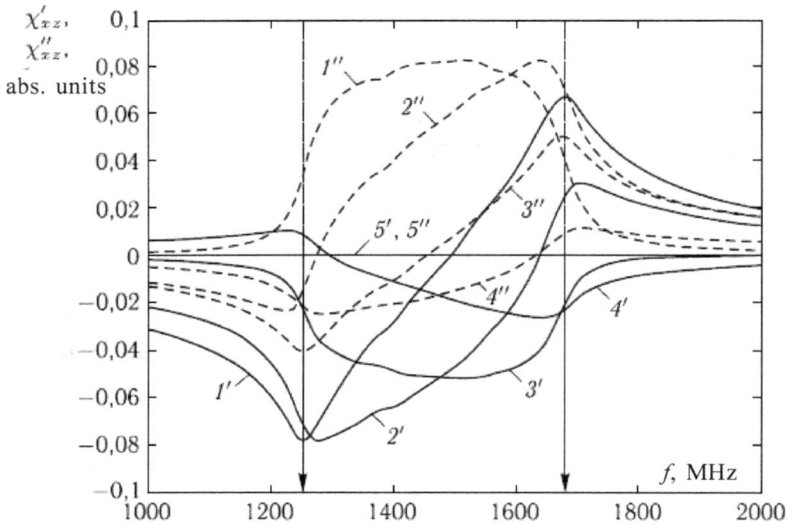

Fig. 8.15. Dependences of real χ'_{xz} and imaginary χ''_{xz} parts of the susceptibility tensor component χ_{xz} versus frequency at different intervals of ordering in the angle φ_a: $1 - \varphi_a = 0°$; $2 - 0° \leq \varphi_a \leq 90°$; $3 - 0° \leq \varphi_a \leq 180°$; $4 - 0° \leq \varphi_a \leq 270°$; $5 - 0° \leq \varphi_a \leq 360°$. The parameters are the same as those taken when constructing Fig. 8.13.

Conclusions in chapter 8

This chapter is devoted to obtaining and studying the properties of the dynamic magnetic susceptibility tensor for a composite medium, consisting of spherical anisotropic magnetic particles, in cases of both an arbitrary and ordered orientation of the axes of anisotropy in individual particles. The main results obtained in this chapter are as follows.

1. A compositional medium is proposed, which is a lattice of anisotropic ferrite spheres interspersed in a nonmagnetic matrix. The task is to study the parameters of such an environment, and first of all the dynamic magnetic susceptibility.

2. It is noted that due to the absence of interaction between particles, the complete problem can be solved in two stages: first, finding the susceptibility tensor for one particle in the case of arbitrary orientation of its anisotropy axis; the second is the summation of the susceptibilities of individual particles, followed by averaging over all possible orientations of the anisotropy axes of individual particles.

3. It has been established that to solve the problem of the susceptibility of a single particle, it is necessary to introduce five auxiliary coordinate systems. The first is the laboratory, connected with the direction of the external field; the second is associated with the anisotropy axis, corresponding to the zero value of the azimuthal angle of the EMA; the third is associated with the equilibrium magnetization, also corresponding to the zero value of the azimuthal angle of the EMA; the fourth is associated with the anisotropy axis, corresponding to an arbitrary value of the azimuthal angle of the EMA; the fifth one is related to the equilibrium magnetization, also corresponding to an arbitrary value of the azimuthal angle of the EMA,

4. It has been established that as the first stage of solving the problem, it is necessary to find the equilibrium orientation of the magnetization vector, determined by the interaction of the magnetization with the anisotropy field and the constant field. To solve this problem, two auxiliary coordinate systems were introduced: one connected with the field and the other connected with magnetization. By minimizing the energy density an equation is obtained for determining the polar angle of the equilibrium position of the magnetization vector. The solution of this equation, carried out by a numerical method, characterizes the behavior of the polar angle of the magnetization vector, which is characteristic of a prolonged orientational transition.

5. A general procedure has been established for calculating the susceptibility of a single particle, which consists in successively transforming the susceptibility obtained in a system related to magnetization to a system associated with a field. The necessity of using a three attic of the transition for such a transformation is revealed. The required transition matrices are obtained, the components of which are expressed in terms of the polar and azimuthal angles of the equilibrium position of the magnetization vector and the specified orientation of the anisotropy axis.

6. The complete course of solving the problem is presented in the form of nine successively performed steps, including solving the equation of magnetization in a system related to the equilibrium magnetization, linearizing the solution obtained and converting it to a system associated with the field. For such a transformation, transition matrices obtained in the previous stage were used. The result of the complete sequence of steps was the dynamic magnetic susceptibility tensor for a single particle, recorded in the laboratory coordinate

438

system associated with the field. The components of the susceptibility
tensor are presented in a complex form and are expressed in terms
of the parameters of the material and the constant field. The polar
and azimuthal angles of orientation of the anisotropy axis of this
particle are distinguished as the main directors of the parameters of
the components of the susceptibility tensor.

7. It was noted that in order to obtain susceptibility for a
composite structure consisting of a large number of particles with
axes of anisotropy randomly oriented in space, the averaged tensor
components should be averaged over all possible polar and azimuthal
angles of such a system. A procedure is presented for averaging the
components of the tensor over the polar and azimuthal angles with
a constant step along the angular coordinates. In order to ensure
a uniform spatial distribution of the fixed charges of the polar
and azimuthal angles, an auxiliary sphere was constructed, on the
surface of which a grid of parallels and meridians was imposed.
As a measure that ensures the uniform distribution of the angular
coordinates, we used the invariance of distances along the grid
parallels between its neighboring nodes both at the equator and as it
approaches the poles of the sphere. It is noted that finding the total
averaged susceptibility consists in calculating the susceptibility for
all angular coordinates obtained on the sphere of possible origins,
which determine the orientations of the anisotropy axes, followed
by summing the obtained values and dividing by the total number
of possible orientations. The scheme of the algorithm for numerical
calculation of the components of the susceptibility tensor is given.

8. The properties of this, obtained by averaging, the susceptibility
tensor of the composite medium are considered. The focus is on
its frequency properties. It is shown that the frequency range
of all components is the same and is between two frequencies,
corresponding to the frequencies of ferromagnetic resonance when
the particle is magnetized along and across the anisotropy axis. Near
these extremes, the components have emissions that are limited in
height, especially since the damping of the ferromagnetic resonance
is higher. Between these emissions there is a more or less smooth
area, having a level several times lower than the maximum emission
amplitude.

9. The dependence of the formation of a flat area on the step
size along the polar coordinate is considered. It is shown that
if at one step the frequency of the ferromagnetic resonance of a
particle changes by an amount exceeding the width of the resonance

line in this particle, then the flat portion of the susceptibility component acquires a sawtooth discrete character. It is noted that this character is due to the superposition of resonant individual particles corresponding to different directions of the anisotropy axes, that is, due to the discrete orientation of the axes, the arrangement of resonant frequencies is also discrete.

10. The complete dependences of all the coefficients of the tensor on frequency are investigated. It is noted that in the case of averaging over only one polar coordinate, all nine components of the tensor are non-zero. The tensor is not diagonal, symmetric or antisymmetric. All its components are complex, that is, they have non-zero real and imaginary parts. In amplitude, only those components of the tensor that in their location correspond to the components of the tensor for an isotropic medium, that is, located on the main diagonal and antisymmetric with respect to it in close proximity, are significant. All other components, although different from zero, but have values smaller than those mentioned by approximately two orders of magnitude. The dependences of the components of the tensor on the field and the attenuation values are considered. It is shown that a change in the field leads mainly to a shift in the extremes of the tensor components in accordance with the usual laws of ferromagnetic resonance. The change in attenuation changes the amplitude of the components of the tensor, first of all, emissions at the edge of the astrotic range, and to a somewhat lesser, although quite noticeable, degree of flat parts.

11. The formation of the components of the tensor is considered when averaged over the azimuthal angle of orientation of the axes of the anisotropy of particles. It is shown that with full uniform averaging over the full azimuthal range from zero to 360 degrees, by the symmetry properties, the tensor takes on the form that completely coincides with the tensor for an isotropic medium. That is, those components that, with incomplete averaging over the azimuthal angle or its absence, were two orders of magnitude smaller than the main ones, now vanish with high accuracy (up to five orders of magnitude). The characteristics of the components of the tensor with partial averaging over the polar and azimuthal angles are considered. A violation of the symmetry of the components of the tensor in frequency with a change in the degree of ordering, manifested in a change in the mutual height and width of emissions at the edge of the frequency range, is revealed. The possibility of analytic calculation of the formation of the components of the tensor is shown in the

particular case of a uniform distribution of the orientation of the anisotropy axes in the full range of changes in the azimuthal angle from zero to 360 degrees.

12. The dynamic susceptibility of a composite medium consisting of specially interacting interacting anisotropic ferrite particles is considered. It is shown that the partial ordering of the orientation of the anisotropy axes of individual particles provides the possibility of the formation of boundary frequencies and the overlapped frequency interval of the susceptibility tensor. The conditions under which the formation can be carried out independently by both parameters are revealed. Thus, a change in the ordering interval of the orientation of the axes of anisotropy of the particles at the polar angle allows changing the frequency range without changing the upper operating frequency, and changing the angular position of the same ordering interval allows changing the working frequency without changing the width of the interval of overlapping frequencies. It is shown that a change in the ordering of the orientation of the anisotropy of the particles at the azimuthal angle leads to the appearance of additional susceptibility tensors, which are absent in a completely ordered or disordered medium. The importance of the obtained results for the expansion of functional capabilities of microwave devices is noted.

9

Precession of positions of equilibrium of magnetization in the conditions of orientational transitions

The consideration of ferromagnetic resonance under the conditions of the orientational transition in previous chapters of this monograph was limited to purely linear problems, that is, those for which the amplitude of oscillations of the magnetization vector was assumed to be small compared to its full length. In this and following chapters, this limitation is absent, the amplitude of oscillations of the magnetization vector can be arbitrary, so that the phenomena considered here are obviously non-linear.

This chapter is devoted to the description of the forced non-linear precession of the magnetization vector occurring under the conditions of the orientational transition. The equilibrium orientation of the magnetization is assumed to be due to the balance of forces between the external field and the demagnetization field. It is shown that when such a system is acted upon by an alternating magnetic field of a certain polarization, the precessional motion of the equilibrium position of magnetization can take place. Such a motion is superimposed on the usual precession of magnetization, that is, it is, as it were, a precession of the second order magnetization. Various characteristics of such a precession are considered, several fundamentally different non-linear modes of its excitation are revealed. The influence of the asymmetry of the constant and variable fields is investigated and the possible variants of their jointly mutually compensating action are revealed.

When considering the problems listed in the surveys, we will mainly follow the works [266, 297–332], as well as the monographs [3]. The remaining necessary references are indicated in the text.

9.1. General illustration of the precession equilibrium positions

In order to quickly introduce the reader to the course of the matter, let us first consider the general picture of the manifestation of the precession of the equilibrium state, as well as its simplest qualitative interpretation based on the vector model.

9.1.1. Problem geometry and basic equations

Consider an infinite isotropic ferrite plate magnetized along the normal to its plane. We will use the geometry of the problem introduced in [3, 264].

In such a geometry, the resonance frequency of oscillations of magnetization corresponds to the bottom of the exchange spin-wave spectrum, when their parametric excitation is excluded, and the precession of the magnetization vector can occur with a deviation from the normal to the film plane by angles of tens of degrees [3,119–130,264]. The choice of such a condition makes it possible to correctly consider the precession of magnetization at orientational transition, consisting in the rotation of the magnetization vector at angles up to 90 degrees, which is done later in this work. So, the overall geometry of the problem is shown in Fig. 9.1. It is based on a plane-parallel magnetic plate of thickness d. The plate material has cubic crystallographic symmetry, the plane (100) of which coincides with the plane of the plate (on the left is shown the orientation diagram of the cubic cell). In addition to the cubic plate, there is a uniaxial magnetic anisotropy with the easy magnetization axis (EMA) perpendicular to the plane of the plate. An external constant magnetic field $\mathbf{H}_0$ is applied perpendicular to the plane of the plate. An alternating magnetic field $\mathbf{h}$ acts in the plane of the plate.

The problem is solved in the Cartesian coordinate system $Oxyz$, the plane Oxy of which coincides with the plane of the plate, and the axes Ox, Oy and Oz are parallel to the edges of the cube of the crystallographic cell. The centre of the coordinate system O is in the

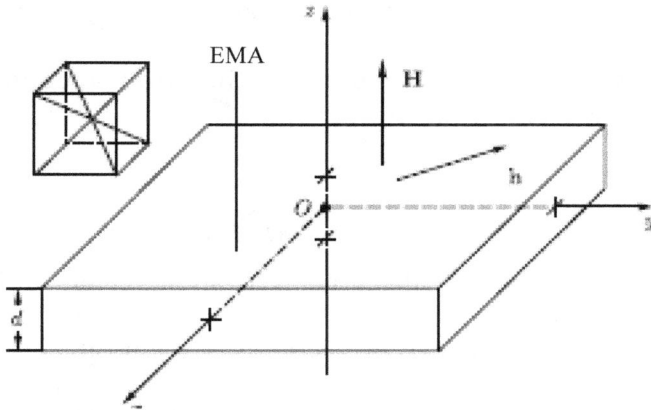

Fig. 9.1. Geometry of the precession problem for the magnetization equilibrium position.

centre of the plate, so that its planes correspond to the coordinates $z = \pm d/2$.

Comment. It should be noted that in the following consideration the precession over the plate thickness is assumed to be homogeneous, therefore the exact value of this thickness does not participate in the calculation, only the demagnetizing factor equal to 4π is used. Accounting for the exact thickness of the plate becomes important when boundary conditions are introduced into consideration, which, for example, is necessary when calculating elastic vibrations [3, 264]. The authors intend to carry out such consideration in one of the subsequent onographies of the present series devoted to magnetoelastic phenomena. Here, the thickness is introduced only for generality, in order not to depart very much from the geometry of the problem adopted in [3, 264].

We assume that the external field has the form

$$\mathbf{H} = \left\{ H_{0x} + h_x; H_{0y} + h_y; H_{0z} \right\}.$$

(9.1.)

where $H_{0x,y,z}$ is a constant magnetizing field, $h_{x,y}$ are the components of the variable field, and we assume that $h_{x,y} < H_{0z}$, as well as $H_{0x,y} < H_{0z}$.

The energy density of the plate taking into account the demagnetizing factor is equal to

$$U = -M_0 h_x m_x - M_0 H_{0x} m_x - M_0 h_y m_y - M_0 H_{0y} m_y -$$
$$- M_0 H_{0z} m_z + 2\pi M_0^2 m_z^2, \quad (9.2.)$$

where $\mathbf{m} = \mathbf{M}/M_0$ is the normalized magnetization vector, M_0 is the saturation magnetization.+

At $H_{0z} = 0$, the magnetization vector in the equilibrium state is oriented in the plane of the plate, and at $H_{0z} \geq 4\pi M_0$ and $H_{0x,y} = 0$ is perpendicular to this plane. In the field interval $0 \leq H_{0z} \leq 4\pi M_0$ with an initial orientation of the magnetization vector other than the equilibrium one, a quasistatic determination of the magnetization occurs through the orientational transition [266, 297]. In equilibrium, at $h_x = h_y = 0$, the orientation of the magnetization vector is determined by minimizing the energy of the components of the magnetization, resulting in:

$$m_z = \frac{H_0}{4\pi M_0}, \quad (9.3)$$

whence, setting $m_y = 0$, which, by virtue of the symmetry of the problem, can always be done by rotating the coordinate system around the Oz axis, which

$$m_x = \sqrt{1 - m_z^2} = \sqrt{1 - \left(\frac{H_0}{4\pi M_0}\right)^2}. \quad (9.4)$$

The angle θ_0 of the deviation of the magnetization vector from the axis Oz is determined by the formula

$$\theta_0 = \arccos(m_z) = \arccos\left(\frac{H_0}{4\pi M_0}\right). \quad (9.5)$$

To solve the problem of the dynamic behaviour of the magnetization vector under conditions of the orientational transition, we use the Landau–Lifshitz magnetization motion equations with the dissipative term in the Hilbert form (2.97)–(2.99):

$$\frac{\partial m_z}{\partial t} = -\frac{\gamma}{1+\alpha^2}\left[\left(m_y + \alpha m_x m_z\right)H_{ez} - \left(m_z - \alpha m_y m_x\right) - \alpha\left(m_y^2 + m_z^2\right)H_{ex}\right];$$

(9.6)

$$\frac{\partial m_y}{\partial t} = -\frac{\gamma}{1+\alpha^2}\left[\left(m_z + \alpha m_y m_z\right)H_{ex} - \left(m_x - \alpha m_z m_x\right) - \alpha\left(m_y^2 + m_x\right)H_{ey}\right];$$

(9.7)

$$\frac{\partial m_z}{\partial t} = -\frac{\gamma}{1+\alpha^2}\left[\left(m_x + \alpha m_z m_y\right)H_{ey} - \left(m_y - \alpha m_x m_z\right) - \alpha\left(m_x^2 + m_y\right)H_{ez}\right];$$

(9.8)

where γ is the gyromagnetic constant ($\gamma > 0$), α is the damping parameter of the precession of magnetization. The effective fields entering into these equations in accordance with (9.2), have the form:

$$H_{ex} = H_{0x} + h_x;$$

(9.9)

$$h_x = h_0 \sin\left(2\pi f t\right),$$

(9.10)

$$H_{ez} = H_{0z} - 4\pi M_0 m_z.$$

(9.11)

In [306–308], it was shown that under the conditions of an orientational transition when a system is excited by an alternating field of circular polarization

$$h_x = h_0 \sin\left(2\pi f t\right);$$

(9.12)

$$h_y = -h_0 \cos\left(2\pi f t\right),$$

(9.13)

where f is the frequency of the alternating field, h_0 is its amplitude, in certain circumstances a precession of the equilibrium position is possible, consisting in that the equilibrium position, around which the magnetization vector with the excitation frequency precesses, itself precesses around the constant field direction with a frequency much lower. Precession of equilibrium is not the only kind precession in the conditions of the orientational transition, however, in view of the complexity of such a precession and importance for further consideration, we will dwell on it in more detail.

9.1.2. Illustration of the equilibrium precession

Let us give the simplest illusration of the precession of the equilibrium state under the conditions of the instrumental transition based on the solution of the system of equations (9.6)–(9.8). The main condition for the implementation of such a precession there will be an establishment of a constant field value of a somewhat smaller magnitude of the demagnetization field, so that the magnetization vector in the equilibrium state is oriented not exactly along the field, but is deflected from it at some angle.

We will proceed from the numerical solution of the system (9.6)–(9.8), performed by the Runge–Kutta method. Consider the development of forced oscillations of magnetization in time. The calculation results are illustrated in Fig. 9.2.

Figures 9.2, a, b show the dependence of the transverse magnetism m_x (solid line) and m_y (dashed line) on time, in Fig. 9.2, c, d there are similar dependences for the longitudinal component of magnetization m_z. Figures 9.2, d, e shows the precession portraits as dependences of the m_y component on the m_x component, when time is a parameter.

The left column (Fig. 9.2, a, c, d) corresponds to a constant field $H_0 = 265$ Oe, that is 15 Oe less than the demagnetization field of the form $H_m = 4\pi M_0 = 280$ Oe, while the magnetization vector in the equilibrium state deviates Oz axis by $19.46°$. The horizontal dashed lines in Fig. 9.2, a and c, correspond to the equilibrium values of the components of magnetization: $m_{xc,yc} = \pm0.34$, $m_{zc} = 0.94$, while: $m_{xc}^2 + m_{yc}^2 + m_{zc}^2 = 1.00$.

The right column (Fig. 9.2, b, d, e) corresponds to the field $H_0 = 295$ Oe, which is 15 Oe more than the demagnetization field of the form, that is, the equilibrium magnetization vector is oriented along the Oz axis.

Comment. It should be noted that in this figure and everywhere further, within the limits of this and subsequent chapters, the authors, in order to reduce the cumbersome designations, do not give the names of units of 'norm' units with components of magnetization $m_{x,y}$, since it is implied that these are full components of the magnetization vector $M_{x,y}$ normalized to saturation magnetization M_0, that is, $m_{x,y} = M_{x,y}/M_0$, which are dimensionless.

From Fig. 9.2 a–d, illustrating the development of oscillations in time, it is clear that the natural oscillations of the magnetization caused by the initial impulse attenuate at times of the order of $(4–5) \cdot 10^{-8}$ s, after which the forced oscillations take a stationary character.

The same applies to Fig.. 9.2, d, e, where after several primary turns, caused by the individual oscillations, the trajectories become stationary. A comparison of the left and right columns of the figure shows that in the fields smaller and larger than the demagnetization field, the shape of the oscillation of the magnetization are fundamentally different.

We first consider the case $H_0 < 4\pi M_0$ (left column). Figure 9.2 shows that after establishing the stationary mode (at $t > 0.5 \cdot 1$

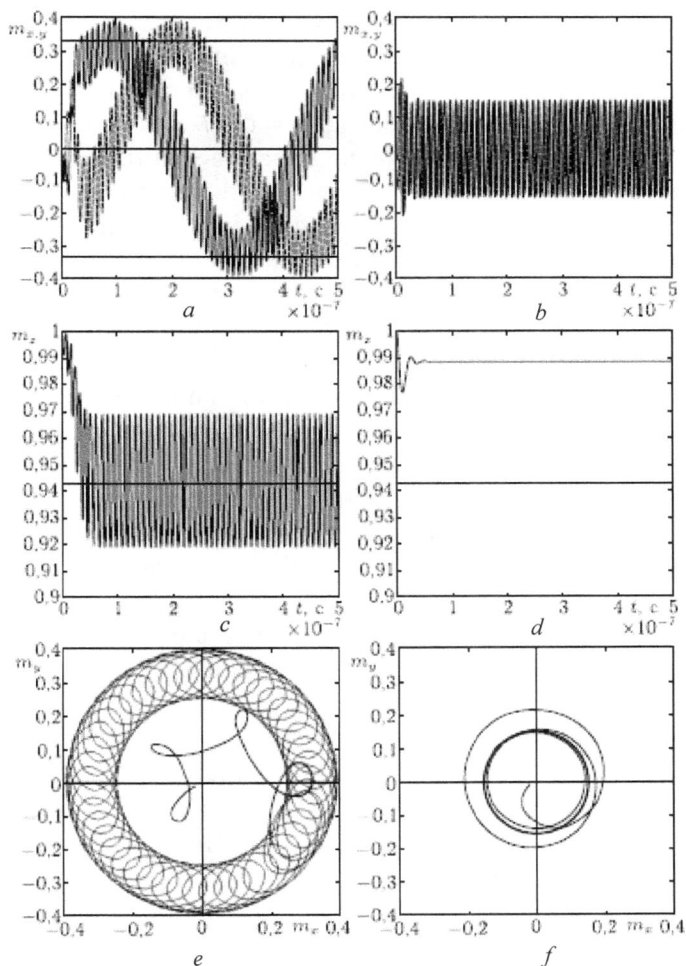

Fig. 9.2. Illustration of the precession of the equilibrium position. Parameters: $4\pi M_0 = 280$ G; $\alpha = 0.3$ (for the sake of clarity, we used the parameters typical for terbium ferrite garnet); the frequency of the alternating field is $f = 108$ Hz; the amplitude of the alternating field $h0_x = h_{0y} = 3$ Oe (circular field), initial values of magnetization components $m_x = m_y = 0$, $m_z = 1$

the average positions of the transverse magnetism components m_x and m_y make relatively slow large amplitude oscillations with a period $T_c = 4.50 \cdot 10^{-7}$ s on which are superimposed faster oscillations of small amplitude with a frequency of driving force, having a period $T_0 = 10^{-8}$ s.

The slow oscillations of the components m_x and m_y are shifted in phase relative to each other by $90°$. From a comparison of fast oscillations with dashed lines, it can be seen that the amplitude of the slow oscillations is determined by the equilibrium position of the magnetization. An additional check shows that the amplitude of the fast oscillations is determined by the amplitude of the alternating field. The magnitude of the oscillations m_x at the maximum of the slow oscillations (for example, at $t = 0.97 \cdot 10^{-7}$ s) is up equal to 0.39, down to 0.24. Considering that the equilibrium value of m_x at this point is equal to 0.34, we find that the magnetization vector in the process of fast free oscillations deviates upwards by 0.061, downwards by 0.097. The deviations m_y at the maximum of the slow-response (for example, at $t = 2.09 \cdot 10^{-7}$ s) are the same. The deviations of both components at the minimum of the emitted oscillations are also the same, but symmetrical about the Oz axis (that is, the same with the opposite sign). Thus, the magnetization vector in the process of fast oscillations approaches the Oz axis more strongly than deviates from it. Considering that the equilibrium deviation angle is $19.46°$, we get the greatest distance from the Oz axis: $23.22°$, the deviation from equilibrium is $3.76°$, the closest approximation to the Oz axis: $13.66°$, the deviation from equilibrium is $5.79°$.

From Fig. 9.2 it is seen that in this case the longitudinal component of the magnetization m_z in the stationary mode also oscillates with a constant amplitude around the stationary equilibrium value (dotted line) with a period of driving force $T_0 = 10^{-8}$ s. The amplitude of oscillations of this component is also determined by the amplitude of the alternating field. The extreme values of the component m_z are 0.97 and 0.92, and the deviations from the equilibrium value of 0.94 are 0.029 and 0.024, respectively.

From Fig. 9.2, e it is clear that with such a motion the equilibrium position of the magnetization vector precesses in a large circle, and the vector itself precesses around this precessing position in a small circle. In this case, the radius of the large circle is determined by the equilibrium deviation of the magnetization vector from the Oz axis, and the small radius is determined by the amplitude of the alternating field. The precession of the equilibrium position along a large circle occurs at a constant speed, as can be seen from the uniformity of filling with a small circle of the space of its movement, and also follows from the sinusoidal character of the slow oscillations of the transverse components of magnetization in Fig. 9.2, a. The constancy of the precession rate of the equilibrium position is due to the fact that the equilibrium position of the magnetization vector is indifferent within the cone, whose axis coincides with the Oz axis, and the angle of aperture is determined by the equilibrium deviation of the magnetization vector from this axis (here 19.46°), i.e. the magnetization vector over the surface of this cone does not change the energy of the system. The phenomenon resembles the precession of a gyroscope in a gravitational field with the difference that the role of rotation of the gyro flywheel around its own axis here is played by the precession of the magnetization vector around the equilibrium position, that is, there is a kind of precession of the equilibrium position or the second precession order. "

Comment. Within this monograph, the authors prefer to use the first of the two last names, that is, the 'precession of the equilibrium state. It should be noted, however, that the expression 'precession of the second order' also occurs quite often in the literature. However, according to the authors, the name 'precession of the equilibrium state' reflects the physical meaning of the phenomenon more specifically, which determines their appreciation.

To more clearly identify the features of the phenomenon in the right column of Fig. 9.2 for comparison, similar dependences are given for a constant field exceeding the anisotropy field of the form, that is, when $H_0 > 4\pi M_0$, when the equilibrium position of the magnetization vector coincides with the axis Oz. From Fig. 9.2, b and 9.2, d it is clear that the components m_x and m_y in a stationary mode oscillate at a frequency of the exciting force with a constant amplitude of 0.15 around the zero value, and the component m_z takes a stationary value equal to 0.99. From Fig. 9.2, f it can be seen that the precession of the magnetization vector in the stationary mode is purely circular (a thick line with a centre at zero). In this case, the angle of deviation of the magnetization vector from the Oz axis during the precession process remains constant and equal to 8.74°. Thus, it can be assumed that in a field larger than the shape anisotropy field, the character of the precession of the magnetization vector fully corresponds to the classical concepts [6–8].

9.1.3. Scheme of formation of precession of the equilibrium state

Let us now consider the general scheme of the formation of the precession of the equilibrium state at a qualitative level. This scheme is illustrated in Fig. 9.3. The left shows the orientation of the vectors of constant $\mathbf{H}_0$ and variable $\mathbf{h}$ fields. The constant field is directed along the Oz axis, the variable field has circular polarization and rotates in the Oxy plane.

In the static state, the magnetization vector $\mathbf{M}$ is oriented at a polar angle θ_0 to the axis Oz. This direction is given by the vector $\mathbf{L}$, and in the absence of a variable field, the orientation of this vector along the azimuth angle φ is arbitrary.

When the alternating field is turned on, the magnetization vector $\mathbf{M}$ starts to precess around its equilibrium position, determined by the vector $\mathbf{L}$. The precession occurs along a small ellipse with the centre at the point O'' defined by the end of the vector $\mathbf{L}$. Due to the difference in gyroscopic fields, the actual magnetization vector is in the positions of the largest and smallest distance from the axis Oz

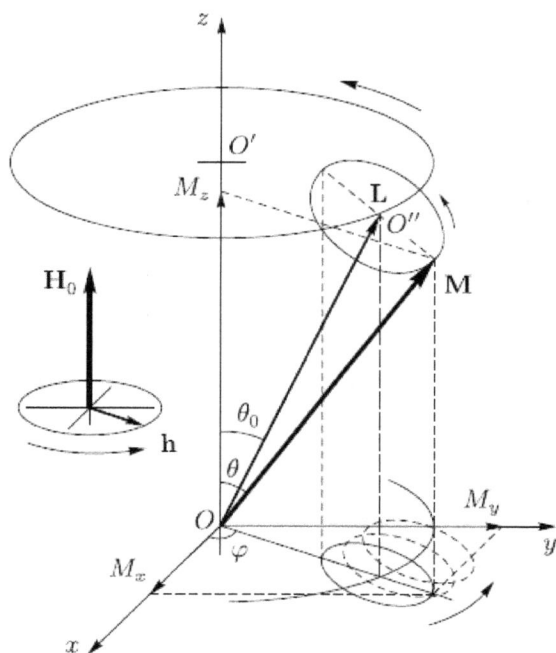

Fig. 9.3. General diagram of formation of precession by equilibrium position.

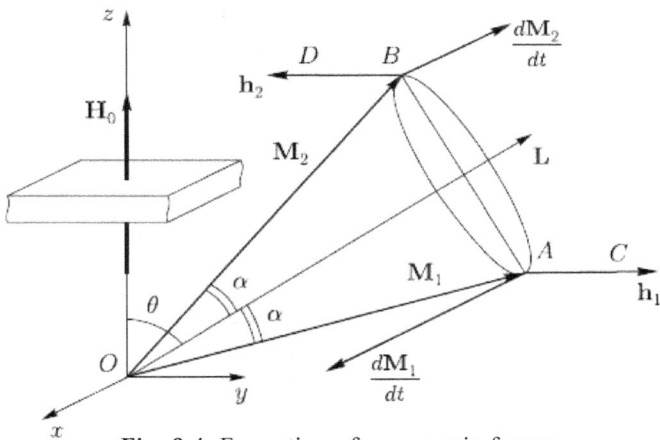

Fig. 9.4. Formation of gyroscopic forces.

along the polar angle θ, the equilibrium position shifts to the side, and the vector **L** starts to precess along a large circle with a centre at the point O', that is, the precession of the position is equal to weight.

The end of the projection of the magnetization vector **M** on the plane Oxy, having a length $\sqrt{M_x^2 + M_y^2}$ describes a small ellipse whose centre moves around point O along a circular path, and the diameter of this path is determined by the projection of the vector **L** on the plane Oxy. As an example, some consecutive positions of a small ellipse on a circular path are shown by dotted lines. Thus, on the OM_xM_y plane, a precession portrait $M_y(M_x)$ is formed, consisting of small ellipses arranged in the form of a ring along a generator of a large circle.

Due to the invariance of the length M_0 of the magnetization vector **M**, it is convenient to normalize its component: $m_x = M_x/M_0$, $m_y = M_y/M_0$, with the result that the precession portrait takes the form $m_y(m_x)$.

9.1.4. Vector model

Let us now consider the scheme of the formation of gyroscopic forces acting on the precessing magnetization vector in the position shown in Fig. 9.3. Consideration will be conducted on the basis of a vector model similar to that constructed in [396, 397].

The general scheme of the formation of gyroscopic forces is shown in Fig. 9.4. This figure on the left shows the magnetic plate and the constant magnetic field **H₀** oriented along its normal and the coordinate system $Oxyz$, whose plane Oxy is parallel to the plane of the plate. It is assumed that the constant field is less than the demagnetization field of the plate shape, that is, $H_0 < 4\pi M_0$. The equilibrium direction of the magnetization vector in this case is determined by the vector **L**, which makes the angle θ with the Oz axis, defined by formula (9.5). The variable field **h** is applied in the Oxy plane, and it is assumed that this field is circularly polarized, that is, the length of the vector **h** is kept constant. Under such excitation, due to the conservation of the length of the magnetization vector, the precession is circular, that is, the end of the vector **M** describes a circle in a plane perpendicular to the vector **L**, and the vector **M** itself moves along the surface of the cone with angle α at the vertex. The figure shows two extreme positions of the magnetization vector M_1 and M_2, where M_1 is the position furthest from the constant field vector H_0, M_2 is the least distant from the same vector. For precession buildup, the variable field must be directed outward relative to the precession cone, as shown by the corresponding vectors h_1 and h_2. According to the Landau–Lifshitz equation, time-dependent derivatives of the magnetization (rate of change of magnetization) $dM1/dt$ and dM_2/dt in extreme positions of the magnetization vector are directed in opposite directions, that is, the resulting rate of change of magnetization modulo is:

$$\left|\frac{d\mathbf{M}}{dt}\right| = \left|\frac{d\mathbf{M}_1}{dt}\right| - \left|\frac{d\mathbf{M}_2}{dt}\right|.$$

(9.14)

The right-hand side of the Landau–Lifshitz equation contains the vector product of the magnetization vector and the field vector, whose modulus is determined by the sine of the angle between these vectors. The angles between the magnetization vectors and the field in the extreme positions of the magnetization vector are different. A diagram of the formation of these angles is shown in Fig. 9.5.

Figure 9.5 a corresponds to the vectors M_1 and h_1, the angle $\angle OAC$ between which is $90° + (\theta + \alpha)$. Figure 9.5 b corresponds to the vectors M_2 and h_2, the angle $\angle OBD$ between which is $90°-(\theta - \alpha)$. Thus, the resultant rate of change of magnetization is obtained proportional

$$v_m \alpha \left|\frac{\Delta \mathbf{M}}{dt}\right| \alpha M_0 h \left\{\sin\left[90° + (\theta + \alpha)\right] - \sin\left[90° - (\theta - \alpha)\right]\right\}.$$

(9.15)

Transforming this expression, we get:

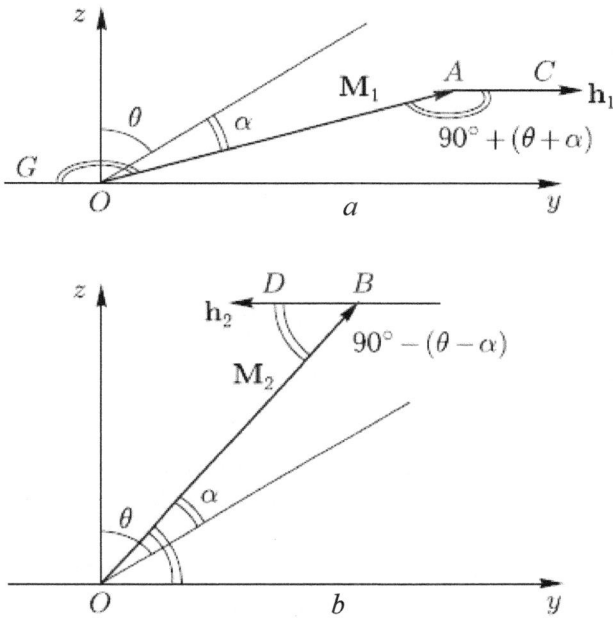

Fig. 9.5. Angles between magnetization vectors and alternating field

$$v_m \infty M_0 h\left[\cos(\theta+\alpha)-\cos(\theta-a)\right]=-2M_0 h\sin\theta\sin\alpha. \quad (9.16)$$

From the general precession, with small α values, at least in the linear approximation [6–8], it follows that: $\alpha \sim M_{x,y}\, h$. Substituting this expression into the formula for speed and omitting a factor proportional to the magnetization (since it is determined by the properties of the static equilibrium magnetization), we obtain

$$v_m \infty h^2 \sin\theta. \quad (9.17)$$

The fact that the velocity v_m is nonzero can be regarded as confirming that the entire cone of the precession is displaced as a whole in the direction perpendicular to the plane passing through the vector of the constant field **H**$_0$ and the vector of the static position of magnetization **L**. This means that the equilibrium position of the magnetization vector precesses around the direction of the constant field.

The precession period, as the time for a complete rotation of a cone around the direction of a constant field, is inversely proportional

454

to the velocity v_m, that is, up to a constant factor A, depending on the material parameters, is determined by the formula

$$T = \frac{A}{h^2 \sin \theta}.$$
(9.18)

Given (9.5), we get:

$$T = \frac{A}{h^2 \sqrt{1 - \left(\dfrac{H_0}{4\pi M_0}\right)}}.$$
(9.19)

From the review we can see that the vector model satisfactorily explains the main features of the motion of the vector magnetization shown in the left column of Fig. 9.2, namely: the participation of the transverse m_x and m_y components in two motions – fast in a small circle with a frequency of an alternating field and slow in a large circle with a much lower frequency, and oscillations of the component m_z with a frequency of an alternating field. At the same time, due to the preservation of the length of the magnetization vector, the slow oscillations m_x and m_y are shifted in phase relative to each other by 90°. It is also seen that the amplitude of the slow oscillations m_x and m_y is determined by the equilibrium value of the magnetization in a constant field, and the amplitude of the fast values is proportional to the amplitude of the alternating field.

It can be seen that the precession of the equilibrium position is non-threshold, that is, it occurs at any arbitrarily small amplitude of the alternating field, but when this amplitude tends to zero, the precession period tends to infinity, that is, the possibility of observing it is determined by the duration of the experiment.

9.2. The main properties of the precession equilibrium positions

Let us now consider the main properties of the precession of the equilibrium position by the method of machine simulation, for which we will numerically solve the system of equations (9.6)–(9.8) for various parameters of the problem.

9.2.1. Period dependences on constant and variable fields

We first consider the dependence of the precession period of the equilibrium position on the values of the constant and variable fields. Some of the results are illustrated in Figs. 9.6 and 9.7.

Figure 9.6 shows the dependences of the precession period of the equilibrium position on the amplitude of the alternating field $h_{0x,y}$ for different values of the stationary field H_0, calculated by formula (9.19) (lines) and obtained by computer experiment (points). The inset points show the values of the parameter A for different values of the constant field, obtained from the lengths of the precession period of the equilibrium position measured during a computer experiment. The solid horizontal line corresponds to the value $A = 1.45$, which is obtained as the arithmetic average of the measured values of A in the field interval from 0 to 240 Oe. The obtained value of $A = 1.45$ is used in the construction of curves 1–4. It can be seen from the figure that the curves 1–4 coincide well with the experimental points. Curve 5, corresponding to the field 272 Oe, was built with $A = 1.03$. Such a choice of the value of the parameter A is necessary because at $A = 1.45$ and the field 272 Oe, the calculated

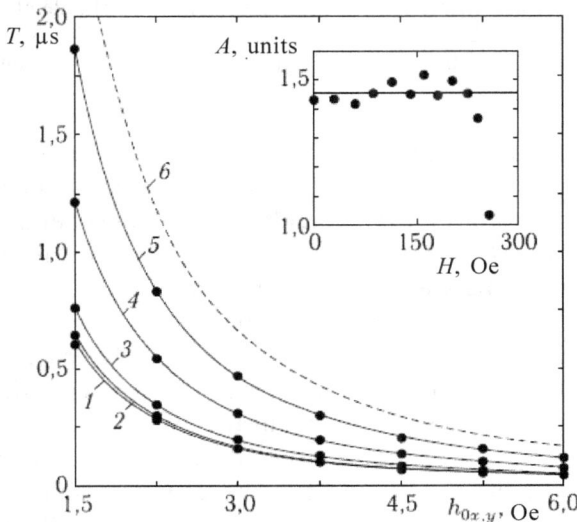

Fig. 9.6. Dependences of the precession period of the equilibrium position on the amplitude of the alternating field for different values of the constant field H_0: 1 – 0 Oe; 2 – 80 Oe; 3 – 160 Oe; 4 – 240 Oe; 5 – 272 Oe; 6 – 272 Oe. Parameters: $4\pi M_0 = 280$ G; $\alpha = 0.3$; $f = 108$ Hz; $h_{0x} = h_{0y} = 3$ Oe

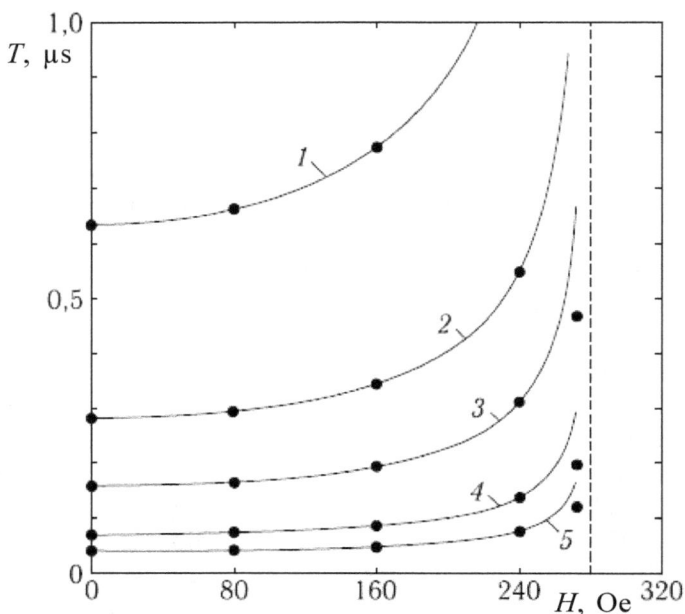

Fig. 9.7. Dependences of the precession period of the equilibrium position on the magnitude of the constant field for different values of the amplitude of the alternating field $h_{0x,y}$: 1 – 1.50 Oe; 2 – 2.25 Oe; 3 – 3.00 Oe; 4 – 4.50 Oe; 5 – 6.00 Oe. The remaining parameters are the same as those taken in the construction of Fig. 9.6

dependence goes higher than the experimental results, as shown by curve 6. The corresponding value of the parameter A, which requires that formula (9.19) coincides with experiment, is noted by the rightmost point at the bottom of the sidebar in Fig. 9.6.

The agreement between the experimental and calculated curves suggests that all dependences of the period on the alternating field decrease inversely with the square of this field, as follows from the general structure of formula (9.19). However, this formula describes quite well (with an accuracy of no less than 5%) the dependence of the precession period of the equilibrium position on the amplitude of the alternating field in constant fields, up to 0.86, on the form anisotropy field ($4\pi M_0 = 280$ Oe), and then, reflecting the same accuracy the overall course of the curves, gives a somewhat overestimated absolute values.

Figure 9.7 shows the dependences of the precession period of the equilibrium position on the magnitude of the constant field for different values of the permeable field $h_{0x,y}$, calculated by formula (9.19) (lines) and obtained by computer experiment (points). All calculated curves are plotted at $A = 1.45$, which corresponds to the dependences shown in Fig. 9.6. The points at $H_0 = 272$ Oe, obtained by experiment, satisfy the formula (9.19) with $A = 1.03$. The vertical dashed line corresponds to $H_0 = 280$ Oe (i.e. the field shape anisotropy).

It can be seen from the figure that all dependences of the period on the constant field increase the faster, the closer H_0 approaches the shape anisotropy field. Such a behaviour completely corresponds to the inverse dependence of the deviation angle of the magnetization vector from a constant field from the sine. At the same time, formula (9.19) describes the dependence of the precession period of the equilibrium position on the constant field also quite well (no worse than 5%) in technical constant fields (to 0.86 on the form anisotropy field), and then gives values, also somewhat overestimated (as can be seen from the curves 3–5).

9.2.2. The ratio of the driving frequency with its own

We now consider the more subtle features of the precession of the equilibrium state, which go beyond the vector model.

First of all, let us pay attention to the ratio of the frequencies of the driving force and the intrinsic precession of the magnetization vector in the absence of excitation. If the excitation frequency is more than 20–30% higher than the eigenfrequency frequency, then with a small amplitude of the alternating field, the precession period of the equilibrium position is inversely proportional to the square of this amplitude, in complete agreement with the vector model. However, if the amplitude of the excitation increases, then when a certain threshold is reached, the precession of the equilibrium state is disrupted and it is replaced with a simple circular precession of the magnetization vector with a large amplitude around the direction of the constant field. The breakdown threshold of the precession of the equilibrium position in the amplitude of the alternating field is quite sharp and is less than 1–2% in the amplitude of the alternating field. If we now fix the amplitude of the alternating field and change its frequency, starting from the lowest frequencies, then first there is the usual precession with a large amplitude, but when a certain critical frequency is reached, the usual precession is replaced by the precession of the equilibrium position, and the threshold in frequency is also quite sharp (units MHz).

The reason for this behaviour of the magnetization vector is as follows. Suppose that the frequency of the alternating field is higher than the resonant frequency of the linear intrinsic precession of magnetization. Then, at a small amplitude of excitation, oscillations of the magnetization vector are linear and, since the constant field is less than the demagnetization field of the plate, the precession of the equilibrium state is excited. With an increase in the amplitude of the alternating field, the resonant curve of forced oscillations of magnetization due to the destructing mechanism expands and leans towards high frequencies. At the same time, its high-frequency edge becomes sharp, acquires an abrupt appearance, and shifts upward with increasing field. When the sharp edge of the resonance curve reaches the frequency of the alternating field, this frequency falls into resonance and conditions are created for the excitation of a simple resonant circular precession of magnetization. The amplitude of this precession can exceed the equilibrium deviation of the magnetization from the constant field. In this case, the precession of the equilibrium

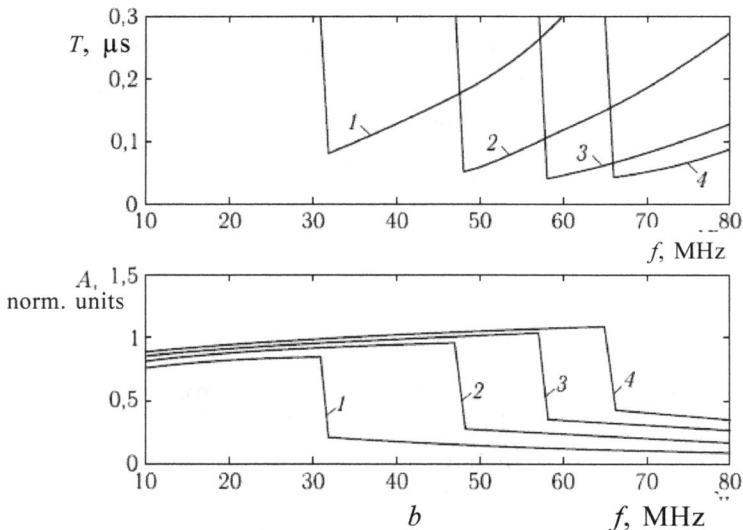

Fig. 9.8. Dependences of the precession period of the equilibrium position (a) and the magnitude of the oscillations of the magnetization (b) on the frequency of the alternating field for different values of its amplitude $h_{0x,y}$: 1 – 1.5 Oe; 2 – 3.Oe; 3 – 4.5 Oe; 4 – 6.0 Oe. Parameters: $4\pi M_0 = 280$ G, $H_0 = 265$ Oe, $\alpha = 0.3$

position is broken and replaced by a simple purely circular precession of magnetization with a large amplitude.

This is illustrated in Fig. 9.8, which shows the dependences of the period of oscillations of the equilibrium position (*a*) and the magnitude of the oscillations of the magnetization (*b*) on the frequency of the alternating field at various values. The oscillation range considered here is the sum of the maximal declination of the magnetization with the frequency of the exciting force in the *Oxy* plane in both directions from the precessing or stationary equilibrium position. Since these deviations differ in magnitude, it makes sense to consider the amplitude of oscillations instead of the traditionally used amplitude. In a certain approximation, it can be assumed that in the absence of the precession of the equilibrium position, this range is governed by the diameter of the magnetization movement around the direction of the constant field. When the precession of the position takes place, the scale is predetermined by the diameter of the precession around the precessing equilibrium position. That is, for example, in Fig. 9.2, and the first swing corresponds to the amplitude of the large sinusoid, and the second to the small one.

The curves in Fig. 9.8, and below a certain critical frequency F_c, different for each curve, go to infinity, which means a steady state of equilibrium. That is, in these cases there is no equilibrium at $F < F_c$, and instead there is the usual circular precession with an equilibrium position of the magnetization along the field. Above the critical frequency, the equilibrium position precesses, and its precession period increases with increasing frequency.

It can be seen from the figure that the value of the critical frequency is the higher, the larger the amplitude of the alternating field.

The curves in Fig. 9.8 b, to a certain extent, represent the resonant characteristics of forced oscillations of magnetization in a nonlinear mode. It can be seen that these resonance characteristics on the side of high frequencies have a sharp decrease due to the tuning mechanism for limiting the amplitude of nonlinear oscillations [6–8, 398]. Although I must say that the detuning here is manifested not so much in a simple change in frequency as in replacing the quiet precession with a vortex-like, corresponding to the second order precession. The figure shows that the decay frequency is the higher, the larger the amplitude of the alternating field.

Comparison of Figs. 9.8, a and b shows that the frequency at which the normal precession replaces the precession of the equilibrium position (Fig. 9.8, a) exactly coincides with the decay frequency of the resonance curve of forced oscillations (Fig. 9.8, b), which confirms the described above picture of the breakdown of the precession of the equilibrium state with an increase in the amplitude of the alternating field or with a decrease in its frequency.

Comment. It should be noted that the thresholds for the breakdown of the precession with a change in both the amplitude and frequency of the alternating field are not infinitely sharp. In the nearest neighbourhood of both thresholds, a transitional regime is observed, in which the equilibrium position performs complex, irregular, close to chaotic movements, abruptly changing from one oscillation state to another. Both in amplitude and frequency, such a neighbourhood is no more than 2–3% of the threshold value itself.

9.2.3. Critical excitation frequency

From the above consideration it follows that the excitation of the precession of equilibrium takes place only when the excitation frequency exceeds a certain critical value, determined by the minimum of the curves in Fig. 9.8, a. It can be seen that as the level of excitation increases, the critical frequency value increases (this is shown further in Fig. 9.10 and described by formula (9.20)).

Thus, if at a given level the frequency of the excitation is below critical, then there is a simple circular precession of the magnetization around the direction of the constant field along a cone-shaped trajectory. In this case, the equilibrium position always remains in place, in full accordance with the direction of the constant field. The amplitude of the precession can be limited by the tuning mechanism, as was noted when considering the left curves of the curves in Fig. 9.8, b.

If at the same excitation level the excitation frequency exceeds the critical one, then such an even calm precession movement of the magnetization along the cone around the stationary equilibrium state is disturbed, the equilibrium position itself begins to precess over its own cone, and the magnetization vector continues to describe its primary cone equilibrium position.

That is, the stationary motion of the magnetization vector undergoes a perturbation of the vortex nature, which is manifested as the appearance of small sheep around a large one in the precession portrait (Fig. 9.2 e. Note that the existence of such a disturbance is determined by both the amplitude of the exciting signal and its frequency, which at a given level must be higher than the critical one. However, for the amplitude of excitation, the critical nature of its level is not visible here. With a small amplitude of excitation, the precession period of the equilibrium state tends to infinity (as noted when considering formula (9.19) and can be seen from Fig. 9.6), that is, the possibility of observing such a precession is determined by the actual experiment duration available to the experimenter. Let us consider some more details on the qualitative side of the formation of the critical frequency value. From Fig. 9.8, a it can be seen that the region of existence of the precession of the equilibrium position with increasing frequency shifts towards large amplitudes of the alternating field.

Such a shift can be explained by the fact that the existence of a precession of the equilibrium position is determined by the motion of the magnetization vector along the energy surface, that is, it depends only on the amplitude of oscillations of the magnetization vector, but not on the frequency. It can be assumed that as the frequency increases, the magnetization vector does not have time to follow the alternating field, therefore, to achieve the same oscillation range, it is required to increase the amplitude of the alternating field. That is, to achieve the same range and a higher frequency requires a greater amplitude of the alternating field. It can be assumed that the inverse relationship holds. That is, at a fixed amplitude of the alternating field, the magnetization vector at a low excitation frequency quite well manages to follow exactly behind a rotating alternating field, forming a regular cone with smooth walls, corresponding to a large precession ring. However, as the frequency increases, the magnetization vector, due to its inertia, begins to lag behind the rotating field, and the lag becomes larger, the higher the frequency. Finally, the frequency becomes so high that the magnetization vector does not have time to complete even half a full revolution during the period of the variable field, so that even before this half passes, the alternating field again starts to pull it in the opposite direction, returning to the previous position. Under the action of such a thrust, the vector of magnetization makes an incomplete coil that does not capture the axis corresponding to the direction of the constant field. Such a coil is manifested in the precessional portrait as a small ring on the periphery of a large one.

Due to the cyclical nature of the action of the variable field, the first small coil is followed by the second, then the third, and so on. The magnetization vector still retains a certain tendency to move along a large ring, that is, each 'jerk' of an alternating field still pushes him to some degree. However, such translational motion now occurs slowly, since the main role is acquired by the vortex-like motion of the magnetization vector along small turns following each other. These small turns give small rings on the precession portrait, and a large ring is formed due to the now slow motion of the magnetization vector along that cone, the solution of which is determined by the deviation of the equilibrium position of the magnetization from the direction of the constant field. Thus, the frequency at which the magnetization vector during the period of change of the variable field does not have time to go through the large cone half of the full circle of precession, so that forms a kind of "twist" in its movement, is just the critical frequency at which simple precession turns into precession of equilibrium.

With an increase in the amplitude of the alternating field, it to some extent overcomes the inertia of the magnetization vector and 'drives' it more successfully over a large circle, as a result of which the vector lags behind the rotating field decreases and the twist that started to form disappears. That is, now the formation of turbulence requires a faster rotation of the field, that is, a higher frequency.

Thus, the critical frequency value with increasing field amplitude also increases, which reflects the shift of the curves to the right side according to Fig. 9.8, a. The fall in the swinging range observed in Fig. 9.8, b is due to the fact that, before the precession of the equilibrium state is excited, this range is the diameter of the large rings of the precession portrait, and after the excitation it is already the diameter of the small ones.

Comment. It should be noted that the analytical analysis of the critical frequency of excitation in the framework of the models presented here seems to require more in-depth improvement or even the creation of some new models that the authors of this monograph leave as a subject for future research. Here we will further consider other properties of the precession of the equilibrium state under the assumption that the excitation frequency will certainly exceed its critical value.

9.2.4. The effect of dissipation on the precession parameters of equilibrium positions

Let us now consider the influence of the dissipation of magnetization oscillations on the main characteristics of the precession of the

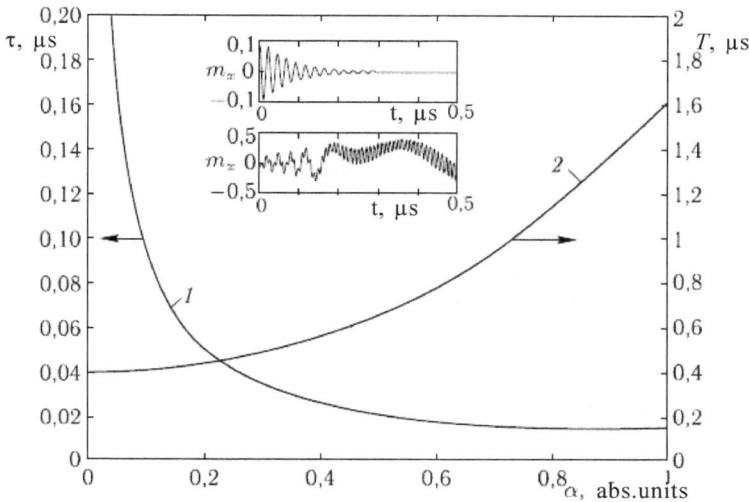

Fig. 9.9. Dependences of the decay time of natural oscillations τ (curve 1, left scale) and the precession period of the equilibrium state T (curve 2, right scale) on the value of the damping parameter α. Parameters: $4\pi M_0 = 280$ G, $H_0 = 265$ Oe, $h_{0x,y} = 3$ Oe, $f = 100$ MHz. The inset image is explained in the text.

equilibrium state. Turn to Fig. 9.9, which show the dependences of the decay time of the characteristic oscillations τ and the precession period of the equilibrium state T on the value of the attenuation parameter α, which is included in the equations (9.6)–(9.8).

The insert to this figure illustrates the scheme for determining the decay time of the natural oscillations of magnetization. The dependences of the magnetization component m_x on time are shown (the dependence for m_y is analogous with a phase shift). The upper picture of the inset corresponds to the natural oscillations of the magnetization at $H_0 = 295$ Oe, that is, in the case of $H_0 > 4\pi M_0$ (where $4\pi M_0 = 280$ Oe). The lower sidebar picture corresponds to the forced oscillations of the magnetization at $H_0 = 265$ Oe, that is, in the case of $H_0 < 4\pi M_0$. It can be seen that in both of the rays, the natural oscillations cease at a time $t \approx 0.2$ μs, after which in the first case, the magnetization is established along the direction of the field, and in the second, the precession of the equilibrium position begins.

It can be seen from the main figure that the precession period of the equilibrium position for values of attenuation parameter α less than 0.2 does not depend on the value of this parameter, and with further increase in attenuation parameter it increases the faster, the larger the attenuation parameter. At the same time, the damping time of the natural oscillations, as the attenuation parameter increases with its values less than 0.2, decreases sharply, after which the decline slows down and, starting with the value of the parameter 0.6, enters the horizontal section. The test showed that, starting with $\alpha \approx 0.2$, the own precession performs no more than one full turn, and starting with $\alpha \approx 0.4$, the precession becomes completely aperiodic, that is, in the process of self-oscillation, the signs of the components of magnetization $m_{x,y}$ do not change. Thus, it can be seen that the precession period of the equilibrium position with a small attenuation ($\alpha < 0.2$), when the precession is periodic, does not depend on the value of the attenuation parameter, but with increasing attenuation ($\alpha > 0.2.. 0.4$), when the precession becomes aperiodic, the more rapidly the larger the attenuation parameter increases. This behaviour is analogous to an increase in the period of a simple oscillatory system with increasing attenuation, that is, the more difficult it becomes for the pendulum to move, the slower it moves [398–400].

9.2.5. Effect of magnetization on the parameters of the precession equilibrium positions

Let us now consider how the properties of the precession of the equilibrium state depend on such an important parameter of the medium as the saturation magnetization M_0.

All the results described above relate to the magnetization equal to $M_0 = 22$ G ($4\pi M_0 = 280$ G), which is close to that of the terbium ferrite garnet (TBFG) [259]. This choice was due to the convenience of machine counting in real time and the relatively small amount of machine memory required. However, from the applied side, the iron yttrium garnet (IYG) seems to be the most interesting, for which $M_0 = 139.26$ G ($4\pi M_0 = 1750$ G) [6–8.259,296]. Obviously, when the magnetization changes, the resonance frequency of the own oscillations also changes, which should lead to a change in the ratio between the critical values of the frequency and amplitude of the alternating field necessary for the transition from ordinary circular precession to precession of the equilibrium position.

The obtained dependences of the critical frequency f_c on the amplitude of the alternating field $h_{0x,y}$ for different saturation magnitudes M_0 are shown in Fig. 9.10. When building these lines, H_0 was chosen such that the equilibrium angle of deviation of the magnetization from the Oz axis was the same, equal to $19.1° \pm 0.4°$. The parameters of curve 1 are close to those for TBFG, curve 4 is for IYG. For each curve, above it is the region where the precession of the equilibrium state is excited, below — the region with the usual circular precession.

It can be seen from the figure that the critical frequency with increasing amplitude of the alternating field increases the stronger, the larger the saturation magnetization. It can be noted that the dependences of the critical frequency on the amplitude of the alternating field with an accuracy of about 10% are proportional to the square root of the amplitude of the variable field:

$$f_c \infty \sqrt{h_{0x,y}}. \tag{9.20}$$

The curves shown in Fig. 9.10 are plotted with the damping parameter $\alpha = 0.3$. However, for single-crystal IYG, this parameter can be more than an order of magnitude less. Additional research has shown that when the attenuation parameter α changes, the critical frequency f_c also changes somewhat. To illustrate this effect, the inset shows the dependence of the critical frequency on the attenuation parameter at $M_0 = 140$ G ($4\pi M_0 = 1760$ Gs), $H_0 = 1660$ Oe, $h_{0x,y} = 10$ Oe. As the α tends to zero, the critical frequency takes a constant value, and with increasing α decreases the faster, the more α. So, for $\alpha = 0.01$, the critical frequency for the same value of $h_{0x,y}$ is 267 MHz, for $\alpha = 0.10$ it is 260 MHz, and for $\alpha = 0.40$ 170 MHz.

9.3. Multimode precession character of equilibrium positions

In the process of studying the precession of the equilibrium state, it was found that as the amplitude of the variable field h_0 increases, five fundamentally different precession modes are observed successively: low-amplitude circular precession, precession of the equilibrium state without centre coverage, continuous precession of the equilibrium state with centre coverage, damped precession of the state equilibrium with centre coverage, unfolded circular precession.

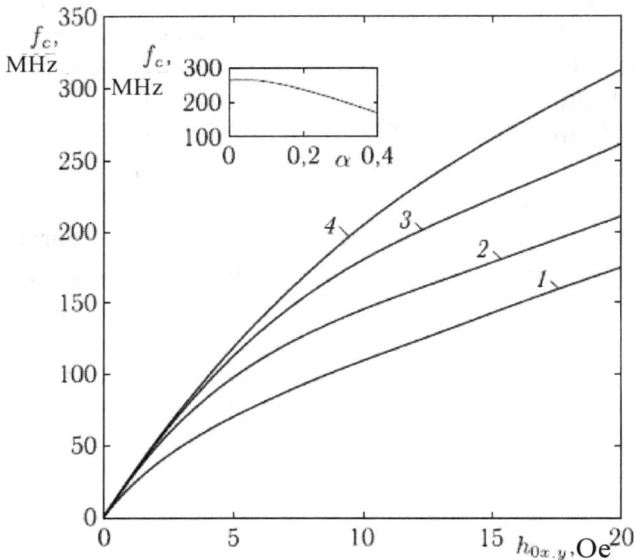

Fig. 9.10. Dependences of the critical frequency, above which the precession of the equilibrium state is observed, on the amplitude of the alternating field at different values of saturation and constant field magnetization: $1 - M_0 = 20$ G ($4\pi M_0 = 251$ Gs), $H_0 = 237$ Oe; $2 - M_0 = 60$ G ($4\pi M_0 = 754$ G), $H_0 = 711$ Oe; $3 - M_0 = 100$ G ($4\pi M_0 = 1257$ G), $H_0 = 1186$ Oe; $4 - M_0 = 140$ G ($4\pi M_0 = 1760$ G), $H_0 = 1660$ Oe. Damping parameter: $\alpha = 0.3$. The inset image is explained in the text.

9.3.1. The development of fluctuations in time and precession por-portraits for different modes

The development of oscillations in time and the corresponding precession portraits m_y (m_x) for all five modes are illustrated in Figs. 9.11–9.15, at construction of which the material parameters typical for terbium ferrite garnet were used. The constant field providing the conditions of the orientational transition was 20 Oe below the demagnetization field. The amplitude of the alternating field of circular polarization h_0 varied from zero to 100 Oe, its frequency was higher than the critical one and was $f = 200$ MHz. The figures refer to the simplest case, symmetrical about the Oz axis, when the constant field is directed exactly perpendicular to the plane of the

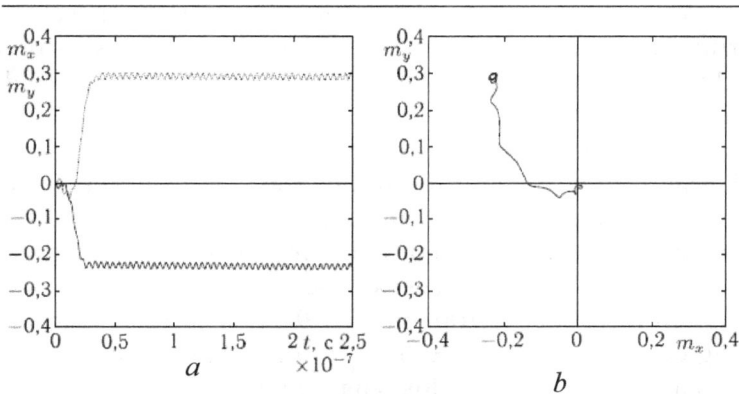

Fig. 9.11. The development of oscillations in time (*a*) and the precession portrait (*b*) with the amplitude of the exciting field $h_0 = 1$ Oe. Mode No. 1 – low-amplitude circular precession. Parameters: $4\pi M_0 = 280$ Gs, $\alpha = 1,2$; $H_{0z} = 260$ Oe; $f = 200$ MHz.

plate, that is, $H_{0x,y} = 0$ Oe.

On all Figs. 9.11–9.15, the letter '*a*' (left figure) denotes the development in time of oscillation of the magnetization components $m_x(t)$ (solid lines) and $m_y(t)$ (dotted lines). In this case, the fast oscillations reflect the precession of the magnetization vector with the frequency of the alternating field, the slow ones reflect the precession of the equilibrium position, the frequency of which is significantly lower. On the technical drawings, the '*b*' letter (right figure) indicates the corresponding precession portraits m_y (m_x), built in the same time interval. In Fig. 9.11 the scale of the components of the magnetization m_x and m_y is enlarged for the purpose of clarity.

Consider the different precession modes separately, following Figs. 9.11–9.15.

Mode number 1 – low-amplitude circular precession (Fig. 9.11). This mode is the limiting case of the precession of the magnetization with the amplitude of the alternating field approaching zero as $h_0 \to 0$. Figure 9.11 was plotted at $h_0 = 1$ Oe (at $h_0 \ll H_0$)

The magnetization vector precesses in a small circle around the equilibrium position, which is determined by the end of the establishment process with a relaxation time of about $0.4 \cdot 10^{-7}$ s. The resulting orientation of the equilibrium position at the azimuth angle is determined by the initial conditions and can be arbitrary. In the absence of a constant field component in the plate plane, this orientation is degenerate, that is, not fixed at all, therefore the equilibrium position of the magnetization vector begins to precess at an arbitrarily small amplitude of the alternating field, and the precession period of the equilibrium position as $h_0 \to 0$ tends to infinity. For a visual illustration of this mode when building Fig. 9.11, the amplitude of the alternating field was chosen so small (1 Oe) that during the observation time ($2.5 \cdot 10^{-7}$ s) the ring in the precession portrait did not have time to shift by a noticeable amount (the equilibrium precession period at $h_0 = 1$ Oe is $1.4 \cdot 10^{-4}$ s). Therefore, the portrait in Fig. 9.11, b is represented by a small ring centreed at $m_x = -0.23$, $m_y = 0.29$. After a sufficient time (commensurate with 10^{-4} s), this small ring moves in a circle with a radius of $m_s = \sqrt{m_x^2 + m_y^2} = 0.37$, which corresponds to mode No. 2. Thus, the mode under consideration is unstable and represents a limiting case of mode No. 2 as $h_0 \to 0$.

However, in the more complex asymmetric case, when the constant field has a component in the plane of the plate, the degeneracy of the orientation of the equilibrium position of the magnetization is removed, and in the absence of a variable field the orientation of the magnetization vector becomes quite definite. Due to this, the mode of low-amplitude circular precession No. 1 ceases to be the limiting case of mode No. 2, becomes stable and is realized in a certain non-zero interval of amplitudes of the alternating field, the larger, the larger the constant field component in the plane of the plate.

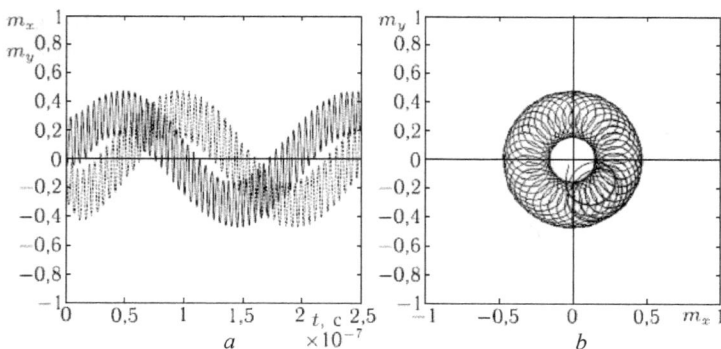

Fig. 9.12. The development of oscillations in time (*a*) and the precession portrait (*b*) with the amplitude of the exciting field $h_0 = 20$ Oe. Mode No. 2 – precession of the equilibrium position without centre coverage Parameters – the same as taken in the construction of Fig. 9.11

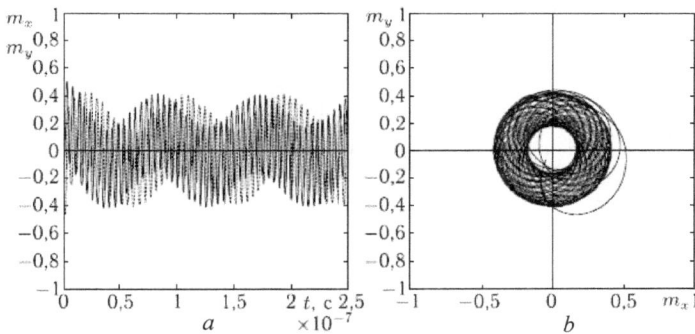

Fig. 9.13. The development of oscillations in time (*a*) and the precession portrait (*b*) with the amplitude of the exciting field $h_0 = 35$ Oe. Mode number 3 - the continuous precession of the equilibrium position with centre coverage. Parameters - the same as taken in the construction of Fig. 9.1

Mode 2 – precession of the equilibrium position without centre coverage (Fig. 9.12). This mode is realized with the amplitude of the alternating field in the interval: $0 < h_0 < 28$ Oe. Figure 9.12 corresponds to $h_0 = 20$ Oe.

The equilibrium position precesses around the field direction, the middle line of fast magnetization oscillations periodically oscillates around the zero line, and with maximum extensions, no rapid oscillations of the zero line reach. The precession portrait has the form of a wide circular ring filled with small rings, the radii of which are smaller than the outer radius of the main ring. A trajectory-free empty core of the main ring, formed by a tangent to the small rings, is located outside the small rings.

Mode number 3 – the precession of the equilibrium position with the centre coverage (Fig. 9.13). This mode is implemented with the amplitude of the alternating field in the interval: 28 Oe $< h_0 < 37$ Oe. Figure 9.13 corresponds to $h_0 = 35$ Oe.

The amplitude of precession oscillations is stationary in time. The equilibrium position precesses around the direction of the constant field, the middle line of fast magnetization oscillations periodically oscillates around the zero line, and the fast oscillations always cross the zero line. The precession portrait has the form of a wide main circular ring filled with small rings forming the radii of which are more than the inner radius of the main ring. The trajectory-free unfilled core of the main ring, formed by a tangent to small rings, is located inside small rings.

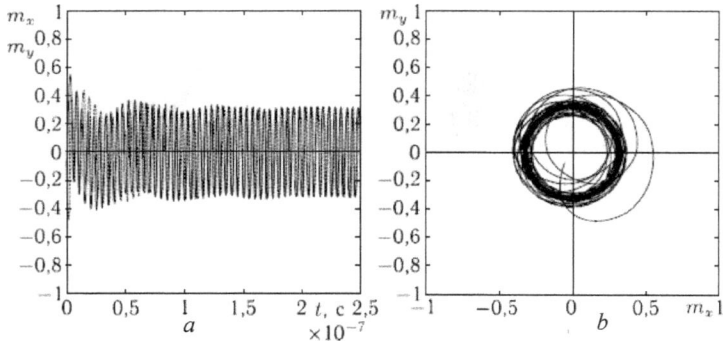

Fig. 9.14. Development of oscillations in time (a) and precessional portrait (b) at the amplitude of the exciting field $h_0 = 38$ Oe. Mode No. 4 – damped precession of the equilibrium position with coverage of the centre. The parameters are the same as taken when constructing Figure 9.11

Mode number 4 – precession of equilibrium with coverage centre, the amplitude of oscillations of which fades in time (Fig. 9.14).

This mode is implemented with the amplitude of the alternating field in the interval: 37 Oe $< h_0 <$ 43 Oe. Figure 9.14 corresponds to $h_0 = 38$ Oe.

The equilibrium position at the initial moment of time after switching on the alternating field first precesses with attenuation, after which it gradually stops, the middle line of fast magnetization oscillations coincides with the zero line. The precession portrait of the oscillation stage has the form of a wide circular ring filled along the generator with small rings whose radii are larger than the inner radius of the main ring and approach it, with the result that the main ring has a kind of smeared form. Over time, the width of the smeared generatrix of the main ring decreases and the precession portrait tends to a stationary state in the form of a large main ring with a small smearing. Trajectory-free unfilled core of the main ring, formed by a tangent to small rings, is located inside small rings.

Mode No. 5 is the unfolded circular precession of the magnetization vector (Fig. 9.15). This mode is implemented with the amplitude of the variable field in the interval: 43 Oe $< h_0 <\infty$. Figure 9.15 corresponds to $h_0 = 70$ Oe.

The equilibrium position is stationary and oriented along the Oz axis, the middle line of fast oscillations coincides with the zero line. The precession portrait has the form of a narrow circular ring, the width of the generator of which tends to zero, and the

$n_0 = 70\ \text{Э}.$

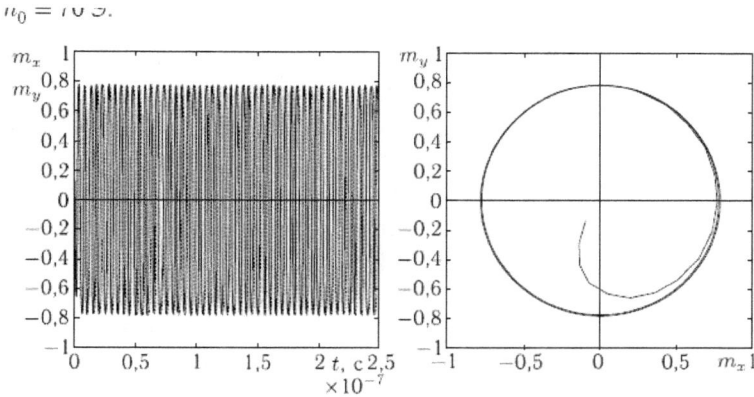

Fig. 9.15. The development of oscillations in time (*a*) and the precession portrait (*b*) with the amplitude of the exciting field $h_0 = 70$ Oe. Mode No. 5 is the unfolded circular precession of the magnetization vector. Parameters are the same as taken in the construction of Fig. 9.11

diameter smoothly increases as the amplitude of the alternating field increases. The limiting case of this mode, realized with amplitudes of a variable field of more than 150–200 Oe, is a fully developed circular precession in which the magnetization vector precesses in a circle near the plane of the plate.

9.3.2. Some features of individual modes

The main results described in the previous sections 9.1 and 9.2 relate to the modes 2 and 3, and the difference between them, although mentioned, is not considered in detail, mode 4 is not considered at all, and mode 5 is characterized as 'disruption precession of equilibrium'.

Add a small note to more accurately describe the relationship between the modes No. 2 and No. 3. In both cases, the recession portrait is a large ring filled with a spiral formed by successively shifting small rings. Depending on the ratio of the amplitudes of the precession of the equilibrium position and the precession of the magnetization vector, two different cases are possible:

a – the diameter of the small rings is smaller than the radius of the large ring, while the inner side of the large ring touches the small rings from the outside;

b – the diameter of the small rings is larger than the radius of the large ring, while the inner side of the large ring touches the small rings from the inside. In the case of a precession portrait, in the middle of a large ring there is a round unfilled core, which, with the exact equality of the diameter of the small rings to the radius of the large ring, degenerates into a point. The expressions 'without centre coverage' and 'with centre coverage' are introduced here for convenience, but there is little fundamental difference between these modes. That is, to a large extent, the difference between the modes corresponds to the difference in the amplitudes of the precession along small rings, while the physics of the process remains the same. Therefore, the transition from one mode to another when the amplitude of the alternating field changes smoothly, there are no jumps and hysteresis. This transition is explained in more detail later in the section on the potential model.

9.4. Mechanical analogy of precession equilibrium positions

The above material allows us already in general to understand the basic properties of the precession of the equilibrium position. At first glance, the phenomenon from the physical side seems somewhat unusual, so we make a slight digression and turn to the interpretation of a similar mechanical model, which allows us to understand the basic properties of precession on an intuitive level.

9.4.1. Flat pendulum with side spring suspension

We first consider a mechanical pendulum, which is a load suspended on a rigid rod, one end of which is hinged. The scheme of the pendulum is shown in Fig. 9.16. The load is indicated by the letter M, the rigid rod is fixed at point A with the possibility of swinging in one plane. The coordinate system Oxz is chosen in this plane and oriented in such a way that the axis Oz coincides with the direction of the action of gravity. A spring hanger is attached to the load, the other end of which is pivotally fixed at point B, located above point A, with line AB parallel to axis Oz.

In the free state, two forces act on the pendulum: the gravitational force $\mathbf{P}$, directed downward along the Oz axis and the tension force of the spring hanger $\mathbf{T}$, directed upward along the line connecting the points B and M, the rest of the pendulum causes its deviation from the axis Oz by an angle θ. Due to the symmetry of the geometry of the system, there are two symmetric, relative to the Oz axis, stable equilibrium positions – at the points M and M' (the second is shown

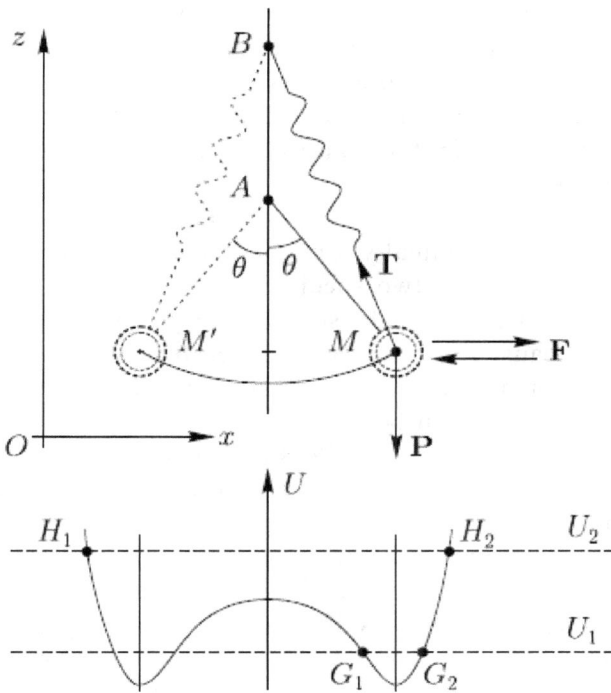

Fig.9.16. Flat mechanical pendulum with lateral spring suspension.

on the left by the dotted line). Such a deviation of the axis of the pendulum from the direction of gravity is similar to the deviation of the magnetization vector from the anisotropy axis under the action of a field perpendicular to this axis. That is, the situation from the physical side is completely similar to the state of the orientational transition (section 5.1).

In the process of oscillation of the pendulum in the transition from position M to position M' must pass through the orientation of the rod AM, parallel to the axis Oz, that is, the suspension spring BM at this intersection should stretch a little. That is, the force **T** in this case increases slightly. This stretching of the spring causes an increase in its potential energy. So the potential U of the system as a whole, depending on the x coordinate, has the form shown below in the same figure. Thus, to the usual increase in potential on both sides with distance from the axis of symmetry of the system, associated with lifting the load M in the field of gravity, a central maximum is added, due to the additional tension of the suspension spring. So the potential takes on the form of two minimum, located symmetrically from the central maximum. At rest, the pendulum occupies one of two possible equilibrium positions, so that the value of the potential corresponds to one of two minimums.

We now assume that the force **F** acts on the pendulum in the direction of the Ox axis, the sign of which periodically changes in time. This force pushes the pendulum out of one or another position of stable equilibrium, so that the potential in time with the action of force increases periodically. Such a force, depending on its size, can cause oscillations of two species.

First, if the force is weak, so that the periodic growth of the potential level caused by it does not reach the height of the central maximum, such as the level U_1 shown in the figure, then oscillations occur in the potential well between points G_1 and G_2. The amplitude of such oscillations remains quite small and the system oscillates only around one of the only two possible equilibrium positions. Secondly, if the force is so great that the potential level exceeds the height of the central maximum, such as, for example, U_2, then oscillations occur in a wider potential well between points H_1 and H_2. The amplitude of such oscillations is much larger than in the first case, so that the system varies from one equilibrium position to another.

Both of these types of oscillations are similar to the precession modes of the equilibrium position of magnetization described above (Section 9.3). Thus, the first type is similar to mode No. 1 – low-amplitude circular precession, and the second is similar to mode No. 5 – unfolded circular precession. The modes No. 2–No. 4, representing the actual precession of the equilibrium position, are not realized in the plane pendulum under consideration, which is due to its purely planar character. To implement these modes, the pendulum must be out of the plane, similar to the precessing magnetization vector out of the plane. That is, the pendulum should have cylindrical symmetry. Consideration of such a pendulum will be performed in the next section.

9.4.2. Cylindrical pendulum with side spring suspension

Let us now consider a mechanical pendulum, similar to the previous one, but, in contrast, allowing free rotation around the vertical axis, on which the ends of the rod and the spring hanger are fixed. A diagram of such a pendulum is shown in Fig. 9.17. The basic designations coincide with those given in Fig. 9.16. The coordinate

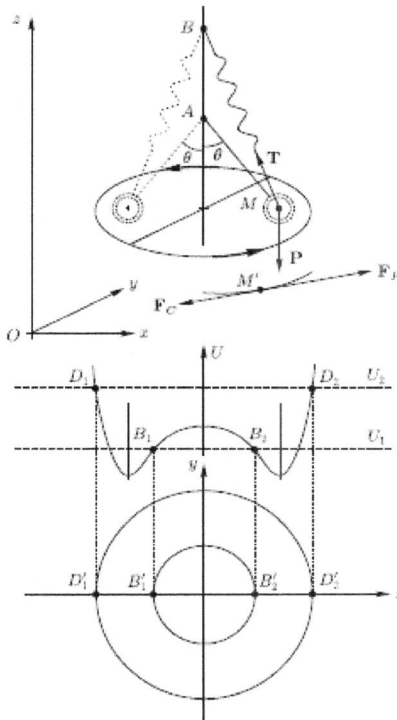

Fig. 9.17. Cylindrical mechanical pendulum with side spring suspension.

system $Oxyz$ is now three-dimensional, that is, it has a third axis Oy, perpendicular to the first two.

The main difference from the previous case is that the tirus of the pendulum can freely rotate around the axis Oz. Moreover, in the free state, it has not two stable equilibrium positions, but an infinite set located on a circle whose plane coincides with the Oxy plane, and the radius is equal to the equilibrium deflection of the pendulum in Fig. 9.16 from the Oz axis.

Accordingly, the potential is now not planar, but volumetric, having cylindrical symmetry, that is, it is formed by rotating the planar potential in Fig. 9.16 around the axis Oz. Such a potential has not two minima, but a whole annular 'ditch' of minima of the same depth, surrounding a single axisymmetric 'hill', rising in the middle of the ring formed by such a ditch. The top view of the potential (viewed along the Oz axis) is shown in Fig. 9.17 below. It can be seen that this species is a system of concentric rings, corresponding to different levels of potential.

The most important difference from the plane case here is the orientation of the exciting force $\mathbf{F}_p$ not along the axis Ox, as in Fig. 9.16, and tangential to the ring formed by the rotation of the equilibrium position of the weight of the pendulum around the axis Oz. This force now has a constant value, but its direction changes in accordance with the change in the direction of the tangent to the said ring. In the absence of attenuation, such a force would cause an accelerated motion of the load of the pendulum, therefore, to ensure uniform motion (necessary for analogy with the motion of magnetization), a friction force $\mathbf{F}_C$ should be introduced, which in general should be proportional to the speed of the motion of the load. The layout of these forces is shown somewhat below the main orbit in the figure, which is done for clarity. It should be borne in mind that the point M' in its physical essence, it is identical to point M. Thus, a continuous force acting on the load is equal to the difference $\mathbf{F}_p - \mathbf{F}_C$, which keeps a constant value and rotates around the axis Oz together with the load. In the case of a magnetic problem, such a force is created perpendicular to a constant alternating magnetic field of circular polarization. Such a field, being directed along the ring radius, due to the gyrotropy of the magnetization vector (in accordance with the Landau–Lifshitz equation), causes the end of this vector to move along the generatrix of the ring trajectory, which is the precession of magnetization in the most general case.

If we further assume that the circular motion of the force is rather slow, as if quasi-static, that is, the load moves beyond the force at the same speed as the force itself turns, then the same properties as in the previous case remain for the potential. To do this, it is necessary to add to the rotating force a variable in time component, perpendicular to the generator of the ring, directed along its radius. Then, at a low potential level, low-amplitude oscillations are possible in a narrow potential circle moving in a circle. With a high level of potential – fluctuations in the full wide well, the amplitude of which would be quite high. Figure 9.17 shows a diagram of these two potential levels, U_1 and U_2. Unlike Fig. 9.16, here the designations of the points $B_{1,2}$ and $D_{1,2}$ do not make sense of the oscillation range along the axis Ox, but simply reflect the concentric nature of the potential in three dimensions.

Comment. It should be noted that in the situations described in the present monograph of the recession of magnetization, any periodic force acting on the vector of magnetization along the radius of the generator of the large ring was not considered, so the authors leave this question for the future investigators.

On the possible realization of the exciting force in the experiment. In the above consideration, the nature of the exciting force FP directed tangentially to the large ring is ignored. On the other hand, it is obvious that the implementation of such a force in the experiment is quite possible. It could be created, for example, by an air screw or a jet engine, similar to an aircraft one, mounted on a load in such a way that the thrust force of this engine would be directed just tangentially to a large ring. However, it can be assumed that an even closer approximation to the magnetic case would be to consider the mechanical pendulum not as a simple load on a rigid rod, but rather a similar load that performs a rotational movement along the axis of this rod. That is, instead of a simple linear inertia, the pendulum would also have gyroscopic properties similar to a mechanical top (gyro [401–403]) with an axis coinciding with the axis of a rigid rod (*AM* in Fig. 9.17). In this case, the action on the top of a force directed along the radius of a large ring would cause the top to move along a tangent to this ring, that is, it would be equivalent to the action of the $\mathbf{F}_p$ force in Fig. 9.17. That is, as a driving force, one could use not a force $\mathbf{F}_p$ tangential to a large ring, rotating together with the load, but a force radial to the same ring, also rotating, but already around the axis of the pendulum and always directed along the radius of the large ring, i.e. remaining perpendicular to this axis. It can be assumed that from the constructive side, the creation of such a rotating radial force would be simpler than the realization of a tangential force using a jet engine. A radial rotating force could be created, for example, by using a system of stationary electrodes that are located radially, fed by a phase delay relative to each other, just as is done in widely used electromotors. At this stage, the authors of this monograph prefer to limit themselves to the above remark, and a more detailed consideration of the mechanical model based on the gyroscope is left for future research.

9.4.3. The oscillation of the pendulum at different frequencies of the exciting force

An important circumstance is the fact that a cargo, as a physical body, having a mass, has a certain inertia, that is, its movement may not keep pace with the driving force of the exciting force. Such a lag can be manifested the more, the greater the speed of a circular

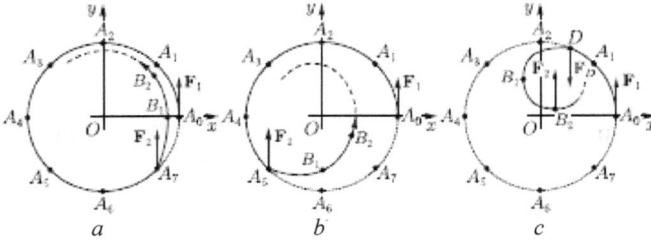

Fig. 9.18. Formation of the modes of motion of the pendulum at different frequencies of the exciting force: a – low frequency; b – the frequency is average; c – high frequency

movement of force.

Thus, the frequency of rotation of the driving force acquires a significant role in the formation of the trajectory of the pendulum. To illustrate the nature of the motion of the pendulum when the frequency of the exciting force changes, consider Fig. 9.18.

Here are the trajectories of the pendulum's load in the Oxy plane at different frequencies of rotation of the exciting force, that is, at different speeds when moving in a circle centred at point O.

Suppose that the movement of the load begins at point A_0. Let the force tangent to the trajectory force $\mathbf{F}_1$ act on the load at this point. Here it is directed upwards by the pattern, that is, along the axis Oy. In the quasistatic case, that is when the load does not lag behind the rotation of the force at all, it moves in a circle, successively passing points A_1, A_2, ... , A_7, after which it returns to the point A_0, whence the movement repeats along the same circular trajectory. The frequency of rotation of the force is given by an independent source (in the magnetic case, the generator of field rotation), the speed of the movement of the load is determined by its inertia in combination with friction. If the speed of rotation of the force is small, then the load can follow it with sufficient accuracy (quasistatic mode). With a high speed of rotation of force, the load may not keep pace with the movement of force, that is, lag behind it. In Fig. 9.18 three possible variants of such a lag are shown, corresponding to different frequencies of rotation of the force.

Consider first Fig. 9.18, a. This figure corresponds to a low frequency of rotation of the force, that is, a small load lag. The force makes a complete revolution in the plane at the moment when the load reaches point A_7. The force $\mathbf{F}_2$ acting on it at this moment is already directed again along the axis Oy, while the load did not reach point A_0. The tangent to the trajectory at this point is somewhat deviated from the Oy axis, so that a component of the force $\mathbf{F}_1$ appears, directed towards the centre of the trajectory (to the point O). This component causes the load to deviate to the centre, as a result of which it does not fall at the point A_0, but instead comes at the point B_1, then at the point B_2 and continues to move along the path shown by the dotted line. However, the lag is still very small, so that the new trajectory differs from the quasistatic also a little, so the load moves in a circle, the diameter of which is less differs from the quasistatic, the lower the frequency.

In the magnet case, this mode is practically no different from the simple circular precession of the magnetization vector along a cone with smooth walls, that is, it can be assumed that it is similar to mode No. 5 – the unfolded circular precession.

Turn now to Fig. 9.18, b. This figure corresponds to the average frequency of rotation of the force, that is, the load is already moving with a noticeable lag. The force makes a complete revolution at the moment when the load reaches point A_5. The force $\mathbf{F}_2$ acting on it at this moment is also directed along the axis Oy, that is, the component of this force, directed toward the centre, is rather large. As a result, the trajectory of the movement of the load noticeably deviates towards the centre, passing successively points B_1, B_2 and then following the curve shown by a dotted line. It is important to note that here the centre (point O) remains on the left side of the new trajectory, so that the diameter of such a new trajectory exceeds the radius of the quasistatic circle. After completing a full turn, the trajectory makes another similar turn, shifted by the azimuth angle relative to the first and so on. As a result, along the generatrix of a quasistatic ring, the plane is covered by successively displaced annular trajectories relative to each other, with the centre point O always remaining inside each such trajectory.

In the magnetic case, such a motion corresponds to mode No. 3 – precession of the equilibrium position with centre coverage.

Comment. It should be noted that there is compliance here also with mode No. 4, which differs from mode No. 3 only by attenuation in time. However, the difference between these modes in the framework of the mechanical model proposed here was not considered by the authors of this monograph, although, quite possibly, a similar analogy can also be found.

Consider now Fig. 9.18, c. This figure corresponds to a high frequency of rotation of the force, that is, the load moves behind the force with a significant lag. The force makes a complete revolution at the moment when the load reaches point A_3. The force $\mathbf{F}_2$ acting on it at this moment is still directed along the Oy axis. However, still far from reaching point A_3, halfway between points A_1 and A_2, that is, when the load is in the vicinity of point D, the force makes half a turn, so that it becomes directed down the axis Oy. In the figure, this force is designated as $\mathbf{F}_D$. The force acting in such a direction deflects the trajectory of the load towards the centre, so that it hits point B_1. Upon further movement of the load, the force rotates even further, so that when the load arrives at point B_2 (which corresponds to point A_3 in time in the semiclassical case), force $\mathbf{F}_2$ again becomes upward. As a result, the trajectory, passing sequentially the points D, B_1 and B_2, is wrapped in a ring, the diameter of which is now smaller than the radius of the quasistatic circle, so that the central point O remains on the right side of such a circular trajectory. After the trajectory returns to the quasistatic circle, the turning force forms a new ring advanced along the azimuth angle a little further than the first one. As a result, the plane along the generatrix of a quasistatic ring is covered by successively displaced ring paths relative to each other, with the centre point O always remaining outside each ring of such a path.

In the magnetic case, such a motion corresponds to mode No. 2 – ring precession of the equilibrium position without centre coverage. It can be assumed that a further further increase in the frequency of rotation of the force will result in the load generally moving very little, that is, experiencing a 'crush' in the region of the equilibrium position, that is, near the generator of the quasistatic ring, with a weak tendency to shift along this generatrix, which with increasing frequency of rotation of force will weaken, tending to zero. In the magnetic case, such a motion will be similar to mode No. 1 – low-amplitude circular precession, which is the limiting case of mode No. 2 – ring precession without centre coverage.

9.4.4. Features of the multi-mode nature of oscillations of cylindrical pendulum

So, from the review we can conclude that a cylindrical pendulum with a side spring suspension has the same basic oscillation modes as the precession of the equilibrium position in the magnetic case.

However, a certain difference should be noted. In section 9.3, these modes replace each other in order from No. 1 to No. 5. Here, the order is reversed, that is, first comes mode No. 5, then No. 3 (or

No. 4), followed by modes No. 2 and No. 1. This difference is due to the fact that in section 9.3 the change of modes was considered as the amplitude of the exciting field increased. Here, the modes replace each other as the frequency of the exciting force increases. The characteristics of the frequency properties of the precession of the equilibrium position of the magnetization vector are discussed in sections 9.2.2, 9.2.3 and are illustrated in Fig. 9.8. The fact of the existence of a critical frequency, above which the precession of the equilibrium state takes place and is absent below, is also noted there. In the mechanical case, you can also talk about the presence of the critical frequency located somewhere between the frequencies corresponding to Fig. 9.18, a and Fig. 9.18, b.

We also add that the equivalent of increasing the amplitude of excitation in the magnetic case is an increase in the force acting on the pendulum in the mechanical case. A preliminary qualitative study showed that in this case the same regimes are also observed, but now, as the force increases, they follow the order from No. 1 to No. 5. Thus, it can be assumed that here the analogy between the magnetic and mechanical cases becomes even more complete.

Comment. Within the framework of this monograph, the authors did not set themselves the tasks of a detailed analytical or numerical calculation of the characteristics of oscillation modes of a cylindrical pendulum, since the main subject of consideration was considered to be the physics of magnetic phenomena. Therefore, the authors restricted themselves to a rather superficial qualitative description of the behaviour of the mechanical model, primarily focusing on compliance with the magnetic case. Bringing the mechanical model here, according to the authors, is the most appropriate, since the further presentation will be devoted to the much more complex aspects of the precession of magnetization. Applying a mechanical model to such cases, apparently, is also possible, but it would require significant complication, which, according to the authors, is hardly advisable at this stage. Nevertheless, the authors believe that a more detailed solution of a purely mechanical problem would also be quite interesting, therefore, they leave such a solution as a subject for individual investigations.

9.5. The specific properties of the multimode nature of the precession of the equilibrium state of magnetization

Let us now return back to the study of the precession of the equilibrium state of magnetization and turn to models that allow us to interpret its properties taking into account the specificity of the magnetic nature of the phenomenon.

9.5.1. Transitions between the magnetization precession modes based on the potential model

First of all, we will consider transitions between different precession modes based on a model that takes into account the dependence of the potential on the amplitude of the alternating field. The general scheme of transitions is illustrated in Fig. 9.19, a, b. The curves are removed at steady-state amplitude values, that is, after the transient oscillations of the precession of equilibrium have already attenuated. In time, this corresponds to approximately $(2...4) \cdot 10^{-7}$ s.

Figure 9.19, a shows the dependences of the maximum (1) and minimum (2) values of the potential on the amplitude of the variable field.

The vertical dashed lines correspond to the boundaries of the transitions between the different precession modes of the equilibrium position: at 0 Oe $< h_0 <$ 28 Oe – mode No. 2 – precession without centre coverage; at 28 Oe $< h_0 <$ 37 Oe – mode No. 3 – precession with centre coverage without attenuation; at 37 Oe $< h_0 <$ 43 Oe

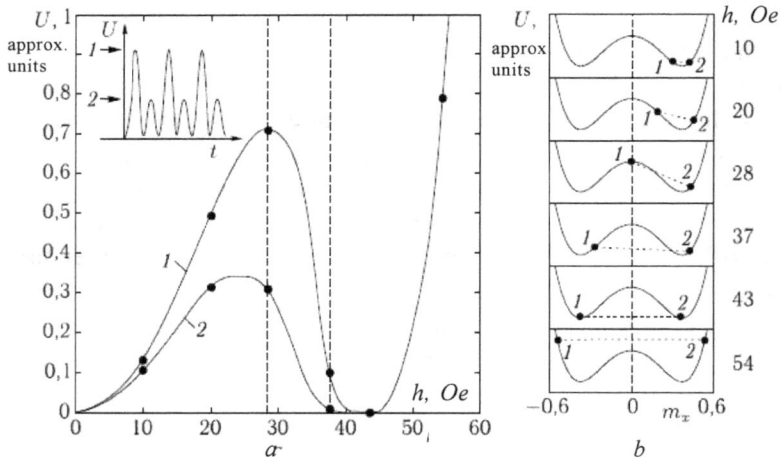

Fig. 9.19. The dependence of the potential on the amplitude of the alternating field (a) and the dependency diagrams for different types of alternating field (b). Explanation of boxes and numbers of curves in the text. Parameters: $4\pi M_0 = 280$ G, $H_{0z} = 260$ Oe; $\alpha = 1.2$; $f = 200$ MHz.

–mode No. 4 – precession with centre coverage with attenuation; at 43 Oe $< h_0 <$ 60 Oe – mode number 5 – deployed circular precession. The aggregated points correspond to the potential values at characteristic light values of the alternating field used in plotting the curves in Figs. 9.11–9.15. The inset at the top left shows the dependence of the potential on time and the scheme of reference for the maximum and minimum values.

Figure 9.19, b shows the dependences of the potential, built in the Oxz plane, on the transverse component of the magnetization m_x (with $m_y = 0$), with some actual peak values of the permeable field shown in the column to the right of the corresponding dependences (these amplitudes of the fields correspond to the values of the enlarged current in Fig. 9.19, a). The choice of m_x (for $m_y = 0$) as a transverse component of magnetization does not limit the generality, since the construction has cylindrical symmetry about the axis Oz. Potential in Fig. 9.19, b calculated by the formula

$$U = -H_0 \sqrt{1 - m_x^2 - m_y^2} - 2\pi M_0 \left(m_x^2 + m_y^2 \right), \qquad (9.21)$$

which is accurate to the constant term and the normalization to M_0 is obtained from the formula

$$u = -\mathbf{MH} - \frac{1}{2} \mathbf{MH}_p, \qquad (9.22)$$

where: $\mathbf{H}_p = -4\pi M_z \mathbf{n}_z$ is the demagnetization field of the plate, $\mathbf{n}_z$ is the unit vector along the Oz axis. Non-integrated check in Fig. 9.19, b correspond to the numbers of curves in Fig. 9.19, a. The dashed lines connecting the enlarged points in Fig. 9.19, b, correspond to the potential values when moving from one point to another (between curves 1 and 2 in Fig. 9.19, a). In this case, the transition does not take place in the Oxz plane, but with access to three-dimensional space, that is, the precessing magnetization does not jump over x ohm corresponding to the Oz axis (where $m = 0$), but bypasses it laterally through the Oy axis, and the magnetization in the process of such bypass acquires the my component, which is reflected in the precession portrait in the form of rings.

The potential of $U(m_x, m_y)$ relative to the plane Oxy is three-dimensional and has the form of a high central peak (hill) with a smooth top, elongated along the Oz axis, corresponding to $m_x = 0$, $m_y = 0$, surrounded by a ring ditch, the outer edge of which tends to infinity. Due to the perpendicularity of the constant field to the plane of the plate, the ditch has axial symmetry about the axis Oz, that is, the three-dimensional form of the potential is formed by rotating a flat curve corresponding to $U(m_x, 0)$ around the axis Oz. With a small amplitude of the alternating field (10 Oe), the magnetization precesses in a small circle that completely fits inside the adjacent opposite slopes of the ditch, and this small circle rolls along the ditch itself, repeating its annular shape. Such a movement corresponds to mode No. 2 – precession of the equilibrium state without centre coverage. As the amplitude of the alternating field (20 Oe) increases, the small circle of precession gradually expands, its internal point fits on the central peak of the potential, the precession without centre coverage (mode No. 2) turns into precession with coverage (field 28 Oe, transition to mode No. 3), after which a small circle of precession, more and more expanding, is put on the central peak of the potential, slides down it (mode No. 4 up to the field 37 Oe) and occupies a symmetrical position relative to the Oz axis (transition point from mode No. 4 to mode No. 5, field 43 Oe). A further increase in the amplitude of the alternating field leads to the precession of the magnetization corresponding to the movement on the outer slope of the potential ditch, which, due to the correct ring shape of the ditch, occurs in the right circle (mode No. 5, field greater than 43 Oe).

9.5.2. Conditions for the excitation of precession modes at different attenuation parameters and amplitudes of an alternating field

From the above consideration, it can be seen that the amplitude of the potential oscillations depends on the amplitude of the alternating field h_0, but, like any oscillation of the dissipative system, it must also depend on the attenuation parameter α. This two-parameter relationship is illustrated in Fig. 9.20, where on the plane 'attenuation parameter – amplitude of the alternating field' $O\alpha h_0$, the boundaries of the regions with different precession modes for different values of h_0 and α for two values of the alternating field frequencies are presented.

Thus, with a symmetric field ($H_{0x} = 0$ Oe): in the limit as $h_0 \to 0$ on curve 1, as well as 5, mode No. 1 is implemented, between the curves 1 and 2, and 5 and 6 - mode No. 2, between the curves 2 and 3, as well as 6 and 7 – mode number 3, between the curves 3 and 4, as well as 7 and 8 – mode number 4, above the curve 4,

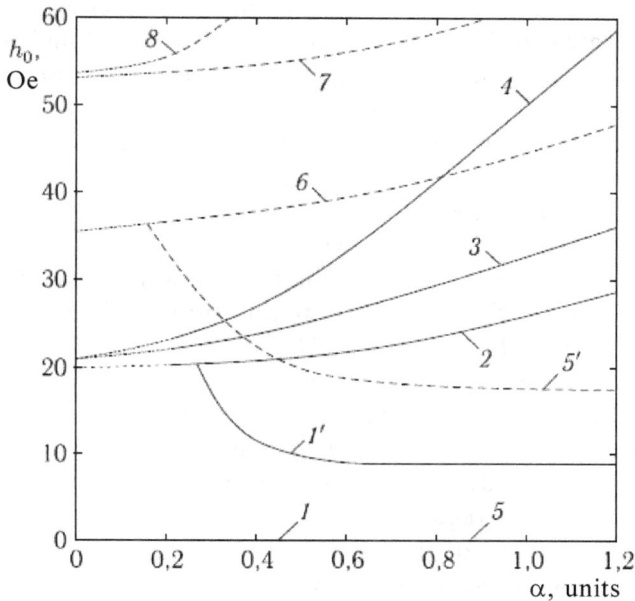

Fig. 9.20. Dependences of the boundaries of regions with different precession modes on the attenuation parameter at various frequencies f and the values of the transverse component of the constant field H_{0x}. The parameters of the curves with the following: 1, 1', 2–4 (solid lines) – $f = 200$ MHz; 5, 5', 6–8 (dotted lines) – $f = 400$ MHz; 1–8 – symmetric field: $H_{0x} = 0$ Oe; 1,1' - the field is asymmetric: $H_{0x} = 0.20$ Oe; 1, 1', 5, 5' – boundaries between areas with modes No. 1 and No.; 2, 6 – borders between areas with modes No. 2 and No. 3; 3, 7 – the boundaries between the areas with modes № 3 and № 4; 4, 8 – the boundaries between regions with modes No. 4 and No. 5. Parameters: $4\pi M_0 = 280$ G, $\alpha = 1,2$; $H_{0z} = 260$ Oe.

and 8 – mode number 5. With an asymmetric field ($H_{0x} = 0.20$ Oe): between curves 1 and 1', as well as 5 and 5' mode 1 is implemented, between curves 1' and 2 as well as 5' and 6 – mode No. 2, between curves 3 and 4, and also 7 and 8 – mode No. 4, above curve 4, and also 8 – mode No. 5.

When $\alpha > 0.2$, different modes are clearly expressed and the boundaries between areas with different modes are sharp. These boundaries are shown by solid and dashed lines. When $\alpha < 0.2$, the difference between the modes is smoothed, the boundaries between areas with different modes are not sharp, the transitions between modes are smooth. These borders are indicated by dotted lines. It can be seen from the figure that when the frequency of an alternating field increases, all the boundaries between areas with different modes increase (when going from 200 MHz to 400 MHz – one and a half to two times), which is apparently due to a change in the resonant words of the nonlinear precession excitation frequency change. The boundaries between the regions with the modes No. 2 and No. 3, No. 3 and No. 4, No. 4 and No. 5 (curves 2–4 and 6–8) do not depend on the magnitude of the constant-field component H_{0x}. When H_{0x} changes, only the boundaries between regions with modes No. 1 and No. 2 (curves 1 and 1′, 5 and 5′) change. At $H_{0x} = 0$ Oe, the region with mode No. 1 is absent (curves 1 and 5). This means that the equilibrium position precesses even at the very smallest amplitudes of the alternating field, that is, in fully symmetrical conditions, the precession of the equilibrium position is thresholdless. At $H_{0x} = 0.20$ Oe, a region with mode No. 1 exists and is bounded above the curve 1′ or 5′, that is, the precession of the equilibrium position (mode No. 2) has a certain threshold in the amplitude of the variable field. As the attenuation parameter increases, the boundaries between the regions with modes No. 2 and No. 3, No. 3 and No. 4, No. 4 and No. 5 increase, regardless of the presence or absence of the transverse component of the constant field. Such an increase in the boundaries seems to be connected with the requirement of increasing the energy consumption for the excitation of oscillations of sufficient amplitude with increasing attenuation. The inclusion of the transverse component of the constant field with the attenuation parameter $\alpha <$ 0.25 does not change the border between areas with modes No. 1 and No. 2, and for $\alpha > 0.25$ this border sharply increases (curves 1 and 5 are replaced by curves 1′ and 5′). Such a dependence on the attenuation parameter is apparently associated with an increase in the role of resonance words with decreasing attenuation.

9.5.3. Properties of the transverse component of magnetization with different precession modes

Let us now dwell on some transitions of transitions between the precession modes when the amplitude of the alternating field changes. First of all, we note that all transitions in a variable field are quite sharp with an accuracy of at least 1% of the transition field. The transitions between modes No. 1 and No. 2, No. 2 and No. 3, and also No. 3 and No. 4 are smooth, that is, they are not accompanied by a sharp change in the transverse component of magnetization, while the transition between modes No. 4 and No. 5 occurs with a sharp jump.

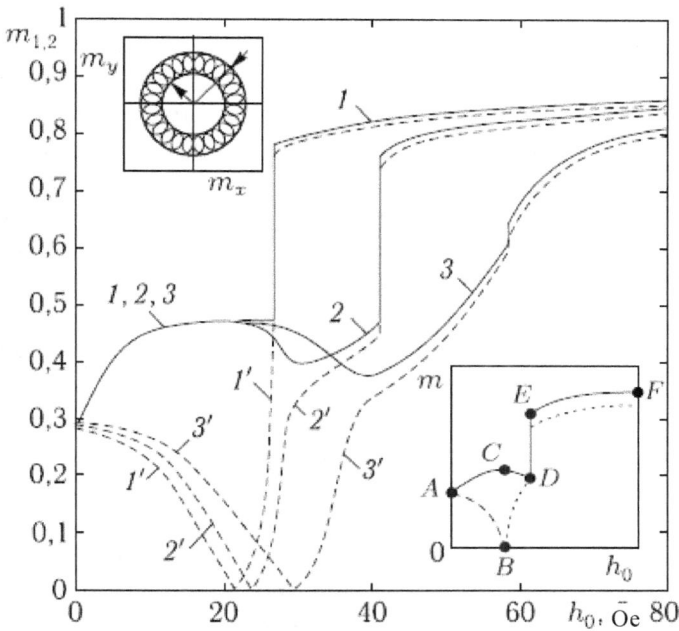

Fig. 9.21. Dependences of the maximum and minimum values of the transverse components of the magnetization of the amplitude of the alternating field at different damping parameters α: 1, 1 – 0.4; 2, 2 – 0.8; 3, 3 - 1.2. Explanation of the sidebars – in the text. Parameters: $4\pi M_0$ = 280 G, H_{0x} = 0 Oe, H_{0y} = 0 Oe, H_{0z} = 260 Oe, f = 200 MHz

The general character of the change in the transverse component of the magnetization with a change in the amplitude of the alternating field is illustrated in Fig. 9.21, which shows the dependences of the maximum and minimum values of the transverse component of the magnetization on the amplitude of the alternating field for different attenuation parameters.

The inset at the top left shows the reference scheme of the minimum m_1 and maximum m_2 values of the transverse component of magnetization. The inset at the bottom right shows the flow diagram of the dependences $m_{1,2}(h_0)$ with different precession modes. Solid lines and numbers without dashes correspond to the maximum values of the transverse component of magnetization $m_2(h_0)$; dotted lines and numbers with strokes the minimum values of the transverse component of magnetization $m_1(h_0)$. The values of the transverse component of the magnetization are measured by the outer and inner radii of the main ring in the precession portrait in the steady state, as shown in the inset in the upper left.

The precession modes shown in the inset at the bottom right are the following: at point A - mode number 1, between the points A and B, and also A and C – mode number 2, between points B and D, and C and D – alternately other modes 3 and 4, from point D to point E - jump in transition from mode 4 to mode 5, from point E to point F – mode 5.

The figure shows the following. The curves without strokes (1, 2, 3) and with strokes (1′, 2′, 3′), Corresponding to the maximum and minimum values of the transverse component of magnetization, for $h_{x,y} = 0$ begin at the point $m_1 = m_2 = 0.30$, which corresponds to the equilibrium position of the magnetization vector in the absence of an alternating field. As the amplitude of the alternating field increases, the equilibrium position begins to precess, and the precession amplitude of the magnetization vector around the precessing equilibrium position increases, which is manifested in a decrease in the inner radius of the ring and an increase in the outer radius and leads to an increase in the distance between curves 1, 2, 3, and 1′, 2′, 3′. In this case, the mode No. 2 is carried out, that is, the precession of the equilibrium position occurs without centre coverage. When the precession radius of the magnetization vector becomes equal to the deviation of the equilibrium position from the Oz axis, the inner ring tightens to a point corresponding to $m_{r1} = 0$, as a result of which mode No. 2 without centre coverage is replaced by mode No. 3 with centre coverage. With a further increase in the amplitude of the alternating field, the inner radius of the ring m_{r1} increases, and the outer radius of m_{r2} slightly decreases until the transition to mode No. 4, when the precession of the equilibrium position becomes damped. Further, the decay of the precession of the equilibrium position increases up to the transition to mode No. 5, when the equilibrium position of the magnetization stops precessing and is aligned along the Oz axis, that is, the precession of the equilibrium position is replaced by the unfolded circular precession. The difference between the inner and outer radii of the rings tends to zero $(m_2 - m_1) \rightarrow 0$.

Thus, with increasing amplitude of the alternating field, the width of the generatrix of the main ring of the precession portrait increases from zero to a value slightly larger than the radius of this ring in mode number 3, decreases in mode 4 and number 5, and to zero. At the same time, the radius of the middle of the generatrix of the main ring in modes No. 2, No. 3 and No. 4 remains almost constant, corresponding to the equilibrium position of the magnetization vector in the absence of an alternating field, and in mode No. 5 it gradually increases, tending to unity. From the general course of the curves in Fig. 9.21 one can see that the transition from mode No. 4 to mode No. 5, unlike other transitions, is quite dramatic, occurs when the amplitude of the alternating field changes by less than 0.1% and is accompanied by a significant change in the precession amplitude (up to one and a half to two times). This means that with a smooth change in the excitation in a very small extent, the opening of the cone of precession changes, that is, we can assume that under these conditions a pre-transitional magnetization vector occurs with a jump-like dynamic orientational transition.

9.5.4. The dependence of the precession period of the equilibrium position on the amplitude of the variable field

From Figs. 9.11–9.15 it is clear that the precession period of the equilibrium position significantly exceeds the oscillation period of the applied alternating field. The same circumstance was noted in section 9.1.4, where, on the basis of the vector model, a connection was obtained between the precession period of the equilibrium position and the amplitude of the variable field in the form (9.19):

$$T = \frac{A}{h_0^2 \sqrt{1-\left(H_0 / 4\pi M_0\right)^2}}.$$
(9.23)

where A is a constant factor. It was also noted there that the reduced formula in mode No. 2 with an accuracy of ~5% describes the dependence of the precession period of the equilibrium state on the amplitude of the variable field h_0 in constant fields H_{0z} to 0.86 on the anisotropy field of the form $4\pi M_0$. For the considered case of different patterns, the dependence of the precession period of the equilibrium position on the amplitude of the variable field is illustrated in Fig. 9.22. In this figure, the points correspond to the period measured by the dependence of the transverse component of magnetization on time, the solid lines are constructed according to the formula (9.23). Curve 1 corresponds to $\alpha = 0.4$, with $A = 10.4$ Oe, the transition field from mode No. 2 to mode No. 3 is $h_{c23} = 21.0$ Oe, the transition field from mode No. 3 to mode No. 4 is $h_{c34} = 23.8$ Oe; curve 2 corresponds to $\alpha = 0.8$, with $A = 20.0$ Oe² μs,

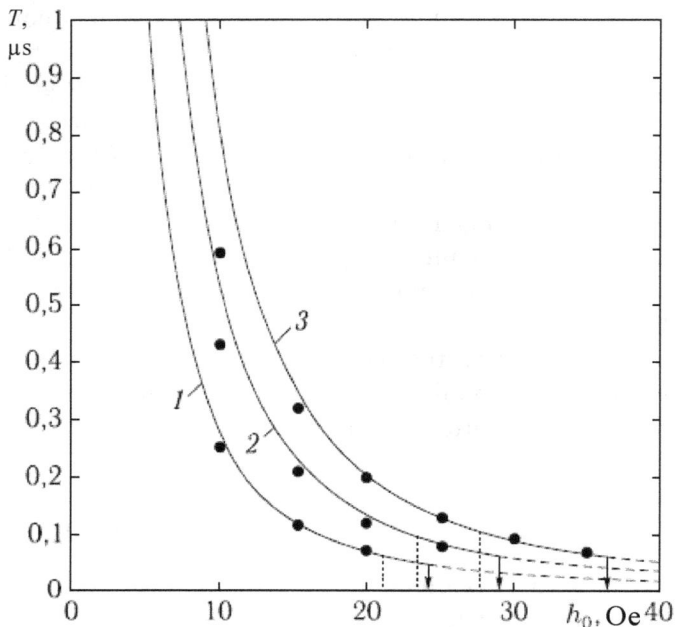

Fig. 9.22. The dependence of the precession period of the equilibrium position on the amplitude of the variable field. The main parameters: $4\pi M_0 = 280$ G, $H_{0x} = 0$ Oe, $H_{0y} = 0$ Oe, $H_{0z} = 260$ Oe, $f = 200$ MHz. Specific parameters are indicated in the text.

$h_{c23} = 23.0$ Oe, $h_{c34} = 29.2$ Oe; curve 3 corresponds to $\alpha = 1.2$, with $A = 30.4$ Oe^2 μs, $h_{c23} = 27.6$ Oe, $h_{c34} = 36.0$ Oe. Vertical dotted lines mark the fields of transition between modes No. 2 and No. 3, arrows on the horizontal axes – transition fields between the modes No. 3 and No. 4. Thus, to the left of the dotted lines (for $h_0 < h_{c23}$), mode 2 is observed for each corresponding curve, between dotted lines and arrows (for $h_{c23} < h_0 < h_{c34}$) is mode No. 3, to the right of arrows (for $h_0 > h_{c34}$) is mode No. 4. Here, the precession of the equilibrium state dies out, therefore the portions of the curves constructed by the formula (9.23) are depicted by a dotted line.

The figure shows the following. The precession period of the equilibrium position in the absence of an alternating field is infinitely large, since the equilibrium position is at rest. When an alternating field is turned on, the equilibrium position begins to precess, the faster the larger the field amplitude, that is, the precession period of the equilibrium position decreases. In this case, the inverse proportionality of the period marked by formula (9.23) to the square of the amplitude of the alternating field is likely to be performed in the interval from ~10 Oe to the transition field h_{c34}. The transition from mode number 2 to mode number 3 with h_{c23} has no features. In the $h_0 \leq 10$ Oe area for any values of the damping parameter α, the actually observed values of the precession period of the equilibrium position become less than the differential formula (9.23), and this difference is the greater, the larger the damping parameter. With an increase in α, the period increases, which is connected with the difficulty of buildup of the oscillatory system with increasing attenuation.

9.6. Precession of equilibrium in terms of symmetry breaking

All the above-described properties of the precession of the equilibrium state are manifested in a pure form only with a fully symmetrical result and implementation. This means, firstly, the symmetry of the excitation, that is, the constant field should be perpendicular to the plane of the plate, and the variable should be oriented in the plane of the plate and should have circular polarization. Secondly, all possible orientations of the magnetization vector in the plane of the plate should be equally probable, that is, any magnetic anisotropy in the plane of the plate should be absent.

9.6.1. General notes on symmetry breaking

Before investigating this issue in more detail, we present some general remarks concerning the violation of such symmetry. If the symmetry of the orientation of the constant field relative to the plate is violated, that is, it is deflected from the plate normal by a certain angle, then on the precession portrait there appears a condensation of small sheep, whose position depends on the azimuth angle of the orientation of the constant field. The development of precession in time ceases to be exactly sinusoidal, so that rather long flat areas appear on the sinusoid, corresponding to a time delay in the formation of condensations of small olec.

If the symmetry of the excitation is violated, that is, the polarization of the alternating field is not purely circular, but elliptical, that is, the components h_x and h_y are different, then the development in equilibrium position during the precession also ceases to be purely sinusoidal and approaches the pulsed one with disruptions. At the same time, the precession period with increasing asymmetry of the excitation increases, that is, the distance between individual pulses and breakdowns becomes longer.

The precession portrait of $m_y(m_x)$ (Fig. 9.2, e) retains the appearance of a regular circular ring, but the filling of the large ring with small rings becomes uneven, on the large ring there appear two condensations of small rings, which are diametrically opposed to each other. With a further increase in the asymmetry of the excitation, the precession of the equilibrium state breaks down and is replaced by the usual purely circular precession around the direction of the constant field. The permissible limits of asymmetry, in which the precession of the equilibrium state is possible, are rather small. Thus, the components of the variable field h_x and h_y can differ from each other by no more than 1.2...1.3 times. With a greater difference, the precession of the equilibrium position is replaced by the usual circular precession. As for anisotropy, first of all it should be noted that the introduction of normal uniaxial anisotropy in the properties of the equilibrium state does not change anything much, except for the addition to the form anisotropy field. Oblique uniaxial anisotropy produces a precession transformation similar to the asymmetry of the constant field. If both the constant field and the excitation remain symmetric, but there is anisotropy in the plane of the plate, then the development of the equilibrium state in the precession time ceases to be purely sinusoidal and approaches the pulsed one with disruptions. At the same time, the precession period also increases with anisotropy. The overall picture is close to that mentioned above for asymmetrical excitation. However, now the condensations of small rings in the precession portrait are determined by the orientation of the anisotropy axes.

As an example, we note the case of cubic anisotropy with the orientation of the [100] type crystallographic axis normal to the plane of the plate and two anisotropy constants of the first K_1 and second K_2 orders, typical of IYG films, both of which are negative. The introduction of cubic anisotropy with a constant K_1 gives the asymmetry of the precession, similar to that caused by the asymmetry of the alternating field, that is, the sinusoidal mode is replaced by a pulsed one. At the same time, in the precession portrait four thickening of small olec are observed, whose positions are determined by the orientation of the anisotropy axes. These condensations are shifted 90° relative to each other in a large circle of precession, but their polarizations do not coincide with the positions of the projections of the axes of cubic anisotropy on the film plane, but are rotated along the azimuth angle by several tens of degrees. Condensations are expressed the better, the larger the amplitude of the variable field. Adding the second cubic anisotropy constant K_2 leads to a stronger expression of condensations. An increase in the constant K_1 leads to the breakdown of the precession of the equilibrium position and its replacement with the usual precession around the direction of the constant field.

Now, after a general overview of the nature of the precession of the equilibrium state in the case of symmetry breaking, we consider this question in more detail and begin with a change in the orientation of the constant field.

9.7. Symmetry violation on a constant field

We first consider a simpler case of violation of the symmetry of the system, namely, a violation by a constant field. Such a violation corresponds to the deviation of the direction of the constant field from the initial one, provided that the same polarization plane of the alternating field is maintained. Therefore, remaining within the same coordinate system, where the alternating field is polarized in the Oxy plane in a circle, we add to the main constant field oriented along the Oz axis, a small addition directed along the Ox axis.

9.7.1. Influence of asymmetry of the constant field on the nature of the precession of the equilibrium state

So, let us consider the general nature of the change in the precession of the equilibrium state in the case of violation of the symmetry of a constant field. As before, we will consider the development of oscillations in time and the resulting precession portrait. We assume that the main constant field, providing the conditions for the orientational transition, as before, is directed along the Oz axis, but now a relatively small field is added to it in the perpendicular direction. That is, there is a field component along the Ox axis, while the Oy component is still zero. The calculation results are illustrated in Fig. 9.23. Figure 9.23, a and c shows the dependences of the oscillations of the transverse directions of the magnetization m_x (solid lines) and m_y (dashed lines) on time, and in Fig. 9.23, b and Fig. 9.23, d – the corresponding precession portraits my (mx). Figures 9.23, a and b correspond to the fully symmetric case (H_{0x} = 0.000 Oe), Fig. 9.23, c and Fig. 9.23, g – asymmetric: H_{0x} = 0.055

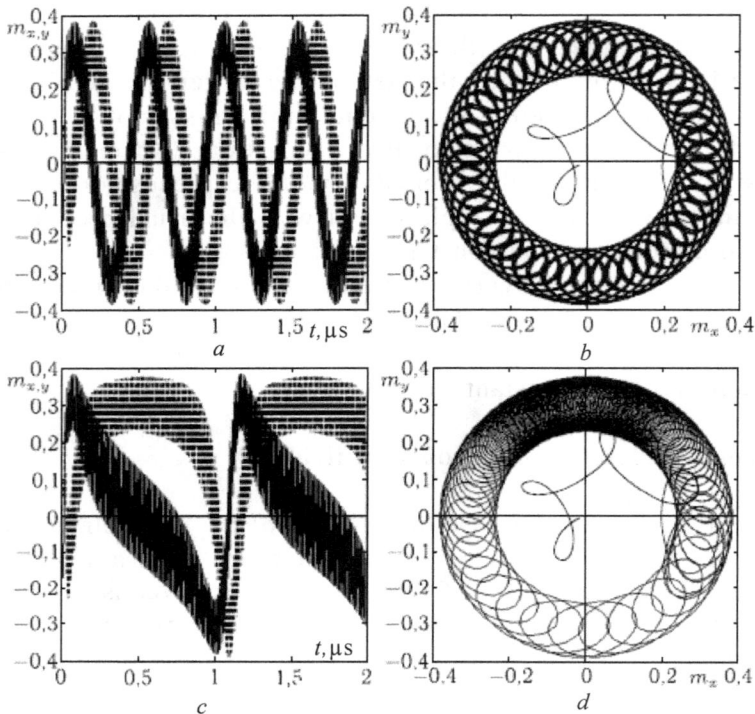

Oe (everywhere H_{0y} = 0 Oe).

Fig. 9.23. Dependences of transverse oscillations of magnetization components on time (*a, c*) and precession portraits (*b, d*) in the symmetric case H_{0x} = 0 Oe (*a, b*) and with asymmetrical constant field H_{0x} = 0.055 Oe (*c, d*). Parameters: $4\pi M_0$ = 280 G; H_{0y} = 0 Oe; h_0 = 3 Oe; f = 100 MHz.

From Fig. 9.23, b and Fig. 9.23, g it is clear that both in symmetric and in asymmetrical cases, the equilibrium position precesses in a circle around the *Oz* axis; however, with an asymmetric constant field, the mean values of the dependences of the transverse magnetization components m_x and m_y on time cease to be purely sinusoidal, as in Fig. 9.23, a, take the form of an irregular pulse shape, shown in fig. 9.23, c. It can also be seen that the period of such pulses significantly (in this case twice) exceeds the period of a sinusoid with symmetric excitation.

The precession portrait of m_y (m_x) (Fig. 9.23, d) retains the appearance of a regular circular ring, similar to a ring with a symmetric constant field (Fig. 9.23, b), but the filling of a large ring with small rings becomes uneven, on a large ring there appears a thickening of small rings, located in the positive direction of the Oy axis.

The position of the condensation observed in the precession portrait is determined by the nature of the asymmetry of the constant field. When $H_{0x} > 0$, $H_{0y} = 0$, the condensation is located at the top on the vertical axis, as shown in Fig. 9.23, b. When $H_{0x} < 0$, $H_{0y} = 0$, the condensation is located below on the same axis. When $H_{0x} = 0$, $H_{0y} > 0$, the condensation is located on the left on the horizontal axis. When $H_{0x} = 0$, $H_{0y} < 0$, the condensation is located on the right on the same axis. This arrangement of the thickening means that it is rotated relative to the direction of the transverse constant field 90° counterclockwise.

9.7.2. Criticality of asymmetry of a constant field

The fact that the precession of the equilibrium state is excited is very critical to the degree of asymmetry of the constant field. If the transverse component of the constant field $H_{0x,y}$ exceeds a certain critical value $H_{0x,y}^{(c)}$, then the equilibrium position when turning around the Oz axis does not make a complete revolution, but stops somewhere on its part, and the magnetization vector continues to precess along an this stopped direction of equilibrium position. It can be said that when the field is greater than the critical one, the precession of the equilibrium state fails, therefore we will call this field the 'critical field of failure' of the precession of the equilibrium position. The permissible limits of the asymmetry of the transverse field, in which the precession of the equilibrium state is possible, are rather small. So, with the selected parameters, the critical stall field $H_{0x}^{(c)}$ is 0.060 Oe (with $H_{0y} = 0$ Oe).

The nature of the precession in the transverse field above the critical is illustrated in Fig. 9.24, where in Fig. 9.24a the corresponding oscillations of the transverse magnetism m_x (solid lines 1) and m_y (dashed lines 1′) corresponding to $H_{0x} = 0.070$ Oe are shown on time, and Fig. 9.24, b shows the precession portrait of m_y (m_x) (curve 1). For comparison, the technical figures show the dependence of the components of magnetization (Fig. 9.24, a) m_x (solid line 2) and m_y (dashed line 2′) on time, and the precession portrait (Fig. 9.24, b) m_y (m_x) (thickened line 2) in the case of free for $h_{0x} = h_{0y} = 0$ Oe of establishing the magnetization of the values $m_x = m_y = 0$, $m_z = 1$.

496

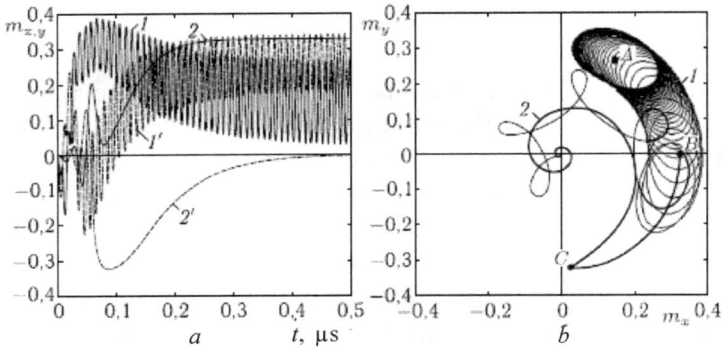

Fig. 9.24. Dependences of vibrations of transverse components of magnetization in time (*a*) and precessional portraits (*b*) in the case of asymmetry of the constant field greater than critical: H_{0x} = 0.070 Oe. Other parameters – the same as adopted in the construction of Figure 9.23

In Fig. 9.24, b point *A* is the centre of an elliptical cycle of forced simple precession, its coordinates are: m_x = 0.147, m_y = 0.262, point *B* is the final point of establishing the free precession of magnetization, its coordinates: m_x = 0.325, m_y = 0.000. The axis of the cone of steady precession is deflected from the coordinate axis *Oz* by an angle of 17.5°, and the angle of deviation from the same axis of the magnetization vector in the static state is 18.9°, that is, the deviation of the precession axis from the normal to the plane of the plate is mainly determined by the ratio between a constant field and a plate-shaped demagnetization field. However, in the plane of the plate, the axis of the cone of the established precession is rotated with respect to the direction of the transverse constant field by 60.7° counterclockwise, which is similar to that observed in Fig. 9.23, d to the rotation of concentration on the precession portrait with the field H_{0x}, less critical.

9.7.3. Left precession

We note in passing one common feature of the precession of magnetization under conditions of the polarization transition. From the course of curve 2 (thickened) in Fig. 9.24 b it can be seen that the free precession without a driving force is first left – in the area from the origin to point *C*, whose coordinates are: m_{0x} = 0.030, m_{0y} = −0.321. At this point, the deviation of the magnetization vector from the Oz axis is 18.8°, which is close to the deviation in the static state. After passing point C, the precession becomes right and remains so in the area from point C to point B. When forced, the curve (curve 1 in Fig. 9.24, b) is in the initial part, where the precession trajectory is a loop-like developing curve, a precession in each small loop is right, but the precession of the trajectory as a whole takes place along the left circle. After the angle of deviation of the magnetization vector reaches the value of the equilibrium static value, the precession becomes right, and the equilibrium position also precesses along the right circle. The same features are visible in the precessional portraits of Fig. 9.23, b and Fig. 9.23, d.

The verification shows that the presence in the precession portrait of a section with a left precession is in no way connected with the symmetry of the field and is due only to the initial position of the magnetization vector. If the initial position is inside a circle whose radius is equal to the transverse component of the static equilibrium position, then there is a section with the left precession, if outside this circle, there is no section with the left precession. The reason for this behaviour of the magnetization is that a constant effective field, which is formed by the sum of the external field and the demagnetization field of the plate, acts on the magnetization vector. If the deviation of the magnetization vector is less than the equilibrium, then the z-component of this field is directed in the negative direction of the Oz axis, if the deviation is greater than the equilibrium, then in the positive direction of the same axis. In this case, in accordance with the Landau–Lifshitz equation, in the first case the magnetization vector moves in the left circle, and in the second – in the right. That is, the replacement of the right precession with the left one occurs as if the direction of the constant field has changed to the opposite.

9.7.4. Mechanisms of formation of the precession portrait

Now, after describing the general range of phenomena accompanying the violation of the symmetry of a constant field, we proceed to consider the physical mechanisms responsible for changing the nature of the precession of the equilibrium state. Consideration will be based on two-do complementary models: energy and vector.

Energy model. From (9.2) with allowance for (9.5), it can be seen that the position of the minimum of the potential well for the magnetization vector, which determines its equilibrium position,

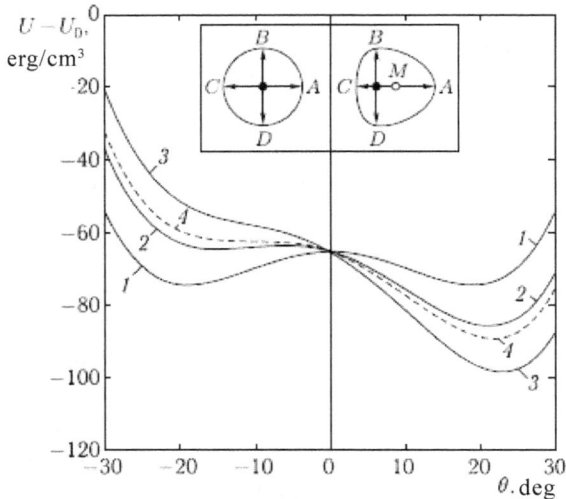

Fig. 9.25. Dependences of the energy density on the polar angle of deviation of the vector of magnetization from the normal of the plane of the plate for different field values H_{0x}: 1 – 0 Oe; 2 – 1.5 Oe; 3 – 3.0 Oe; 4 – 1.91 Oe. Parameters: $4\pi M_0 = 280$ G; $H_{0y} = 0$ Oe; $H_{0z} = 265$ Oe. Energy density is calculated from the value $U_0 = -2720.0$ erg·cm^{-3}. Explanation of the box in the text

depends on the transverse component of the constant field H_{0x}. An illustration of this change is Fig. 9.25, where the dependences of the energy density U on the angle θ between the magnetization vector and the normal to the plane of the plate are shown for different values of H_{0x}. The inset shows the consecutive in time $A \rightarrow B \rightarrow C \rightarrow D \rightarrow A$ positions of the transverse component of the alternating field with an amplitude of 3 Oe during one oscillation period. The left shows the rotation of the field in the absence of the constant component H_{0x}; it can be seen that it has a circular character, that is, $h_A = h_B = h_C = h_D = 3$ Oe. On the right is shown the field rotation with $H_{0x} = 1.5$ Oe, whose constant value corresponds to point M. It can be seen that in this case the field is not circular the h_B and h_D components are kept the same $h_B = h_D = 3$ Oe, while the h_A component increases to $h_A = 4.5$ Oe, and the h_C component decreases to $h_C = 1.5$ Oe.

Curve 1 in the main figure reflects the energy density in the absence of the field H_{0x}, which corresponds to the magnetization of the plate exactly normal to its plane. This curve has two symmetric minima at $\theta_{1.2} = \pm 18.8°$, corresponding to $U - U_0 = 54.2$ erg×cm^{-3}. All orientations of the magnetization along the azimuth angle φ are equal in energy, that is, the magnetization vector can move freely in the angle φ.

Suppose now that the transverse alternating field of circular polarization is included, the amplitude of which is insufficient for the development of precession at angles comparable with the magnitude of the static deviation of the magnetization from the normal. In this case, the magnetization vector precesses with a small amplitude around the equilibrium position, and, due to the difference in gyroscopic forces, the magnetization vector at the positions of maximum and minimum polar angles, the equilibrium position shifts to the side. Thus, the precession of this position occurs in a circle, the radius of which is determined by the magnitude of the equilibrium deviation from the normal, and the period by the amplitude of the alternating field (as shown in [306]). Due to the independence of the energy from the orientation of the equilibrium position along the azimuth angle φ, the movement of the equilibrium position in a circle is uniform. In the precession portrait, this looks like a uniform filling of a large ring with small rings (Fig. 9.23, b).

When specifying the transverse component of the field $H_{0x} = 1.5$ Oe, curve 1 is replaced by curve 2, which has two asymmetrical minima: the first is for $\theta_1 = 21.1°$ and $\varphi = 0°$, corresponding to $U - U_0 = -65.6$ erg cm^{-3}, the second at $\theta_2 = -14.7°$ and $\varphi = 180°$, corresponding to $U-U_0 = -44.3$ erg cm^{-3}. As we see, the first minimum is much deeper than the second, therefore the precession of the equilibrium position is delayed in the region of this minimum for a longer time than near the second. Such a time delay leads to a condensation of the small rings of the precession portrait shown in Fig. 9.23 d.

When specifying the transverse component of the field $H_{0x} = 3.0$ Oe, curve 1 is replaced by curve 3, which has only one minimum at $\theta = 22.9°$ and $\varphi = 0°$, corresponding to $U-U_0 = -98.2$ erg cm^{-3}. In this case, the precession of the equilibrium position can occur only in the region of this single minimum and the large ring in the precession portrait opens, like the one shown in Fig. 9.24, b, that is, there is a breakdown of the precession of the equilibrium position.

With the chosen parameters, the critical breakdown field is $H_{0x}^{(c)} = 1.91$ Oe, and the dependence $U(\theta)$ is represented by curve 4 (dashed line). This curve has a single minimum at $\theta = 21.7°$ and $U-U_0 = -89.0$ erg cm^{-3}. Instead of another minimum, there is an inflection corresponding to $\theta = -10.5°$, $U-U_0 = -62.2$ erg cm^{-3}. The

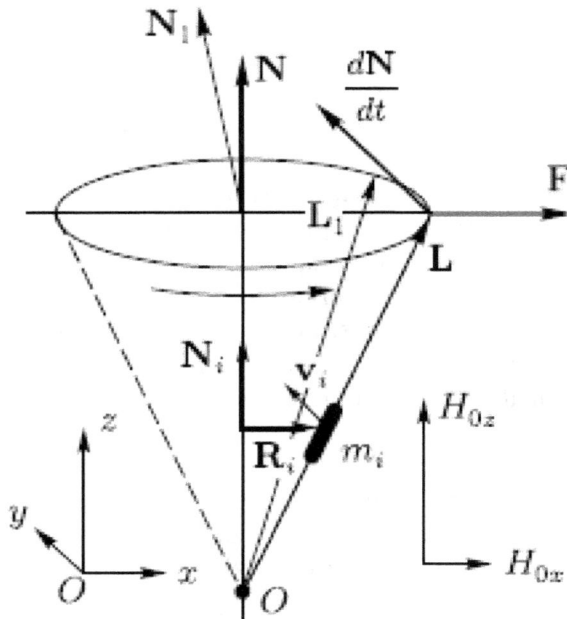

Fig. 9.26. The scheme of formation of a thickening of the rings in the precession portrait.

flat region in the vicinity of the bend is located in θ from $-13.1°$ to $-8.0°$ and corresponds to $U-U_0 = -(62.2 \pm 0.1)$ erg cm^{-3}.

Vector model. Based on Fig. 9.25, condensation should be formed where the energy surface has the deepest minimum, that is, along the axis Ox, where the field H_{0x} acts. However, as already noted, when considering Fig. 9.23, d, the thickening corresponds to the Oy axis, that is, it is shifted along the azimuth angle by $90°$. Such a shift in the condensation position can be explained on the basis of a vector model similar to that constructed in [307, 396].

Consider Fig. 9.26, which shows a diagram of the formation of thickening shear. We will consider the equilibrium position as an independent vector that precesses around the direction of the field. This can be considered if the precession period of the equilibrium position is much longer than the alternating field period, in time with which the magnetization vector precesses, that is, the position of the magnetization vector over several periods can be averaged, and then the resulting averaged vector will move as an equilibrium position vector. In the figure, the averaged equilibrium position vector is denoted by **L**. This vector precesses along a cone, the axis of which coincides with the axis Oz (or close to this axis), along which the main component of the constant field H_{0z} is directed. By analogy with mechanics [401], we can consider the vector **L** as a rigid rod with a distributed mass, for which one can enter the angular momentum using the formula

$$\mathbf{N} = \sum_i m_i \left[\mathbf{R}_i \times \mathbf{v}_i \right],$$

$$(9.24)$$

where m_i, $\mathbf{R}_i$, $\mathbf{v}_i$ is the mass, radius-vector and speed of the i-th element of the rod. The moment of amount of motion introduced in this way will be directed along the Oz axis, as shown in the figure. Next, we take into account that the constant field component H_{0x} acts on the magnetization vector and, therefore, on the average equilibrium position vector **L** in the direction that coincides with the direction of this component, that is, along the axis Ox, forming force **F**. The moment of this force relative to the point O of fixing vector **L** is proportional to the vector product:

$$\mathbf{P} \quad [\mathbf{L} \times \mathbf{F}] \qquad (9.25)$$

In this case, according to the basic law of the dynamics of a solid body, the change in the angular momentum **N** is determined by

$$d\mathbf{N}/dt = \mathbf{P}, \qquad (9.26)$$

or

$$d\mathbf{N}/dt \propto [\mathbf{L} \times \mathbf{F}], \qquad (9.27)$$

that is, directed along the Oy axis. As a result, the angular momentum **N** under the action of the field component H_{0x} leans towards the axis Oy, assuming the position $\mathbf{N}_1$. Such a change in the orientation of the moment of momentum **N** leads to a similar change in the orientation of the equilibrium position vector **L** also in the direction of inclination to the axis Oy to the position $\mathbf{L}_1$, which, with a sufficient value of H_{0x}, leads to the transfer of the small rings concentration on the precession portrait from the axis Ox to the axis Oy counterclockwise in relation to the arrows. Thus, the consideration carried out on the basis of the energy model makes it possible to qualitatively explain the appearance of condensation of small rings in the precession portrait, as well as the phenomenon of the breakdown of the precession of the equilibrium position in a field that is more critical. The proposed vector model also makes it possible to qualitatively explain the rotation of the condensed region counterclockwise. Note, however, that the value of the stall field obtained on the basis of the energy model is somewhat overestimated in comparison with that observed in the dynamic mode.

9.7.5. Period formula check

As mentioned above (section 9.1.4), the formula (9.19) was proposed in [306], which relates the precession period of the equilibrium state T with the constant field H_0 and the amplitude of the alternating field $h_{0x,y}$:

$$T = \frac{A}{h_0^2 \sqrt{1 - \left(\frac{H_0}{4\pi M_0}\right)^2}}. \tag{9.28}$$

In order to determine the applicability of this formula, in the asymmetrical case, a test was carried out for $4\pi M_0 = 280$ G on two parts 100 and 500 MHz with a constant field of 80 to 240 Oe and an amplitude of the alternating field from 2 to 70 Oe. from above, the constant field varied from 0.02 to 0.20 Oe. The test showed that the transverse component of the constant field $H_{0x,y}$, being small compared to the longitudinal component H_{0z}, does not have a significant effect on the dependence of the period on H_{0z}, and the reduced formula (9.28) was satisfied in the whole range of parameters change in a constant field with an accuracy of at least 9.4%, in a variable field – at least 6.5%.

9.7.6. Areas of precession equilibrium positions

Let us now consider how the asymmetry of a constant field, determined by its transverse component H_{0x}, affects the x-ray region of the existence regions of the precession of the equilibrium position along the constant H_{0x} and alternating $h_{0x,y}$ fields. The general character of the areas of existence is illustrated in Fig. 9.27, where the dependences of the critical field of breakdown of the precession of the equilibrium position on the amplitude of the alternating field at different frequencies are shown. Each curve 1–5, corresponding to a certain frequency, consists of solid and dashed sections. In

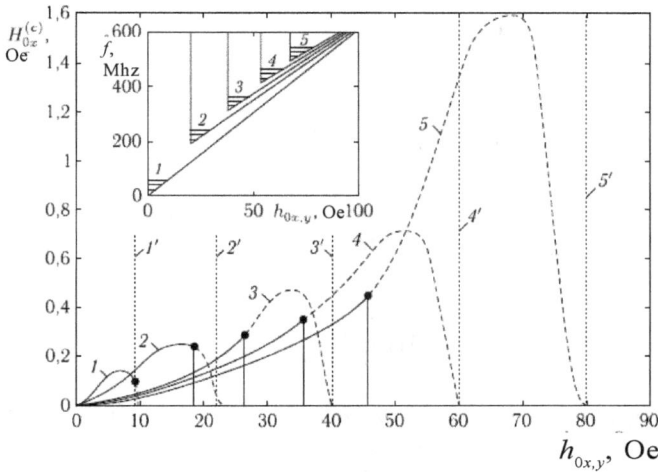

Fig. 9.27. The dependences of the critical field of the precession and the equilibrium position of the amplitude of the alternating field at different frequencies f: 1, 1' – 100 MHz; 2, 2' – 200 MHz; 3, 3' – 300 MHz; 4, 4' – 400 MHz; 5, 5' – 500 MHz. Parameters – $4\pi M_0$ = 280 G; α = 0.3; H_{0y} = 0 Oe, H_{0z} = 265 Oe. Cutting parameters given in the text

this case, the solid lines correspond to the mode without centre coverage, and the dotted one – with centre coverage. Large points and vertical straight dotted lines emanating from them correspond to the boundaries of the transition from one mode to another. For each frequency, the precession of the equilibrium position exists only below the corresponding curve of line 1–5, and to the left of the enlarged point with a vertical dotted line the mode without centre coverage is implemented, to the right is the mode with centre coverage.

It can be seen from the figure that at each frequency for any given amplitude of the alternating field $h_{0x,y}$ there is a critical value of the breakdown field $H_{0x}^{(c)}$, below which the precession of the position exists, and no higher. All curves 1–5 are similar to each other, with each curve having a maximum, and consists of smoothly ascending and sharply falling sections. The maximum corresponds to the amplitude of the variable field $h_{0x,y}^{(m)}$ and the stall field $H_{0x}^{(cm)}$. As the frequency increases, the maximum becomes larger and shifts toward a larger alternating field. All curves begin near the amplitude of the field $h_{0x,y}$ = 0.00 Oe and end at the field value $h_{0x,y}^{(e)}$, its own for each frequency. When a certain critical value

$h^{(e)}_{0x,y}$ is exceeded by a variable field, the mode without centre coverage is replaced by the mode with coverage.

The critical values of the maximum stall field and the amplitudes of the variable field are as follows:

at frequency f = 100 MHz: $h^{(m)}_{0x,y}$ = 7.0 Oe; $H^{(cm)}_{0x}$ = 0.141 Oe; $h^{(c)}_{0x,y}$ = 8.4 Oe; $h^{(e)}_{0x,y}$ = 9.65 Oe;

at frequency f = 200 MHz: $h^{(m)}_{0x,y}$ = 16.0 Oe; $H^{(cm)}_{0x}$ = 0.245 Oe; $h^{(c)}_{0x,y}$ = 17.0 Oe; $h^{(e)}_{0x,y}$ = 23.25 Oe;

with frequency f = 300 MHz: $h^{(m)}_{0x,y}$ = 34.0 Oe; $H^{(cm)}_{0x}$ = 0.460 Oe; $h^{(c)}_{0x,y}$ = 0.27 Oe; $h^{(e)}_{0x,y}$ = 40.55 Oe;

with frequency f = 400 MHz: $h^{(m)}_{0x,y}$ = 51.0 Oe; $H^{(cm)}_{0x}$ = 0.710 Oe; $h^{(c)}_{0x,y}$ = 35.0 Oe; $h^{(e)}_{0x,y}$ = 61.25 Oe;

with frequency f = 500 MHz: $h^{(m)}_{0x,y}$ = 69.0 Oe; $H^{(cm)}_{0x}$ = 1,600 Oe; $h^{(c)}_{0x,y}$ = 46.0 Oe; $h^{(e)}_{0x,y}$ = 85.05 Oe.

The sparse dotted lines 1'–5' shows the boundaries between the two modes when the field is greater than the critical field of failure. These boundaries are vertical straight with an accuracy of at least 5%. On both sides of these gates, the precession of the equilibrium position is absent and the precession occurs, which is close to circular, but the characters of the precession on either side of the boundary differ. The precession portrait to the left of these lines has the form of a ring, slightly smeared partially elongated along the Oy axis, the centre of which is close to the point of static equilibrium. The precession portrait to the right of these lines has the form of a narrow well-defined ring, the shape of which approaches a circular one, and the centre coincides with the origin. From different heights of the maxima of curves 1–5, it follows that for a certain value of the transverse constant field, the precession of the equilibrium position can exist only at frequencies exceeding some critical value. This circumstance is additionally illustrated in the sidebar, where the dependences of the critical frequency on the amplitude of the alternating field are schematically shown for various transverse constant fields. The numerical designations of the curves correspond to the values of the transverse field $H^{(c)}_{0x}$: 1 – 0.0 Oe; 2 – 0.2 Oe; 3 – 0.4 Oe; 4 – 0.6 Oe; 5 – 0.8 Oe. The precession of the equilibrium position exists only inside the triangles, the angles at the lower edges of which, formed by the inclined and vertical lines, are shaded.

Comment. Thus, from the sidebar in Fig. 9.27 it can be seen that as the amplitude of the alternating field increases, the region of existence of the precession of the equilibrium position shifts higher in frequency. It should be noted that a similar shift has already been noted in section 9.2.2 when considering Fig. 9.8, which shows that as the amplitude of the alternating field increases from 1.5 Oe to 6.0 Oe, a sharp drop in the curves 1–4, to the right of which the precession of the equilibrium state takes place, shifts from 32 MHz to 66 MHz. That is, despite the difference in the intervals of variation of the field and frequency in Fig. 9.27 and Fig. 9.8, the fact of the existence of a critical frequency corresponding to a given field value, in both cases, is completely preserved. Thus, it can be assumed that Fig. 9.27 serves as an additional confirmation of the position expressed in section 9.2.2 about a certain lag of the magnetization vector from the rotation of the alternating field vector, which is the reason for the existence of the critical frequency, which can only be increased by increasing the amplitude of the alternating field.

9.7.7. Explanation of the form of existence regions based on the energy model

The general character given in Fig. 9.27 curves can be qualitatively explained using the energy model described above (section 9.7.4). Consider Fig. 9.28, where the dependences of the energy density U on the polar angle θ of the deviation of the magnetization vector from the normal to the plane of the plate are shown for three-fold values of the transverse component of the constant field H_{0x}. Figure 9.28 a corresponds to $H_{0x} = 0$. In this case, the dependence $U(\theta)$ with respect to the line $\theta = 0$ is symmetric and has two minima of the same depth for $U = U_1$ with a maximum between them at $U = U_2$. In the absence of an alternating field, the energy level of the system as a whole corresponds to minima at $U = U_1$ and there is no

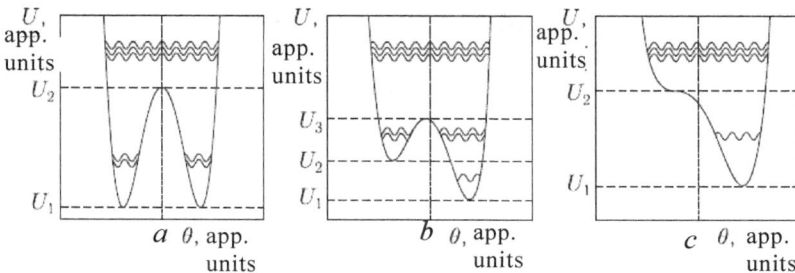

Fig. 9.28. Dependences of the energy density on the polar angle of deviation of the magnetization vector from the normal to the plane of the plate, for different values of the transverse component of the constant field.

precession of magnetization. As the amplitude of the alternating field increases, the energy level of the system as a whole increases, and a precession of magnetization occurs near the minima. In the interval $U_1 < U < U_2$, due to the equality of energy minimums, the symmetric precession of the equilibrium position occurs (conventionally, the level for this energy case is shown by a double wavy line). In the precession portrait, the diameter of small rings is determined by the width of the minima of the $U(\theta)$ dependence, and the diameter of the large ring is determined by the distance between them.

With further increase in the alternating field, the energy level increases and, at $U > U_2$, the magnetization vector begins to precess along a large circle, determined by the distance between the side branches of the $U(\theta)$ dependence (triple wavy line). On the precession portrait, the small rings disappear and only one narrow large ring remains, which means the breakdown of the precession of the equilibrium position. Thus, the precession of the equilibrium position exists in the range of the variable field from zero to the final value corresponding to $U = U_2$. In Fig. 9.27, this corresponds to the lower edge of the regions of existence of the precession of the equilibrium position, depicted by the curves 1–5.

Consider now Fig. 9.28, b, corresponding to $0 < H_{0x} < H^{(c)}_{0x}$. In this case, the dependence $U(\theta)$ is asymmetric and has two minima at $U = U_1$ and $U = U_2$ with a maximum between them at $U = U_3$. In the absence of an alternating field, the energy level of the system as a whole corresponds to the minimum at $U = U_1$ and there is no precession. With an increase in the amplitude of the alternating field, the energy level of the system as a whole increases, and precession of magnetization occurs near the first minimum. In the interval $U_1 < U < U_2$, this precession has a circular character around the position of the static equilibrium of the magnetization vector, determined by an asymmetric constant field (single wavy line). In the precession portrait there is one small ring, the shift of its centre relative to the origin of coordinates is determined by the asymmetry of the constant field. In the interval $U_2 < U < U_3$, conditions arise for the precession of the equilibrium position (the energy level for this case is shown by a double wavy line). The precession portrait has the form of a large ring filled with small rings with condensation, the intensity of which is determined by the difference between the minima of U_1 and U_2. At $U > U_3$, the precession of the equilibrium position is broken and in the precession portrait there remains one narrow large ring (in accordance with the triple wavy line), whose centre is near the origin. In Fig. 9.27 this behaviour of the precession of the equilibrium state corresponds to the interval of the alternating field determined by the intersection of the curves type 1–5 with the horizontal lines corresponding to a given value of the transverse component of the constant field in the interval $0 < H_{0x} < H_{0x}^{(c)}$.

Figure 9.28 c corresponds to $H_{0x} > H_{0x}^{(c)}$. In this case, the dependence $U(\theta)$ has only one minimum at $U = U_1$ and the precession of the equilibrium position is absent. In the interval $U_1 < U < U_2$, precession occurs along a small circle, the centre of which in the precession portrait is located near the equilibrium position of the magnetization. When $U > U_2$, the precession occurs in a large circle, the centre of which in the precession portrait is located near the origin.

9.7.8. Dependence of the precession period of the equilibrium position on the asymmetry of the constant field

Let us now consider how the asymmetry of the constant field is reflected in the precession period of the equilibrium state. Figure 9.29 shows the dependences of the precession period of the equilibrium position T on the transverse component of the constant field H_{0x} for various times of the variable field $h_{0x,y}$.

The vertical dotted lines correspond to the critical values of the breakdown field of the precession of the equilibrium state $H_{0x}^{(c)}$ with technical values of the magnitudes of the variable field $h_{0x,y}$. The enlarged points correspond to the values of the precession period of the equilibrium position when the field is less than the breakdown field by 0.001 Oe. For the curves 1 and 2, these points are out of the picture frame and correspond to:

curve 1 – $T = 3.14$ μs with $H_{0x} = 0.039$ Oe (with $H_{0x}^{(c)} = 0.040$ Oe);
curve 2 – $T = 3.00$ μs with $H_{0x} = 0.136$ Oe (with $H_{0x}^{(c)} = 0.137$ Oe).

For the rest, the values of the integrated checks are as follows:
3 – $T = 0.300$ μs with $H_{0x} = 0.22326$ Oe (with $H_{0x}^{(c)} = 0.22327$ Oe);
4 – $T = 0.105$ μs with $H_{0x} = 0.300$ Oe (with $H_{0x}^{(c)} = 0.350$ Oe);
5 – $T = 0.080$ μs with $H_{0x} = 0.440$ Oe (with $H_{0x}^{(c)} = 0.450$ Oe);
6 – $T = 0.038$ μs with $H_{0x} = 0.600$ Oe (with $H_{0x}^{(c)} = 0.660$ Oe).

FThe dashed curve passing through the enlarged points corresponds to the boundary of the region of existence of the precession of the equilibrium position. Below and to the left of this curve, the precession of the equilibrium state is present; above and to the right,

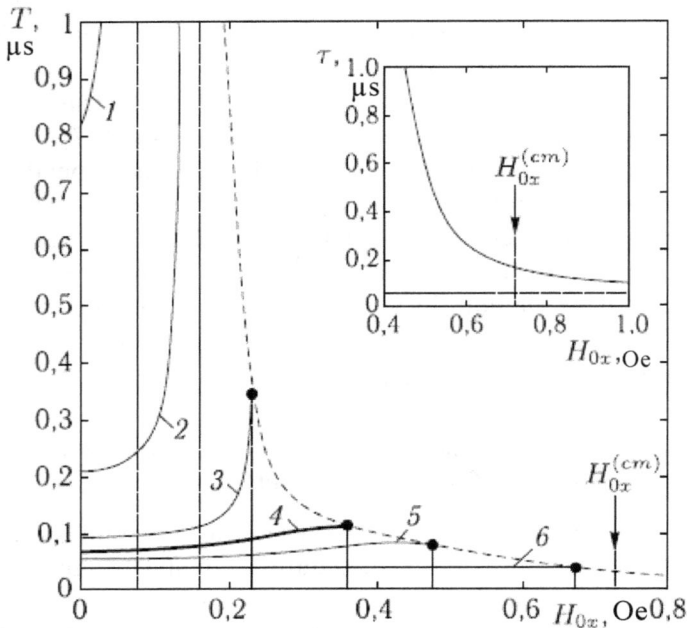

Fig. 9.29. Dependences of the precession period of the equilibrium position on the transverse component of the constant field at different amplitudes of the permeable field $h_{0x,y}$: 1 – 10 Oe; 2 – 20 Oe; 3 – 30 Oe; 4 – 35 Oe; 5 – 40 Oe; 6 – 50 Oe. Parameters: $4\pi M_0 = 280$ G, $\alpha = 0.3$, $H_{0y} = 0$ Oe, $H_{0z} = 265$ Oe, $f = 400$ MHz. Inset parameters are given in the text.

it is absent. When $H_{0x} \to 0$, the dotted curve tends to infinity. Only in this unique case, the precession period of the equilibrium state tends to infinity, that is, when $H_{0x} > 0$, it is always limited by the stronger, the more H_{0x}. When $H_{0x} \to \infty$, the dashed curve tends to the value of the period of the exciting alternating field, equal to $T_h = 0.0025$ μs (at the excitation frequency $F = 400$ MHz). Curves 1–3 correspond to the mode without centre coverage, curves 5, 6 correspond to the mode with centre coverage. Curve 4 (thickened line) corresponds to the transition from the mode without centre coverage to the mode with centre coverage, that is, it is the boundary between areas with different modes: above this curve mode without centre coverage is implemented, below – mode with centre coverage.

It can be seen from the figure that under the regime without centre coverage (curves 1–3), the period of precession of the equilibrium state increases slowly, first as the field component H_{0x} increases, and then rises sharply as it approaches the breakdown field. The increase in the period compared with the value at $H_{0x} = 0$ Oe is greatest at $h_{0x,y} = 20$ Oe (curve 2) – about 15 times, slightly less - at $h_{0x,y} = 10$

Oe (curve 1) – 4 times and even less when $h_{0x,y} = 30$ Oe (curve 3) – about 3 times. With a further increase in the amplitude of the variable field $h_{0x,y}$, the increase in the precession period of the equilibrium state with increasing field H_{0x} slows down, and to the border between areas with different modes (curve 4) the period increases by less than 1.5 times, and at the right end of the T dependence on H_{0x} appears flat horizontal area, which then turns into a slightly falling (curve 5). With a further increase in $h_{0x,y}$, starting from $h_{0x,y} = 50$ Oe, the dependence of period T on the field H_{0x} becomes extremely small or even completely absent (curve 6). The transition from precession of the equilibrium position to simple precession under the regime without centre coverage (curves 1–3) occurs with a sharp change in the period and amplitude of oscillations of the magnetization. Slow oscillations corresponding to the precession of the equilibrium state completely disappear, only the fast ones remain, the period of which is equal to the exciting field of the alternating field, and the amplitude of the fast oscillations jumps twice and more, usually somewhat exceeding the amplitude of the slow oscillations. Any traces of the precession of the equilibrium position completely disappear, the magnetization vector precesses in a stationary circular (or close to circular) orbit without any deviations.

The transition from the precession of the equilibrium position to the simple precession under the regime with centre coverage (curves 5, 6) proceeds quite smoothly. The amplitude of fast oscillations occurring with the exciting field period increases slightly, but the amplitude of the precession of the equilibrium state ceases to be stationary in time and gradually decreases, that is, the precession of the equilibrium position becomes damped, and the equilibrium state gradually assumes a new stationary position. As a result, after some time, only the stationary circular (or close to circular) precession of the magnetization vector around the new stationary equilibrium state remains. In the process of damping, the period of oscillation of the precession of the equilibrium state remains close to the value that was in the continuous mode, and the amplitude decreases the faster, the larger the field H_{0x}.

As an example, the inset shows the dependence of the decay time τ (in amplitude up to 0.1 of the initial value) on the field H_{0x} at $h_{0x,y} = 40$ Oe. At the bottom, the horizontal dotted line shows the decay time of the natural oscillations of the magnetization in the absence of excitation equal to 0.05 ms It can be seen from the figure that the precession of the equilibrium position decays much slower than free oscillations, and this excess is greater, the smaller the field H_{0x} is separated from its critical value. The continuous precession of the equilibrium position at any amplitude of the variable field $h_{x,y}$ exists only when the field $H_{0x} < H_{0x}^{(cm)}$, where the value $H_{0x}^{(cm)} = 0.710$ Oe is indicated by the arrow. The decaying precession of the equilibrium position after the excitation is turned on can also exist with fields H_{0x} greater than $H_{0x}^{(cm)}$, and the decay time decreases with increasing H_{0x} in accordance with the dependence shown in the inset. The decaying precession of the equilibrium position disappears completely only in the H_{0x} field, such that its decay time becomes less than the decay time of natural oscillations, which at $\alpha = 0.3$ corresponds to approximately $H_{0x} \sim 1.60$ Oe.

Note that the decay time of the precession of the equilibrium position at the constant field H_{0x} also decreases with increasing field $h_{x,y}$. So, with $H_{0x} = 0.4$ Oe, the critical value of the variable field $h_{x,y}$, above which the precession of the equilibrium state occurs, is 54 Oe. Near the field, the precession decay time is $\tau = 0.50$ μs. At $h_{x,y} = 56$ Oe, the decay time is 0.24 μs, at 58 Oe it is 0.16 μs, and at 60 Oe 0.10 μs. Here, too, the precession of the equilibrium state fades much slower than free oscillations.

9.7.9. Frequency properties of the precession period in case of asymmetry of a constant field

Consider now Fig. 9.30, where the dependences of the precession period of the equilibrium state T on the transverse component of the constant field H_{0x} are shown for different values of the variable field $h_{x,y}$. The vertical dotted lines correspond to the critical values of the breakdown field of the precession of the equilibrium position $H_{0x}^{(c)}$ with the same values of the height of the variable field. To the left of these lines, the precession of the equilibrium position is stationary in time, to the right – for the curves 2–4 and 8 – is absent, for the curves 1 and 5–7 it is damped (over 2–5 periods), and the faster, the higher the frequency, decays. Large points on the curves 1, 6 and 7 correspond to a change in the modes of precession of the equilibrium position from stationary to decaying. For curve 5, the critical field of stall and regime change is located behind the right edge of the figure and corresponds to 1.10 Oe.

It can be seen from the figure that, depending on the ratio between the frequency and amplitude of the alternating field, two types of dependence of the precession period on the magnitude of

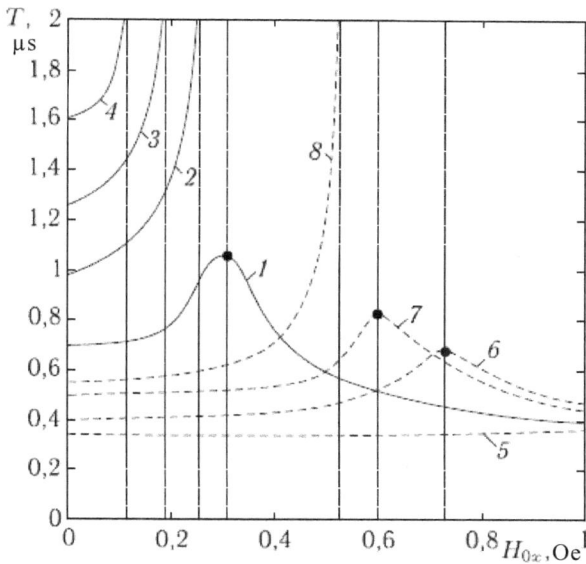

Fig. 9.30. Dependences of the precession period of the equilibrium position on the transverse component of the constant field for different frequencies of the ac field $h_{x,y}$: 1 – 300 MHz; 2 – 400 MHz; 3 – 500 MHz; 4 – 600 MHz; 5 – 500 MHz; 6 – 600 MHz; 7 – 700 MHz; 8 – 800 MHz. The solid curves 1–4 correspond to $h_{0x,y}$ = 30 Oe. The dotted curves 5–8 correspond to $h_{0x,y}$ = 60 Oe. Parameters: $4\pi M_0$ = 280 G, α = 0.3, H_{0y} = 0 Oe, H_{0z} = 265 Oe.

the transverse component of the constant field are possible. The dependence of the first type (curves 1, 5–7) – with an increase in the transverse component of the constant field, the period increases to a certain final value, after which it decreases. The maximum period corresponds to the critical field of failure, below which the precession is stationary, and above – damped. The dependence of the second type (curves 2–4, 8) – with an increase in the transverse component of the constant field near the breakdown field, the period tends to infinity, below the breakdown field, the precession is stationary, and above – not.

For a given amplitude of the alternating field, there is a critical frequency, below which the dependences of the first type are observed, and above – the second. With amplitude $h_{0x,y}$ = 30 Oe (curves 1–4), the critical frequency is approximately equal to $f_c \approx 350$ MHz, with amplitude $h_{0x,y}$ = 60 Oe (curves 5–8) the critical frequency is $f_c \approx 750$ MHz. From the general nature of the dependences shown in Figs. 9.29 and 9.30, we can conclude that when the transverse field approaches the breakdown field, the precession period of the equilibrium state increases (in some cases, to infinity). This can be explained by the fact that the approximation of the transverse field to the field of breakdown means an increase in the asymmetry of the constant field. Such an increase in asymmetry leads to an increase in the delay time of the main precession, which in the precession portrait is manifested in an increase in the concentration of small sheep in the condensation region. An increase in the delay time leads to an increase in the precession period of the equilibrium position, which is observed in Figs. 9.29 and 9.30.

9.8. Violation of symmetry in a transverse alternating field

In the previous section, the precession of the equilibrium state was studied under conditions of non-symmetry in a constant field. We now turn to a more detailed consideration of asymmetric excitation, which consists in the inequality of the transverse components of the alternating field, as well as in the presence of its longitudinal component. As an additional factor in these words, we also take into account the asymmetry of the constant field. We will limit consideration to the case of mode 2 – the precession of the equilibrium position without centre coverage.

9.8.1. Symmetrical excitation conditions

As a starting point for the following analysis of asymmetric excitation, we turn to Fig. 9.31, which shows the development in time of the component of magnetization m_x (a) and m_y (b), as well as the corresponding precession portrait $m_y(m_x)$ (c). In Figs. 9.31, a and 9.31, b frequent oscillations correspond to the precession of the magnetization with the frequency of excitation, rarely the precession of the equilibrium position. It is seen that the precession of the equilibrium position has a sinusoidal character with a phase shift between the components by 90°. The precession portrait is a large ring corresponding to the precession of the equilibrium position, uniformly filled with small rings corresponding to the precession of the magnetization around the equilibrium position with the frequency of excitation. Further, the parameters of the excitation and the constant field will vary with respect to the parameters adopted in this figure.

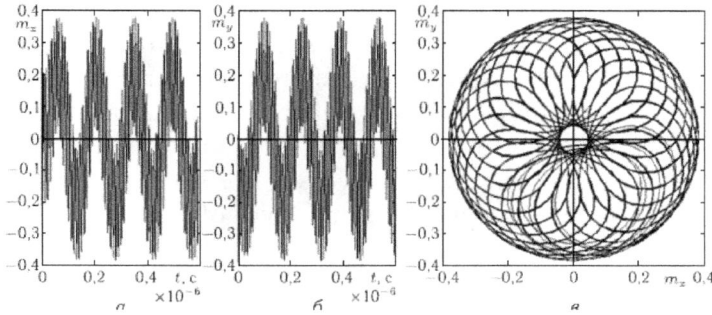

Fig. 9.31. The precession of the equilibrium position under symmetric excitation: h_{0x} = 7.00 Oe, h_{0y} = 7.00 Oe. Parameters: $4\pi M_0$ = 280 G, H_{0z} = 267 Oe, α = 0.3, f = 100 MHz

9.8.2. Asymmetrical transverse ac field

In the course of these studies, it was found that asymmetrical excitation, in comparison with the symmetric case, dramatically changes the nature of the precession of the equilibrium position. We first consider the case of transversely asymmetric excitation, when the alternating field has not circular, but elliptical polarization. Let the variable field be:

$$h_x = h_{0x} \sin(2\pi f t), \tag{9.29}$$

$$h_y = h_{0y} \cos(2\pi f t), \tag{9.30}$$

where F is the frequency of the alternating field, $h_{0x,y}$ is its amplitude. Symmetric excitation corresponds to circular polarization of an alternating field at which h_{0x} = h_{0y}. Transversely asymmetric excitation, we will call the case of an elliptical polarization field, that is, when h_{0x} = H_{0y}. The verification shows that the nature of the oscillations at different levels of excitation, even if the same degree of asymmetry is maintained, differs markedly. A preliminary check shows that a significant difference in the nature of the oscillations occurs when passing through the excitation amplitude $h_{0x,y}$ of order 2...4 Oe.

Comment. A more precise meaning, as well as its criticality, that is, the sharpness of change, leading to a significant change in the nature of the oscillations, is not known to the authors of this monograph at the time of its writing, since such articles are not available in the articles on which this presentation is based. Thus, the authors leave this question as a subject for further research.

So, at first we will consider as a more characteristic case in which the level of excitation is already sufficient for the manifestation of the strongest difference of oscillations from the symmetric case. Turn to Fig. 9.32, where the same processes are shown as in Fig. 9.31, occurring at approximately the same level of excitation, but with a significant asymmetry.

It can be seen from the figure that in this case the precession of the equilibrium state has the character of pulses of irregular shape (*a, b*), with a phase shift between the components by the same 90°. In the precession portrait (*c*), the filling of the large ring with small ones is sharply uneven and there are two ring-shaped dampings near the axis *Ox*.

In the example, the violation of the symmetry of the variable field is about 12.5%. However, a preliminary check shows that here also (as with the change of amplitude) there are two clearly distinct oscillation modes. The boundary between them corresponds to a violation of symmetry by about 15...20%. To reveal another, different from the illustrated Fig. 9.32 mode, we now consider a more significant degree of symmetry breaking at the same level of excitation amplitude. Turn to Fig. 9.33, corresponding to the violation of symmetry about 25%. It can be seen that in this case the periodic precession of the equilibrium position is absent, and only oscillations of the magnetization with the excitation frequency (*a, b*) occur, that is, from the initial moment of time the equilibrium position is established in a stationary state, and the magnetization

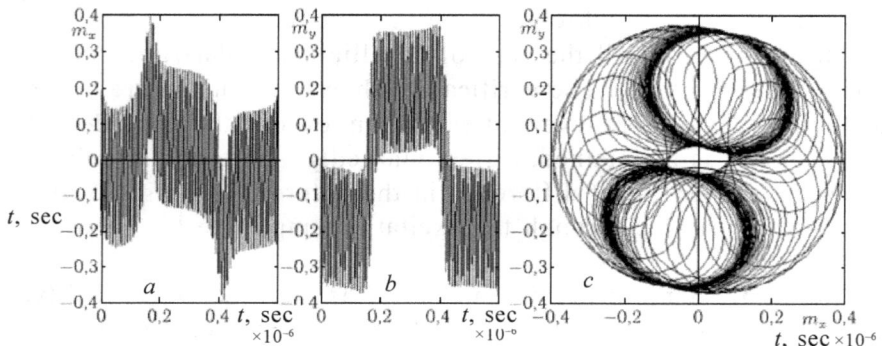

Fig 9.32. The precession of the equilibrium position with asymmetric excitation: $h_{0x} = 7.00$ Oe, $h_{0y} = 6.12$ Oe. The remaining parameters are the same as those adopted in construction of Fig. 9.31

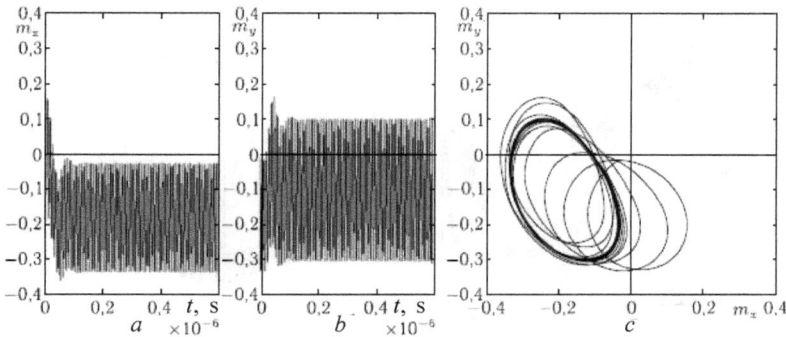

Fig. 9.33. Magnetization precession with a strong asymmetry of the alternating field: $h_{0x} = 7.00$ Oe, $h_{0y} = 5.25$ Oe. The remaining parameters are the same as those used in the construction of Fig. 9.31

further precesses around this position with frequency excitement. In the precession portrait (*c*) it looks like the formation of a stable cycle, the centre of which is relatively zero shifted.

So, in contrast to the case of symmetric excitation, the above picture is sharply critical to the value of the asymmetry of the alternating field. So in the interval $0 < h_{0y} < 0.86 h_{0x}$, the character of the precession corresponds to Fig. 9.33, in the interval $0.86 h_{0x} < h_{0y} < 0.90 \ h_{0x}$ – fig. 9.32, in the interval $0.90 \ h_{0x} < h_{0y} < 1.00 \ h_{0x}$ – Fig. 9.31.

9.8.3. Cases of small and large amplitudes

Finally, let us consider the nature of the precession of the equilibrium position with a small degree of asymmetry of the excitation (less critical) in cases of scarlet and large levels. Consider Fig. 9.34, where the dependences of the transverse oscillations of the magnetization on time and the corresponding precession portraits at two levels of excitation are shown.

It can be seen from the figure that, depending on the amplitude of excitation, oscillations occur in two substantially different modes. The most characteristic features that distinguish both of these modes from the symmetric case (Fig. 9.31), apparently, can be considered the change in the oscillation period (Fig. 9.34, a, c), as well as the condensation of small rings, as more or less uniform at low levels excitement (Fig. 9.31, b), and characteristic forked at its high level (Fig. 9.31, d). Consider these signs separately in more detail.

516

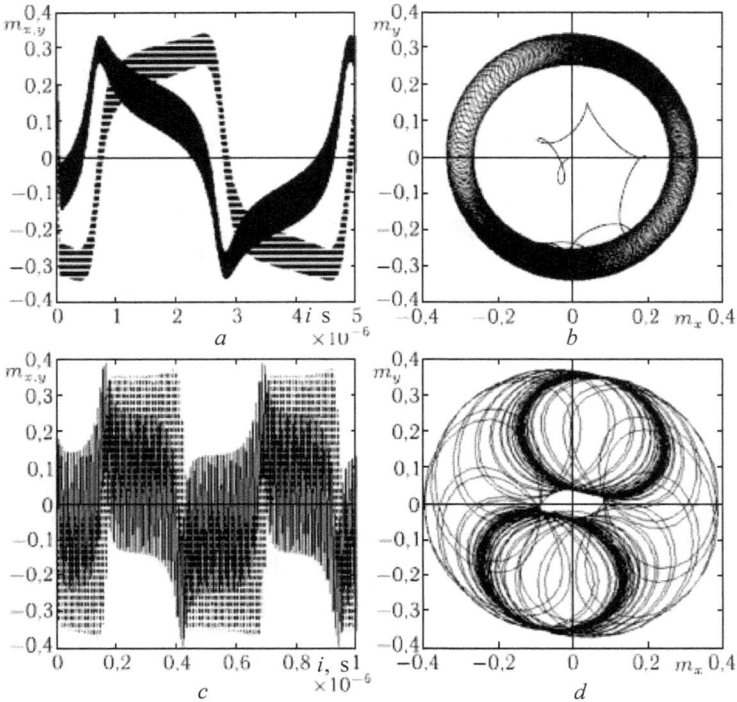

Fig. 9.34. Dependences of vibrations of transverse components of magnetization on time (*a, c*) and precessional portraits (*b, d*) bring symmetric excitation (*c, d*) of small (*a, b*) and large (*c, d*) amplitudes. (*a, b*) corresponds to: h_{0x} = 1.75 Oe, h_{0y} = 1.43 Oe. Large amplitude (*c, d*) corresponds to: h_{0x} = 7.00 Oe, h_{0y} = 6.12 Oe. The remaining parameters are the same as those taken by construction of Fig. 9.31

Period of oscillation. From time sweeps in the left-hand portions of the figure (*a, c*), it can be seen that the period of oscillation of the precession of the equilibrium state decreases with increasing excitation amplitude: at small amplitude it is T_1 = 4.4 • 10^{-6} s, at large T_2 = 0.52 • × 10^{-6} s. For the dependence of the period on the amplitude of the variable field in [306], the formula (9.19) was proposed:

$$T = \frac{A}{h_0^2 \sqrt{1 - \left(\dfrac{h_0}{4\pi M_0}\right)^2}}.$$

(9.31)

For symmetric excitation, the value of A = 1.45 Oe² s was obtained in this work, whence with small amplitude, based on the average value of the asymmetric field of 1.59 Oe and its extreme values of 1.43 Oe and 1.75 Oe, the period is equal to: $T_1 = (1.90 \pm 0.04) \cdot 10^{-6}$ s, and with a large amplitude with an average field of 6.56 Oe and extreme values of 6.120 Oe and 7.000 Oe, we get: $T_2 = (0.112 \pm 0.014) \cdot 10^{-6}$ s.

Thus, it can be seen that at any amplitude, with asymmetric excitation, the period of oscillations of the precession of the equilibrium position is much longer (2...5 times) than with the symmetric one. Find the corresponding values of the parameter A. From the formula (9.31) we obtain

$$A = Th_0^2 \sqrt{1 - \left(\frac{H_0}{4\pi M_0}\right)^2}. \qquad (9.32)$$

For a small amplitude up to rounding, we get: $A_1 = (3.35 \pm 0.13)$ Oe^2 s. Similarly, for a large amplitude, we obtain: $A_2 = (6.76 \pm 0.16)$ Oe^2 s. That is, the value of the constant A here also turns out to be much larger.

Precession portrait. It can be seen from the precessional portraits that at a small amplitude the small rings are smeared, overlapping each other, forming a continuous smeared area in the first coordinate quarter between $20°$ and $90°$ relative to the axis Ox. The centres of the rings are distributed more or less evenly between the angles from $30°$ to $80°$. In the third quarter there is the same smeared area, which is symmetrical to the first relative to the rotation of $180°$ (that is, between $200°$ and $270°$ from the same axis) with the centres of the rings between $210°$ and $260°$. At large amplitudes, small rings are sharp and are concentrated mainly in the first, with a partial approach to the second, coordinate area between $10°$ and $210°$ (tangent to the edges of the rings) relative to the axis Ox. The centres of the rings are concentrated in a narrow region of $80° \pm 3°$. In the third quarter there is the same region of sharp butts, symmetrical to the first one relative to the rotation of $180°$ (that is, between $190°$ and $300°$ from the same axis). The centres of the rings are concentrated in the region of $260° \pm 3°$.

From Fig. 9.34 d it can also be seen that with a large amplitude the precession portrait is obtained slightly flattened vertically ($|m_y|_{\max} < |m_x|_{\max}$). This flattening is especially pronounced on the inner envelope of a large ring. So, on the outer envelope: $|m_y|_{\max} = 0.370$, $|m_x|_{\max} = 0.395$, from where: $|m_y|_{\max} / |m_x|_{\max} = 0.94$.

At the same time on the inner envelope: $|m_y|_{\max} = 0.45$, $|m_x|_{\max} = 0.10$, from where: $|m_y|_{\max} / |m_x|_{\max} = 4.50$.

We also note that from the sweeps in time (a, c) one can see an additional feature of the oscillations: the pulses at any amplitude are highly asymmetrical, contain smooth, almost flat sections, separated by sharp steep drops. The flat areas of the pulses on the timescales (a, c) correspond to the condensations of small rings on the precession portraits (b, d), and the sharp differences between the pulses – to the rarefaction sites of technical rings.

9.8.4. Diagram of the formation of condensations of small rings in a precession portrait

Let us now consider the qualitative scheme for the formation of condensations of small rings in the precession portrait with transverse asymmetrical excitation, illustrated in Fig. 9.35.

Figure 9.35 shows a diagram of the formation of condensations of small rings on a precession portrait with asymmetric excitation in the case of $h_x > h_y$. Figure 9.35, a shows the general scheme of the formation of a precession portrait. The plane of the drawing coincides with the planes Oxy and $Om_x m_y$, that is, with the plane of the precession portrait. The constant field is oriented perpendicular to the plane of the pattern towards the observer, the alternating field with the components h_x and hy rotates in the plane Oxy. The dashed lines show the boundaries of the large precession ring under symmetric excitation. Points A, B, C, D are consecutive positions of

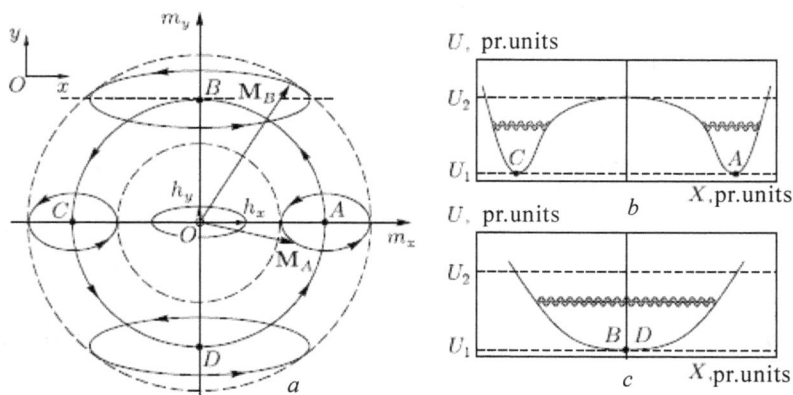

Fig. 9.35. Scheme of the formation of condensations of small rings in the precessional portrait with asymmetric excitation ($h_x > h_y$). a – education scheme of the precession portrait; b – the dependence of the potential on the coordinate x at $y = 0$; c – dependence of the potential on the coordinate x at $y = y_B$.

the centres of small rings of the precession portrait, shown by solid thick lines. The $\mathbf{M}_A$ and $\mathbf{M}_B$ vectors are the instantaneous positions of the projections of the magnetization vector $\mathbf{M}$ on the Oxy plane during its precession around points A and B. The thin solid ring between the dotted (dashed) lines is the locus of the end of the magnetization vector in the absence of excitation. This line passes through points A, B, C, D. The arrows indicate the directions of precession of the magnetization vector and the equilibrium position.

Figures 9.35, b and c show the dependence of the static potential on the x coordinate at $y = 0$ (b) and $y = y_B$ (c). Double wavy lines show equal potential levels corresponding to the maximum possible magnitude of the precession of the magnetization vector. Static potential is a figure of rotation around the axis Oz (perpendicular to the plane of the drawing), the top view of which corresponds to Fig. 9.35, a. The solid line in Fig. 9.35, a, passing through points A, B, C, D is the line on which the potential has a minimum (the potential has the form of an 'annular ditch' with a 'hill' in the middle).

Due to the fact that both potential wells in the cross section of the potential surface at $y = 0$ (b) are much narrower than the potential well in the section at $y = y_B(c)$, the magnitude of the magnetization vector along the axis Ox at the point B is much larger than at point A (the ellipse in the vicinity of point B is horizontally strongly extended). This means that the kinetic energy of precession in the vicinity of point B is significantly greater than in the vicinity of point A. In order for the precessing magnetization vector to move from point A to point B, it needs to gain additional energy that it receives from the exciting alternating field. Due to the identical nature of the potential in the vicinity of points A and C, when the magnetization moves from point B to point C, this energy is released and leaves through relaxation processes into the lattice. Thus, the total energy of oscillation of magnetization at points A and C is less than at points B and D.

Figuratively speaking, in the process of moving from point A through point B to point C, the system under the action of an exciting alternating field, coming out of point A, first climbs onto the energy hill at point B, after which, having crossed the top, rolls down from it to point C under the action of the same field. Since during the ascent the exciting and rolling forces act in different directions, and during the descent their action is unidirectional, the ascent occurs much more difficult and longer than the descent. Thus, advancing in the area between points A and B takes much longer time than advancing between points B and C, as a result, much more small circles accumulate on section AB than on section BC, which manifests itself as a condensation of small plot AB precession portrait. Next, the system rises again on the hill at point D and descends from it to point A, after which the process repeats, giving a thickening of small rimgs in areas AB and CD with rarefaction in the areas BC and DA. Some flattening of the precession portrait along the Oy axis is due to the ellipticity of the small rings in the vicinity of points B and D. On the time interval, ascent to the hill correspond to smooth sections of impulses, to a rapid descent – sharp differences between the pulses.

9.8.5. Difference between low and high levels of excitation

From Fig. 9.34 it follows that with low (a, b) and high (c, d) in the excitation levels, the nature of the vibrations and the precession portrait differs markedly. As shown in the previous section, the condensation of small rings is due to the difference in the speed of ascent and descent from the energy hill. However, it can be assumed that as the level of arousal increases, the system will start moving faster. That is, the ascent and the descent from the hill will occur in less time. This explains the reason for the decrease in the magnitude of the oscillation period in Fig. 9.34,c compared with the period in Fig. 9.34, a. On the other hand, since the passage of the slide will be carried out in a shorter period of time, the entire full turn of the precession along the large ring will also require less time. Therefore, successive small rings on the precession portrait will be placed along the generatrix of a large ring over long distances. Figure 9.34, d shows much more rare arrangement of small rings than on Fig. 9.34, b.

Further, with a higher level of excitation, the amplitude of oscillations corresponding to the excitation rate will increase, as can be seen from a comparison of Fig. 9.12, a and Fig. 9.13, a. In Figs. 9.34, a and c, there is also a significant increase in the amplitude of such oscillations, that is, the width of the shaped lines increases. Such an increase corresponds to an increase in the diameter of the little small rings, as can be seen from the comparison of Figs. 9.34, b and d,

Small rings of such a large diameter as in Fig. 9.34, d already cease to fit within the extremes of the plane of the precession portrait, now only their midpoints are more or less symmetrically arranged. The small ring thickening is now happening around these centres, as a result of which the thickening is now not a smearing of a part of a large ring, as in Fig. 9.34, b, and collected in two groups, spread to a rather small extent. It is noteworthy that these centres of the azimuthal coordinate (about 70°) are quite close to the middle of the condensations of small rings in Fig. 9.34, b (about 50°). A certain forward angle shift (approximately by 20∘), apparently, is due to the faster movement of the magnetization vector along the large ring due to the higher level of excitation.

Thus, it can be assumed that the difference between the modes shown in Fig. 9.34, a, b and Fig. 9.34, c, d is not fundamental, that is, it does not contain any other mechanisms other than those presented in the previous section (Fig. 9.35). That is, the observed difference is due only to the higher velocity of the magnetization vector at a higher level of excitation. At the same time, the transformation of Fig. 9.34, a, b in Fig. 9.34, c, d should also occur in a continuous manner.

Comment. Generally speaking, the statement about the completely unified nature of the regimes shown in Fig. 9.34, a, b and Fig. 9.34, c, d, to the authors of this monograph is not completely certain. However, in the literature known to the authors, this question is ignored, so, in the authors' opinion, it may also be of interest for more detailed research.

9.8.6. The precession period of the equilibrium position for varying degrees of asymmetry of an alternating field

We now consider some properties of the precession of the equilibrium position with asymmetry of the transverse field. Figure 9.36 shows the dependence of the precession period of the equilibrium position on the ac field component h_{0y} at $h_{0y} = 7$ Oe, and the transition to the centre-coverage mode corresponds to $h_{0y} = 8$ Oe.

In the figure, the points are the period values obtained by machine experiment. The solid line is built on the empirical formula of inverse proportionality.

$$y = \frac{1}{30(x-6)} + 0.1 \qquad (9.33)$$

It can be seen from the figure that as the degree of asymmetry of an alternating field increases (that is, as h_{0y} decreases as compared with the constant h_{0x}), the precession period of the equilibrium

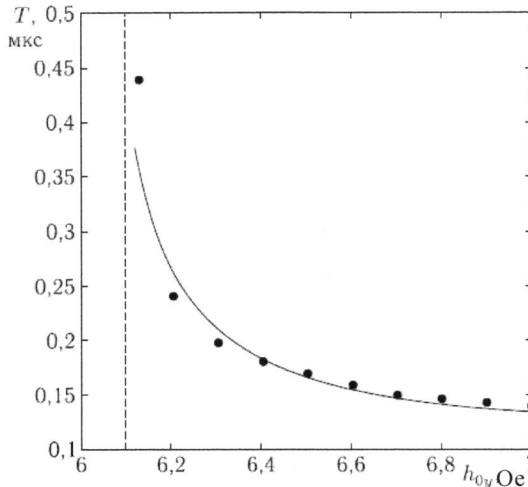

Fig. 9.36. The dependence of the precession period of the equilibrium position on the degree of asymmetry of the alternating field. Parameters: $4\pi M_0$ = 280 G, H_{0z} = 267 Oe, α = 0.3, f = 100 MHz

state increases, and the rate of the period increases with increasing asymmetry (decrease h_{0y}). At h_{0y} = 6.1 Oe (vertical dashed line), the period tends to infinity, which corresponds to the break of the large ring of the precession portrait. It can be seen that the empirical formula reflects the real dependence with an accuracy of about 10%. An increase in the period with an increase in asymmetry corresponds to an increase in the rise time of the oscillatory system on a potential slide, when the precessional portrait is a condensation of small olets.

9.8.7. Critical degree of asymmetry of an alternating field

The precession of the equilibrium state is critical to the degree of asymmetry of the alternating field. So, if the asymmetry exceeds a certain critical value, then the precessing magnetization vector no longer describes the complete circle around the Oz axis, since it cannot overcome the energy 'hill' formed by the sum of potential and kinetic energy, because its kinetic energy is not enough. In the precession portrait in this case, the big circle is broken. You can add kinetic energy by either increasing the amplitude of the alternating field or increasing its frequency. The phenomena occurring at the same time are illustrated by Fig. 9.37, where the dependences of the critical degree of asymmetry of the alternating field on the relative

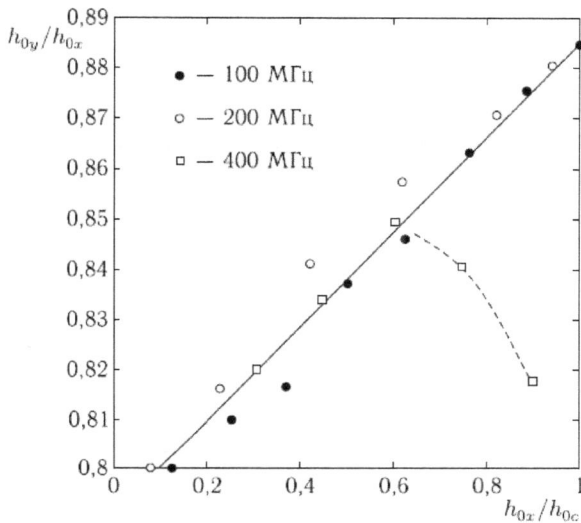

Fig. 9.37. Dependence of the critical degree of asymmetry of the alternating field h_{0y}/h_{0x} on the relative amplitude of the field h_{0x}/h_{0c} at different frequencies f: 1 - 100 MHz; 2 – 200 MHz; 3 – 400 MHz. Parameters: $4\pi M_0 = 280$ Gs, $H_{0z} = 267$ Oe, $\alpha = 0.3$.

amplitude of the field h_{0x} at various frequencies are shown. Points obtained during the machine experiment. The solid straight line is based on the empirical formula:

$$\frac{h_{0y}}{h_{0x}} = 0.0934\frac{h_{0x}}{h_{0c}} + 0.7916. \tag{9.34}$$

where h_{0c} is the critical value of the alternating field of circular polarization, in which the precession mode of the equilibrium position without centre coverage (No. 2) goes into the centre-covering mode (No. 3). According to the above experimental picture, the precession of the equilibrium state describes a full circle, below – this circle is broken. The horizontal axis shows the relative values in order to combine the dependences for different frequencies. Moreover, the higher the frequency, the higher the h_{0c} value. So, for a frequency of 100 MHz, $h_{0c}^{(100)} = 8.0$ Oe, for a frequency of 200 MHz, $h_{0c}^{(200)} = 16.0$ Oe; for a frequency of 400 MHz, $h_{0c}^{(400)} = 33.0$ Oe.

The figure shows that the line constructed by the formula (9.34) limits the region of existence of the precession of the equilibrium position for the frequencies 100 and 200 MHz from below with an accuracy of no worse than 1%. The same can be said about the frequency of 400 MHz to $h_{0x}/h_{0c} \leq 0.6$. This behaviour fits into the framework of the qualitative model discussed above. However, at a frequency of 400 MHz to the right of $h_{0x}/h_{0c} = 0.6$, the experimental points go down quite sharply (dotted line), which means that the precession of the equilibrium state expands to an asymmetry of the alternating field up to $h_{0y}/h_{0x} = 0.82$. It can be assumed that such a decrease is associated with a rather long relaxation time of the magnetization precession, that is, the magnetization vector with a large amplitude at such a high frequency does not have time to completely follow the change in the alternating field, as a result of which it is somewhat delayed and the large precession circle does not have time to break before how it should spin again. This assumption is also supported experimentally (by machine) the established fact that the lower limit of the region of existence of the precession of the equilibrium state with an increase in the attenuation parameter also decreases at any frequencies studied. So, even at a frequency of 100 MHz, an increase in the attenuation parameter α by 10% (from 0.30 to 0.33) leads to a decrease in the critical value of h_{0y}/h_{0c} by 1...2% (That is, for example, the experimental point at $h_{0x}/h_{0c} = 0.53$ instead of 0.836 takes the value 0.825).

Additional measurements show that the area of existence expands further with increasing frequency. So at a frequency of 1000 MHz with $h_{0c}^{(1000)} = 210.0$ Oe, the value of $h_{0x} = 50$ Oe corresponds to $h_{0y}/h_{0x} = 0.68$, and the value of $h_{0x} = 100$ Oe – the value of $h_{0y}/h_x = 0.54$. On the other hand, with such large variable fields, the precession mode without centre coverage (No. 2) becomes unstable and has a tendency to switch to extended circular precession mode (No. 5) after several orbital periods, that is, the critical value of the degree of asymmetry becomes ambiguous.

Comment. All the effects described above relate to the mode without centre coverage (No. 2 by [308]). With an increase in the amplitude of the ac field up to the value of h_c, the precession period of the equilibrium state increases up to the tendency toward infinity. However, with a further increase in the amplitude of the alternating field, the mode without centre coverage (No. 2) is replaced by a mode with centre coverage (No. 3), in which the precession period decreases again until the transition to a mode with centre coverage with attenuation (No. 4). At the same time, the asymmetry of the alternating field also causes the thickening of the small rings of the precession portrait, however, now the diameter of these small rings exceeds the radius of the large ring, as a result the two diametrically oppositely arranged thickening circles partially overlap each other. In this mode, the condensation is less clearly expressed than in the mode without centre coverage, however, it shows quite similar properties, up to mirror reflection, relative to the transition field between modes.

9.9. Violation of symmetry in a longitudinal variable field

Let us now consider the case of longitudinal asymmetric excitation, when a longitudinal variable component is added to the transverse variable field of circular polarization, parallel to the constant field.

9.9.1. General picture of magnetization oscillations with longitudinal asymmetry of an alternating field

We assume that the time dependence of a variable field is:

$$h_x = h_{0x} \sin(2\pi ft),$$ (9.35)

$$h_y = -h_{0y} \cos(2\pi ft),$$ (9.36)

$$h_z = -h_{0z} \sin(2\pi ft).$$ (9.37)

The overall picture of the observed phenomena is illustrated in Fig. 9.38. Here the left shows the development of oscillations in time (solid lines – m_x, dotted line – m_y), on the right – the resulting precession portrait.

The given pair of Figs 9.38, a and b corresponds to the longitudinal asymmetry of the alternating field while maintaining the symmetry constant: $h_{0x} = h_{0y}$, but $h_{0z} \neq 0$, subject to the condition $H_{0x} = H_{0y} = 0$. The parameters compared to the previous figures are chosen somewhat different to increase visibility. It can be seen that in this case, the sinusoidal oscillations are replaced by irregular pulses, and on the precession portrait in the lower part of it there is a strongly pronounced thickening of the small rings.

9.9.2. The mechanism of formation of the condensation of small rings

Figure 9.39 illustrates the mechanism of formation of the condensation of small rings of a precession portrait with a longitudinal asymmetry of an alternating field, which shows the dynamic scheme of the mutual orientation of the constant and variable fields for this case.

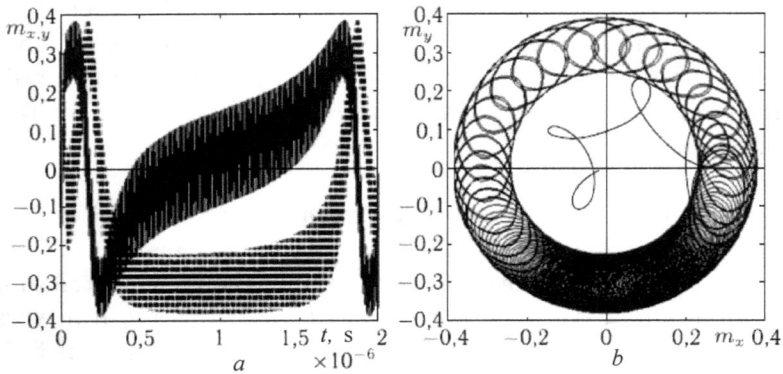

Fig. 9.38. Development of oscillations in time (*a*) and the corresponding precession portrait (*b*) with longitudinal asymmetry of the alternating field. Parameters of constant fields: $H_{0x} = 0$ Oe, $H_{0y} = 0$ Oe, $H_{0z} = 267$ Oe; variable: $h_{0x} = 3$ Oe, $h_{0y} = 3$ Oe, $h_{0z} = 1.5$ Oe. Other parameters: $4\pi M_0 = 280$ G, $\alpha = 0.3$, $f = 100$ MHz.

The general scheme of the relative orientation of the fields is shown in Fig. 9.39, a. The constant field $\mathbf{H}_0$, directed along the axis Oz, corresponds to the segment OM. A transverse alternating field with components h_x and h_y rotates around the axis Oz in such a way that the end of the vector h moves along a circle $ABCD$ lying in the $Mx'Y'$ plane parallel to the coordinate plane Oxy. The longitudinal alternating field is directed along the Oz axis, and its intensity in time varies in antiphase with the component of the symmetric field h_x. In this case, the vector of the total variable field $\mathbf{h}$ (the MF segment) rotates in the $EBFD$ plane, which makes the angle α with the $ABCD$ plane, where $\alpha = \arctan(h_{0z}/h_{0x})$.

It can be seen that the vector of the constant field makes the same angle α with the $EBFD$ plane. That is, the situation of the relative position of the plane of rotation of the alternating field and the vector of the constant field is completely similar to the case of asymmetry of the constant field considered in [307]. It has been shown that the asymmetry of the constant field causes the thickening of the small ring precession portrait, the formation and position of which is explained on the basis of the energy and vector models. Thus, it can be assumed that the reason for the formation of a condensation of small rings with a longitudinal asymmetry of an alternating field is the same as the reason for the formation of a similar condensation with asymmetry of a constant field [307], discussed in detail in Section 9.7, that is, there is no need to dwell on it.

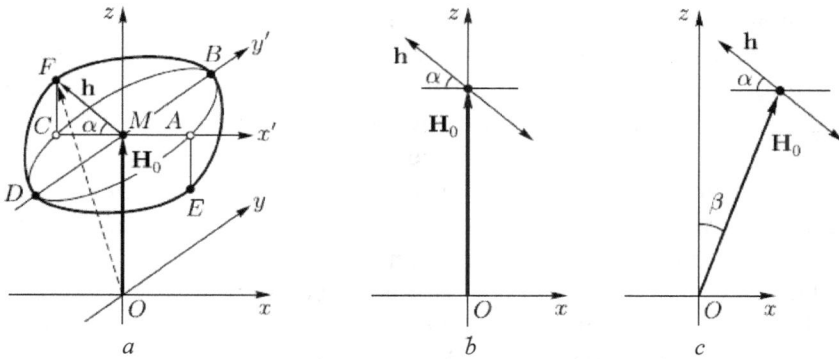

Fig. 9.39. General scheme of mutual orientation of constant and variable fields with longitudinal asymmetry of the variable field (*a*), as well as mutual arrangement of the vectors of constant and variable fields with symmetric (*b*) and asymmetric (*c*) orientation of the constant field.

Figure 9.39, b shows the mutual orientation of the constant and variable fields, when the constant field is directed exactly along the axis *Oz*. This case corresponds to Fig. 9.38.

Figure 9.39, c shows the mutual orientation of the constant and variable fields, when the constant field is deflected from the axis *Oz* by angle β. This case is considered in the next section.

9.9.3. Some common properties of precession in case of asymmetry of an alternating field

The basic properties of the precession of the equilibrium position with the longitudinal asymmetry of the alternating field are similar to those of the precession with the asymmetry of the transverse field. Thus, as the asymmetry increases, there is a certain critical value of the alternating field, above which the large ring of the precession portrait breaks, and the precession period also increases as the field approaches this value, tending to infinity.

The critical value of asymmetry with increasing amplitude of the alternating field also increases, and with increasing frequency it tends to decrease. The qualitative explanation of these properties is the same as in the case of transverse inhomogeneity.

9.9.4. Compensating effect of asymmetrical constant field

The identical nature of the formation of the condensation of small-fisted precessional portrait in the case of asymmetry of constant and longitudinal asymmetry of variable fields raises the question of the possible interaction of these asymmetries. An illustration of this interaction is Fig. 9.40, where the left shows the development of oscillations in time (solid lines – m_x, dotted lines – m_y), on the right – the resulting precession portraits. The main parameters of the task are the same as the previous ones.

The first pair of Figs. 9.40, a and b corresponds to a constant field with an asymmetric component H_{0x}. It is seen that in this case, the sinusoidal oscillations are also replaced by irregular pulses, and

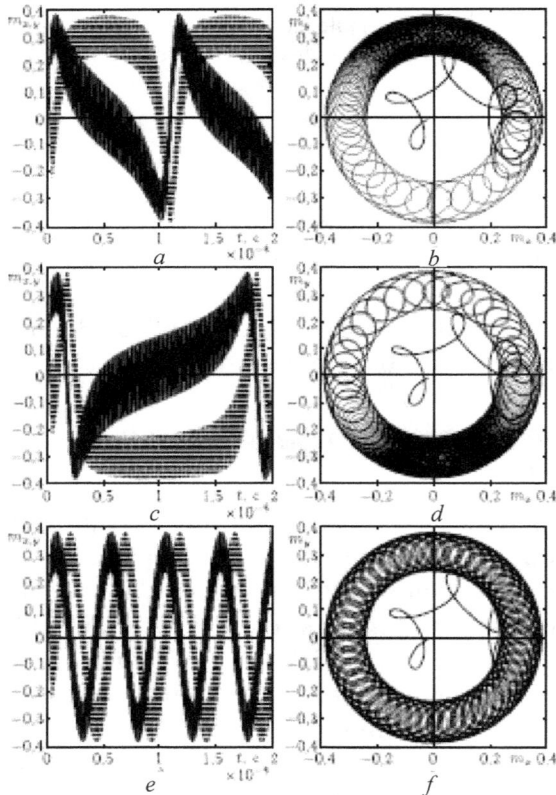

Fig. 9.40. The development of oscillations in time and the corresponding precession portraits with asymmetry of constant (*a, b*), variable (*c, d*) and both at once (*d, e*) fields. The parameters of the fields: constant: H_{0z} = 265 Oe, H_{0y} = 0 Oe; *a, b, d, e* – H_{0x} = 0.055 Oe, *c, d* – H_{0x} = 0 Oe; variable: H_{0z} = 3 Oe, h_{0y} = 3 Oe, *a, b* – h_{0z} = 0 Oe, c, d, e, f – h_{0z} = 1.5 Oe. Other parameters: $4\pi M_0$ = 280 Gs, α = 0.3, f = 100 MHz.

in the precession portrait, as in Fig. 9.23, d, there is a condensation of small rings in its upper part.

The second pair of Figs. 9.40, c and d corresponds to a longitudinal variable field with an asymmetric component h_{0z}. In this case, the sinusoidal oscillations are also replaced by irregular pulses, and on the precession portrait, as in Fig. 9.38, b, there is a condensation of small rings in its bottom.

Thus, under the action of a constant field (Fig. 9.40, b) the condensation of small rings is located at the top, and under the action of a longitudinal alternating field (Fig. 9.40, d) – at the bottom of the precession portrait. It can be expected that with the simultaneous action of both fields, both thickening will be mutually compensated.

This situation is realized in the third pair of drawings – Figs. 9.40, e and Fig. 9.40, f, which correspond to the simultaneous asymmetry of constant and variable fields, that is, with non-zero fields H_{0x} and h_{0z}. It can be seen that in this case the pulses are again replaced by sinusoidal oscillations (Fig. 9.40, e), and there are no condensation on the precession portrait, and the large ring is filled with small rings uniformly (Fig. 9.40, f), similar to the case of symmetric fields, considered in the previous sections 9.1–9.5 (Fig. 9.2 or Fig. 9.12), as well as in [306]. Thus, it can be said that in this case there is a mutual compensation of the features of the precession, due to the asymmetry of both fields.

The scheme of mutual orientation of vectors of constant and variable fields is illustrated in Figs. 9.39, b and c. In this figure. 9.39, b corresponds to the precession portrait shown in Fig. 9.38, b, and Fig. 9.39, c – the portrait shown in Fig. 9.40, e. It can be seen that in the second case, the vectors of constant and variable fields are almost mutually perpendicular (perpendicularity corresponds to $\alpha = \beta$), as in the case of a fully symmetric field arrangement (that is, with $H_{0x} = 0$ and $h_{0z} = 0$), which explains condensation compensation in the precession portrait. A more detailed additional study shows that the criticality of such compensation is very high and amounts to hundredths of a percent of the total value of the constant field. So, the compensation shown in Fig. 9.40, d occurs at a constant field $H_{0x} = 0.055$ Oe, which is about 0.02% from the field along the Oz axis, equal to $H_{0z} = 265$ Oe. The criticality of compensation in a variable field is also high and amounts to a few percent of its full value (as with h_{0z} decreasing from 1.5 Oe to 1.4 Oe, the compensation is already completely violated).

Conclusions for chapter 9

This chapter is devoted to a specific form of precession of magnetization under conditions of a transitional transition — precession of a second-order equilibrium state or precession, in which the magnetization vector precesses around its equilibrium position, which in turn precesses around a constant field direction. Consideration is limited to the case of an isotropic medium, but the problems of breaking the symmetry of the system are considered in some detail.

The main results of this chapter are as follows.

1. Considered the forced nonlinear precession of the magnetization vector in a normally magnetized magnetic plate under the conditions of the orientational transition. It is shown that in a field smaller than the demagnetization field of the form, when the orientation of the stationary equilibrium position of the magnetization vector is deflected from the direction of the constant field, the alternating field of circular polarization causes a precession of the equilibrium position of the magnetization vector. In this case, the magnetization vector precesses with the frequency of the alternating field along the small cone around the equilibrium position, and the equilibrium position itself precesses along the large cone with a frequency significantly (up to several orders of magnitude) less.

2. The time dependences and precession portraits of magnetization oscillations were constructed, it was shown that the precession portrait in its most general form is a large ring filled with small rings forming, and the large ring reflects the precession of the equilibrium position, and the small rings represent the precession of the magnetization vector around this equilibrium position.

3. A vector model of the phenomenon under consideration was constructed, it was shown that the precession period of the equilibrium state is inversely proportional to the square of the amplitude of the alternating field and the sine of the angle of deviation of the equilibrium position of the magnetization from the constant field. The method of machine modeling has been used to verify the obtained dependency. It is shown that the theory based on the vector model describes well (with accuracy no worse than 5%) a computer experiment in the range of fields of orientational transition from zero to 0.86 from the form anisotropy field, and then gives slightly overestimated values for the precession period.

4. It is shown that the critical parameters of the precession excitation of the equilibrium state, are the amplitude and frequency of the alternating field. For the precession to be excited, it is necessary that the amplitude of the alternating field is less than the critical value, and the frequency of the alternating field exceeds the critical value. If critical conditions are not fulfilled, the precession of the equilibrium state is replaced by the simple circular precession of the magnetization vector around the direction of the constant field. It is shown that when changing the parameters of the variable field, the transition from one type of precession to another occurs abruptly and has a threshold character. It has been established that the reason for this behavior of the magnetization precession is the excitation of a nonlinear resonance of the magnetic oscillatory system in combination with an amplitude limiting detuning mechanism, and the threshold of the precession changes the frequency of the nonlinear resonance of the magnetic system and the alternating field.

5. On the plane in the magnitude – frequency coordinate of the alternating field, diagrams were constructed defining the regions of existence of the precession of the equilibrium position and simple circular precession for different values of the saturation magnetization of the magnetic medium. It is shown that with an increase in the amplitude of the alternating field, the critical frequency increases the stronger, the larger the saturation magnetization.

6. The role of dissipation of magnetization oscillations is considered. It is shown that the precession period of the equilibrium position with a small attenuation, when the precession is periodic, does not depend on the magnitude of the attenuation parameter, and as the attenuation increases, when the precession becomes aperiodic, the faster the attenuation parameter increases. The critical frequency of the precession excitation of the equilibrium state with a small attenuation parameter does not depend on its value, and with a large one, when the oscillations become aperiodic, the faster the larger the attenuation parameter decreases.

7. Based on the analysis of time dependencies and precession portraits, five different precession modes are revealed, replacing each other when the amplitude of the alternating field changes: mode No. 1 – low-amplitude circular precession, mode No. 2 – precession of the equilibrium position without centre coverage, mode No. 3 – continuous precession equilibrium positions with centre coverage, mode No. 4 – damped precession of an equilibrium position with centre coverage, mode No. 5 – developed circular precession.

8. To interpret the multimode nature of the precession of the equilibrium position, a mechanical model of the pendulum with a spring side suspension is proposed. The analogy between the equilibrium position of the pendulum under the action of a lateral force with the equilibrium position of the magnetization vector during the orientational transition is noted. An interpretation of this behavior is proposed on the basis of a potential model having two minima of equal depth located symmetrically with respect to the central maximum.

9. Based on the model of a flat mechanical pendulum with a side spring suspension, the possibility of the existence of two mechanical pressure oscillations similar to the modes of oscillation of magnetization is illustrated: No. 1 – low-amplitude circular precession and No. 5 – unfolded circular precession. It is noted that the implementation of modes No. 2 – No. 4 is hindered by the planar nature of the pendulum. In order to overcome this drawback, a model of a cylindrical pendulum is proposed, which is a flat pendulum that can freely rotate around a vertical axis passing through the points of the main and lateral suspension. It is shown that the potential of such a pendulum has the form of an annular 'ditch' surrounding the central maximum in the form of an axisymmetric 'hill'. An action of a force directed tangentially to the circular trajectory of the load of a cylindrical pendulum has been proposed as an analog of the variable field of circular polarization.

10. It is noted that with circular motion caused by a circular tangential force, the load of the pendulum can move around the ring both synchronously and due to its inertia, lag behind the rotation of the force to a greater or lesser extent. It is shown that the growth of the lag with increasing rotational speed of the force leads to a sequential change of oscillation modes, similar to the magnetization precession modes, namely: from mode No. 5 – deployed circular precession through modes No. 4 and No. 3 – precession with centre coverage to mode No. 2 – precession without centre coverage and further to mode No. 1 – low-amplitude circular precession.

11. The formation of etir regimes is illustrated on the basis of the analysis of the trajectories of the movement of cargo under the action of a circular force of various frequencies. The difference from the magnetic case, which consists in the reverse order of the modes, is revealed. It was found that the reverse order of the modes in the mechanical case is due to a change in frequency, while the direct order in the magnetic case is due to a change in the amplitude of the

excitation. The equivalence of both approaches is noted, confirming the validity of the use of the mechanical model for the interpretation of magnetic phenomena.

12. Transitions between the magnetization precession regimes are considered on the basis of a three-dimensional potential model, which has the form of an axisymmetric central peak surrounded by a ring ditch, the outer edge of which goes to infinity. On the basis of a dynamic model of the motion of the end of a precessing magnetization vector along a three-dimensional potential surface, the need to take into account the dependence of the potential on the amplitude of an alternating field has been revealed and the mechanism of the formation of precession portraits for various modes is clearly demonstrated. The conditions for the excitation of all precession modes with different attenuation parameters and amplitudes of the permeable field are considered. On the plane 'attenuation parameter – amplitude of the alternating field', the boundaries of the regions with separate precession modes are defined for different frequencies and values of the transverse component of the constant field. The properties of the transverse component of the magnetization under conditions of transitions between different modes of precession with varying amplitude of the variable field are considered.

13. It is shown that in the process of transitions between modes No. 1 to No. 4, the amplitude of oscillations of the transverse component of the magnetization changes smoothly, and the transition from mode No. 4 to mode No. 5 is very sharp, occurring when the amplitude of the alternating field changes less than 0.1% and is accompanied by a significant change in the amplitude of the precession (up to one and a half to two times), that is, it represents an abrupt dynamic orientational transition. The dependence of the precession period of the equilibrium position on the amplitude of the alternating field is considered, an abrupt change in the period in modes 2 and 3 is found, corresponding to the inverse proportionality of the square of the amplitude of the alternating field.

14. The forced nonlinear precession of the magnetization vector in a normally magnetized magnetic plate under conditions of orientational transition with an asymmetric orientation of a constant field is considered. The conditions under which the variable field field of circular polarization causes the precession of the equilibrium state of the magnetization vector are revealed. It is shown that with an asymmetric constant field, the average values of the dependences

of the transverse magnetization components on time cease to be purely sinusoidal, but take the form of irregular pulse shapes, the period of which significantly exceeds the sinusoid period with symmetric excitation. The precession portrait retains the appearance of a regular circular ring, similar to a ring with a symmetric constant field, but the filling of a large ring with small rings becomes uneven;

15. A very high criticality of the excitation of the precession of the equilibrium state to the degree of asymmetry of the constant field was found. The existence of the critical value of the transverse component of the constant field, above which the precession of equilibrium is disrupted and replaced by a simple circular precession of large amplitude, is revealed. The magnitude of the critical field of failure does not exceed a few percent of the saturation magnetization of the plate material.

16. The areas of existence of the precession of the equilibrium position in constant and alternating fields were studied at various frequencies of the alternating field. It is shown that on the plane 'the amplitude of the alternating field is the transverse component of the constant field' of the existence region have the form of irregular three angles with a maximum, and consisting of smoothly growing and sharply falling sections. With increasing frequency, the region of existence increases and shifts towards the high-magnitude litters of the alternating field.

17. The dependences of the precession period of the equilibrium position on the transverse component of the constant field are investigated for different amplitudes of the frequencies of the alternating field. It is shown that near the field of breakdown, the precession period of the equilibrium state tends to infinity or to a sufficiently large final value, depending on the frequency of the alternating field. The main features of the observed phenomena are qualitatively explained on the basis of mutually complementary energy and vector models of each other.

18. Considered the precession of the equilibrium position with asymmetric orientation of the transverse and longitudinal alternating fields. It is shown that in this case the sinusoidal nature of the precession of the equilibrium position is replaced by a pulsed one, and the precessional portrait of oscillations, retaining the appearance of a regular circular ring filled with small rings, acquires non-uniform thickening of small lobsters. The position of the condensations is determined by the nature of the asymmetry of the field. In the case of transverse asymmetry of an alternating field, in the precession

portrait there are two condensations located diametrically opposite to each other, in the case of longitudinal asymmetry, only one condensation located on the periphery of the large ring.

19. It is shown that the application of a longitudinal constant field allows one to compensate for the anomalous nature of the precession with the longitudinal asymmetry of the variable field, leading to the restoration of the sinusoidal nature of the oscillations in combination with the uniform filling of the large ring with small rings. A very high criticality of compensation in a constant field is noted, amounting to hundredths of a percent of its full value. The observed phenomena are explained on the basis of a dynamic vector model.

10

Precession of the position of the equilibrium magnetization in an anisotropic medium

This chapter is devoted to the further development of the material presented in the previous chapter. The main attention is paid to the precession properties of the equilibrium position in a medium with oblique uniaxial and cubic anisotropy for a number of the most characteristic crystallographic orientations. Recommendations are given for observing the precession of the equilibrium position of the magnetization in the experiment, and also some application possibilities are discussed. In conclusion, some possible questions are presented for further investigation.

When considering the surveys listed in the polls, we will mainly follow the same works as in the previous chapter [266, 297–332], as well as the monographs [3]. The remaining necessary references are indicated in the text.

10.1. The precession of the equilibrium state of magnetization in a medium with uniaxial anisotropy

In Chapter 9, the precession of the equilibrium state was considered in an isotropic medium with axial symmetry created by a uniform constant field in combination with an alternating field of circular polarization. However, it follows from [307] that even a small violation of axial symmetry leads to a strong change in the nature of the precession. It can be expected that in an anisotropic medium the nature of the precession will also be different from the isotropic case. Consider this question in more detail.

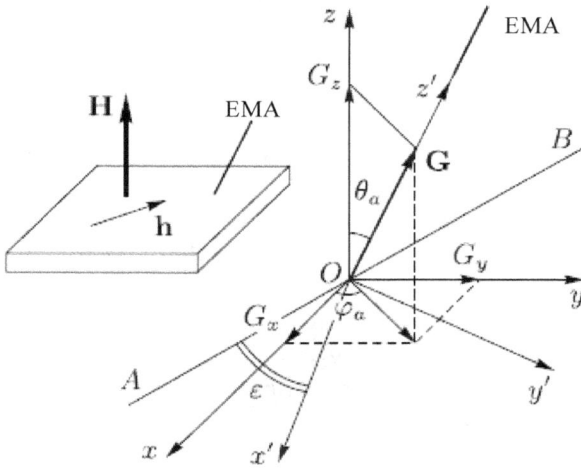

Fig. 10.1. General geometry of the problem of precession of an equilibrium position in an anisotropic medium.

10.1.1. Task geometry

Consider a normally magnetized ferrite plate with uniaxial magnetic anisotropy, whose axis is deflected from the normal to the plane of the plate by an arbitrary angle.

Comment. Note that the choice of ferrite as a material for a plate is of no fundamental importance here and is adopted only for the convenience of terminology, based on its most frequent use in practice. For further consideration, only the properties of the anisotropy of the magnetic plate are important, including the anisotropy of its shape.

The overall geometry of the problem is illustrated in Fig. 10.1. The ferrite plate is magnetized by a constant field **H** perpendicular to its plane, the alternating field **h** is applied in the plane of the plate. The Oxy plane of the Cartesian coordinate system $Oxyz$ coincides with the plane of the plate, the axis Oz is perpendicular to it. The direction of the anisotropy axis or the easy magnetization axis (EMA) is given by the unit vector **G**, the projections of which on the axes of the $Oxyz$ system are equal to G_x, G_y, G_z, respectively. The vector **G** is with the axis Oz angle θ_a. The projection of the vector **G** on the plane of the plate is the angle φ_a with the axis Ox. The straight line AB passing through the point O lies in the plane Oxy and forms an angle of 90° with the anisotropy axis.

To describe the uniaxial anisotropy, an auxiliary Cartesian coordinate system, $Ox'y'z'$, the Oz' axis of which is introduced. which is directed along the anisotropy axis, and the plane $Ox'y'$, crossing the plane $Ox'y'$

along the straight line AB, is rotated around the axis Oz' so the angle between the Ox' axis and the line OA is equal to ε.

10.1.2. Transition matrix

The problem of the dynamic behaviour of the magnetization vector will be solved in the coordinate system associated with a constant field, that is, $Oxyz$. On the other hand, the energy density of uniaxial anisotropy has the simplest form in the system $Ox'y'z'$, with the axis Oz' directed along the anisotropy axis. Thus, in the initial state, the anisotropy energy density is written through the components of the unit magnetization vector $m_{x'}$, $m_{y'}$, $m_{z'}$ normalized to the saturation magnetization M_0, whereas the problem should be solved in the components of the analogic unit vector m_x, m_y, m_z, through which the energy density should be written. It is enough to express the components $m_{x'}$, $m_{y'}$, $m_{z'}$ through the components m_x, m_y, m_z, and then substitute them into the expression for the energy density in the system $Ox'y'z'$. As a result, we obtain the required expression for the energy density of anisotropy, written through the components m_x, m_y, m_z.

So, we need the components of the normalized magnetization vector $\mathbf{m'}$, Defined in the system $Ox'y'z'$, expressed through the components of the magnetization vector $\mathbf{m}$, defined in the system $Oxyz$, that is, to carry out the transformation:

$$\mathbf{m'} = \ddot{A}^{-1}\mathbf{m}. \tag{10.1}$$

where $\ddot{A}^{-1}$ is the transition matrix of the system $Ox'y'z'$ to the $Oxyz$ system. To obtain the transition matrix, we use the apparatus described in detail in Chapter 3. Given that the components of the transition matrix are the cosines of the angles between the respective axes of the original and rotated coordinate systems, as well as using the Euler angle system [261], for which the main axis is Oz' and the line of nodes is the straight line AB, we obtain the transition matrix in the form

$$
\ddot{A}^{-1} =
\begin{pmatrix}
\cos\theta\cos\varphi\sin\varepsilon + \sin\varphi\cos\varepsilon & \cos\theta\sin\varphi\sin\varepsilon - \cos\varphi\cos\varepsilon & -\sin\theta\sin\varepsilon \\
\cos\theta\cos\varphi\cos\varepsilon - \sin\varphi\sin\varepsilon & \cos\theta\sin\varphi\cos\varepsilon\_\# + \cos\varphi\sin\varepsilon & -\sin\theta\cos\varepsilon \\
\sin\theta\cos\varphi & \sin\theta\sin\varphi & \cos\theta
\end{pmatrix} \tag{10.2}
$$

It can be seen that the resulting matrix of the transition, up to notation, is similar to (4.61), which is due to the identity of the geometry of both parts (Figs. 4.5 and Fig. 10.1).

10.1.3. Total energy density, effective fields and equations of motion in the original (laboratory) coordinate system

Energy density of anisotropy and external field. The energy density of uniaxial anisotropy in the $Ox'y'z'$ system has the form

$$U_a = K\left(m_{x'}^2 + m_{y'}^2\right),$$
(10.3)

where in the case of a light axis $K > 0$.

In accordance with formula (10.1), taking into account (10.2), we find:

$$m_{x'} = \left(\cos\theta_a \cos\varphi_a \sin\varepsilon + \sin\varphi_a \cos\varepsilon\right)m_x +$$
$$+\left(\cos\theta_a \sin\varphi_a \sin\varepsilon - \cos\varphi_a \cos\varepsilon\right)m_y - \sin\theta_a \sin\varepsilon m_z;$$
(10.4)

$$m_{y'} = \left(\cos\theta_a \cos\varphi_a \cos - \sin\varphi_a \sin\varepsilon\right)m_x +$$
$$+\left(\cos\theta_a \sin\varphi_a \cos\varepsilon + \cos\varphi_a \sin\varepsilon\right)m_y - \sin\theta_a \cos\varepsilon m_z;$$
(10.5)

Substituting these expressions into (10.3), we obtain the energy density of uniaxial anisotropy in the $Oxyz$ system, written through the components of the vector $\mathbf{m}$:

$$U_a = K\left\{m_x^2\left(\cos^2\theta_a \cos^2\varphi_a + \sin^2\varphi_a\right) + m_y^2\left(\cos^2\theta_a \sin^2\varphi_a + \cos^2\varphi_a\right) + \right.$$
$$+m_z^2\sin^2\theta_a - 2m_x m_y \sin^2\theta_a \sin\varphi_a \cos\varphi_a -$$
$$\left. -2m_x m_z \sin\theta_a \cos\theta_a \cos\varphi_a - 2m_y m_z \sin\theta_a \cos\theta_a \sin\varphi_a \right\},$$
(10.6)

where the angles θ_a and φ_a determine the orientation of the anisotropy axis in the $Oxyz$ system.

Let us now assume that the external magnetic field is

$$\mathbf{H} = \{H_{0z} + h_x, H_{0y} + h_y, H_{0z}\},$$
(10.7)

where $H_{0x,y,z}$ is a constant field, h_x, h_y are the components of the variable field of frequency f:

$$h_x = h_0 \sin(2\pi f t), \tag{10.8}$$

$$h_y = -h_0 \cos(2\pi f t). \tag{10.9}$$

We assume that $h_{x,y} < H_{0z}$, as well as $H_{0x,y} < H_{0z}$.

Total energy density. The total energy density of the plate taking into account the demagnetization and anisotropy takes the form

$$U = -M_0 h_x m_x - M_0 H_{0x} m_x - M_0 h_y m_y -$$
$$-M_0 H_{0x} m_z + 2\pi M_0^2 m_z^2 +$$
$$+K \left\{ m_x^2 \left(\cos^2 \theta_a \cos^2 \varphi_a + \sin^2 \varphi_a \right) + m_y^2 \left(\cos^2 \theta_a \sin^2 \varphi_a + \cos^2 \varphi_a \right) + \right.$$
$$+ m_z^2 \sin^2 \theta_a - 2m_z m_y \sin^2 \theta_a \sin \varphi_a \cos \theta_a -$$
$$\left. -2m_x m_z \sin \theta_a \cos \theta_a \cos \varphi_a - 2m_y m_z \sin \theta_a \cos \theta_a \sin \varphi_a \right\}. \tag{10.10}$$

where M_0 is the saturation magnetization.

Equations of motion of magnetization. To calculate the magnetization precession, similarly to Chapter 2, we use the Landau–Lifshitz equations with a dissipative term in the Hilbert form (2.97)–(2.99):

$$\frac{\partial m_x}{\partial t} = -\frac{\gamma}{1+\alpha^2} \left[\left(m_y + \alpha m_x m_z \right) H_{ez} - \right.$$
$$\left. -\left(m_z - \alpha m_y m_z \right) H_{ey} - \alpha \left(m_y^2 + m_z^2 \right) H_{ex} \right]; \tag{10.11}$$

$$\frac{\partial m_y}{\partial t} = -\frac{\gamma}{1+\alpha^2} \left[\left(m_z + \alpha m_y m_x \right) H_{ex} - \right.$$
$$\left. -\left(m_x - \alpha m_z m_y \right) H_{ez} - \alpha \left(m_z^2 + m_x^2 \right) H_{ey} \right]; \tag{10.12}$$

$$\frac{\partial m_z}{\partial t} = -\frac{\gamma}{1+\alpha^2} \left[\left(m_z + \alpha m_z m_y \right) H_{ey} - \right.$$
$$\left. -\left(m_y - \alpha m_z m_z \right) H_{ex} - \alpha \left(m_x^2 + m_y^2 \right) H_{ez} \right]; \tag{10.13}$$

where γ is the gyromagnetic constant ($\gamma > 0$), α is the damping parameter of the magnetization precession. Calculating the effective fields from (10.10) using the formula (1.2) or (2.81), (2.84):

$$H_{ei} = -\frac{\partial U}{\partial M_i} = -\frac{1}{M_0}\frac{\partial U}{\partial m_i}.$$
(10.14)

we get:

$$H_{ex} = H_{0z} + h_x + \frac{2K}{M_0}[-m_x\left(\cos^2\theta_a\cos^2\varphi_a + \sin^2\varphi_a\right) +$$
$$+m_y\sin^2\theta_a\sin\varphi_a\cos\varphi_a + m_z\sin\theta_a\cos\theta_a\cos\varphi_a];$$
(10.15)

$$H_{ey} = H_{0y} + h_y + \frac{2K}{M_0}[m_x\sin^2\theta_a\sin\varphi_a\cos\varphi_a -$$
$$-m_y\left(\cos^2\theta_a\sin^2\varphi_a + \cos^2\varphi_a\right) + m_z\sin\theta_a\cos\theta_a\sin\varphi_a];$$
(10.16)

$$H_{ez} = H_{0z} - 4\pi M_0 m_z + \frac{2K}{M_0}[m_z\sin\theta_a\cos\theta_a\cos\varphi_a +$$
$$+m_y\sin\theta_a\cos\theta_a\sin\varphi_a - m_z\sin^2\theta_a].$$
(10.17)

As before, the numerical solution of the system (10.11)–(10.13) with allowance for (10.15)–(10.17) allows to find the development of forced magnetization oscillations in time, to the consideration of which we proceed further.

10.1.4. Precession portraits

The results of the calculation of the precession of the equilibrium position of the magnetization in a medium with uniaxial anisotropy are illustrated in Fig. 10.2, which shows the development of fluctuations in time and the corresponding precession portraits.

Figure 10.2, a, b corresponds to the absence of anisotropy ($K = 0$) and is given here for comparison. In this case, the oscillations of the average positions of the components of the magnetization m_x and m_y are identical sinusoids with a period of $0.50 \cdot 10^{-6}$ s, shifted in phase by 90°. The precession portrait is a large ring, which is completely uniform along the small ring.

Figure 10.2, c, d corresponds to the case of uniaxial anisotropy with a constant $K = 3.9$ erg $\times$ cm^{-3}, whose axis is oriented at angles: $\theta_a = 10°$, $\varphi_a = 0°$, that is, the projection of the anisotropy axis on the plane of the ferrite plate is parallel to the coordinate axis Ox and the axis itself makes an angle of 10° with the normal to the plate. In this case, the oscillations of the average positions of the components of magnetization have the character of stretched edges with a period

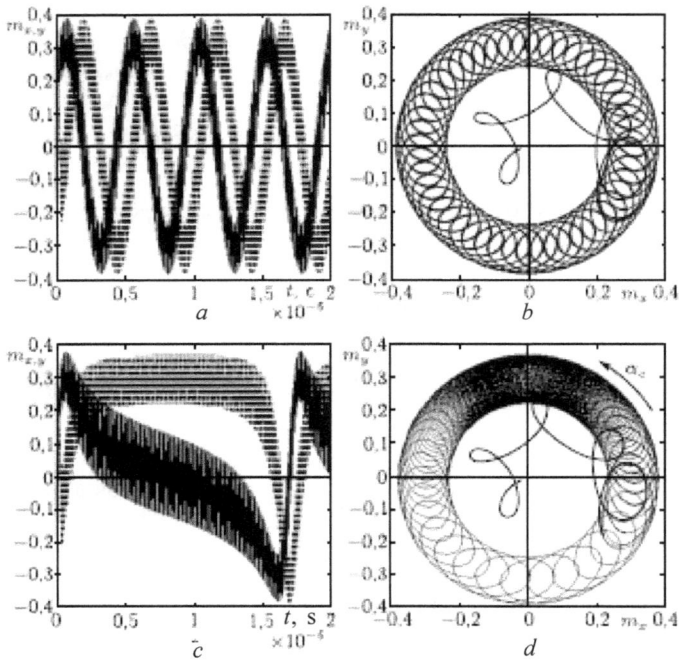

Fig. 10.2. Development of fluctuations in magnetization in time (*a, c*) and the corresponding precessional portraits (*b, d*). *a, b* no anisotropy: *c, d* - uniaxial anisotropy with an inclined axis: $K = 3.9$ erg cm^{-3}, $\theta_a = 10°$, $\varphi_a = 0°$. In the figure: *a, b* - solid lines - m_x, dotted lines – m_y. Parameters: $4\pi M_0 = 280$ G, $H_{0z} = 265$ Oe, $\alpha = 0.3$, $f = 100$ MHz, $h_0 = 3$ Oe

of $1.72 \cdot 10^{-6}$ s, and the oscillations m_x are close to sawtooth, and m_y – to rectangular pulses with rounded fronts. The large ring in the precession portrait is filled with small, very uneven: there is a significant thickening of the small rings, rotated in the positive direction relative to the *Ox* axis by the angle α_c, measured from the O_{mx} axis, close to 90°, and diametrically opposite to this thickening, that is, in the vicinity of $\alpha_c \approx 270°$, small rings are highly sparse.

Comment. The value of the anisotropy constant $K = 3.9$ erg $\times$ cm^{-3} adopted here seems to be extremely small at first glance, especially in comparison with the anisotropy constants in mixed ferrite–garnets, where they are about 10^4 erg $\times$ cm^{-3} (sections 5.3, 7.6).

In this case, the anisotropy field is about 0.35 Oe, whereas in mixed garnets, this field reaches 1000 Oe and more. It should be noted, however, that the phenomenon of precession of the equilibrium position considered here to the symmetry breaking of the system as a whole is extremely sensitive. So for example, from Fig. 9.23, it can be seen that a strong condensation of small rings is observed when the transverse field H_{Ox} is only 0.055 Oe. An increase in such a field at least two to three times leads to a breakdown of the precession of the equilibrium position. Therefore, here too, the order of magnitude of the anisotropy constant is chosen so that the corresponding anisotropy field for breaking the precession is insufficient. In more detail the question of the criticality of the magnitude of the anisotropy is discussed below.

10.1.5. Model of thickening and rarefaction

To interpret the observed phenomena, we will use the energy model of the potential introduced in Section 9.7.7.

We first consider how the uniaxial anisotropy affects the energy density of the system as a whole. We will characterize the orientation of the magnetization vector in the same spherical coordinate system as for the anisotropy axis by the polar and azimuthal angles θ_m and φ_m. Thus, we will assume that the components of the magnetization vector, have the form:

$$m_x = \sin\theta_m \cos\varphi_m; \tag{10.18}$$

$$m_y = \sin\theta_m \sin\varphi_m; \tag{10.19}$$

$$m_z = \cos\theta_m. \tag{10.20}$$

The static state of magnetization corresponds to the minimum of the potential (10.10) with $H_{Ox,y} = 0$ and $h_{x,y,z} = 0$. The corresponding dependences of the potential U on the polar angle of the magnetization vector θ_m at different azimuth angles are illustrated in Fig. 10.3. The figure shows that the potential has a minimum in the polar angle, whose position is almost independent of the azimuth angle. The energy density is minimal at azimuth angle $\varphi_m = 0°$ (curve 1) and maximum at $\varphi_m = 270°$ (curve 4). The values of the energy density at $\varphi_m = 90°$ (curve 2) and $\varphi_m = 180°$ (curve 3) coincide. Thus, with the precession of the equilibrium position along a large circle of the precession portrait, the magnetization vector successively passes from the low potential region (1) through the medium potential region (2) to the large potential region (4), and then again goes to the medium potential region (3) and returns to the low potential region (1).

Let us now consider the smooth variation of the potential with a change in the azimuthal angle of the magnetization vector φ_m and the associated features of the motion of this vector. In particular, let us pay attention to the dependence of the speed of motion of the precessing equilibrium position on the nature of the potential change.

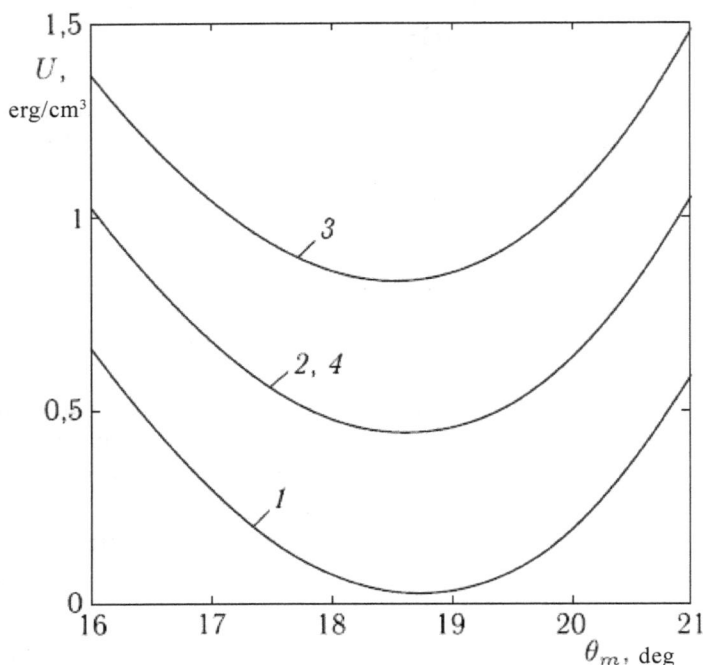

Fig. 10.3. The dependence of the potential (energy density) on the polar angle of the magnetization vector θ_m at various azimuth angles φ_m: 1 – 0°; 2 – 90°; 3 – 180°; 4 - 270°. Parameters: $4\pi M_0$ = 280 Gs, H_{0z} = 265 Oe, K = 3.9 erg × cm^{-3}, θ_a = 10°, φ_a = 0°. Initial potential value: −2794.1 erg × cm^{-3}.

Figure 10.4 shows the corresponding dependences of the potential (*a*) and the precession velocity along a large circle (*b*) on the azimuthal angle of the magnetization vector.

The scatter of points in Fig. 10.4, b is due to the difficulty of accurately measuring the precession rate over a large circle because of the imposition of a small precession on it. The dotted curve is drawn on average over the measured points. From Fig. 10.4, a it can be seen that the potential increases in the range of angles from 0° to 180°, and decreases in the range of angles from 180° to 360°. The rate of increase in potential is maximum near 90°, and the rate of decrease in potential is maximum near 270°. From Fig. 10.4, b it is clear that the precession rate along a large circle is minimal near 90° and maximum near 270°. Thus, the precession rate over a large circle decreases when the potential along the azimuth angle increases and increases when this potential decreases. That is, the minimum of the precession rate falls on the maximum growth of the potential, and its maximum falls on the maximum of its decrease.

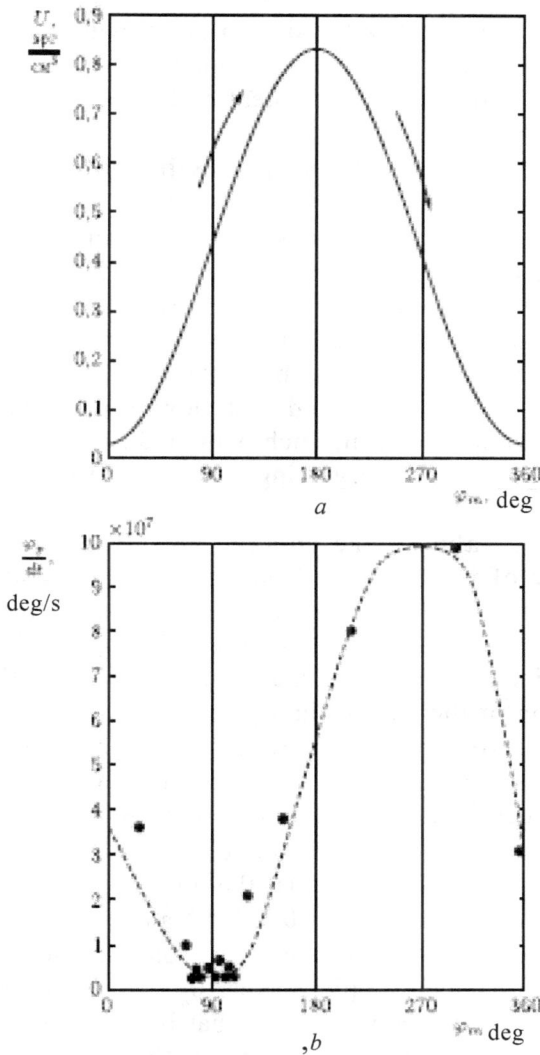

Fig. 10.4. Dependences of the potential (a) and the precession velocity along a large circle (b) on the azimuthal angle of the magnetization vector. Parameters: $4\pi M_0 = 280$ Gs, $H_{0z} = 265$ Oe, $\alpha = 0.3$, K = 3.9 erg $\times$ cm^{-3}, $\theta_a = 10°$, $\varphi_a = 0°$, $f = 100$ MHz, $h_0 = 3$ O. Initial value of the potential: -2794.1 erg $\times$ cm^{-3}.

By analogy with mechanics, it can be said that precession slows down when the system has to climb a 'slide' of potential and accelerates when the system rolls down from this 'slide', as shown by arrows in Fig. 10.4, a. When the precession rate slows down in a large circle, a precessional portrait appears a condensation of small rings, and when it is accelerated, a rarefaction occurs. Thus, the condensation of small rings in the precession portrait is due to the growth of the potential in the path of movement of the precessing magnetization vector, and their reduction is due to the decline in potential during the same movement.

10.1.6. Criticality of system symmetry breaking

Oblique uniaxial anisotropy violates the axial symmetry of the system, with the result that when its magnitude is sufficient, the circular precession of the equilibrium state is disrupted, that is, the equilibrium position in the precession process does not describe the full circle, but tends with time to a stationary stable position, and the magnetization vector continues to precess around it stationary position.

In the precessional portrait, such a breakdown manifests itself in the form of a rupture of a large ring and the exit of small rings into a stationary orbit.

The larger the value of the uniaxial anisotropy constant and the larger the angle of its deviation from the normal to the plane of the plate, the more significantly the symmetry of the system is broken.

Therefore, it is of interest to trace the influence of the value of the constant and orientation angle of the anisotropy axis on the breakdown of the precession of the equilibrium position.

Figure 10.5 shows the dependence of the critical value of the anisotropy constant K_c, corresponding to the breakdown of the equilibrium precession, on the angle θ_a between the anisotropy axis and the normal to the plane of the plate.

It can be seen from the figure that the minimum value of the critical value of the constant takes near $\theta_a = 45°$ and $\theta_a = 135°$. These two angles differ by 90° and correspond to the same tilt of the axis relative to the plane of the magnetic plate. Near $\theta_a = 90°$, the anisotropy axis lies in the plane of the magnetic plate, that is, the system is symmetric with respect to this plane, therefore the critical value of the constant increases. Near $\theta_a = 0°$, the violation of the axial symmetry of the system relative to the normal to the plane of the plate disappears, as a result of which the influence of uniaxial anisotropy decreases and the critical value of the constant also increases. In this case, the critical value of the constant increases as the angle θ_a decreases very significantly. So, at $\theta_a = 10°$, the value of K_c is 3.9 erg×cm^{-3}, at $\theta_a = 1°$ it becomes equal to 30 erg×cm^{-3}, and at $\theta_a = 0.1°$ it reaches 101 erg×cm^{-3}.

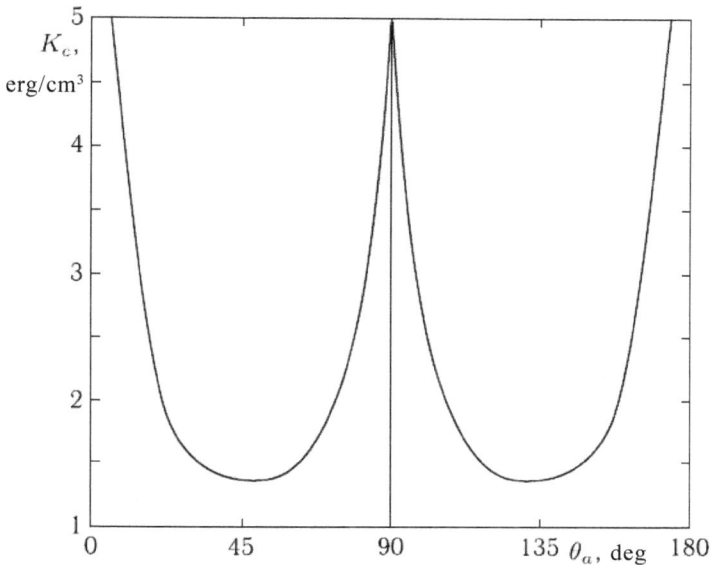

Fig. 10.5. The critical value of the uniaxial anisotropy constant, at which the precession of the equilibrium state occurs, depending on the angle of deviation of the anisotropy axis from the normal to the plane of the plate. Parameters: $4\pi M_0 = 280$ G, $H_{0z} = 265$ Oe, $\alpha = 0.3$, $K = 3.9$ erg×cm^{-3}, $\varphi_a = 0°$, $f = 100$ MHz, $h_0 = 3$ Oe.

With a further decrease in the angle θ_a, the critical value of the constant K_c becomes so significant that the anisotropy field $H_a = 2K/M_0$, adding to the constant field H_{0z}, begins to exceed the demagnetization field $4\pi M_0 = 280$ Oe, with the result that the equilibrium position of the magnetization vector is aligned along the direction of the constant field, that is, the conditions of the orientational transition are violated and the precession of the equilibrium state transforms into a simple circular precession of the magnetization vector around the field direction.

10.1.7. The transformation of thickening rings when the inclination of the anisotropy axis changes

The change in the symmetry of the system with a change in the angle of inclination of the anisotropy axis is manifested in a certain change in the position of the condensation of small rings in the precession portrait. Figure 10.6 illustrates the rotation of the thickening of the small-fringed precession portrait when the angle of inclination of the axis of uniaxial anisotropy is changed.

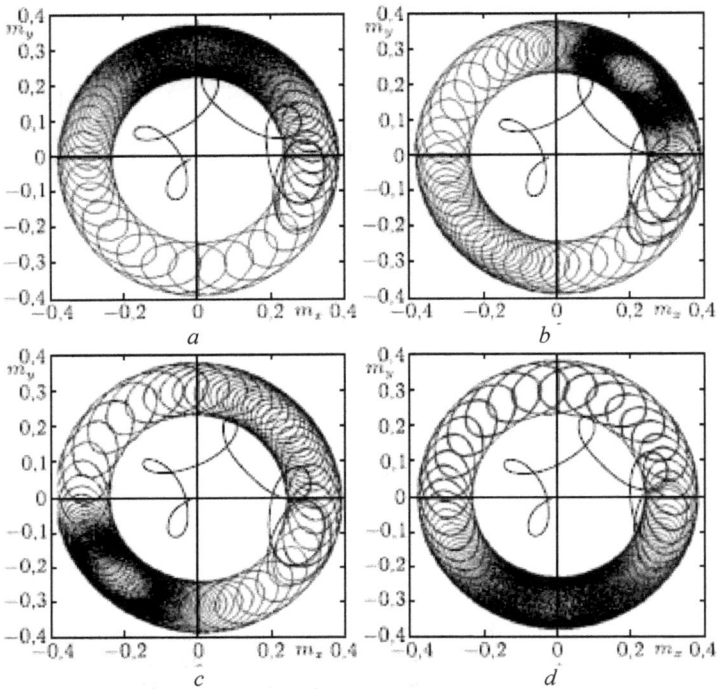

Fig. 10.6. The change in the position of the condensation of the small rings when the inclination of the axis of uniaxial anisotropy θ_a changes: $a - 10°$; $b - 87°$; $c - 92.5°$; $d - 170°$. Parameters: $4\pi M_0 = 280$ G, $H_{0z} = 265$ Oe, $\alpha = 0.3$, K = 3.9 erg×cm^{-3}, $\phi_a = 0°$, $f = 100$ MHz, $h_0 = 3$ Oe.

Figure 10.6 illustrates the sequential transformation of a precession portrait with increasing polar angle θ_a of the axis uniaxial anisotropy.

Figure 10.6, a corresponds to a small deviation of the anisotropy axis from the normal to the plane of the plate: $\theta_a = 10°$. In this case, the condensation of small rings corresponds to $\alpha_c \approx 90°$. Fig. 10.6, b corresponds to $\theta_a = 87.5°$, that is, the anisotropy axis almost lies in the plane of the plate. In this case, there is a well-pronounced main condensation at $\alpha_c \approx 45°$ and a weakly pronounced small condensation near $\alpha_c \approx 225°$.

Figure 10.6 corresponds to $\theta_a = 92.5°$, that is, the anisotropy axis has already crossed the plane of the plate and that part that was in the positive half-space relative to the Oxy plane has passed into the negative half-space, and that part that was in the negative half-space has moved to positive. At the same time, the condensations, as it were, are reversed relative to the centre of the precession portrait, that is,

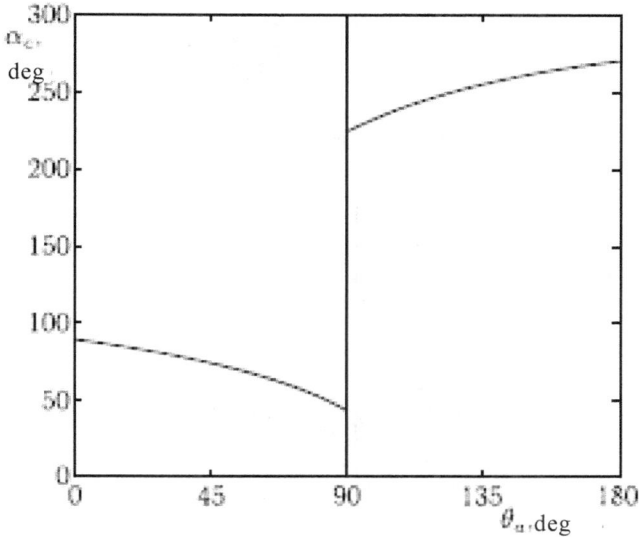

Fig. 10.7. The angle of rotation of the thickening of the small rimgs of the precession portrait αc depending on the angle of deviation of the anisotropy axis from the normal to the plane of the plate θ_a. Parameters – the same as taken in the construction of Fig. 10.6.

there is now a well-pronounced main condensation at $\alpha_c \approx 225°$ and poorly pronounced at $\alpha_c \approx 45°$.

Figure 10.6, d corresponds to $\theta_a = 170°$, that is, again, a significant exit of the anisotropy axis from the *Oxy* plane and its approximation to the normal to the plate plane. Now, however, parts of the anisotropy axis that were originally located in the positive and negative half-spaces, relative to the *Oxy* plane, are reversed. Therefore, the condensation of small rings is observed at $\alpha_c \approx 270°$.

Figure 10.7 illustrates in more detail the smooth change in the position of the condensations of small rings when the polar angle of the anisotropy axis changes, where, in the form of continuous curves, the dependence of the angle of rotation of the centre of condensation of small rings α_c on the angle of deviation of the anisotropy axis from the normal to the plane of the plate θ_a is shown.

It can be seen from the figure that near $\theta_a = 0°$, the angle of rotation of the condensation of small rings is 90° and with θ_a increasing to 90°, it gradually decreases to approximately 45°. When θ_a exceeds the angle of 90° the angle of rotation of the condensation abruptly jumps from 45° to 225°, that is, the condensation in the precession portrait jumps by 180° relative to its centre, that is, its position changes to the

550

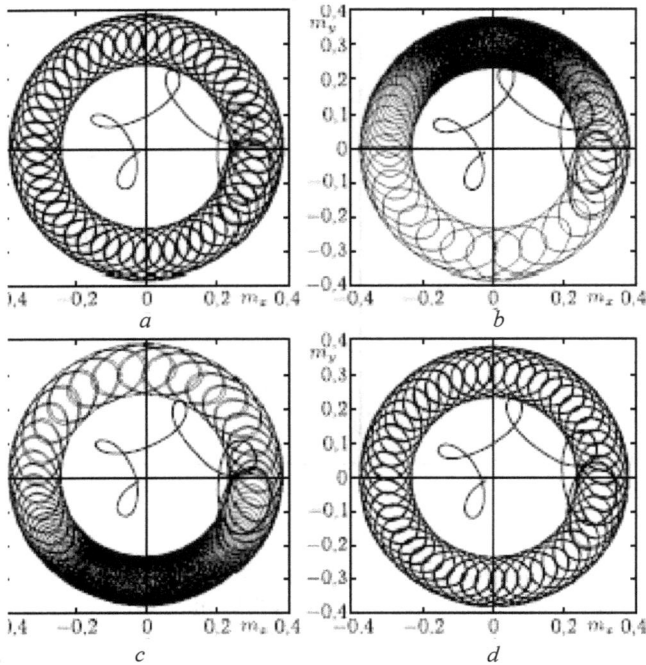

Fig. 10.8. Illustration of the possibility of compensation caused by anisotropy thickening of rings by switching on a transverse constant field: $a - K = 0$, $H_{0x} = 0$ Oe; $b - K = 3.9$ erg cm^{-3}, $H_{0x} = 0$ Oe; $c - K = 0$, $H_{0x} = -0.057$ Oe; $d - K = 3.9$ erg cm^{-3}, $H_{0x} = -0.057$ Oe. Other parameters: $4\pi M_0 = 280$ G, $H_{0z} = 265$ Oe, $\alpha = 0.3$, $\theta_a = 10°$, $\varphi_a = 0°$, $f = 100$ MHz, $h_0 = 3$ Oe

diametrically opposite. Such a jump corresponds to the transition of the anisotropy axis through a position parallel to the plane of the plate. At the same time, the positive and negative parts of the anisotropy axis, located above and below the plate plane, change places, as a result of which the positions of the minima of the surface of the potential of the energy density U are also changed. With a further increase in the angle θ_a up to 180°, the angle of rotation of the condensation tends to 270°, which is diametrically opposite to the initial position at $\theta_a = 0°$ and corresponds to the mutual change of places of the areas of increasing and decreasing potential U.

10.1.8. Constant field axis tilt compensation

When the plate is magnetized by a constant field exactly perpendicular to its plane, the inclined axis of anisotropy introduces asymmetry into the system, which leads to a condensation of the small finger in the

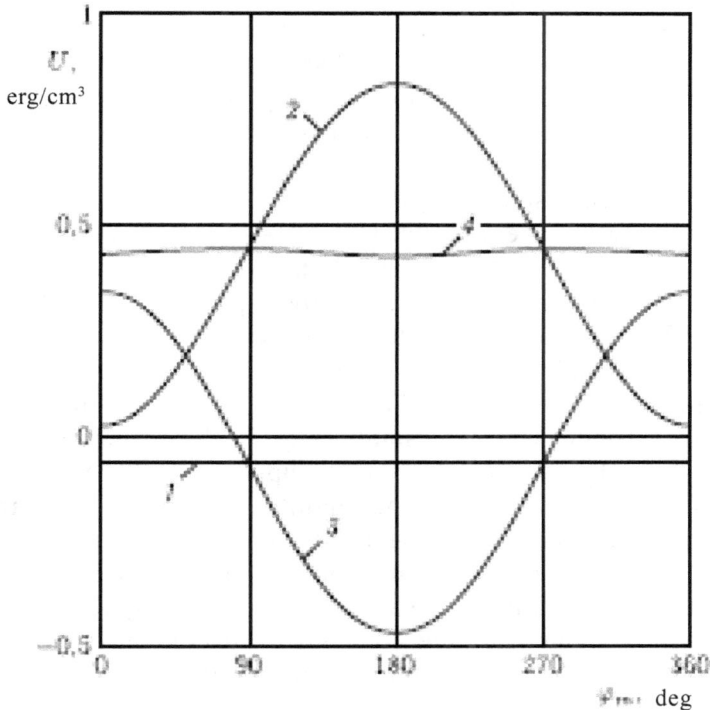

Fig. 10.9. Dependences of the potential on the azimuthal angle of the magnetization vector at different symmetry values: 1 – $K = 0$, $H_{0x} = 0$ Oe; 2 – K = 3.9 erg×cm^{-3}, $H_{0x} = 0$ Oe; 3 – $K = 0$, $H_{0x} = -0.057$ Oe; 4 – K = 3.9 erg×cm^{-3}, $H_{0x} = -0.057$ Oe. The initial value of the potential is -2794.1 erg×cm^{-3}. The remaining parameters: $4\pi M_0 = 280$ G, $H_{0z} = 265$ Oe, $K = 3.9$ erg×cm^{-3}, $\theta_a = 10°$, $\varphi_a = 0°$

precession portrait. Similar condensation in the absence of uniaxial anisotropy can be created by introducing asymmetry into a constant field, in particular, slightly deviating it from the normal to the plane φ of the plate, which is equivalent to introducing its transverse component (section 9.7).

This component should be oriented along the projection of the anisotropy axis on the plane of the plate and have a sign opposite to the positive direction of this projection, that is, at $\varphi_a = 0°$, the orientation of the transverse field should coincide with the negative direction of the Ox axis.

Thus, if the asymmetry created by the transverse constant field is opposite to the sign of asymmetry created by the inclination of the anisotropy axis, then mutual asymmetry can be compensated, as a result of which the small rings in the precession portrait will be no longer present.

The possibility of such compensation is illustrated in Fig. 10.8, where precession portraits are shown for various asymmetry.

Figure 10.8, a corresponds to the absence of both anisotropy and transverse constant field, that is, a completely symmetric case. As you can see, in this case, any condensation of small rings is completely absent.

Figure 10.8, b corresponds to the presence of uniaxial anisotropy in the absence of a transverse field. In this case, the asymmetry of the system leads to a condensation of small rings in the upper part of the portrait ($\alpha_c \approx 90°$) and a vacuum in the lower part ($\alpha_c \approx 270°$).

Figure 10.8, c corresponds to the presence of a transverse field in the absence of uniaxial anisotropy. Here, the asymmetry of the system, due to the negative component of the transverse field along the axis Oxz, leads to a condensation of small rings in the lower part of the portrait ($\alpha_c \approx 270°$) and rarefaction in the upper ($\alpha^c \approx 90°$).

Figure 10.8, d corresponds to the simultaneous action of both the anisotropy axis and the transverse field component. It can be seen that in this case, as a result of the opposite action of both factors mentioned, any condensation of small rings is completely absent. Thus, the transverse component of the constant field compensates for the asymmetry of the system, due to the inclination of the axis of uniaxial anisotropy.

The features of compensation caused by the anisotropy of the thickening of the rings when the transverse constant field is turned on are due to the nature of the potential behaviour, which for the cases considered is illustrated in Fig. 10.9. Here, the dependences of the potential on the azimuthal angle of the magnetization vector are presented for various degrees of asymmetry of the system. Curve 1 corresponds to the absence of both anisotropy and transverse constant field. At the same time, the potential with a change in the azimuthal angle of the magnetization vector φ_m remains constant, as a result of which the rate of change of this angle also does not change and the rings do not thicken (Fig. 10.8, a)

Curve 2 corresponds to the presence of anisotropy in the absence of a constant field. In this case, the potential has a bell-shaped appearance with a maximum at $\varphi_m = 180°$, with the result that the rate of change of the angle φ_m near 90° decreases (the system climbs the hill), and near 270° increases (the system rolls down the hill). This change in velocity leads to the observed thickening of the rings near $\alpha_c = 90°$ and a rarefaction near $\alpha_c = 270°$, as can be seen in Fig. 10.9, b. Curve 3 corresponds to the presence of a transverse field in the absence of anisotropy. In this case, the potential also has a bell-shaped appearance, but now it has a negative sign with a minimum at $\varphi_m = 180°$, that is, as φ_m increases, it first decreases and then increases, which leads to a rarefaction of the rings near $\alpha_c = 90°$ and thickening near $\alpha_c = 270°$, as can be seen in Fig. 10.8, c.

Curve 4 corresponds to the simultaneous presence of anisotropy and the transverse field. In this case, due to the difference in the signs of potential 2 and 3, their influence is destroyed and the potential remains almost constant, which leads to the absence of thickening of the rings in the precession portrait (Fig. 10.8, d).

Thus, the reason for compensating for the thickening of the rings of the precession portrait is the opposite direction of the anisotropy and the constant field, which leads to the equalization of the potential dependence on the azimuth angle of the magnetization vector.

10.1.9. Criticality ot the anisotropy constant

The above quantitative consideration of the properties of the precession of the equilibrium position in an anisotropic medium was made using the example of rather small quantities of the anisotropy constant, on the order of a few erg×cm^{-3}, whereas in some frictions this value can reach tens, hundreds, or even thousands of erg×cm^{-3} [38, 40 , 259, 296]. Therefore, it is of interest to consider the possibility of exciting the precession of the equilibrium position in a wider range of the anisotropy constant. To clarify this possibility, the dependence of the critical angle of the polar angle of the anisotropy axis θ_c on the magnitude of the uniaxial anisotropy constant K was studied. This critical value means that at $\theta_a < \theta_c$ the large ring in the precession portrait is completely closed, and at $\theta_a > \theta_c$ – open. The resulting dependence, built on a double logarithmic scale, is shown in Fig. 10.10.

It should be taken into account that, since the anisotropy axis is almost perpendicular to the plane of the plate in the considered cases, the anisotropy field, adding to the external field, gives a positive additive to the effective field. Therefore, so that at a constant frequency of the alternating field and different quantities of the anisotropy constant, it is possible to ensure equal conditions of excitation when constructing Fig. 10.10, the external field H_{0z} decreased by the magnitude of the anisotropy field H_a, that is, the z-component of the constant field was set equal to

$$H_{0z} = H_{00z} - H_a, \tag{10.21}$$

where $H_{00z} = 265$ Oe; $H_a = 2K/M_0$. In the figure, the dots indicate

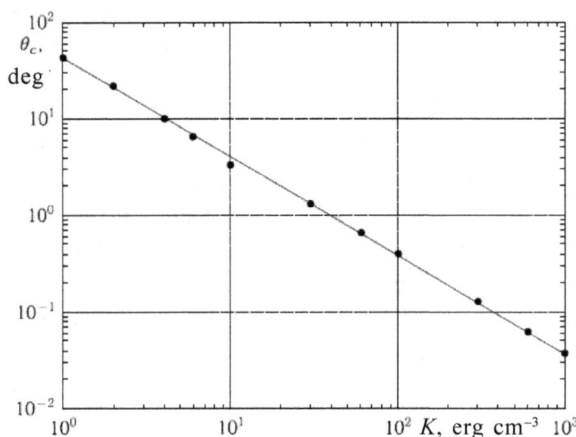

Fig. 10.10. The dependence of the critical value of the polar angle of the anisotropy axis on the value of the anisotropy constant. Parameters: $4\pi M_0 = 280$ GS, $H_{0z} = 265$ Oe, $H_{0x} = 0$ Oe, $\alpha = 0.3$, $\varphi_m = 0°$, $f = 100$ MHz, $h_0 = 3$ Oe.

the results of a machine experiment for given anisotropy values, the solid line is a straight line drawn through these points. It can be seen from the figure that all the points fall quite well on the straight line described by the equation:

$$\log \theta_c = 1.6532 - 1.0293 \log K, \qquad (10.22)$$

which corresponds to a power dependence

$$\theta_c = 10^{(1.6532 - 1.0283 \log K)}, \qquad (10.23)$$

or the same dependence in exponential form

$$\theta_c = \exp(3.8067 - 1.0283 \ln K), \qquad (10.24)$$

Thus, it can be seen that as the anisotropy constant increases, the criticality of the precession of the equilibrium state along a closed ring increases markedly. It should be noted that this circumstance imposes rather strict requirements on the observation of the precession of equilibrium in the experiment. This issue is discussed in more detail in section 10.6.

10.2. Precession of the equilibrium state of magnetization of a medium with cubic anisotropy

In the preceding sections, the consideration of the precession of the equilibrium position of the magnetization in an anisotropic medium is limited only to the case of uniaxial anisotropy. However, in many ferrites that are of interest to practice, including iron–yttriumferrite garnet, there is a noticeable cubic anisotropy. We now turn to the consideration of the effect of such anisotropy on the properties of the precession of the equilibrium position in the case of various orientations of the cubic axes of easy magnetization of the [111] type with respect to the plate plane.

10.2.1. Task geometry

Consider a normally magnetized ferrite plate with cubic magnetic anisotropy. The overall geometry of the problem is illustrated in Fig. 10.11, where the diagram of the main orientations of the cubic cell relative to the plane of the ferrite plate (similar to that shown in Fig. 3.7) is shown. The ferrite plate is magnetized by a constant field **H** perpendicular to its plane, the alternating field **h** is applied in the plane of the plate. The Oxy plane of the Cartesian coordinate system $Oxyz$ coincides with the plane of the plate, the axis Oz is perpendicular to it.

Three schematically shown in Fig. 10.11 variants of orientation of the cubic crystallographic cell of the plate material relative to its plane were investigated:

1) **orientation [001]**: along the normal to the plane of the plate, that is, along the Oz axis, one of the [001] axes, which is the edge of

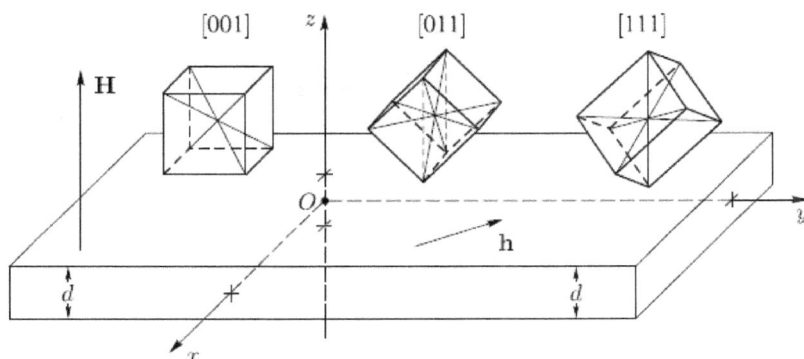

Fig. 10.11. The scheme of the main orientations of the cubic cell relative to the plane of the ferrite plate.

556

the cube, is directed, while the axes of Ox and Oy are oriented along the two edges of the cube;

2) **orientation [011]:** along the normal to the plane of the plate or axis Oz, one of the [011] axes is directed, that is, the diagonal of the cube face, while the Ox axis is oriented along one of the cube edges, and the Oy axis is perpendicular to this edge;

3) **orientation [111]:** along the normal to the plane of the plate or axis Oz, one of the axes of the [111] type is directed, that is, the spatial diagonal of the cube, while the axis Ox is oriented along the projection of one of the edges of the cube onto the Oxy plane, and the axis Oy is perpendicular this projection.

It is assumed that the axes of the easy magnetization are spatial diagonals of the cube, that is, axes of the type [111], shown by thin lines inside the cells.

10.2.2. Basic equations

Similarly to the case of the uniaxial anisotropy analyzed above, it is assumed that the constant field H_0 is not enough to orient the magnetization vector in the equilibrium state perpendicular to the plane of the plate, and the system is excited by the alternating field of circular polarization (10.8)–(10.9):

$$h_x = h_0 \sin (2\pi f t), \tag{10.25}$$

$$h_y = -h_0 \cos (2\pi f t). \tag{10.26}$$

To calculate the magnetization precession, similarly to the previous one, we use the Landau–Lifshitz equations with the dissipative term in the Hilbert form (2.97)–(2.99):

$$\frac{\partial m_x}{\partial t} = -\frac{\gamma}{1+\alpha^2}[(m_y + \alpha m_x m_z)H_{ez} - (m_z - \alpha m_y m_x)H_{ey} - \alpha(m_y^2 + m_z^2)H_{ex}]; \tag{10.27}$$

$$\frac{\partial m_y}{\partial t} = -\frac{\gamma}{1+\alpha^2}[(m_z + \alpha m_y m_x)H_{ex} - (m_x - \alpha m_z m_y)H_{ex} - \alpha(m_z^2 + m_x^2)H_{ey}]; \tag{10.28}$$

$$\frac{\partial m_z}{\partial t} = -\frac{\gamma}{1+\alpha^2}[(m_x + \alpha m_z m_y)H_{ey} - (m_y - \alpha m_x m_z)H_{ex} - \alpha(m_x^2 + m_y^2)H_{ey}]. \tag{10.29}$$

where γ is the gyromagnetic constant ($\gamma > 0$), α is the damping parameter of the magnetization precession.

Effective fields put in the form:

$$H_{ex} = h_x + H_{ax} \tag{10.30}$$

$$H_{ey} = h_y + H_{ay} \tag{10.31}$$
$$H_{ez} = H_{0z} - 4\pi M_0 m_z + H_{az}. \tag{10.32}$$

where H_{ax}, H_{ay}, H_{az} are the components of the anisotropy field, determined by the orientation of the crystallographic cell. With such a record, it is assumed that in addition to anisotropy in the plane of the plate, external variable fields h_x and h_y act, and along the normal to the plate there is a constant field H_{0z} and a demagnetization field $4\pi M_0$ directed in opposite directions. The demagnetization field tends to place the magnetization vector in the plane of the plate, and the normal constant field H_{0z} pulls this vector out of the plane in the direction normal to the plane of the plate, creating an orientational transition situation. The interaction of the external field with the demagnetization field has already been considered in sufficient detail, so we will focus here primarily on the influence of cubic anisotropy.

In the general case, the effective fields of cubic anisotropy are determined using formula (1.2) or (2.19):

$$H_{ai} = -\frac{\partial U}{\partial M_i} = -\frac{1}{M_0}\frac{\partial U_a}{\partial m_i}, \tag{10.33}$$

where U_a is the energy density of cubic anisotropy for a given orientation of the crystallographic cell. We note that we are talking here about effective fields in the sense of (10.33), and not about the 'cubic anisotropy field' (4.119), which is a material constant and is equal to K_1/M_0, where K_1 is the first cubic anisotropy constant.

The basic equations (10.27)–(10.29) used to solve the problem are written in the $Oxyz$ system, associated with the direction of the constant field and the plane of the plate. However, the energy density of cubic anisotropy has the simplest form in the coordinate system $Ox'y'z'$, whose axes coincide with the edges of the cube, which in the general case is somehow rotated relative to the system $Oxyz$.

For the $Ox'y'z'$ system we used the form of the energy density of the cubic anisotropy in the form similar to (4.120) (at $K_1 > 0$, $K_2 = 0$), i.e. with the accuracy to notations, it is assumed that

$$U_a = -K\left(m_{x'}^2 m_{y'}^2 + m_{y'}^2 m_{z'}^2 + m_{z'}^2 m_{x'}^2\right).$$

(10.34)

where for light axes of the type [111], it is assumed that $K > 0$ (here the index '1' at the constant is omitted to simplify recording). The preliminary step in solving the problem is the conversion of the anisotropy energy density (10.34) specified in the $Ox'y'z'$ system to the type of record in the $Oxyz$ system, for which we use the apparatus of the transition matrices described in Chapter 3. In this case, the transformation of the components of the magnetization vector **m** according to the formula similar to (3.14):

$$\mathbf{m'} = \ddot{A}^{-1}\mathbf{m}.$$

(10.35)

where $\ddot{A}^{-1}$ is the matrix of transition from the system $Oxyz$ to the $O'x'y'z'$ system whose components, like (3.8), are cosines of the anglex between the axe of these systems:

$$\ddot{A}^{-1} = \begin{pmatrix} \cos(\angle x'Ox) & \cos(\angle x'Oy) & \cos(\angle x'Oz) \\ \cos(\angle y'Ox) & \cos(\angle y'Oy) & \cos(\angle y'Oz) \\ \cos(\angle z'Ox) & \cos(\angle z'Oy) & \cos(\angle z'Oz) \end{pmatrix}$$

(10.36)

In Chapter 3, it is noted that the most important for practice in the case of a cubic cell are orientations of the type [001], [011] and [111].

Since in order to obtain effective fields (10.33) entering the equations of motion of magnetization (10.27)–(10.29), knowledge of the anisotropy energy U_a is necessary, we will dwell in detail on the derivation of energy density expressions for the above basic orientations.

10.2.3. Orientation [001]

We first consider the orientation [001], illustrated in Fig. 10.12.

Here in Fig. 10.12, a in the $Oxyz$ coordinate system, a cubic cell is shown in the orientation position [001], oriented so that the edges of the cube coincide with the axes of the $Oxyz$ system. These edges are shown by thick solid and dashed lines. The vertices of the cube are marked with the letters O (this vertex is placed at the origin), A, B, C, D, E, F, G. The intersection point of spatial diagonals is indicated by the letter S.

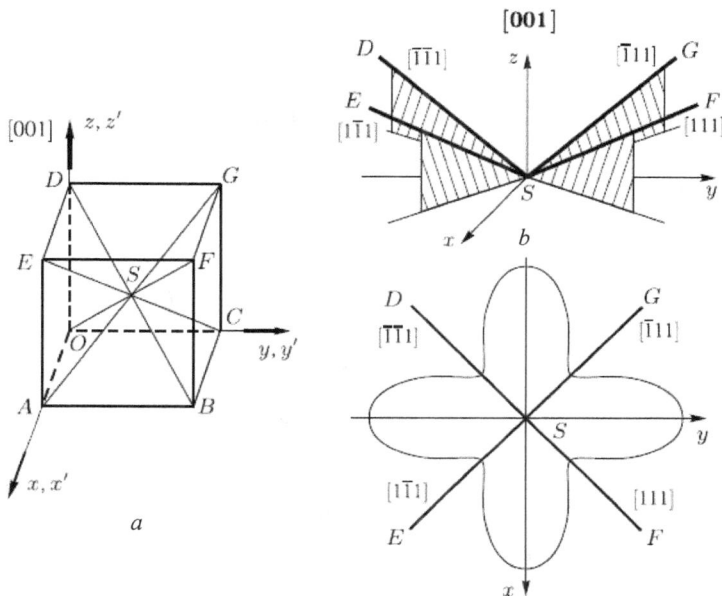

Fig. 10.12. The orientation of the cubic cell and anisotropy axes of type [111] at orientation [001]. *a* – cubic cell and coordinate systems *Oxyz* and *Ox'y'z'*; *b* – axes of anisotropy and in the *Sxyz* system; *c* – projections of the anisotropy axes onto the *Sxy* plane and the azimuthal portrait of the energy density.

Figure 10.12, b in the system of the coordinates *Sxyz*, the beginning of which is placed at the intersection point of spatial cube diagonal, and the axes *Sx*, *Sy*, *Sz* parallel to the axes *Ox*, *Oy*, *Oz*, respectively, show the location of the parts of the [111] axes in the upper half-space relative to the plane *Sxy*. The letters *D, E, F, G* correspond to the vertices of the cube in Fig. 10.12, a. All these parts of the axes *SD, SE, SF, SG* lie above the plane *Sxy*. In Fig. 10.12, a view is shown of the projection of the [111] axes on the *Sxy* plane from the positive direction of the *Sz* axis. Hereinafter, the *Sx* axis is directed downwards, and the *Sy* axis to the right is for ease of visual comparison with Fig. 10.12, b. The projections of the technical axes from the negative direction of the *Sz* axis coincide with the first projections, that is, on the Sxy plane or parallel to the *Oxy* plane, there are four dedicated easy magnetization directions parallel to the *SF, SG, SD, SE* axes located at 45° angles relative to the axes *Ox* and *Oy*.

Since the axes of the *Oxyz* and *Ox'y'z'* systems coincide in the given orientation, the transformation matrix (10.36) is unique and the energy density of anisotropy in the *Oxyz* system have the form similar ot (10.34):

$$U_a^{(001)} = -K\left(m_x^2 m_y^2 + m_y^2 m_z^2 + m_z^2 m_x^2\right). \tag{10.37}$$

In accordance with (10.33), the effective anisotropy fields are:

$$U_{ax}^{(001)} = \frac{2K}{M_0} m_x \left(m_y^2 + m_z^2\right); \tag{10.38}$$

$$U_{ay}^{(001)} = \frac{2K}{M_0} m_y \left(m_z^2 + m_x^2\right); \tag{10.39}$$

$$U_{az}^{(001)} = \frac{2K}{M_0} m_z \left(m_x^2 + m_y^2\right); \tag{10.40}$$

We introduce a spherical coordinate system whose polar axis coincides with the axis Oz, and the azimuthal axis with the axis Ox. In this case, the components of the normalized vector of magnetization, expressed through the polar and azimuth angles θ and φ, have the form:

$$m_x = \sin\theta\cos\varphi, \tag{10.41}$$

$$m_y = \sin\theta\sin\varphi; \tag{10.42}$$

$$m_z = \cos\theta. \tag{10.43}$$

The energy density (10.37) takes the form

$$U_a^{(001)} = -K\left(\sin^4\theta\sin^2\varphi\cos^2\varphi\sin^2\theta\cos^2\theta\right). \tag{10.44}$$

In Fig. 10.12, c the solid line shows the dependence of the normalized energy density $U^{(001)}/K$ on the azimuthal angle φ, constructed by the formula (10.44) at $\theta = 20°$. The value of the angle θ is chosen from the condition of proximity of the direction of the magnetization vector to the equilibrium orientation, which takes place later in the construction of the precession portraits. For clarity, the figure to the value of the energy density by the formula (10.44) added a constant value of 0.109, not changing the essence of construction, since the potential is determined with an accuracy of an arbitrary constant.

It can be seen that the energy density has four minima, which coincide in position with the projections of the [111] axes on the Sxy plane or the Oxy plane parallel to it.

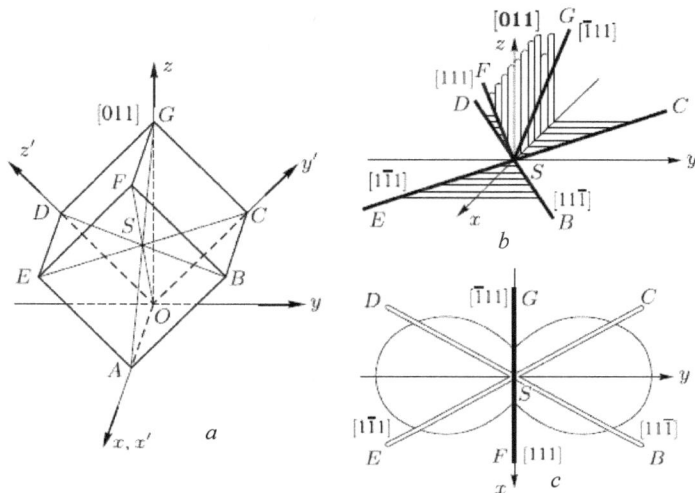

Fig. 10.13. The orientation of the cubic cell and anisotropy axes of the [111] type with the [011] orientation. *a* – cubic cell and coordinate systems *Oxyz* and *Ox'y'z'*; *b* – anisotropy axes in the *Sxyz* system; *c* – the projections of the anisotropy axes onto the *Sxy* plane and the azimuthal portrait of the energy density.

10.2.4. Orientation [011]

We now consider the orientation [011], illustrated in Fig. 10.13.

Here in Fig. 10.13, and in the coordinate system *Oxyz*, a cubic cell is shown in the orientation position [011], oriented so that one of the edges of the cube coincides with the axis *Ox*, and one of the diagonals of the face, namely *OG*, coincides with the axis *Oz*. The image and designations of the edges and vertices of the cube, as well as the spatial diagonals and points of their intersection coincide with those shown in Fig. 10.12, a.

Figure 10.13, b in the system of the coordinates *Sxyz*, the beginning of which is placed at the intersection point of spatial cube diagonals, and the axes *Sx, Sy, Sz* parallel to the axes *Ox, Oy, Oz*, respectively, shows the location of parts of the [111] axes in the upper half-space relative to the plane *Sxy*. The letters *B, C, D, E, F, G* correspond to the vertices of the cube in Fig. 10.13, a. Two axes of the [111] type, *BD* and *CE*, lie in the *Sxy* plane, parts of the *SF* and *SG* axes lie above the *Sxy* plane.

In Fig. 10.13, a view is shown of the projection of the [111] axes on the *Sxy* plane from the positive direction of the *Sz* axis. The projections of the parts of the axes *SF* and *SG* lying above the *Sxy* plane are shown by thick solid lines. The *BD* and *CE* axes lying in the *Sxy* plane, coinciding with their projections, are shown by double thin lines.

In the accepted orientation, from the axes of the coordinate system $Ox'y'z'$ the axis Ox' coincides with the axis $Ox,$ and the axis Oy' and Oz' make up with the axes Oy and Oz angles of $45°$, respectively. The transformation matrix formed by the cosines of the angles between the axes of the systems $Ox'y'z'$ and $Oxyz,$ has the form

$$
\ddot{A}_{(111)} = \begin{pmatrix} 1 & 0 & 0 \\ 0 & \dfrac{\sqrt{2}}{2} & \dfrac{\sqrt{2}}{2} \\ 0 & -\dfrac{\sqrt{2}}{2} & \dfrac{\sqrt{2}}{2} \end{pmatrix}
\tag{10.45}
$$

The transformation of the components of the magnetization vector, is produced by the same formula (10.35), from which we obtain:

$$
m_{x'} = m_x;
\tag{10.46}
$$

$$
m_{y'} = \frac{\sqrt{2}}{2}\left(m_y + m_z\right);
\tag{10.47}
$$

$$
m_{z'} = \frac{\sqrt{2}}{2}\left(-m_y + m_z\right);
\tag{10.48}
$$

Substituting these components into (10.34), we obtain the anisotropy energy density, expressed through the components of the magnetization vector in the $Oxyz$ system, in the form

$$
U_a^{(001)} = -K\left(\frac{1}{4}m_y^4 + \frac{1}{4}m_z^4 + m_x^2 m_y^2 + m_x^2 m_z^2 - \frac{1}{2}m_y^2 m_z^2\right).
\tag{10.49}
$$

Effective anisotropy fields are:

$$
H_{ax}^{(011)} = \frac{2K}{M_0} m_x \left(m_y^2 + m_z^2\right);
\tag{10.50}
$$

$$
H_{ay}^{(011)} = \frac{K}{M_0} m_y \left(2m_x^2 + m_y^2 - m_z^2\right);
\tag{10.51}
$$

$$
H_{az}^{(011)} = \frac{K}{M_0} m_z \left(2m_x^2 - m_y^2 + m_z^2\right).
\tag{10.52}
$$

In the spherical coordinate system with components in the magnetization vector of the form (10.41)–(10.43), the energy density takes the form:

$$U_a^{(011)} = -K\left(\frac{1}{4}\sin^4\theta\sin^4\varphi + \frac{1}{4}\cos^4\varphi + \sin^4\theta\sin^2\varphi\cos^2\varphi + \right.$$

$$\left. + \sin^2\theta\cos^2\theta\cos^2\varphi - \frac{1}{2}\sin^2\cos^2\theta\sin^2\varphi\right). \tag{10.53}$$

In Fig. 10.13, the solid line shows the dependence of the normalized energy density $U_a^{(011)}/K$ on the azimuthal angle φ, constructed by the formula (10.53) at $\theta = 20°$. The value of the angle θ is chosen similarly to Fig. 10.12, c. A constant addition to the energy density is also 0.109. It can be seen that in this case the energy density has two minima, which coincide in position with the projections of the [111] -type network on the Sxy or Oxy plane, which lie in the Oxz plane perpendicular to it.

10.2.5. Orientation [111]

Finally, we consider the orientation [111], illustrated in Fig. 10.14. Here in Fig. 10.14, and in the $Oxyz$ coordinate system, a cubic cell is

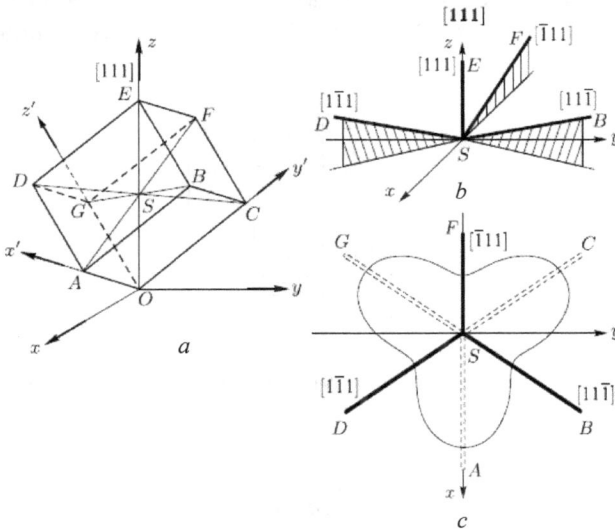

Fig. 10.14. Orientation of the cubic cell and anisotropy axes of the [111] type at orientation [111]. a - cubic cell and coordinate systems $Oxyz$ and $Ox'y'z'$; b – axes of anisotropy in the $Sxyz$ system; c – projections of the axes of anisotropy on the plane Sxy and azimuthal energy density portrait

shown at the [111] orientation, oriented so that the projection of the cube edge OA on the Oxy plane coincides with the Ox axis, and the spatial diagonal of the cube OE coincides with the Oz axis. The image and designations of the edges and vertices of the cube, as well as the spatial diagonals and their points of intersection coincide with those shown in Fig. 10.12, a.

Figure 10.14, b in the coordinate system $Sxyz$, the beginning of which is placed at the intersection point of spatial cube diagonals, and the axes Sx, Sy, Sz are parallel to the axes Ox, Oy, Oz, respectively, shows the location of parts of the [111]-type axes in the upper half-space relative to the plane Sxy. The letters B, D, E, F correspond to the vertices of the cube in Fig. 10.14, a. The OE axis is perpendicular to the Sxy plane, and parts of the SB, SD, SF axes lie above the Sxy plane and form an angle with it equal to $\pi/2 - 2$ arctg $\sqrt{2}/2 \approx 19.47°$.

Figure 10.14, shows a view on the projection of axes of the [111] type on the Sxy plane from the positive direction of the Sz axis. The projections of the parts of the axes SB, SD, SF lying above the plane Sxy are shown by thick solid lines, the projections of the parts of the axes SA, SC, SG are lying below the plane Sxy are shown by double dashed lines.

In the accepted orientation the Ox' axis lies in the plane Oxz and makes an angle with the axis Ox equal to arctg $\sqrt{2}/2$. The Oy' and Oz' axes lie in a plane that forms an angle with the axis Ox equal to arctg $\sqrt{2}$ and are located symmetrically relative to the plane Oxz.

.

$$\ddot{A}_{(111)} = \begin{pmatrix} \dfrac{\sqrt{6}}{3} & 0 & \dfrac{\sqrt{3}}{3} \\ \dfrac{\sqrt{6}}{6} & \dfrac{\sqrt{2}}{2} & \dfrac{\sqrt{3}}{3} \\ \dfrac{\sqrt{6}}{6} & \dfrac{\sqrt{2}}{2} & \dfrac{\sqrt{3}}{3} \end{pmatrix} \tag{10.54}$$

Transforming the components of vector **m** using formula (10.35), we get:

$$m_{x'} = \frac{\sqrt{6}}{3} m_x + \frac{\sqrt{3}}{3} m_z; \tag{10.55}$$

$$m_{y'} = -\frac{\sqrt{6}}{6} m_x + \frac{\sqrt{2}}{2} m_y + \frac{\sqrt{3}}{3} m_z; \tag{10.56}$$

$$m_{z'} = -\frac{\sqrt{6}}{6}m_x - \frac{\sqrt{2}}{2}m_y + \frac{\sqrt{3}}{3}m_z;$$

(10.57)

Substituting these components into (10.34), we obtain the anisotropy energy density, expressed through the components of the magnetization vector in the $Oxyz$ system, in the form

$$U_a^{(111)} = -K\left(\frac{1}{4}m_x^4 + \frac{1}{4}m_y^4 + \frac{1}{3}m_z^4 + \frac{1}{2}m_x^2m_y^2 - \frac{\sqrt{2}}{3}m_x^3m_z + \sqrt{2}m_xm_y^2m_z\right).$$

(10.58)

Effective anisotropy fields are:

$$H_{ax} = \frac{K}{M_0}\left(m_x^3 + m_xm_y^2 - \sqrt{2}m_x^2m_z + \sqrt{2}m_y^2m_z\right);$$

(10.59)

$$H_{ay} = \frac{K}{M_0}\left(m_y^3 + m_x^2m_y + 2\sqrt{2}m_xm_ym_z\right);$$

(10.60)

$$H_{az} = \frac{K}{M_0}\left(\frac{4}{3}m_z^3 - \frac{\sqrt{2}}{3}m_x^3 + \sqrt{2}m_xm_y^2\right);$$

(10.61)

In a spherical coordinate system with components in the magnetization vector of the form (10.41)–(10.43), the energy density takes the form:

$$U_a^{(111)} = -K\left(\frac{1}{4}\sin^4\theta\cos^4\varphi + \frac{1}{4}\sin^4\theta\sin^4\varphi + \frac{1}{3}\cos^4\varphi + \right.$$

$$\left. + \frac{1}{2}\sin^4\theta\sin^2\varphi\cos^2\varphi - \frac{\sqrt{2}}{3}\sin^3\theta\cos\theta\cos^3 + \right.$$

(10.62)

$$\left. + \sqrt{2}\sin^3\theta\cos\theta\sin^2\varphi\cos\varphi\right).$$

In Fig. 10.14, the solid line shows the dependence of the normalized energy density $U_a^{(111)}/K$ on the azimuth angle φ, constructed by the formula (10.62) with $\theta = 20°$. The value of the angle θ is chosen similarly to Fig. 10.12, c.

A constant additive to the energy density is 0.210. It can be seen that in this case the energy density has three minima occurring on the projection onto the Sxy or Oxy plane of those parts of the [111] axes, which lie above the Sxy plane, that is, at angles φ equal to 60°, 180° and 300°. This position of the minima is due to the positive value of

566

the angle θ, measured from the axis Oz. When this angle is $160°$ (that is, when the magnetization vector is mirror-symmetric with respect to the plane Sxy), the energy density minima are closer to those parts of the $[111]$ axes that lie below the plane Sxy, that is, at angles $0°$, $120°$ and $240°$. At $θ = 90°$, the dependence of the normalized energy density $U_a^{(111)}/K$ on the angle $φ$ is absent.

10.2.6. Precession portraits

Now, when the expressions for the energy density of cubic anisotropy for different orientations of the cubic cell are obtained, we now turn to the study of the properties of the precession of the equilibrium state in these orientations. The total energy density, which includes, besides the anisotropy energy, also the energy of interaction of magnetization with the external field and the anisotropy field, has the form:

$$U = -m_0 h_x n_x - M_0 h_y m_y - M_0 H_{0z} m_z + 2\pi M_0^2 m_z^2 + U_a, \quad (10.63)$$

where U_a is the energy density of cubic anisotropy, defined by the formulas (10.37), (10.49), (10.58) or their equivalents in the spherical coordinate system (10.44), (10.53), (10.62).

We solve the equations of motion of magnetization (10.27)- (10.29) with effective fields determined from the energy density in accordance with formulas (10.14) and (10.33). As an object of study, we consider precession portraits with different orientations of the anisotropy axes constructed by numerically solving Landau–Lifshitz nonlinear equations (10.27)–(10.29) using the fourth order Runge–Kutta method [261], just as it was done in [307, 308, 312, 323, 324].

The resulting portraits are shown in Fig. 10.15. To compare the portraits with the anisotropy character, in Fig. 10.15, b–d thickened lines show projections of axes of the (111) type on the Sxy plane (here the coordinate axes Ox and Oy are oriented in the traditional way, that is, they are rotated relative to their position in Fig. 10.12, c – Fig. 10.14, c by $90°$) . Azimuthal portraits of the energy density, constructed under the technical conditions, as in Fig. 1, are shown on the interior of the tubes in each of these ortrets. 10.12, c – Fig. 10.14, c. The direction of movement of the equilibrium position (along the central line of the large ring) occurs counterclockwise, which corresponds to its right character. The solid lines of the energy density portraits correspond to the growth of the potential as the equilibrium position progresses, the dotted lines to its decrease.

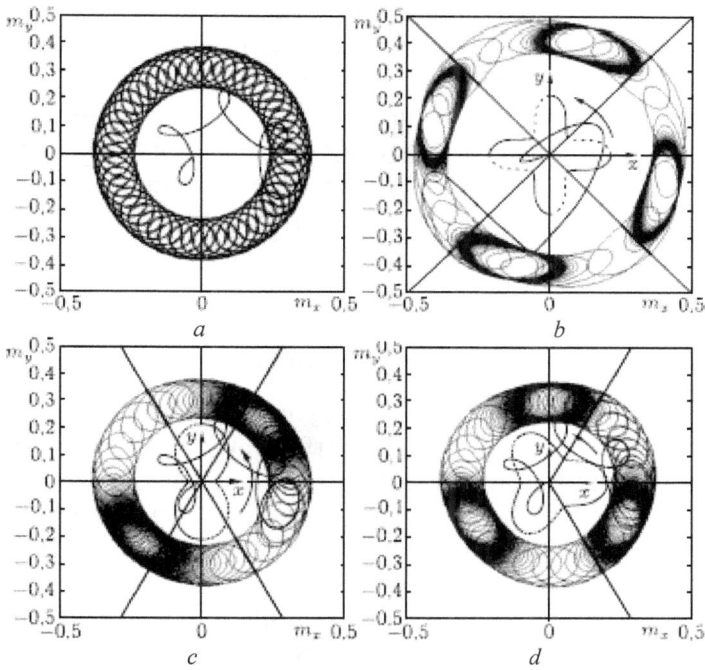

Fig. 10.15. Precession portraits and anisotropy energy densities at various orientations of cubic axes. Parameters: a – no anisotropy: $K = 0$; b — orientation [001], $K = 221$ erg $\times$ cm^{-3}; c – orientation [011], $K = 5.5$ erg$\times$cm^{-3}; (d) orientation [111], K = 12.9 erg $\times$ cm^{-3}. Parameters: $4\pi M_0 = 280$ G, $H_{0z} = 265$ Oe, $\alpha = 0.3$, $f = 100$ MHz, $h_0 = 3$ Oe

Figure 10.15, a corresponds to the absence of anisotropy and is shown here for comparison. It is seen that in this case the large ring is filled with small rings completely uniformly. Figure 10.15, b corresponds to the orientation [001]. It can be seen that in this case the precession portrait has four thickening of small rings with rarefactions located between them, and the positions of those and others are completely correlated with projections of [111] axes. From the comparison with the azimuthal portrait of the energy density, one can see that the condensations fall on the same azimuth angle as the potential growth areas (from 45° to 90° and further between 135° and 180°, 225° and 270°, 315° and 360°).

Figure 10.15, c corresponds to the orientation [011]. In this case, the precession portrait has only two thickenings of small rings with rarefactions located between them, which also correlate well with projections of [111] axes. Here, the condensation centres also fall in the middle of the same azimuth angle intervals as the potential growth areas (from 0° to 90° and from 180° to 270°)

Figure 10.15, d corresponds to the orientation [111]. Now, the precession portrait has three thickenings of small rings with intermediate dilutions, also correlated with projections of [111] axes. And in this case, thickening falls on the same intervals of the azimuth angle as the potential growth areas (from 60° to 120°, from 180° to 240° and from 300° to 360°). Thus, as in the case of uniaxial anisotropy, the condensation of small oleants corresponds to a deceleration of the equilibrium position due to its rise on a potential slide, and a rarefaction to an acceleration of movement due to rolling off this slide.

10.2.7. Criticality of system symmetry breaking

The cubic anisotropy violates the axial symmetry of the system, with the result that when its constant is sufficiently large, the circular precession of the equilibrium state is disrupted, which manifests itself in the precession portrait as a rupture of the big ring and the exit of the small rings to a stationary orbit.

As can be seen from the previous review, with a different orientation of the cube axes, the type of precession portrait is different, that is, the symmetry is broken in various ways and to different degrees.

All precession portraits in Fig. 10.15 are built in the conditions preceding the breakdown, with an accuracy of about 0.01, that is, the breakdown occurs already with an increase in the constant by 1...2%. At the same time, Fig. 10.15, b corresponds to $K = 221$ erg×cm^{-3}, Fig. 10.15, c $K = 5.5$ erg×cm^{-3}, Fig. 10.15, $K = 12.9$ erg×cm^{-3}. The strong difference between the values of the breakdown constants at different orientations of the cubic cell is determined by the sharpness of the orientational dependence of the potential, illustrated in Fig. 10.16. This figure shows the dependences of the potential constant U/K (energy density) normalized to the magnitude of the azimuthal angle of the magnetization vector for different orientations of the axes of cubic anisotropy constructed by formulas (10.44), (10.49) and (10.62). For convenience of comparison, all zeros of zeros are combined by introducing constant potentials to the potential.

It is seen from the figure that with orientation [001], the potential dependence on the azimuth angle is extremely small, has an average value with orientation [111], and reaches the maximum value with orientation [011]. That is, the orientation [001] breaks symmetry to the least extent, and the orientation [011] – to the greatest. This corresponds to the fact that with the orientation [011], the breakdown occurs already with such a small value of the constant as $K = 5.5$ erg×cm^{-3}, that is, with the anisotropy field, defined as $H_a = K/M_0$,

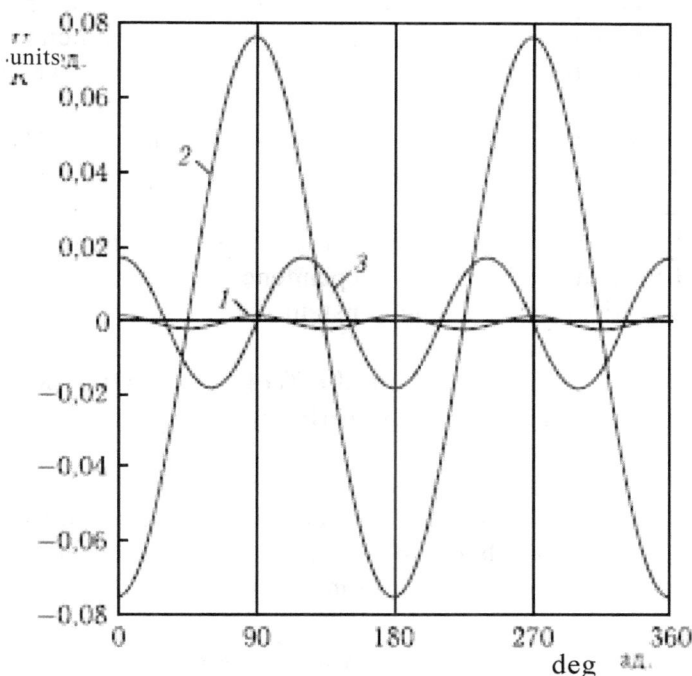

Fig. 10.16. Dependences of the normalized potential (energy density) on the azimuthal angle of the vector on the magnetization at different orientations axes of cubic anisotropy: 1 – orientation [001], shift zero: +0.105; 2 – orientation [011]; shift zero: +0.223; 3 – orientation [111]; shift zero: +0.263. Parameters: $4\pi M_0 = 280$ G, $H_{0z} = 265$ Oe, $\theta_a = 20°$.

equal to $H_a = 0.247$ Oe, whereas for orientation [001], the value of the constant $K = 221$ erg×cm^{-3}, which corresponds to $H_a = 9.92$ Oe, is required for disruption. Thus, the ratio of critical values of the constants corresponding to the disruption for the [001] and [011] orientations is 40.2. From a comparison of the arrangement of the axes of the [111] type with respect to the plane of the plate, illustrated in Fig. 10.12, b, Fig. 10.13, b, Fig. 10.14 b, it can be seen that with the [001] and [111] orientations all axes are inclined from the plane of the plate at the same angles, whereas when [011] are oriented, two axes lie in this plane, and the other two are inclined from it at a very significant angle equal to arctg $\sqrt{2} \approx 54.74°$.

Thus, it can be assumed that the symmetry of the system is broken the most when the angles that make up the anisotropy axis with the plane of the plate are different. On the other hand, with the [001] orientation, the arrangement of the [111] axes can be translated into

570

itself by rotating the cubic cell around the normal to the plane of the plate by 90°, that is, the normal is a fourth-order symmetry axis. With the [111] orientation, such a translation is possible when rotating around the same normal by 120°, that is, in this case, the normal is the axis of symmetry of the third order. The orientation [011] for such a translation requires turning already at 180°, that is, here the normal to the plane of the plate is the axis of symmetry of the entire second order. Thus, it can be seen that as the degree of symmetry decreases, the amplitude of the azimuthal dependence of the potential increases, while the critical value of the constant decreases.

10.3. Features of the additional effects of constant and variable fields in the case of cubic anisotropy

As in the case of uniaxial anisotropy, the properties of the precession of the equilibrium position in a medium with cubic anisotropy also significantly change when exposed to additional constant and variable fields. Consider some of them in more detail.

10.3.1. Constant field in the plane of the plate

As shown in [307, 323, 324], the application of a constant field in the plane of the plate, breaking the symmetry of the system, changes the nature of the precession. In the case of cubic anisotropy under consideration, as in the previous cases, the application of a constant field in the plane of the plate with a sufficient value leads to a breakdown in the precession of the equilibrium position, that is, the large ring of the precession portrait breaks and small rings go to a stationary orbit. In this case, the orientation of the constant field in the plane is not significant, only its magnitude remains important. Before the breakdown, the sweep of the oscillations of magnetization in time takes the form of pulses of unequal duration, and this inequality increases as the field approaches the value sufficient for breakdown. At the same time, in the precessional portrait, the positions of the condensations do not change, only the density of filling up of the condensations, which accounts for a longer pulse, increases somewhat. The breakdown phenomenon is distinguished by a high field criticality: the breakdown occurs at the H_{0x} or H_{0y} fields, very small compared with the H_{0z} field normal to the plane of the plate. So, with $H_{0z} = 265$ Oe and other parameters, corresponding to the precessional portraits in Fig. 10.15, with orientation [001], the breakdown occurs at a constant field in the plane of $8.7 \cdot 10^{-4}$ Oe, with orientation [011] with a field of 2.6 • 10^{-3} Oe and with orientation [111] with field 1.9 • 10^{-3} Oe. As you can see, the precession of the equilibrium position is most critical to the transverse field when the amplitude of the azimuthal potential dependence is minimal (Fig. 10.16, curve 1), that is, with the orientation [001] corresponding to the highest degree of symmetry. The smallest criticality is observed for the [011] orientation, that is, with the greatest amplitude of the azimuthal potential (Fig. 10.16, curve 2) and the smallest degree of symmetry, and the [111] orientation in criticality, as in the degree of symmetry, takes an intermediate position (Fig. 10.16, curve 3).

10.3.2. Variable field in the plane of the plate

The effect of a constant field in the plane of the plate is a static factor that is not connected in time with the precessional motion of the magnetization vector. Let us now consider the application in the plane of the plate in addition to the exciting additional and additional alternating field, which can somehow be consistent with the motion of the precession.

Since during the precession of the equilibrium position, the magnetization vector participates immediately in two-stage precessional character: with the frequency of the exciting field and with the frequency of the precession of the equilibrium position, the effect of the alternating field, depending on its frequency, can be synchronized with both the other movement. Consider these cases separately.

10.3.3. Synchronism with the frequency of excitation

Consider the orientation [011], for which the precession portrait contains only two thickening small rings (Fig. 10.15, c), located diametrically opposite to each other, that is, it has the simplest form. For the case of an isotropic medium, it was shown in [312] that, in a non-symmetric field of excitation, with a sufficient degree of asymmetry, a two-dimensional and oppositely opposite condensations are also possible in a precession portrait. We now consider the simultaneous action of both actors: cubic anisotropy with the [011] orientation and asymmetry of the exciting field. The phenomena occurring at the same time are illustrated by Fig. 10.17, where the sweep of magnetization oscillations in time and precession portraits are shown for different values of the anisotropy and the components of the alternating field. Fig. 10.17, and corresponds to the absence of anisotropy and full symmetry of an alternating field. The oscillations of the magnetization components mx and my are purely sinusoidal with a phase shift of 90°, which

572

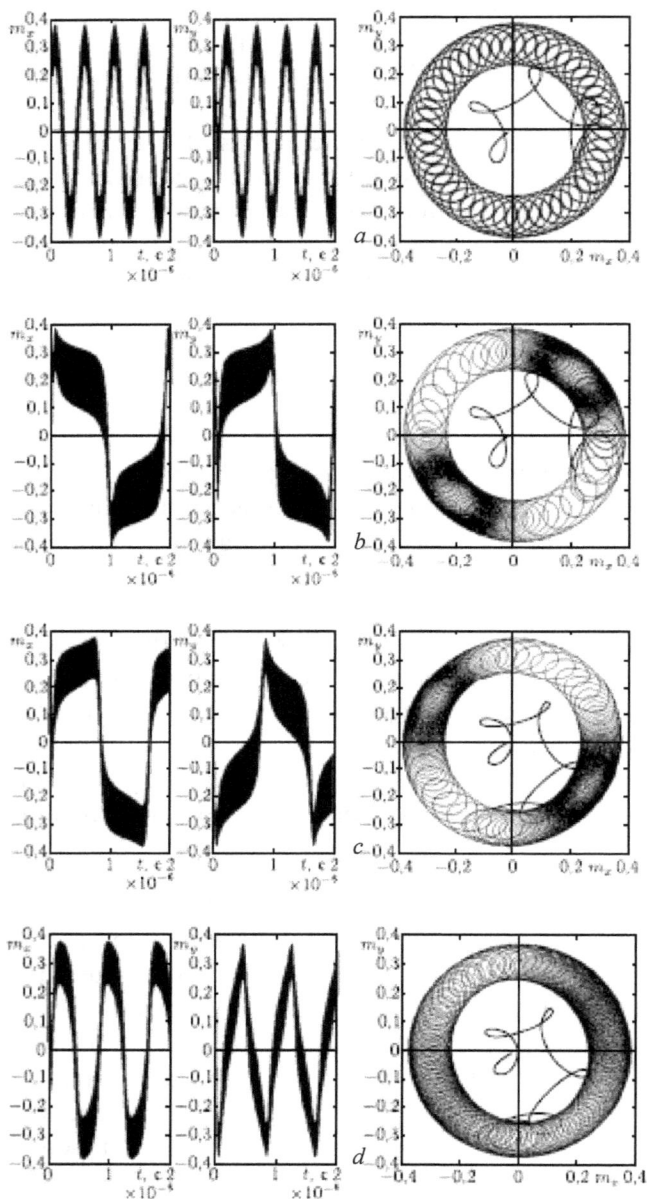

Fig. 10.17. Sweeps of magnetization oscillations in time and precession portraits at various counts of low-drop and variable field components: $a - K = 0$, $h_{0x} = 3.0$ Oe, $h_{0y} = 3.0$ Oe; $b - K = 5.5$ erg×cm^{-3}, $h_{0x} = 3.0$ Oe, $h_{0y} = 3.0$ Oe; $c - K = 0$, $h_{0x} = 2.4$ Oe, $h_{0y} = 3.0$ Oe; $d - K = 5.5$ erg×cm^{-3}, $h_{0x} = 2.4$ Oe, $h_{0y} = 3.0$ Oe; Orientation [011]. Parameters: $4\pi M_0 = 280$ Gs, $H_{0x} = 265$ Oe, $\alpha = 0.3$, $f = 100$ MHz.

corresponds to a strictly circular polarization of the exciting field. The oscillation period is $0.5 \cdot 10^{-6}$ s. In the precession portrait, there are no thickening of small rings.

Figure 10.17, b corresponds to the presence of cubic anisotropy with the [011] orientation, but the alternating field remains symmetric. In these words, the oscillations of the components of magnetization have the form of close-to-square-wave pulses, of in-phase polarity, that is, they do not have a noticeable phase shift. At the same time, however, the amplitudes of extended pulse arrays corresponding to mx decrease somewhat (by about a third) for the duration of each pulse, and the amplitudes of extended pulse arrays corresponding to m_y increase to the same extent. The period of the pulses is $2.0 \cdot 10^{-6}$ s, that is, significantly (four times) exceeds the period of sinusoidal oscillations in the absence of anisotropy. In the precession portrait, two diametrically opposed 'rasping' of small rings are observed, located in the first and third quarters of the azimuthal plane. Accordingly, in the second and fourth quarters of the same plane, rarefaction of small rings is observed.

Figure 10.17, c corresponds to the absence of cubic anisotropy, however, the alternating field is asymmetric. The ratios of its components are chosen in such a way that the locations of the condensations in the precession portrait are the most different from the previous case, namely: the amplitude of the h_{0x} component is 0.8 from the amplitude of the h_{0y} component. In these words, the oscillations of the magnetization are also close to rectangular pulses with an inclined extended peak, but now their field of brightness is antiphase. The nature of the slope of the extended wavelengths of the pulses for both components of the magnetization is similar to the previous case, but now the increase and decrease are interchanged. The period of the pulses is $1.8 \cdot 10^{-6}$ s and is close to the period of the pulses due to the anisotropy (3.6 times the period of the sinusoidal oscillations). The precession portrait shows two diametrically opposed thickening of small rings located in the second and fourth quarters of the azimuthal plane (due to the choice of the ratio of the alternating field components mentioned above). Correspondingly, in the first and third quarters of the same plane, rarefactions of small rings are observed.

Fig. 10.17, d corresponds to the simultaneous combination of cubic anisotropy with asymmetry of the alternating field. It is seen that in this case both of the listed factors act in opposite directions. Thus, the type of magnetization oscillations loses a pulsed character and approaches a sinusoidal, albeit somewhat deformed. The phase shift also tends to 90°. The oscillation period is $0.8 \cdot 10^{-6}$ s, that is, it approaches the oscillation period in the absence of disturbances. In the precession portrait in this case, the large ring is filled with small ones almost evenly, that is, much more uniform than in Fig. 10.17, b and c. At the same time, although some condensations in the intervals of angles from −10° to 80 and from 170° to 260° still remain, however, their intensity is significantly less than the condensations in Fig. 10.17, b and Fig. 10.17, c. At the same time, a significant weakening of the observed rarefaction occurs.

Thus, it can be seen that a proper choice of the asymmetry of the alternating field can largely compensate for the disturbance of the uniform motion of the precession of the equilibrium position caused by anisotropy. The conditions of such compensation are very critical to the choice of both parameters, both the magnitudes of the anisotropy constant and the degree of asymmetry of the alternating field. Thus, a decrease in the anisotropy constant from $K = 5.5$ erg×cm^{-3} to 4.0 erg× cm^{-3}, that is, a decrease in the anisotropy field from 0.24 Oe to 0.17 Oe leads to resorption of condensations, and an increase in the same constant to a value of 5.6 erg×cm^{-3}, that is, by 2%, causes a breakdown of the precession with a corresponding break of the large ring. Reducing the degree of asymmetry of the field by increasing its components h0x from 2.4 Oe to 2.7 Oe does not allow for the visible resorption of thicknesses, and increasing the same degree of asymmetry by reducing h0x to 2.2 Oe (that is, less than 10%) leads again to the breakdown of the precession. That is, to be able to observe compensation, it is necessary to choose the values of the parameters with an accuracy of a few percent.

Recall that the described compensation was considered for the case of orientation [011], when the precession portrait has only two thickenings. Additional consideration shows that in the case of the [001] and [111] orientations, when such a portrait has four or three condensations, only no more than one or two condensations can be partially compensated, while the others, although deformed, still remain.

10.3.4. Synchronism with the precession frequency of the equilibrium state

Consider the same orientation [011], and we assume that the amplitude of the exciting field is such that in the precession portrait the thickening is expressed quite clearly, that is, the sweep of oscillations of the components of magnetization m_x and m_y in time has a pulsed character, similar to that shown in Fig. 10.17, b. We assume that in the plane of the plate an additional alternating field of right circular polarization is applied, the frequency of which is close to the precession frequency of the equilibrium position. That is, a field of the form is added to the exciting field of precession (10.25)

$$h_{px} = h_{p0} \sin\left(2\pi f_p t\right),$$

(10.64)

$$h_{py} = -h_{p0} \sin\left(2\pi f_p t\right),$$

(10.65)

where f_p is the frequency of the additional variable field, h_{p0} is its amplitude. Thus, in the plane of the plate there is a rotating magnetic field, the rotation speed of which can be both more and less than the speed of movement of the precessing equilibrium position. The introduction of such a field leads to the emergence of a 'hill moving in a circle and the 'hollow' potential U that is diametrically opposed to it, which, if it is sufficiently large, can accelerate or slow down the precessional movement of the equilibrium position. In this case, the magnetization vector will be delayed in the condensation region for more or less time, that is, the duration of the sweep pulses with respect to the oscillation time of the components mx and my will change.

The phenomena observed at it are illustrated by Fig. 10.18, which shows the dependence of the duration τ of the second sweep pulse in time of oscillation of the component m_x on the frequency f_p of the additional alternating field in the plane of the plate. The sidebar illustrates the scheme for registering the pulse duration of the magnetization component m_x (the m_y component behaves similarly). Here, solid and dashed lines reflect the transformation of pulses with a change in the phase of the alternating field. For cases shown in two, phase differ by 90°. It can be seen that the phase shift of the alternating field leads to a change in the duration of the first pulse, after which the duration of all subsequent parameters stabilizes and becomes constant. That is, immediately after switching on the alternating field, the system 'adapts' to its phase for some time, after which it moves synchronously with it. To eliminate interference from the transient mode in experiments, the duration of the second pulse τ, shown schematically in the inset, was measured.

The figure shows the solid dots 1 of the machine experiment for measuring the pulse duration, the dotted curve 2 shows the dependence of the half-period of the alternating field on its frequency, the level of the pulse duration in the absence of an alternating field is shown by a dotted line 3. The amplitude of the additional field is set at $h_{p0} = 0.004$ Oe.

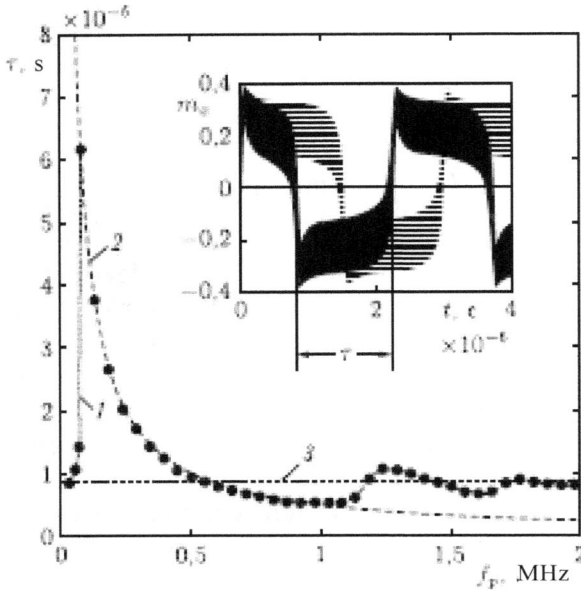

Fig. 10.18. Dependence of the duration of the second sweep pulse during the oscillation of the component on the frequency of the additional alternating field in the plane of the plate. The inset shows the scheme for measuring the duration of the second pulse. 1 – points — the result of a machine experiment; 2 – dependence of the duration of the half-period of the alternating field from the frequency; 3 – duration of the second pulse in the absence of an additional alternating field. Orientation [011], the amplitude of the additional alternating field: $h_{p0} = 0.004$ Oe. Parameters: $4\pi M_0 = 280$ G, $H_{0z} = 265$ Oe, $\alpha = 0.3$, $K = 5.5$ erg cm^{-3}, $f = 100$ MHz, $h_{0x} = 3.0$ Oe, $h_{0y} = 3$,

It can be seen from the figure that in the frequency range from 0.1 MHz to 1.2 MHz, the duration of the second pulse is exactly equal to half the period of the additional field, that is, there is complete synchronism between the precession of the equilibrium position and this alternating field. Below this frequency interval, the duration of the second pulse remains close to that in the absence of an alternating field, that is, to a value of $0.9 \cdot 10^{-6}$ s, after which it abruptly increases to $6.2 \cdot 10^{-6}$ s on the boundary of the interval (there is almost an order of magnitude), then in frequency following half the period of the variable field. By the end of the synchronism interval, the duration of the second pulse, following the half period of the alternating field, becomes less than $0.9 \cdot 10^{-6}$ s almost twice, after which, breaking out from the conditions of synchronicity and experiencing smooth gradual damped oscillations, tends to this value. The coincidence of the

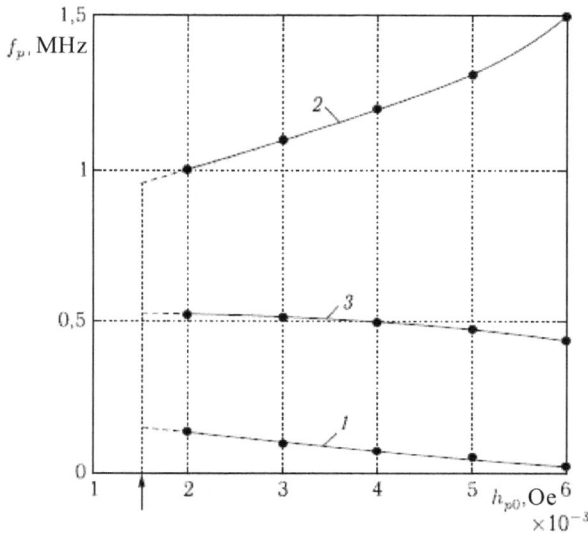

Fig. 10.19. The boundaries of the synchronism region, depending on the amplitude of the additional variable field. 1 – low-frequency boundary of the synchronism region; 2 – high-frequency synchronism boundary; 3 – the line of equality of the pulse duration to its value in the absence of an alternating field. Orientation [011]. Parameters: $4\pi M_0 = 280$ G, $H_{0z} = 265$ Oe, $\alpha = 0.3$, $K = 5.5$ erg×cm^{-3}, $f = 100$ MHz, $h_{0x} = 3$ Oe; $h_{0y} = 3$ Oe.

pulse duration in synchronism with its duration in the absence of an additional alternating field occurs near the frequency of 0.5 MHz, that is, just below the middle of the synchronism interval. From the almost exact symmetry of this frequency with respect to the edges of the sync interval, it can be assumed that the conditions of capture and quitting for the precession of the equilibrium state occur in close conditions, that is, with a close value of the potential slope in both rays.

The curves shown in Fig. 10.18 were obtained with the amplitude of the additional field equal to $h_{p0} = 0.004$ Oe. Figure 10.19 shows how the boundaries of the synchronism region change with a change in the amplitude of this field. The synchronism region lies between curves *1* and *2*. Curve *3* reflects the condition that the pulse duration equals its value in the absence of an alternating field ($0.9 \cdot 10^{-6}$ s). It can be seen from the figure that the frequency interval of synchronism extends down to one order and upwards two to three times with respect to the precession frequency in the absence of an additional field. With an increase in the amplitude of the additional field, the frequency boundaries of the synchronism region expand, which is associated with an increase in the height of the rotating potential slide (or an increase in the depth of the opposing trough) created by an additional field.

The curves in the amplitude region below $h_{p0} = 0.002$ Oe are shown by a dotted line, since more accurate measurements are difficult here because of the significant spreading and deformation of the pulse shape. A more detailed study in the field of small amplitudes showed that the synchronism phenomenon is a threshold one: when the amplitude of the additional alternating field is lower than $h_{p0} = 0.005$ Oe, synchronism is absent. A very high criticality of the synchronism threshold to the field amplitude should be noted: the noted value is $\sim 2 \cdot 10^{-6}$ from the field perpendicular to the plane of the plate.

A detailed study in the field of small amplitudes showed that the synchronism phenomenon is a threshold: when the amplitude of the additional alternating field is lower than $h_{p0} = 0.005$ Oe, synchronism is absent. A very high criticality of the synchronism threshold to the field amplitude should be noted: the noted value is $\sim 2 \cdot 10^{-6}$ from the field perpendicular to the plane of the plate.

In the amplitude region above $h_{p0} = 0.006$ Oe, accurate measurements are also difficult, since, in synchronous regions, the synchronism pulses also spread out, and the boundaries become less defined. The reason for this spreading is that two mechanisms are responsible for the movement of the equilibrium position of the magnetization vector in a large circle of the precession portrait. One is due to the action of forces of a gyroscopic nature, which were interpreted in detail in [306] on the basis of a vector model. The movement of the equilibrium position under the action of this mechanism can be called 'spontaneous', since in this case there is no external force with circular polarization. Another mechanism is due to the action of an additional alternating field, which creates a certain 'potential well' moving in a circle, followed by the equilibrium position. Such a movement of the equilibrium position can be called 'forced'. With a small amplitude of the additional variable field, the first mechanism dominates, that is, the movement of the equilibrium position is spontaneous. In this case, the shape of the pulses is expressed quite clearly, since their frequency is due to a sharp 'jumping off' of the equilibrium position from the potential 'slide' into the potential 'hole'. When the amplitude of the additional field begins to exceed $h_{p0} = 0.006$ Oe, the second mechanism prevails, and the movement of the equilibrium position becomes forced. Such movement follows a potential well formed by an additional field, that is, it occurs quite smoothly, and the fronts of the pulses are smeared. During the study, it was clearly established that, starting from these amplitudes (h_{p0} = 0.006 Oe), the synchronous forced movement of the equilibrium position begins to prevail over the spontaneous one, with the result that at h_{p0} = 0.100 Oe up to the frequency 4.0 MHz any traces of spontaneous the movements disappear, the condensations in the precession portrait dissolve, and the precession of the equilibrium position acquires a purely forced character. Such a forced movement takes place in a much larger (up to several times) range of exciting and constant fields than in the case of spontaneous movement, including in the technology, when spontaneous movement undergoes a breakdown.

10.4. Some additional notes

It should be noted that the review carried out here concerns only three very specific orientations of the axes of the cubic cell, which are the most characteristic and studied. In the most general case, one should consider an arbitrary orientation of the [111] axes with respect to the plane of the plate. In this case, the symmetry of the system will be broken even more than in the case of the [011] orientation, that is, dependencies are possible, such as that given for uniaxial anisotropy in Fig. 10.5, where the critical value of the constant already with a relatively small deviation from the normal to the plane of the plate (by units and fractions of a degree), decreases sharply. On the other hand, as shown in 10.1.8, for uniaxial anisotropy, such a symmetry breaking can be compensated for by a proper deviation of the constant field from the same normal. It can be hoped that in the case of cubic anisotropy, the symmetry violation caused by an imbalance in the orientation of the [111] axes with respect to the plate plane can be compensated by a suitably oriented constant field. The results reported relate to the precession mode without centre coverage [308]. The sequence of manifestation of the rest of the presses with a change in the value of the alternating field in the case of an anisotropic medium is preserved in the main. However, between the mode with the centre coverage and the simple circular precession mode, successively alternating from two four modes are observed, consisting in localizing the precession portrait in the directions of the individual anisotropy axes.

Comment. The properties of the precession of the equilibrium state depend substantially on a number of other actors. The authors of this monograph in no way can claim to be comprehensive and their investigation. However, some features of the phenomenon under consideration, in their opinion, are of particular interest, therefore, the next section is devoted by the authors to a brief review of one of these features — the kinetics of transition between certain regimes.

10.5. Kinetics of transition between different balance precession modes

In the previous sections, various modes of precession of the equilibrium position in a fully steady state were considered. That is, after the end of the initial development, the amplitude and the general nature of the oscillations did not change further. However, of particular interest are not only stationary oscillations, but also the process of transition from one mode to another. Of particular importance here is the time of such transitions, including its connection (or its absence) with the time of the system's own relaxation.

Generally speaking, studies of the dynamics of both transitions and the initial establishment of various modes of precession of the equilibrium state, as far as the authors of this monograph know, have not yet been conducted, that is, is a possible subject for future research. However, in order to shed some insignificant light on the substance of the issue, or at least to show which side we can move to, in this section we will give a brief review of one of these transitions, following [305]. So, we note that the angles of precession of magnetization in a normally magnetized ferrite plate, where parametric excitation of exchangeable waves is excluded, can reach 20–40 degrees or more, which makes it possible to implement complex regular and stochastic modes [127, 130, 297]. It was shown in [306] that at a constant field less than the demagnetization field, that is, under the conditions of an orientational transition, a precession of the equilibrium position of the magnetization is possible, according to the nature of which five different modes are distinguished in [308] (Section 9.3).

This section is devoted to the study of the kinetics of transition between modes No. 3 and No. 4, in particular, the time taken to establish a dynamic stationary state of precession on both sides of the transition point.

The geometry of the problem coincides with that adopted in [130, 297]. The ferrite plate is magnetized by a constant field $\mathbf{H}_0$ directed along the normal to its plane. The magnitude of the constant field is somewhat less than the demagnetization field of the $4\pi M_0$ plate. The variable field $\mathbf{h}$ of circular polarization is oriented in the plane of the plate. The Oxy plane of the Cartesian coordinate system $Oxyz$ is parallel to the plane of the plate, the Oz axis is perpendicular and coincides with the direction of the constant field.

In Fig. 10.20 the precession portraits (a, d) and the development of oscillations in time are shown for the components of magnetization m_x (solid lines) and m_y (dotted line) (b, e) and the transverse component

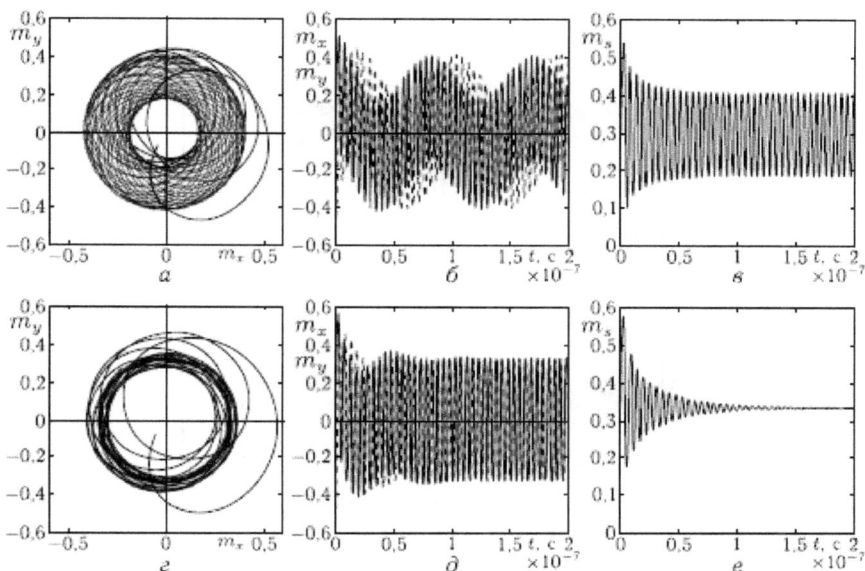

Fig. 10.20. Precessional (a, d) portraits and temporal dependences (b, c, e, d) with various types of letters of the alternating field h_0: a, b, c – 35 Oe; d, e, f – 38 Oe. The transition point from mode No. 3 to mode No. 4 corresponds to $h_0 = 37$ Oe. Parameters: $4\pi M_0 = 280$ G, $H_{0x} = H_{0y} = 0$ Oe, $H_{0z} = 260$ Oe, $\alpha = 1.2$, $f = 200$ MHz

of magnetization $m_s = \sqrt{m_x^2 + m_y^2}$ (c, f). The top row of figures (a, b, c) corresponds to mode No. 3, the lower one (d, e, f) to mode No. 4.

The nature of the establishment of magnetization oscillations is most pronounced in Fig. 10.20, c (mode number 3) and Fig. 10.20, e (mode number 4). In both presses, the establishment of a steady state occurs through the damping of the initial unsteady precession, and in mode No. 3 in steady state after establishment, the equilibrium position continues to precess with constant amplitude (Fig. 10.20, c), and in mode No. 4 in steady state, the equilibrium state is aligned along the constant field and calms down (Fig. 10.20, e).

The settling time of the decaying component of the precession of equilibrium considerably (many times) exceeds the precession relaxation time of the magnetization vector and strongly depends on the amplitude of the variable field, as illustrated in Fig. 10.21. Here the points are the values of the settling time obtained from the time dependences shown in Fig. 10.20, c, e, solid lines – constructed by an empirically found formula

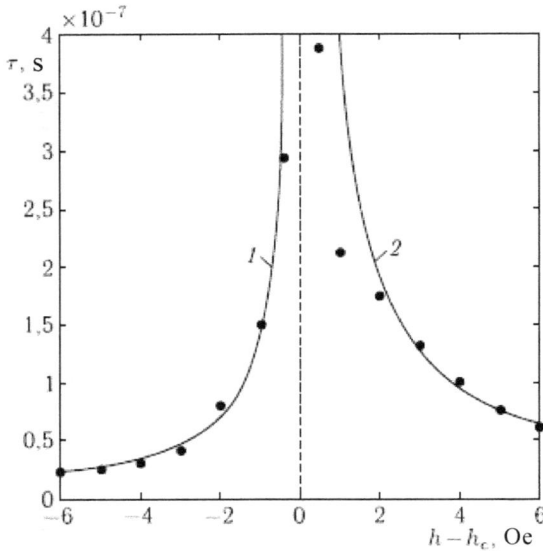

Fig. 10.21. Dependence of the settling time on the amplitude of the ac field. Parameters - the same as taken in the construction of Fig. 10.20.

$$\tau = \frac{A}{\left|h - h_c\right|},$$

$$(10.66)$$

where h_c = 37 Oe, with an alternating field below the transition A = 1.4, above the transition $(h > h_c)$ A = 3.8.

It is seen from the figure that as the amplitude of the exciting signal changes, the settling time changes from a value of the order of the relaxation time of the magnetization vector (far from the transition between modes No. 3 and No. 4) to a value tending to infinity (near the transition point). This phenomenon resembles the kinetics of the second-order phase transition described by the Landau–Khalatnikov equation [404, p. 517, forms. (101.2)].

In the precession problem, the equilibrium position can be considered as the transition point of the amplitude of the excitation signal, in which mode No. 3 is replaced by mode No. 4, and for the order parameter, the transverse component of magnetization deviates from the value corresponding to the transition point between modes No. 3 and No 4.

Consider, on the basis of such an analogy, the establishment of magnetization during the transition between modes No. 3 and No. 4. We assume that the rate of establishment of magnetization is the greater, the stronger the dependence of the potential on the magnetization, that is, the ratio

$$\frac{dm}{dt} = -\delta \left| \frac{\partial U}{\partial m} \right|.$$

(10.67)

where $\delta > 0$ is a constant coefficient. Suppose that near the transition point, where $m = m_0$ (s is omitted for simplicity) and $h = {}_hc$, the potential is

$$U = U_0 + \beta |h - h_c| (m - m_0)^2,$$

(10.68)

where $\beta > 0$ is a constant coefficient. Substituting (10.68) into (10.66), we obtain

$$\frac{dm}{dt} = -2\delta\beta |h - h_c| |m - m_0|,$$

(10.69)

Solving this equation, we get

$$|m - m_0| = Ge^{-2\delta\beta|h-h_c|}$$

(10.70)

where G is a constant factor.

Assuming that the establishment of magnetization occurs according to the law

$$|m - m_0| = Ge^{-t/\tau},$$

(10.71)

where $\tau > 0$ is the settling time, that is, the time during which the magnetization decreases by $e = 2.71828...$ times, we get

$$\tau = \frac{B}{|h - h_c|}$$

(10.72)

where $B = 2\delta\beta > 0$ is a constant coefficient. The resulting formula, up to a constant coefficient, coincides with (10.66).

Thus, from Fig. 10.21 one can see that in both regimes in a not very small neighbourhood of the transition, the dependence of the establishment time on the amplitude of the variable field is well (with an accuracy of about 2%) described by the law of inverse proportionality (10.72). Such a match confirms the validity of the use of the relation (10.67), as well as the decomposition (10.68). In the vicinity of about 1 Oe, which is about 3% of the transition field, the accuracy of formula (10.72) falls to 50% or more, especially above the transition. We note that the Landau theory of second-order transitions [404] has the same loss of accuracy property near the transition, which is traditionally explained by the lack of consideration for fluctuations. In this brief excursion into the region of the kinetics of transitions between various precession modes of the equilibrium state, the authors allow themselves to finish, leaving a virtually untouched field of activity for future investigations.

10.6. Precession observation conditions of the equilibrium positions in the experiment

In the previous sections, numerous theoretical aspects of the precession of the equilibrium state are considered, but the conditions for observing such a precession are not specified in the experiment. We now consider the conditions necessary for such an observation. First of all, let us dwell on cases of complete symmetric fields and anisotropy, as less critical excitation conditions, and also estimate the possibility of observing different precession modes separately. We also give some qualitative considerations concerning the violation of symmetry. As an addition, we will briefly discuss the technique of creating the required fields.

10.6.1. Frequency-field and energy parameters

Section 9.2.5 (Fig. 9.10) shows that with fully symmetric words in constant and alternating fields, the precession mode of the equilibrium position without centre coverage (No. 2) is non-threshold in the alternating field, that is, it is excited for any arbitrarily small values. In this case, the precession period of equilibrium with a variable field is 1.5...6.0 Oe, constant field 10...280 Oe and $\alpha = 0.3$ (TBFG – terbium-ferrite–garnet) is 1.5...6.0 μs (Fig. 9.6, Fig. 9.7).

Further, from Fig. 9.10, it can be seen that the precession of the equilibrium state at $\alpha = 0.3$ (TBFG) is excited at a field $h_c = 10$ Oe at frequencies higher than the critical $f_c = 116$ MHz. When $\alpha = 0.03$ (YIG) with a field $h_c = 10$ Oe, the critical frequency is $f_c = 203$ MHz. When the field $h_c = 20$ Oe, the same frequencies are equal, respectively, to 170 MHz and 310 MHz. An approximate empirical formula (9.20) is also given there.

$$f_c = \beta\sqrt{h_c},$$

$$(10.73)$$

in which at $\alpha = 0.3$ (TBFG): $\beta = 37.4$ MHz Oe$^{-1/2}$, with $\alpha = 0.03$ (YIG): $\beta = 66.8$ MHz Oe$^{-1/2}$. From this formula, it is possible for a given value of frequency f_c to find the critical value of an alternating field, above which the precession of the equilibrium state transforms into the unfolded circular precession

$$h_c = (f_c / \beta)^2.$$

$$(10.74)$$

Moreover, for $f_c = 1000$ MHz, we obtain: for $\alpha = 0.3$, $h_c = 715$ Oe; with $\alpha = 0.03$ $h_c = 224$ Oe. Similarly, with $f_c = 10\ 000$ MHz, we obtain: with $\alpha = 0.3$, $h_c = 71\ 500$ Oe; with $\alpha = 0.03$ $h_c = 22\ 400$ Oe.

This refers to the transition from the damped precession regime with the coverage centre (No. 4) to the deployed circular precession (No. 5). From section 9.7.8 (Fig. 9.29) one can see that transitions between modes without centre coverage (No. 2), with centre coverage without attenuation (No. 3) and with centre coverage with attenuation (No. 4) occur with variable fields, 2...3 times smaller than the above critical values of the fields of transition to the mode of deployed precession (No. 5), that is, they require a lower level of excitation, but no more than 2...3 times. In real experiments with ferrites on microwave, with a power input of no more than a few watts, the alternating field in a rectangular half-wave resonator with a quality factor of about 100–200 at 10 000 MHz does not exceed 1...2 Oe, and at a frequency of 1000 MHz, in a similar resonator, the field is usually an order of magnitude smaller. In a resonator with a field concentration (for example, coaxial with a spiral central conductor), these values can be increased by an order of magnitude, but in any case they remain significantly less than the critical values required for disrupting the precession of the equilibrium position.

So, we can assume that both the frequency-field and energy conditions necessary for observing the precession of the equilibrium state in the non-centre coverage mode in real experiments up to frequencies of 10 000 MHz and more are extremely easy to achieve using standard radio equipment. However, for confident observation of the remaining modes, first of all, to observe the transition of the decaying precession of the equilibrium position with the centre (No. 4) covered in the deployed circular precession (No. 5), the frequencies of the ac field should not exceed 100...200 MHz for which the order is 5...10 Oe are achievable by relatively simple radio equipment.

10.6.2. Geometric parameters

All theoretical results described in chapter 9, as well as in all previous sections of this chapter, relate to the case of a normally magnetized plate, infinite in its plane. However, in a real experiment, the plate always has some edges that create demagnetizing fields; therefore, the very strong criticality of the excitation conditions and the properties of the precession of the equilibrium state to the asymmetry of both the constant and alternating fields requires consideration of this factor. The simplest way to neutralize the demagnetization of the edges of a plate is to choose its shape in the form of a circular disc, the demagnetizing factors of which are completely symmetrical about its plane.

10.6.3. Domain structure

Since the necessary condition for the precession of the equilibrium state is the constant field less than $4\pi M_0$, at which the component of magnetization appears in the plane of the plate, the formation of domains is possible due to the demagnetizing action of the ends. Therefore, the disk diameter must be in such a ratio with the plate thickness so that the formation of domains is energetically unfavorable, for example, if the size of the domains significantly exceeds the plate thickness, then you can choose a disk diameter close to the width of one domain. In typical films of rare-earth ferrite garnets with a through domain structure, having a thickness of 5...20 μm, the size of domains is from 5 to 50 μm [40]. Thus, the diameter of the disk should also be no more than 50...100 μm. A more convenient option is provided by YIG films, where, due to the smallness of the anisotropy fields as compared with the magnetization, the nature of the domain structure is different from that mentioned. Thus, in [333], it was shown that, according to the nature of the domain structure of a YIG film, oriented in the (111) plane, there can be two types: at a film thickness of 5...20 μm width of domains in perforated films is 2...5 μm, and in films of the second type 10...20 μm. The difference between these types of films is determined by the magnitude of the normal uniaxial anisotropy, the critical value of which is 90...120 Oe. In the demagnetized state in films of the first type, the magnetization vectors in domains are oriented at an angle of 70°...90° to the film plane. In films of the second type, the magnetization vectors are oriented near the film plane at an angle of about 20°. Since the total magnitude of the saturation magnetization of a YIG is $M_0 = 140$ G, in the films of the first type the normal component of the magnetization vector is 131.5 G, and in the films of the second type 47.9 G. If we assume that in order to destroy domains, the field perpendicular to the film plane should exceed $4\pi M_\perp$, where $M_\perp$ is the normal component of the magnetization in the demagnetized state, in films of the first type this value is 1653.2 Oe, and in films of the second type 601.7 Oe.

Thus, to be able to observe the precession of the equilibrium position in films of the first type, the field interval remains from 1650 to 1750 Oe, and in films of the second type, from 600 to 1750 Oe to the film plane, and its orientation in the film plane is indifferent (to the accuracy of the smallness of the cubic anisotropy). In general, such a YIG film should have the form of a circular disk, the diameter of which can be arbitrary and is determined by the technical possibilities of creating sufficiently uniform constant and alternating fields over the entire film area.

One more additional opportunity for observing the precession of the equilibrium position is opened by films of submicron-thick iron yttrium garnet. In [405–407], it was shown that, in a fully demagnetized state, in the films of the YIG, a block domain structure is usually observed, which is especially pronounced in films of submicron thickness. In the thick flakes, the blocks are filled with small band domains, and as the film thickness decreases, the transverse dimensions of the blocks increase, and the strip domains inside them tend to disappear. Thus, in the YIG film described in [405] with a thickness of 0.27 μm, the block size reached 10 mm, and there were no domains inside such blocks. Thus, in order to be able to observe the precession of the equilibrium position in such words, you should choose YIG films with a thickness of less than 1 micron, and the diameter of the disk cut from such a film should fit in the size of the unit, that is, it may be about 10 mm.

10.6.4. About the possibility of observing precession equilibrium positions under asymmetric conditions

The previous consideration relates to the observation of the precession of the equilibrium state in completely symmetrical conditions. As shown in Sections 9.6–9.9, for any violation of symmetry, the interval of permissible values of fields, angles, and magnitudes of the anisotropy constants noticeably narrows. So from Fig. 10.10 one can see that with uniaxial anisotropy with a constant of about 10^2–10^3 erg×cm^{-3}, typical for most mixed ferrites–garnets, the angle of deviation of the anisotropy axis from the normal to the film plane should not exceed about 0.1 degree. Although it is quite possible to control such high accuracy in the experiment, it still becomes quite difficult by simple means. Note that this angle is measured from the normal to the film plane, that is, two types of anisotropy are involved in its formation. The first is magnetic anisotropy, determined by the crystal structure of the film. The field of such anisotropy is inversely proportional to the magnitude of the magnetization. The second is the anisotropy of the shape of the film, the field of which is directly proportional to the magnitude of the magnetization. Between two such contributions, a common anisotropy there is a certain competition, that is, there is some kind of minimum, which sets the optimal ratio between the values of the anisotropy constant and the magnetization. That is, to observe the precession of magnetization, one should choose ferrite, in which this ratio is close to the optimum. Further, it should be noted that in a number of mixed garnets, the axis of easy magnetization deviates from the normal by an angle that significantly (up to several times) exceeds the angle of deviation of the corresponding crystallographic axis of the substrate. Such a deviation is discussed in detail in section 5.2. This circumstance imposes increased requirements on the orientation of the crystallographic substrate, which is usually performed by x-ray methods (diffraction).

The most important circumstance playing in favor of observing the precession of the equilibrium position is the possibility of compensating for the deviation of the anisotropy axis by the corresponding introduction of asymmetry in a constant field (section 10.1.8, Fig. 10.8). That is, a constant field can serve as an important tool to ensure the convenience of the realization of the precession of the equilibrium position. Similar compensation due to the constant field can be achieved in the case of asymmetry of the longitudinal variable field (section 9.9.4, Fig. 9.40). It can be assumed that an asymmetric longitudinal alternating field can also help to compensate for the asymmetry caused by the tilt of the anisotropy axis, however, this issue seems to require more detailed study. Thus, summarizing what has been said, we can conclude that for successful observation of the precession of the equilibrium state in the experiment, films with the smallest possible deviation of the anisotropy axis from the normal should be chosen, and it is highly desirable to include a constant or longitudinal alternating field.

10.6.5. Field creation method

Let us now consider the possible configuration of the device for creating magnetic spikes and registering the precession of the equilibrium position. A necessary condition for precession excitation is the presence of a constant field of up to 2000 Oe, perpendicular to the film plane, and also an alternating field of circular polarization in the film plane. The method of creating a constant field of this size on an area up to several tens of square centimeters with a heterogeneity of less than 0.1% is well known and is widely used, for example, in devices for observing FMR and especially EPR (electronic paramagnetic resonance) [408, 409]. The simplest way here is to increase the area of the magnet poles and reduce the gap between them. So in the EPA-2M spectrometer used in the experiments described in Section 7.6 above, a fairly uniform field of strength up to 5000 Oe was achieved using an electromagnet having round poles about 12 cm in diameter with a gap between them of the order of 1 cm. The method of creating such fields is not considered here because of its triviality. An alternating field with a frequency of tens and hundreds (especially thousands) of megahertz, with a strength of several oersteds, more or less homogeneous on a noticeable area (at least several square meters of air) is much more difficult to create, so we'll dwell on this question in a little more detail. Some possible options for creating an alternating field of circular polarization and recording the signal from the precession of the equilibrium state are illustrated in Fig. 10.22.

Figure 10.22, a shows a scheme for creating a variable field, suitable for frequencies up to 100 MHz. The field is created by a pair of crossed coils 2 and 3 fed by alternating current with a phase shift of 90° relative to each other. At frequencies above 100 MHz, the coils will have only 1 each...2 turns, that is, it will be better to replace them with a pair of crossed short-circuited wires or strip lines, similar to those

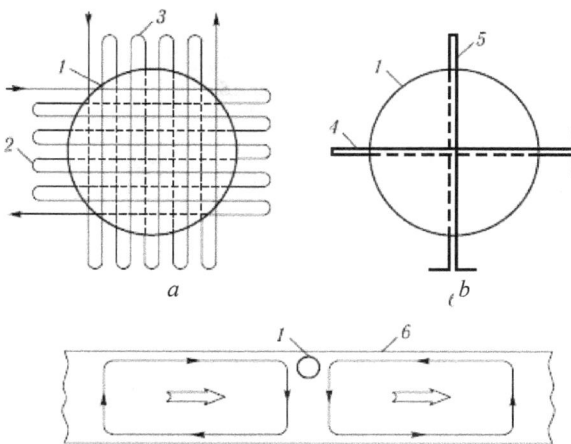

Fig. 10.22. Schemes for creating an alternating field (*a, c*) and registering a signal from the precession of the equilibrium position (*b*). 1 – ferrite film, 2–5 – various coils, 6 – waveguide.

written in [410, 411]. The reading of the signal from the precession of the equilibrium position in two coordinates can be accomplished using crossed 4 and 5 crossed films, shown in Fig. 10.22, b. At the same time, in order to eliminate direct interference from the excitation signal, a sufficiently narrow band filter must be used in the receiving circuit: either bandpassing at the frequency of excitation, or bandpassing at the precession frequency of the equilibrium position. At frequencies above 1000 MHz, a rectangular waveguide can be used to excite an alternating field, the view of which from the side of the wide wall is shown in Fig. 10.22, c. When a travelling wave of type H_{10} (TE_{10}) passes through such a waveguide near its narrow wall, a region with a rotating magnetic field is formed, where the ferrite film 1 should be placed. The signal from the precession of the equilibrium position can be read by the same loop system shown in Fig. 10.22, b.

For sufficient uniformity of the alternating field, the diameter of the ferrite film should be significantly smaller than the width of the waveguide. Assuming that the difference should be no less than an order of magnitude, we find that for frequencies of the order of 10 000 MHz, where the typical waveguide width is about 2 cm, the film diameter should not exceed 2 mm, which is quite convenient for a real experiment. For sufficient homogeneity of an alternating field the diameter of the ferrite film must be significantly less than the width of the waveguide. Assuming that the difference should be no less than an order of magnitude, we find that for frequencies of the order of 10 000 MHz, where the typical waveguide width is about 2 cm, the film diameter should not exceed 2 mm, which is quite convenient for a real experiment.

10.7. Possible technical applications

Let us now consider some possibilities of the technical application of the equilibrium position precession. We group them into main features in order of increasing complexity.

10.7.1. Ferromagnetic resonance

First of all, we note that in all described recession regimes, there are always oscillations with the frequency of the exciting alternating field, that is, the traditional ferromagnetic resonance (FMR) is possible. In accordance with this, the field of application of such oscillations coincides with the field of application of ordinary FMR, for example, in nonreciprocal peripheral devices (valves, circulators) or in analog information processing devices in the microwave range (filters, phase shifters, delay lines, and others). An advantage over traditional devices may be the expansion of the frequency range towards low frequencies, due to the fact that the FMR frequency tends to zero during the orientational transition (soft mode). Another advantage may be the expansion of the dynamic range up to work at high power levels (kilowatts) due to the exclusion of excitation of the exchange of pin waves due to the selected geometry.

10.7.2. Frequency conversion

An important feature of the considered phenomena is the excitation of the precession of the equilibrium state in modes No. 2, No. 3, and No. 4. At the same time, the precession frequency of the equilibrium position is significantly lower than the frequency of the exciting alternating field and very strongly depends on its amplitude. For example, from Fig. 9.22 (Section 9.5.4) it can be seen that at a variable field frequency of 200 MHz, the precession frequency of the equilibrium position can be from 1.7 MHz to 25 MHz and varies between these values when the amplitude of the alternating field changes from 10 to 35 Oe. The maximum gradient of the observed dependence is 0.32 MHz Oe^{-1} (curve 3 in the region of 10 Oe). Such a strong dependence on such low frequencies (as compared with microwave frequencies) can serve as the basis for creating a microwave frequency converter at a low frequency of the order of MHz units, which is useful for processing analog information. You can also talk about the possibility of creating a high-precision sensor of the amplitude of the microwave field based on the measurement of low frequency (since the measurement accuracy at low frequencies, as a rule, is higher than the measurement accuracy in the microwave range).

10.7.3. Asymmetry compensation

Further, it should be noted that the phenomenon of compensating for the anomalous nature of asymmetry of precession by a transverse constant field, being very critical to the value of both constant and alternating fields, can serve as the basis for creating high-precision sensors, meters and precision stabilization systems of the constant and alternating fields, as well as microwave . Indeed, as follows from the example given in section 9.9.4 in Fig. 9.40, the transverse constant field equal to 0.055 Oe fully compensates for the condensation caused by the longitudinal alternating field of 1.5 Oe. Thus, turning the constant field on and off within 0.055 Oe, it is possible to create or destroy condensations, that is, to change the precession period of the equilibrium position in two times from 0.5 to 1.0 μs. Note that the constant field required for switching (0.055 Oe) is more than an order of magnitude smaller than the FMR line width in YIG (0.5...0 Oe), as a result of which the sensitivity of such a sensor can more than an order of magnitude exceed the sensitivity of the instruments, based on the phenomenon of FMR.

10.7.4. Hopping stall

Another very characteristic feature of the observed phenomena can be considered a very sharp dynamic orientational transition between modes No. 4 and No. 5. So from Fic. 9.21 it can be seen that at a frequency of 200 MHz and the attenuation parameter α equal to 0.4 or 0.8 (curves 1 and 2), the change in the amplitude of the alternating field h0 is not more than 0.1 Oe at a value of the order of 25...40 Oe leads to a change in the maximum transverse component of the magnetization m_2 from 0.45 to 0.75. Thus, the change in the amplitude of the variable field is less than 0.2...0.4% leads to a change in the amplitude of magnetization oscillations 1.67 times. With an increase in the frequency of the alternating field, the sharpness of the transition increases. So, with a frequency of 400 MHz and a damping parameter $\alpha = 0.6$, mode No. 4 with $h_0 = 83.2$ Oe and amplitude of oscillation of magnetization $m = 0.50$ switches to mode No. 5 with increasing h_0 not more than 0.1 Oe, which is accompanied by an increase in amplitude jump to the value of $m_2 = 0.92$. At $\alpha = 1.0$, an increase in h_0 from 128.4 Oe to 0.1 Oe leads to a jump in m_2 from a value of 0.69 to 0.92. That is, a change in the amplitude of the alternating field of less than 0.1% leads to a jump of magnetization 1.84 times. In relative terms, one can say that the gradient of change in magnetization in the transition region is about 2000 units of Oe^{-1}. Note that the numbers given here are limited to the step in the amplitude of the variable field used in this work, equal to 0.1 Oe, given by the real time scale of the machine counting. It can be assumed that, in the limit, the magnitude of the above gradient tends to infinity, that is, a jump-like transition occurs by breaking, apparently having a hysteresis character. The observed so sharp dependence of the precession amplitude on the amplitude of the alternating field can be the basis for creating high-precision sensors, gauges, and precision systems for stabilizing the amplitude of the alternating field or microwave frequency. The introduction of such a cell into the feedback circuit may allow the creation of bistable and self-oscillating devices useful in information processing circuits and various automation devices.

10.7.5. Uniaxial anisotropy

The possibilities of use noted above relate to the case of a completely isotropic medium. However, the introduction of anisotropy allows us to see a number of additional features.

So, it is noteworthy that with an increase in the uniaxial anisotropy constant, the interval of allowable values of the polar angle of the anisotropy axis is greatly narrowed (section 10.1.9). So, if at a constant value $K = 4$ erg×cm^{-3}, which corresponds to the anisotropy field $H_a = 0.36$ Oe, the precession of the equilibrium position exists within the deviation of the anisotropy axis from the normal to the plate to 10, then at $K = 100$ erg×cm^{-3}, which corresponds to $H_a = 9.0$ Oe, the admissible value of the angle θ_a is $0.40°$, and at $K = 1000$ erg×cm^{-3}, that is, $H_a = 90.0$ Oe, the value of θ_c decreases to $0.037°$. Such a small value of the angle θ_c means a very high criticality of the excitation of the precession of the equilibrium position to the experimental conditions. It can be assumed that such a high criticality may be useful both for the study of material parameters and for creating a precision sensor of small dimensions of the orientation angle of the magnetic plate. A study of the possibility of compensating for the thickening of the rings of the precession portrait caused by the inclination of the anisotropy axis by a transverse constant field (section 10.1.8) showed that the fields required for compensation are also very small. So, for $K = 4$ erg cm^{-3}, the compensation field is about 0.057 Oe, for $K = 30$ erg cm^{-3} it decreases to 0.040 Oe, and for $K = 1000$ erg cm^{-3} drops to 0.015 Oe. As you can see, the compensation conditions are already satisfied for very small values of the constant field. This circumstance may also be useful for creating a precision sensor for the magnitude or direction of a constant magnetic field.

10.7.6. Cubic anisotropy

An important feature of the described phenomena can be considered an extremely high criticality to the parameters of the cubic anisotropy of the material of the magnetic plate and the magnitudes of the outer plates (Section 10.2.7). Criticality to the size of the anisotropy field is manifested in the fact that the opening of the full large ring (precession breakdown) in the case of magnetization 280 Gs (TBFG) and the field $H_0 = 265$ Oe, with orientation (001) occurs with anisotropy field of about 9.9 Oe, with orientation (011) with a field of only 0.025 Oe, and with orientation (111) with a field of 0.58 Oe. Criticality to the value of the constant field in the plane is also very large: a field of $8.7 \cdot$

594

10^{-4} Oe is sufficient to break the precession under technical conditions. Oe, which is $3.3 \cdot 10^{-6}$ from the field perpendicular to the plane of the plate. The criticality to the synchronism condition, if you look at its threshold, where the duration of the second impulse increases abruptly several times, is $5.0 \cdot 10^{-4}$ Oe, which is $1.9 \cdot 10^{-6}$ from the field perpendicular to the plane of the plate. It can be assumed that such a high criticality may be useful both for studying material parameters and for creating a precision field sensor or small dimensions of the orientation angle of the magnetic plate.

10.8. Questions for further research

At the end of this chapter, we list some questions that could form the subject for further study.

1. The first and foremost question is to set up experiments to directly detect the precession of the equilibrium state. Certain recommendations are given here (section 10.6), however, it is undoubted that in solving real experimental problems they will not be sufficient. That is, it is necessary not only the use of these instructions, but also the participation of living thought and solving a number of inventive problems.

2. An important issue, including one related to the first, is the search for conditions under which the succession of the equilibrium state ceases to be so critical to the parameters of the excitation and symmetry of the system. Here you can vary the magnetization, as well as the field and the nature of the anisotropy in order to find the optimum between them. Another possible way is to search for optimal asymmetry compensation in terms of properly selected and oriented constant and variable fields.

3. In the conducted consideration of the multi-mode nature of the precession of the equilibrium state (Section 9.3), only one mode change path was considered – an increase in the amplitude of the exciting signal. At the same time, using the example of the mechanical model (Section 9.4.4), it is shown that such a change of modes is possible as the frequency increases. It is suggested that in this case the order of mode change is reversed. That is, it is desirable to consider the multi-mode nature of the precession of the equilibrium position depending on the change in the excitation frequency.

4. The study of the properties of asymmetry of the system performed here relates only to one mode – the precession of the equilibrium state without centre coverage (mode No. 2 according to the classification introduced in section 9.3.1). However, it can be expected that for

other modes of manifestation of asymmetry of the system, including the conditions for its compensation, will be excellent. This question is also left without proper attention at the moment and requires more detailed consideration.

5. In most of the examples given here, the excitation frequencies are relatively small and do not exceed 100–200 MHz. This frequency range in the present work (and the references cited in it) was chosen primarily to simplify the calculation and reduce the computer time. At the same time, the higher frequency range is interesting for practice, that is, thousands and tens of thousands of MHz. Preliminary testing showed that there the precession properties of the equilibrium state are approximately similar to those described here, but at the same time they differ somewhat, primarily due to the limitations caused by the relaxation processes. In addition, some nonlinearity properties are not noted here. That is, it is desirable to continue the study at higher frequency ranges.

6. In the study done, the frequency of excitation is taken as noted in the previous paragraph, that is, of the order of 100...200 MHz. In this case, the possibility of ferromagnetic resonance is not taken into account. However, it is known that, under the conditions of a transitional transition, the frequency of ferromagnetic resonance can significantly decrease, up to the realization of a 'soft mode' (Section 6.3.5, Fig. 6.6). It can be expected that if the excitation frequency is close or coincides with the frequency of free ferromagnetic resonance in a given constant field, the character of the precession of the position will somehow change. This question also requires more detailed research.

7. In most of the examples considered here, the precession attenuation parameter (Hilbert) was chosen rather large, of the order of 0.3, which was done on the basis of the convenience of precession development over a rather small period of time, reducing the required machine counting time. At the same time, as noted in Section 9.5.2, the nature of the precession of the equilibrium position, depending on the magnitude of the attenuation parameter, can vary in a noticeable way. However, the study conducted in this section is rather superficial and requires more detailed consideration, especially considering that the most common material in films, such as the iron–yttrium garnet, has a significantly smaller attenuation parameter and is of the order of 0.03 or less.

8. All the results presented here relate to relatively small angles of deviation of the equilibrium position of the magnetization from the normal to the plane of the film, which do not exceed 20°. At the same

time, a preliminary check shows that with large deviations, the nature of the precession of the equilibrium position changes noticeably, so that new modes are observed that are not considered here. Moreover, in some rays, when the angle of deviation approaches 90°, especially at sufficiently high levels of excitation, self-modulation and stochastic modes arise. The authors of this monograph do not undertake to assert this with complete certainty, but they believe that research under conditions of strong nonlinearity is quite interesting.

9. The consideration carried out here mainly relates to the description of the fully established in time precession modes of the equilibrium position. However, Section 10.5 shows that the transition in time between certain modes can be quite unexpected. The analogy of the kinetics of such a transition with the kinetic phenomena that occur with a fairly general second-order phase transition is noted. According to the authors of this monograph, such an analogy can be quite fruitful, that is, a more detailed study of the kinetics of transitions between the precession modes of the equilibrium state can shed additional light on the general physics of phase transitions.

10. The examination carried out here was carried out under the assumption of the purely magnetic nature of the phenomena. However, it is well known that almost all magnetic materials (except, perhaps, specially prepared permalloy) have noticeable magnetoelastic properties. The nature of the precession of the equilibrium position in a magnetoelastic medium was partially investigated in [319, 330.332]. However, the study conducted there is not complete. Thus, various types of orientation of the axes of uniaxial and cubic anisotropy are not considered, the issues of compensation for symmetry breaking of the system are left without proper attention, the issues noted above with respect to the magnetoelastic medium are not mentioned. It can be assumed that a more detailed study of the precession of the equilibrium position in a medium possessing magnetoelasticity would give a number of interesting physical and practical results.

11. As a parallel to the precession of the equilibrium position for magnetization, it would seem interesting to develop in more detail the mechanical analogy, which is partially touched upon in section 9.4. So, the examination carried out there was carried out on a purely qualitative level without building any mathematical model. It can be assumed that the construction of such a model, including taking into account various compensation options, besides numerical goodies, would allow for the precession of the equilibrium position of the magnetization to understand more deeply and to model its most characteristic properties in more detail.

The questions listed, of course, the possibilities of studying the precession of the equilibrium state are far from exhausted. Some additional points that require more detailed consideration are noted in the text. In addition, it should be noted that all the provisions cited here relate to the study of more or less steady-state modes or transitions between quasi-static modes (i.e., rather slow ones). It can be assumed that the study of the precession of the equilibrium state in transient conditions, including pulsed excitation or pulsed effects on steady-state oscillations, would open up a huge new field for both fundamental and applied activity. So the authors of this monograph wish inquisitive researchers on this path all sorts of spikes.

Conclusions for chapter 10

This chapter is devoted to further consideration of the precession of the equilibrium state of magnetization. The focus is on the case of an anisotropic medium. Recommendations for experiments are given, and some application possibilities are discussed. In conclusion, some questions are noted for further research.

The main results of this chapter are as follows.

1. The precession of the equilibrium position in a normally magnetized magnetic plate with uniaxial anisotropy, whose axis of easy magnetization is inclined from the normal to the plane of the plate, is considered. The matrix of energy density conversion from the coordinate system associated with the anisotropy axis to the system associated with the plane of the magnetic plate and the direction of the constant field is obtained. The expression found for the energy density is used to solve the equation of motion of the magnetization vector in the field of the orientational transition. A sweep of forced oscillations in time and their recession portrait, formed by a large precession ring of the equilibrium position, followed by an envelope of the magnetization vector around this equilibrium position along the envelope of the small precession rings, were obtained.

2. It is shown that the deviation of the anisotropy axis from the normal to the plane of the magnetic plate leads to a condensation of the small-flanks of the precession portrait, which is ahead of the position of the projection of the anisotropy axis on the plane of the plate by 90°. It is noted that the thickening of the rings in one place of the portrait is accompanied by their dilution in the diametrically opposite place.

3. The observed phenomenon is explained on the basis of the energy model of the potential, whose growth as the precessing equilibrium advances causes a decrease in the precession rate, leading to thickening

of the rings, and a decrease in the same potential – its increase, leading to their reduction.

4. The possibility of compensating for condensation and rarefaction of rings by applying a small constant field in the plane of the plate along the projection of the anisotropy axis onto this plane is shown. The compensation is explained by the equalization of the potential level due to the transverse constant field, which leads to the equalization of the precession rate of the equilibrium position over the entire large circle of the precession portrait with the corresponding resorption of condensations and the normalization of rarefactions.

5. The presence of a critical value of the angle of deviation of the anisotropy axis from the normal to the plane of the plate, above which the precession of the equilibrium state is disrupted, is revealed, which is manifested in the opening of the large ring of the precession portrait. A relationship is obtained between the value of the mentioned critical angle and the magnitude of the anisotropy constant, which is exponential or exponential.

6. The precession of the equilibrium position in a normally magnetized magnetic plate with cubic anisotropy is considered when the plane of the plate is oriented relative to the cubic cell in the (001), (011) and (111) planes. The matrices for the transition from the coordinate systems related to the axes of the cubic cell to the coordinate system associated with the magnetic plate are obtained. The energy densities and effective anisotropy fields are found for all the reorientations.

7. Based on the numerical solution of the equation of motion of the magnetization vector under conditions of a transition, a precession portrait of forced vibrations is obtained, formed by a large ring, followed by a small ring around the envelope. It is shown that the presence of cubic magnetic anisotropy leads to the formation in the precession portrait of condensations and rarefactions of small rings, the location of which is tied to the spatial orientation of the [111] easy magnetization axes. When the [001] orientation is on the precession portrait, four thickening and the same rarefaction occur, while the [011] orientation has two thickening and two rarefaction, while the [111] orientation is the number of thickening and rarefaction is three. Such an arrangement of small rings is interpreted on the basis of the energy model of the potential, which makes it possible to show that the condensation of small rings corresponds to the motion of the equilibrium position of the magnetization vector in the direction of increasing potential, and the rarefaction to movement in the direction of decreasing it.

8. The high criticality of the orientation of cubic axes and the magnitude of the anisotropy constant to the fact of the existence of a precession of the second order, as well as the presence of condensations and rarefactions of small olec on the precession portrait were revealed. It is shown that the cubic anisotropy violates the axial symmetry of the system, as a result of which, with a sufficient value of its constant, a circular precession of the equilibrium position is disrupted, which manifests itself in the precession portrait as a gap of a large ring. A strong (up to 40 times) difference between the values of the anisotropy constant sufficient for a breakdown with different orientation of the cubic cell, explained on the basis of the orientational dependence of the potential, was revealed. It is shown that the value of the anisotropy constant required for disruption is smaller, the stronger the cubic anisotropy violates the axial symmetry of the system.

9. The effect of a constant field in the plane of the plate on the nature of the precession is considered. It is shown that such a field, even at a very small value, leads to a breakdown of the precession of the equilibrium position. So for the breakdown there is enough field in a plane smaller by six–seven orders of magnitude than a constant field perpendicular to the plane of the plate. Such high criticality is most pronounced with the [011] orientation, somewhat less with the [111] orientation and even less with the [001] orientation, but in this case it is five to six orders of magnitude.

10. The effect of the asymmetry of the alternating field in the plane of the plate on the nature of the precession is considered. It is shown that for the orientation [011], when the precession portrait has two diametrically opposed thicknesses of the small rings, the symmetry breaking of the alternating field exciting the precession in such a way that one of its components is 0.8 from the other allows one to largely compensate for the asymmetry introduced by the anisotropy system, leading to almost complete resorption of condensations of small rings in the precession portrait. With the orientations of [001] and [111], such compensation cannot be achieved.

11. The effect of an additional variable field in the plane of the plate, whose frequency is close to the precession frequency of the equilibrium position in a large circle of the precession portrait, is considered. The phenomenon of spatial synchronism has been discovered, which consists in the fact that, under certain conditions, the equilibrium state precesses completely synchronously with the rotation of the vector of the additional alternating field. In this case, the duration of the stay of the magnetization vector within a single thickening of the precession

portrait corresponds to the duration of one half period of the variable field.

12. The frequency nature of the synchronism phenomenon is investigated. It is shown that the frequency interval of synchronism extends down to one order and upwards two or three times with respect to the precession frequency in the absence of an additional field. With an increase in the amplitude of the additional field, the frequency boundaries of the region of synchronism expand. The threshold nature of the synchronism phenomenon is revealed, it is shown that the threshold value is very small and almost six orders of magnitude smaller than the field perpendicular to the plane of the plate. As the amplitude of the alternating field increases to a value of the order of 4 $\cdot$ 10^{-4} from the field value perpendicular to the plane of the plate, the synchronism becomes complete and the precession of the equilibrium position acquires a purely forced character.

13. The kinetics of the transition between different precession regimes of the equilibrium position is considered. The main attention is paid to the transition between modes No. 3 – continuous precession without coverage of the centre and No. 4 – also precession with coverage of the centre, but damping in time. The time taken to establish a dynamic stationary state of precession on both sides of the transition point was investigated. It was shown that the establishment time significantly (many times) exceeds the relaxation time of the free precession of the magnetization and strongly depends on the amplitude of the alternating field in such a way that it diverges near the transition field, tending to infinity. The analogy of the observed feature of the transition with the kinetics of the second-order phase transition described by the Landau–Khalatnikov equation is established.

14. An analytical model has been constructed, which assumes that the rate of establishment of magnetization is the greater, the stronger the potential of the system depends on the magnitude of the magnetization. On the basis of such a model, it was shown that, near the transition, the dependence of the establishment time on the amplitude of the variable field is described by the law of inverse proportionality. Comparing the analytical dependence with the data of numerical calculation revealed the high accuracy of the proposed model, which is about 2%.

15. The conditions necessary to observe the precession of the equilibrium state in the experiment are considered. It is noted that from the practical side, the fully symmetric case is most convenient, as it is less critical to the conditions of excitation. Geometric, frequency-field and energy excitation parameters are estimated. Practical recommendations are given for the selection of these parameters in

accordance with the parameters of the material and the technical capabilities of the equipment.

16. It is shown that both the frequency-field and energy conditions necessary for observing the precession of the equilibrium position in the non-centre coverage mode, in experiments up to frequencies of 10 000 MHz and more, are quite achievable using standard radio equipment. However, to confidently observe other modes, first of all, to observe the transition of the decaying precession of the equilibrium position with the centre (No. 4) covered in the deployed circular precession (No. 5), the frequencies of the alternating field should not exceed 100...200 MHz, since at higher frequencies, the achievement of variable intensity of about 5...10 Oe and higher requires the use of quite complex hardware.

17. Discussed the impact of the domain structure. It was noted that to eliminate interference with the formation of domains in the case of films of mixed garnet ferrites, the sample size should be chosen less than the width of one domain, which is of the order of fractions of a millimeter. The presence of a homeless range of fields is noted in the films of an iron–yarn garnet, which makes it possible to set the film size to be arbitrary within the framework of the possibility of creating homogeneous variable fields of the corresponding area. It is noted that in films with a small thickness, the size of homeless tumors can reach an area of up to a square centimeter or more.

18. The technique of creating fields required for the experimental excitation of the precession of the equilibrium state of magnetization is considered. For the formation of an alternating field that is uniform over a sufficiently large area, recommendations are given on the use of flat windings attached to the sample, waveguides, and strip lines.

19. The possibilities of the technical application of observables for designing microwave devices, as well as high-precision sensors of constant and variable fields, are discussed. The possibility of using the precession of the equilibrium position to create frequency converters of the microwave range at relatively low frequencies down to units of megahertz and below is indicated. It is noted that the very high sensitivity of the precession of the equilibrium position to the deviation of the anisotropy axis from the normal to the plane of the plate can be useful for studying the parameters of the material and creating precision angle pickers and a constant magnetic field. Compensation of the abnormal nature of the precession caused by the inclination of the anisotropy axis, by applying a transverse constant field, can serve as the basis for creating high-precision sensors, meters and precision systems for stabilizing the constant and alternating fields,

as well as the microwave frequency. The sharp jump-like dependence of the precession amplitude on the amplitude of the alternating field during the transition from mode No. 4 to mode No. 5 can be the basis for creating high-precision sensors, meters and precision systems for stabilizing the amplitude of the alternating field or microwave frequency. The introduction of such a cell into the feedback circuit may allow the creation of bistable and self-oscillating devices that are useful in information processing circuits and various automation devices. Extremely high criticality of the phenomenon to the parameters of cubic anisotropy can be useful both for studying material parameters and for creating a precision field sensor or small values of the orientation angle of the magnetic plate.

20. Briefly lists some issues that, in the opinion of the authors of this monograph, are of particular interest for further research. It is noted that the main thing here is the formulation of real experiments with the aim of detecting and studying in detail the properties of the precession of the equilibrium position. In this case, the most important issue is to identify the conditions for reducing the criticality of precession to various violations of the symmetry of the system, as well as the search for optimal ways to compensate for such violations. The importance of conducting precession studies in a wide range of frequencies, including the identification of a multimode nature, the influence of the role of resonant and dissipative media, was noted. A wish was expressed to conduct research in the full range of deviations of the equilibrium position of the magnetization from the normal to the plane of the plate – from zero to 90°. The perspectives of studying the kinetics of transitions between different precession modes of the equilibrium state, including on the basis of an analogy with the kinetics of phase transitions of the second kind, are noted. A more detailed consideration of the precession of the equilibrium position on the basis of the mechanical model, as well as the study of the properties of precession in a medium with magnetoelastic properties, are proposed.

Bibliography

1.   Shavrov V.G., Shcheglov V.I. Magnetostatic waves in inhomogeneous fields. - Moscow, Fizmatlit, 2016.
2.   Shavrov V.G., Shcheglov V.I. Magnetostatic and electromagnetic waves in complex structures. - Moscow, Fizmatlit, 2017.
3.   Vlasov V.S., Kotov L.N., Shcheglov V.I. Nonlinear vector precession magnetization under conditions of orientation transition. - Syktyvkar , IPO SyktSU, 2013.
4.   Vonsovsky S.V., Shur Ya. S. Ferromagnetism. - Moscow, OGIZ Gostekhizdat,
5.   Landau L.D., Lifshitz E.M., On the theory of the dispersion of magnetic permeability in ferromagnetic bodies Phys. Zs. der Sowjetunion. 1935. V. 8, No. 2. - P. 153.
6.   6. Gurevich A.G. Ferrites at ultrahigh frequencies. - Moscow, State. Ed. Physical-mat. lit., 1960.
7.   Gurevich A.G. Magnetic resonance in ferrites and antiferromagnets. Moscow: Nauka, 1973.
8.   Gurevich A. G., Melkov G. A. Magnetic vibrations and waves. Moscow: Fizmatlit, 1994.
9.   Gilbert T. L. A phenomenological theory of damping in ferromagnetic materials // IEEE Trans. on Magn. - 2004. - V. 40, No. 6. - P. 3443.
10. Monosov Ya.A. Nonlinear ferromagnetic resonance. - Moscow, Nauka, 1971.
11. Zakharov V. E., Lvov V. S., Starobinets S. S. Turbulence of spin waves beyond the threshold of their arametric excitation // Phys. - 1974. -V. 114, No. 4. - P. 609.
12. Lvov V.S. Nonlinear spin waves. - Moscow: Nauka, 1987.
13. Kozlov V.I. Study of inhomogeneity and anisotropy of magnetic films using gyromagnetic effects. - Dissertation for the application of the scientific degree of Doctor of Phys.-Math. sciences. - Moscow, 1997.
14. Sparks M., Tittmann B. R., Mee J. F., Newkirk C. Ferromagnetic resonance in epitaxial garnet thin films // JAP. - 1969. - V. 40, No. 3. - P. 1518.
15. Laulicht I., Suss J. T., Barak J. The temperature dependence of the ferromagnetic and paramagnetic resonance spectra in thin yttrium-iron garnet films // JAP. - 1991. - V. 70, No. 4. - P. 2251.
16. Chen H., De Gasperis P., Marcelli R., Pardavi-Horvath M., McMichael R., Wigen P. E. Wide-band linewidth measurements in yttrium iron garnet films // JAP. - 1990. - V. 67. No. 9. - P. 5530.
17. Pomyalov A.V., Zilberman P.E. Magnetic resonances in small film samples of yttrium iron garnet // RE. - 1986. - V. 31, No. 1. - P. 94.
18. Archer J. L .. Bongianni W. L., Collins J.H. Magnetically tunable microwave bandstop filters using epitaxial YIG film resonators // JAP. - 1970. - V. 41, No. 3. - P. 1360.
19. Hansen P., Krumme J. P. Determination of the local variation of the magnetic properties of liquid-phase epitaxial iron garnet films // JAP. - 1973. - V. 44, No. 6. - P. 2847.
20. Telesnin R.V., Kozlov V.I., Dudorov V.N. Ferromagnetic resonance in epitaxial films $Y_3Fe_{5-x}Ga_xO_{12}$ // Phys. - 1974 .-- V. 16, No. 11. - P. 3532.
21. Algra H.A., Robertson J.M. A FMR study on horizontally dipped LPE grown (La,

Ga): YIG films // JAP. - 1979. - V. 50, No. 3. - P. 2173.

22. Avaeva I. G., Lisovskiy F. V., Osika V. A., Shcheglov V. I. Study of epitaxial films of mixed ferrite-garnets by the ferromagnetic resonance // FTT. - 1975. - V. 17, No. 10. - S. 3045.

23. Avaeva I. G., Lisovskiy F. V., Osika V. A., Shcheglov V. I. Ferelectromagnetic resonance in epitaxial lenses, see data ferrite-garnets // RE. - 1976. - V. 21, No. 9. - S. 1894.

24. Avaeva I. G., Lisovskiy F. V., Osika V. A., Shcheglov V. I. Ferelectromagnetic resonance in epitaxial lenses, see data ferrite garnets // FTT. - 1976. - V. 18, No. 12. - P. 3694.

25. Shcheglov V.I. Investigation of dynamic phenomena and phase transitions in magnetic and electric materials with space-time inhomogeneous noses. - Dissertation for the degree of Cand. physical-mat. sciences. - Moscow, 1980.

26. Hsia L. C., Wigen P. E., De Gasperis P., Borghese C. Enhancement of uniaxial anisotropy constant by introducing oxygen vacancies in Ca-doped YIG // JAP. - 1981. - V. 52. No. 3. - P. 2261.

27. Tsutaoka T., Ueshima M., Tokunaga T., Nakamura T., Hatakeyama K. Frequency dispersion and temperature variation of complex permeability of Ni-Zn ferrite composite materials // JAP. - 1995. - V. 78, No. 6. - P. 3983.

28. Glushchenko A. G., Kurushin E. P., Koshkin L. I. Microwave microwave devices on thin onocrystalline films of Mg-Mn ferrite // Izv. Univ., Radioelectronics. - 1975. - V. 18, No. 11. - P. 93.

29. Glushchenko A. G., Kurushin E. P., Koshkin L. I. Using thin mononcrystalline films of Mg-Mn ferrite in microstrip lines for microwave transmission // RE. - 1974. - V. 19, No. 11. - P. 2397.

30. Avaeva I. G., Lisovskiy F. V., Shcheglov V. I. On the tilt of the magnetic axis of anisotropy in epitaxial films with mixed ferrite-garnets //FTT. - 1975. - V. 17, No. 5. - P. 2102.

31. Smit J., Beljers H. G. Ferromagnetic resonance absorption in $BaFe_{12}O_{19}$ a highly anisotropic crystal // Philips Res. Rep. - 1955. - V. 10, No. 2. - P. 113.

32. Turov E.A. Physical properties of magnetically ordered crystals. - Moscow: Ed. USSR Academy of Sciences, 1963.

33. Gulyaev Yu. V., Dikshtein I. E., Shavrov V. G. Surface magnetic acoustic waves in magnetic crystals in the region of orientational phase transitions // Phys. - 1997. - V. 167, No. 7. - S. 735.

34. Gulyaev Yu. V., Tarasenko S. V., Shavrov V. G. S spin-wave acoustics antiferromagnetic structures as magnetoacoustic metamaterials // Usp. Fiz. Nauk. - 2011. - V. 181, No. 6. - P. 595.

35. Bobeck A.H. Properties and device applications of magnetic domains in orthoferrites // Bell Syst. Tech. J. - 1967. - V. 46, No. 8. - P. 1901.

36. Kooy C., Enz U. Experimental and theoretical study of the domain configuration in thin layers of $BaFe_{12}O_{19}$ // Phil. Res. Rep. - 1960. - V. 15, No. 1. - P. 7.

37. Thiele A.A. Device implications of the theory of cylindrical magnetic domains// Bell Syst. Tech. J. - 1969. - V. 50, No. 3. - P. 725.

38. Lisovsky F.V. Physics of cylindrical magnetic domains. - Moscow: Sov. Radio, 1979.

39. Hubert A. Theory of domain walls in ordered media. - Moscow, Mir, 1977.

40. Malozemov A., Slonzuski J. Domain walls in materials with cylindrical with dry magnetic domains. - Moscow, Mir, 1982.

41. O'Dell T. Ferromagnetodynamics. Dynamics of CMD, domains and domains walls. - Moscow, Mir, 1983.

42. Avaeva I.G., Kopylov Yu.L., Kravchenko V.B., Lisovskiy F.V., Sobolev A.T., Shcheglov V.I. Epitaxial films of mixed ferrites garnets for CMD applications // Microelectronics. - 1975. - V. 4, No. 4. - P. 325.

43. Avaeva I. G., Kravchenko V.B., Lisovsky F.V., Shcheglov V.I. Non-reciprocal effects during the motion of domain boundaries in an inhomogeneous magnetic field // FTT. - 1976. - V. 18, No. 6. - P. 1780.

44. Kraftmakher G.A., Meriakri V.V., Chervonenkis A.Ya., Shcheglov V.I. Natural Resonance Associated with Domain Walls in Orthoferrites on submillimeter in waves / ZhETF. - 1972. - T. 63, No. 10. - P. 1353.

45. Shamsutdinov M.A., Farztdinov M.M., Ekomasov E.G. Dynamic bevel of magnetic gratings in a magnetic field and spin waves in red earth orthoferrites with domain structure // Physics and Technology. - 1990. - V. 32, No. 4. - P. 1133.

46. Dotsh H., Smith H. J., M"uller J. Detection and generation of magnetic bubble domains using ferromagnetic resonance // Appl. Phys. Lett. - 1973. - V. 23, No. 11. - P. 639.

47. Turov E.A., Shavrov V.G. On the energy gap for spin waves in ferro- and antiferromagnets associated with magnetoelastic energy // FTT. - 1965. - T. 7, No. 1. - S. 217.

48. Borovik-Romanov A.S., Rudashevsky E.G. About the influence of spontaneous antiferromagnetic resonance in hematite // ZhETF. - 1964. - V. 47, No. 6 (12). - S. 2095.

49. Shcheglov V.I. Dependence of the speed of sound on the magnetic field in ferro- and antiferromagnets // FTT. - 1972. - V. 14, No. 7. - S. 2180.

50. Shcheglov V.I. Method for smooth reversible frequency tuning of elastic resonance of vibrations of a rigid body // USSR author's certificate No. 386262, M.cl. (3) G 01 h 13/00. MKI: S01h13 / 00; UDC 534.63.09.12.1970. Application No. 1603337.18-10. Publication: Of. bull. "Discoveries, inventions". 1973, No. 26. - P. 141.

51. Seavey M.H. Acoustic resonance in the easy-plane weak ferromagnets α-Fe_2O_3 and $FeBO_3$ // Sol. St. Comm. - 1972. - V. 10, No. 2. - P. 219.

52. P. P. Maksimenkov, V. I. Ozhogin. Study of magnetoelastic inter- and interaction in hematite using antiferromagnetic resonance // ZhETF. - 1973. - V. 65, No. 2 (8). - S. 657.

53. 53. Dikshtein I.E., Tarasenko V.V., Shavrov V.G. The influence of pressure on magnetoacoustic resonance in uniaxial antiferromagnets // ZhETF. - 1974. - V. 67, No. 2. - P. 816.

54. Dikshtein I.E., Tarasenko V.V., Shavrov V.G. Magnetoelastic waves in orthoferrites // Phys. - 1977. - V. 19, No. 4. - P. 1107.

55. Gerus S. V., Tarasenko V. V. Rayleigh waves in magnetic crystals with anizotropy of the "easy plane" type and their amplification // FTT. - 1975. - V. 17, No. 8. - P. 2247.

56. Dikshtein I. E., Tarasenko V. V. Parametric excitation of sound in ferro-, ferri- and antiferromagnets in the vicinity of points of phase transitions // FTT. - 1978. - V. 20, No. 10. - P. 2942.

57. Suhu R. Magnetic thin films. - Moscow, Mir, 1967.

58. Antonov L.I., Mironova G.A., Lukasheva E.V., Malova T.I. Study of the phenomenon of ferromagnetic resonance (FMR) and determination of the nature of magnetic anisotropy of a ferromagnetic single crystal. MSU preprinNo. 5/1999. - Moscow, 1999.

59. Goryunov Yu.V., Khaliullin G.G., Garifullin I.A., Tagirov L. R., Schreiber F., Bodeker P., Brohl K., Morawe Ch., M"uhge Th., Zabel H. FMR studies of magnetic properties Co and Fe thin films on Al_2O_3, and MgO substrates // JAP. - 1994. - V. 76, No. 10. - P. 6096.

60. Dupuis V., Perez J. P., Tuaillon J., Paillard V., Melinon P., Perez A., Barbara B.,

Thomas L., Fayeulle S., Gay M. Magnetic properties of nanostructured thin films of transition metal obtained by low energy cluster beam deposition // JAP. - 1994. - V. 76, No. 10. - P. 6676.

61. Miroshnikov Yu.F., Khramov BV Perpendicular anisotropy in two axis films // Izv. Universities. Physics. - 1974. - No. 11 (150). - O. 119.

62. Gavrilin V.P., Berezin D.G., Miroshnikov Yu.F. Ferromagnetic resonance and magnetic crystallographic anisotropy of single crystals of lithium ferrite flakes // Izv. Univ. Physics. - 1973. - No. 9 (136). - P. 86.

63. Meckenstock R., von Geisau O., Peizl J., Wolf J.A. Conventional and photothermally modulated ferromagnetic resonance investigations of anisotropy fields in an epitaxial Fe (001) film // JAP. - 1995. - V. 77, No. 12. - P. 6439.

64. Komenou K., Zebrowski J., Wilts C.H. Ferromagnetic resonance study of the anisotropy field and nonmagnetic regions in implanted layers of bubble garnet films // JAP. - 1979. - V. 50, No. 8. - P. 5442.

65. Wilts C.H., Zebrowski J., Komenou K. Ferromagnetic resonance study of the anisotropy profile in implanted bubble garnets // JAP. - 1979. - V. 50, No. 9. - P. 5878.

66. Voronenko A.V., Gerus S.V., Krasnozhen L.A. Measurement method of gyromagnetic films // Microelectronics. - 1989. - V. 18, No. 1. - P. 61.

67. Shcheglov V.I. Ferromagnetic resonance in elastically deformed films of iron-yttrium garnet // Microelectronics. - 1987. - V. 16, No. 4. - P. 374.

68. Shcheglov V.I. Ferromagnetic resonance in elastically deformed films of yttrium iron garnet // Collection of abstracts "II All-Union seminar on functional magnetoelectronics ". - Krasnoyarsk, 1986 .-- P. 42.

69. V. I. Shcheglov. Propagation of magnetostatic waves in elastically formed environment // Collection of abstracts "II All-Union seminar on functional magnetoelectronics ". - Krasnoyarsk, 1986 .-- P. 44.

70. McMichael R.D. Method for determining both magnetostriction and elastic modulus by ferromagnetic resonance // JAP. - 1994. - V. 75, No. 10. - P. 5650.

71. Oliver S.A., Harris V.G., Vittoria C. Magnetostriction measurements on thin films by a slot-line ferromagnetic resonance technique // JAP. - 1990. - V. 67, No. 9. - P. 5019.

72. Dikshtein I. E., Maltsev O. A. Ferromagnetic resonance and magnetic static waves in nonuniformly deformed films of ferrite pomegranates // RE. - 1992. - V. 37, No. 11. - P. 2003.

73. Smith A.B., Jones R.V. Magnetostriction constants from ferromagnetic resonance // JAP. - 1963. - V. 34, No. 5. - P. 1283.

74. Morgenthaler F.R. Two-dimensional magnetostatic resonances in a thin film disk containing a magnetic bubble // JAP. - 1979. - V. 50, No. 3. - P. 2209.

75. Surig C., Hempel K.A. Interaction effects in particulate recording media studied by ferromagnetic resonance // JAP. - 1996. - V. 80, No. 6. - P. 3426.

76. Orth Th., Pelzl J., Chantrell R. W., Veitch R., Jakusch H. Ferromagnetic resonance and transverse susceptibility measurements on particulate recording media // JAP. - 1993. - V. 73, No. 10. - P. 6738.

77. Yu Y., Harrell J.W., Doyle W.D. Ferromagnetic resonance spectra of oriented barium ferrite tapes // JAP. - 1994. - V. 75, No. 10. - P. 5550.

78. Vinogradov A.P. Electrodynamics of composite materials. - Moscow, 2001.

79. Kazantseva N.E., Ryvkina N.G., Chmutin I.A. Promising materials for absorbers of electromagnetic waves of ultrahigh frequency range zone // RE. - 2003. - V. 48, No. 2. - P. 196.

80. Pendry J.B., Holden A. J., Stewart W. J., Youngs I. Extremely low frequency plas-

mons in metallic mesostructures // Phys. Rev. Lett. - 1996. - V. 76, No. 25. - P. 4773.

81. Smith D. R., Padilla W. J., Vier D. C., Nemat-Nasser S. C., Schultz S. Composite medium with simultaneously negative permeability and permittivity // Phys. Rev. Lett. - 2000. - V. 84, No. 18. - P. 4184.

82. Vendik I.B., Vendik O. G. Metamaterials and their application in technology ultra-high frequencies // ZhTF. - 2013. - T. 83, No. 1. - P. 3.

83. Strelniker Y.M., Bergman D. J. Theory of magnetotransport in a composite medium with periodic microstructure for arbitrary magnetic fields // Phys. Rev. B. - 1994. - V. 50, No. 19. - P. 14001.

84. Shevchenko V. V. Chiral electromagnetic objects and environments // Sorosovsky educational journal. - 1998. - No. 2. - P. 109.

85. Katsenelenbaum B.Z., Korshunova E.N., Sivov A.N., Shatrov A.D. Electrodynamic objects // Phys. - 1997 .-- V. 167, No. 11. - P. 1201.

86. Tretyakov S.A. Electrodynamics of complex media: chiral, biisotropic and some bianisotropic materials (review) // RE. - 1994. - V. 39, No. 10. - P. 1457.

87. Shuster A. Introduction to theoretical optics. - M.-L .: ONTI, 1935.

88. Veselago VG Electrodynamics of substances with simultaneously negative values of ε and μ // Phys. - 1967. - V. 92, No. 3. - S. 517.

89. Lindell I.V., Tretyakov S.A., Nikoskinen K. I., Ilvonen S. BW media - media with negative parameters, capable of supporting backward waves // Microwave and Optical Technology Letters. - 2001. - V. 31, No. 2. - P. 129.

90. Shevchenko V. V. Forward and backward waves: three definitions, their inter-connection and conditions of applicability // Phys. - 2004. - V. 177, No. 1. - S. 301.

91. Blioh K.Yu., Blioh Yu.P. What are left-handed environments and how are they threading? // UFN. - 2004. - V. 174, No. 4. - S. 439.

92. Byrdin V.M. Backward waves: a century of first work, origins and development of the twist of backwave mechanics and electrodynamics // RE. - 2005. - V. 50, No. 12. - P. 1413.

93. Pendry J.B. Negative refraction makes a perfect lens // Phys. Rev. Lett. - 2000. - V. 85, No. 18. - P. 3966.

94. Agranovich VM, Gartstein Yu.N. Spatial dispersion and negative refraction of light // Phys. - 2006. - V. 176, No. 10. - P. 1052.

95. Vinogradov A.P., Dorofeenko A.V., Zukhdi S. To the question of effective parameters of metamaterials // Phys. - 2008. - V. 178, No. 5. - S. 511.

96. Emkin N.A., Merzlikin A.M., Vinogradov A.P. Deviation of laws refraction from Fresnel in composite materials // RE. - 2010. - V. 55, No. 5. - P. 601.

97. Veselago VG Waves in metamaterials: their role in modern physics // UFN. - 2011. - V. 181, No. 11. - P. 1201.

98. Damon R.W., Eshbach J. R. Magnetostatic modes of a ferromagnet slab // J. Phys. Chem. Solids. - 1961. - V. 19, No. 3/4. - P. 308.

99. V. I. Zubkov, V. I. Scheglov. The conditions for the existence of inverse surfaces of magnetostatic waves in the ferrite – dielectric – metal structure // PZHTF. - 1998. - V. 24, No. 13. - P. 1.

100. V. I. Zubkov, V. I. Scheglov. Propagation of backward surface of magnetostatic waves in the ferrite – dielectric – metal structure, linearly inhomogeneous magnetic field // ZhTF. - 1999. - V. 69, No. 2. - P. 70.

101. V. I. Zubkov, V. I. Shcheglov. Reverse surface magnetostatic waves in the ferrite – dielectric – metal structure, nonuniformly magnetized native field of the "valley" type // RE. - 2000. - V. 45, No. 1. - P. 116.

102. 102. V. I. Zubkov, V. I. Scheglov. Dispersion of magnetostatic waves in a completely

magnetized ferrite plate with an anisotropy axis, perpendicular its dicular plane // RE. - 2000. - V. 45, No. 4. - P. 471.

103. V. I. Zubkov, V. I. Shcheglov. Reverse surface magnetostatic waves in the ferrite – dielectric – metal structure, nonuniformly magnetized native field of the "shaft" type // RE. - 2001. - V. 46, No. 12. - S. 1471.

104. V. I. Zubkov, V. I. Shcheglov. Electromagnetic waves in a bigyrotropic plate propagating in a direction perpendicular to the field bias // RE. - 2002. - V. 47, No. 9. - P. 1101.

105. V. I. Zubkov, V. A. Epanechnikov, V. I. Shcheglov. Dispersion characteristics of surface magnetostatic waves in a two-layer ferromagnetic film // RE. - 2007. - V. 52, No. 2. - P. 192.

106. V. I. Zubkov, V. I. Scheglov. Surface magnetostatic waves in the structure ferrite-dielectric-lattice of metal strips // RE. - 2006. - T. 51, No. 3. - S. 328.

107. Lax B., Button K. Ultrahigh-frequency ferrites and ferrimagnetics. - Moscow, Mir, 1965.

108. Mikaelyan A. L. Theory and application of ferrites at ultra-high frequencies. - Moscow, Gosenergoizdat, 1963.

109. Adam J.D. Analog signal processing with microwave magnetics // Proc.IEEE. - 1988. - V. 76, No. 2. - P. 159.

110. Adam J.D. Analog signal processing using microwave ferrites(Translation [109]) // TIER. - 1988. - V. 76, No. 2. - P. 73.

111. 111. Ishak W. S. Magnetostatic wave technology: a review // Proc. IEEE. - 1988. - V. 76, No. 2. - P. 171.

112. Iskhak VS Application of magnetostatic waves: a review. (Transfer [111]) // TIIER. - 1988. - V. 76, No. 2. - P. 86

113. 113. Schloemann E. F. Circulators for microwave millimeter wave integrated circuits // Proc. IEEE. - 1988. - V. 76, No. 2. - P. 188.

114. Shleman E.F. Circulators for integrated SV Ch-circuits. (Transfer [113]) // TIER. - 1988. - V. 76, No. 2. - P. 105.

115. Adam J. D., Davis L. E., Dionne G. F., Schloemann E. F., Stitzer S. N. Ferrite devices and materials // IEEE Trans. on MTT. - 2002. - V. 50, No. 3. - P. 721.

116. Dikshtein I. Ye., Shcheglov V. I. The receptivity of the compositional environment, consisting of anisotropic spherical ferrite particles // In the book: XVII International School-Seminar "New Magnetic Materials in electronics (NMMM-2000) ". - Moscow: URSS. Moscow State University, 2000 .-- P. 21.

117. V. I. Zubkov, V. I. Scheglov. The susceptibility of the compositional environment from of anisotropic ferrite spheres under the conditions of orientational transition // In the book: VIII International Seminar "Magnetic Phase Transitions". - Makhachkala: publication of the Institute of Physics of the Dagestan Scientific Center RAS, 2007 .-- P. 63.

118. Ferrites in nonlinear microwave devices. Sat. articles ed. A.G. Gurevich. - Moscow, IL, 1961.

119. Temiryasev A.G., Tikhomirova M. P., Zilberman P. E. "Exchange" spin waves in nonuniform yttrium iron garnet films // JAP. - 1994. - V. 76, No. 9. - P. 5586.

120. Gulyaev Yu.V., Temiryazev A.G., Tikhomirova M. P., Zilberman P. E. Magnetoelastic interaction in yttrium iron garnet films with magnetic inhomogeneities through the film thickness // JAP. - 1994. - V. 75, No. 10. - P. 5619.

121. Zilberman P. Ye., Temiryazev A. G., Tikhomirova M. P. Excitation and relaxation of the space of the exchange spin of the outgoing waves in the films of yttrium iron garnet // ZhETF. - 1995. - V. 108, No. 1. - P. 281.

122. Gulyaev Yu. V., Zilberman P. Ye., Temiryazev A. G., Tikhomirova M. P. The new mode of nonlinear spin-wave resonance in the normal films // FTT. - 2000. - T. 42, No. 6. - P. 1062.

123. Alvarez L. F., Pla O., Chubykalo O. Quasiperiodicity, bistability, and chaos in the Landau-Lifshitz equation // Phys. Rev. B. - 2000. - V. 61, No. 17. - P. 11613 (5)

124. Shutiy A. M., Sementsov D. I. Nonlinear effects of precessional motion of magnetization in the region of ferromagnetic resonance // Phys. - 2000. - V. 42, No. 7. - S. 1268.

125. Shutyi A.M., Sementsov D.I. Dynamics of magnetization under conditions nonlinear ferromagnetic resonance in a film of the (111) type // Phys. - 2001. - V. 43, No. 8. - P. 1439.

126. Shutyy A.M., Sementsov D.I. Nonlinear precessional dynamics motion of magnetization in a ferrite-garnet film of type (100) // FTT. - 2002. - V. 44, No. 4. - S. 734.

127. Sementsov D.I., Shutyi A.M. Nonlinear regular and stochastic dynamics of magnetization in thin film structures // Phys. - 2007. - V. 177, No. 8. - S. 831.

128. Sementsov D.I., Shutyy A.M. High-amplitude precession and dynamics magnetic immunity of the two-layer film //FTT. - 2003. - V. 45, No. 5. - S. 877.

129. Gerrits Th., Schneider M. L., Kos A.B., Silva T. J. Large-angle magnetization dynamics measured by time-resolved ferromagnetic resonance // Phys. Rev. B. - 2006. - V. 73, No. 9. - P. 094454 (7).

130. Karpachev S.N., Vlasov V.S., Kotov L.N. Nonlinear relaxation dynamics of magnetic and elastic subsystems of a thin ferrite film near acoustic resonance // Vestnik MGU. Ser. 3. - 2006. -No. 6. - P. 60.

131. Vlasov V.S. Study of relaxation and nonlinear dynamics of mag- and magnetoelastic vibrations of films and particles. - Dissertation on Candidate of Science degree physical-mat. sciences. - Moscow: Moscow State University, 2007.

132. Kadomtseva A.M. Orientational phase transitions // Encyclopedia of science and technology (electronic resource). www.femto.com.ua/articles/part-2/2661.html

133. Belov K.P., Zvezdin A.K., Kadomtseva A.M., Levitin R.Z. On spin reorientation in rare-earth materials // Phys. - 1976. - V. 110, No. 3. - P. 447.

134. Belov K.P., Zvezdin A.K., Kadomtseva A.M., Levitin R.Z. Orientation transitions in rare-earth magnets. - Moscow, Nauka, 1979.

135. Landau L. D., Lifshits E. M. Statistical physics. Part 1. (Theoretical physics. - T. 5). - Moscow: Nauka, 1976.

136. Landau L. D. On the theory of phase transitions. I // Collected Works. - V. 1. - Moscow: Nauka, 1969 .-- P. 234.

137. Landau L. D. On the theory of phase transitions. II // Collected Works. - V. 1. - Moscow: Nauka, 1969 .-- P. 253.

138. White R., Jebell T. Long-range order in solids. - Moscow, Mir, 1982.

139. Patashinsky A. Z., Pokrovsky V. L. Fluctuation theory of phase transitions. - Moscow, Nauka, 1982.

140. Dikshtein I. E., Lisovsky F. V., Mansvetova E. G., Tarasenko, Shapovalov V.I., Shcheglov V.I. Domain structure of uniaxial ferrimagnets with a compensation point in strong magnetic fields // ZhETF. - 1980. - V. 79, No. 2 (8). - S. 509.

141. M.M. Farztdinov Structure of antiferromagnets // Phys. - 1964. - V. 84, No. 4. - P. 611.

142. Pastushenkov Yu. G., Suponev N.P., Skokov K.P., Lyakhova M.B., Semenova L.V. Magnetocrystalline anisotropy, domain structure and orientational phase transitions in intermetallic compounds $Nd_2Fe_{14}B$ and R (Fe, Co) 11Ti (R = Tb, Dy, Er, Ho) // Bulletin of TVGU, Series "Physics". - 2004. - No. 4 (6). - S. 25.

143. Pomyalov A.V., Andreev A.S. Waves of coupled oscillations in periodic system of thin film ferrite resonators. - 1986. - V. 31, No. 9. - P. 1739.

144. Shutyi A.M., Sementsov D.I. Dynamic magnetization reversal and strong states in antiferromagnetic multi-layer structures // FTT. - 2004. - V. 46, No. 2. - P. 271.

145. 145. Shutyi A. M. Controlling orientation transitions in crossed dipole gratings // ZhTF. - 2015. - V. 85, No. 7. - P. 1.

146. 146. A.K. Zvezdin, V.A.Kotov. Magneto-optics of thin lenses. - Moscow, Science, 1988.

147. Hubert A., Sch˝afer R. Magnetic domains: the analysis of magnetic microstructures. - Berlin Heidelberg: Springer, 1998.

148. Deryugin A.A., Tsyrkin V.V., Krasovsky V.E. Application of the integral memory: Handbook. - Moscow, Radio and communication, 1994.

149. Raev V.K. Development of the foundations of the theory and principles of design information sources on cylindrical agnetic domains. - Dissertation for the degree of Doctor of Engineering. sciences. - Moscow, 1989.

150. 150. Raev VK, Khodenkov GE Cylindrical magnetic domains in the element of computing technology. - Moscow, Energoizdat, 1981.

151. Rosenblat M.A. External computer storage devices // Foreign electronic equipment. - Moscow, Central Research Institute "Electronics", 1988. - No. 8. - P. 50.

152. Elsgol'ts L. E. Differential equations and calculus of variations - Moscow, Nauka, 1965.

153. Thiele A.A. Theory of the static stability of cylindrical domains in uniaxial platelets // J. Appl. Phys. - 1970. - V. 41, No. 3. - P. 1139.

154. Thiele A.A. The theory of cylindrical magnetic domains // Bell Syst. Tech. J. - 1969. - V. 48, No. 10. - P. 3287.

155. Boyarchenkov M.A., Prokhorov N.L., Raev V.K., Rosenthal Yu.D. Magnet- domain logical and storage devices. - Moscow, Energy, 1974.

156. 156. Smolenskiy G.A., Boyarchenkov M.A., Lisovskiy F.V., Raev V.K. Cylindrical magnetic domains in magnetically uniaxial materials: physical principles and fundamentals of technical applications // Microelektronika. - 1972. - V. 1, No. 1. - P. 26; No. 2. - P. 99.

157. Bar'yakhtar V. G., Gorobets Yu. I., Filippov B. N. CMD theory. 1. Statical and dynamic properties of an isolated CMD // FMM. - 1977. - V. 43, No. 2. - P. 231.

158. Baryakhtar V.G., Gann V.V., Gorobets Yu.I., Smolensky G.A., Philippov B.N. Cylindrical magnetic domains // Phys. - 1977 .-- V. 121, No. 4. - P. 593.

159. Rosenblat M.A. Magnetic elements of automation and computing technology. - Moscow: Nauka, 1974.

160. Dorleijn J.W.F., Druyvesteyn W. F., Bartles G., Tolksdorf W. Magnetic bubbles and stripe domains subjected to in-plane fields. I. Uniaxial anisotropy // Phil. Res. Rep. - 1973. - V. 28, No. 2. - P. 133.

161. Dorleijn J.W.F., Druyvesteyn W. F., Bartles G., Tolksdorf W. Magnetic bubbles and stripe domains subjected to in-plane fields. II. Contribution of the cubic anisotropy // Phil. Res. Rep. - 1973. - V. 28, No. 2. - P. 152.

162. Bobeck A.H., Bonyhard P. I., Geusic I. E. Magnetic bubbles - an emerging new memory technology // Proc. IEEE. - 1975. - V. 63, No. 8. - P. 1176.

163. Cohen M. S., Chang H. The frontiers of magnetic bubble technology // Proc. IEEE. - 1975. - V. 63, No. 8. - P. 1196.

164. Telesnin R. V., Dudorov V. N. Periodicity of the hexagonal lattice CMD in films of rrit-garnets // Physics and Technology. - 1975. - V. 17, No. 6. - S. 1627.

165. Balbashov A.M., Chervonenkis A.Ya., Cherkasov A.N., Bakhteuzov V.E.Domain

structure in epitaxial films Y-Bi-Ga // FTT. - 1974. - V. 16, No. 10. - S. 3102.

166. Druyvesteyn W. F., DeJonge F.A. A special kind of magnetic domain: hollow bubble with at its center a bubble // Phys. Lett. - 1971. - V. 36A, No. 1. - P. 1.

167. De Bonte W. J. The static stability of half-bubbles // Bell Syst. Tech. J. - 1972. - V. 51, No. 9. - P. 1933.

168. Vaskovsky V.O., Kandaurova G.S., Balbashov A.M., Chervonenkis A.Ya. Oblique CMD in Y Fe O3 crystals // Phys. - 1977. - V. 19, No. 1. -P. 20.

169. Chetkin M. V., Shalygina A. N., Drachev V. A. CMD in the plate, ferrites perpendicular to the optical axis, Mikroelektronika. -1974. - V. 3, No. 1. - P. 71.

170. Lisovskiy FV, Mansvetova EG Intra-volume CMD in epitaxial magnite garnets // PZhTF. - 1978. - V. 4, No. 21. - S. 1268.

171. Chervonenkis A.Ya., Rybak V.I. Bistable cylindrical domains in Bi-containing gram anatom films // PZhTF. - 1978. - T. 4, No. 1. - P. 24.

172. Tabor W. J., Bobeck A.H., Vella-Coleiro G. P., Rosencwaig A. A new type of cylindrical magnetic domains (bubble isomers) // Bell Syst. Tech. J. - 1972. - V. 51, No. 6. - P. 1427.

173. Malozemoff A. P. Interacting Bloch lines: a new mechanism for wall energy in bubble domain materials // Appl. Phys. Lett. - 1972. - V. 21, No. 4. - P. 149.

174. 174. VN Dudorov, VV Randoshkin, RV Telesnin. Synthesis and physical properties of monocrystalline films of rare earth ferrites-granatov // UFN. - 1977. - V. 122, No. 2. - P. 253.

175. Kobayashi T., Nishida H., Sugita J. Statics of extraordinary bubbles // J. Phys. Soc. Jap. - 1973. - V. 34, No. 2. - P. 555.

176. Thiele A.A. Application of the gyrocoupling vector and dissipation dyadic in the dynamics of magnetic domains // J. Appl. phys. - 1974. - V. 45, No. 1. - P. 377.

177. Rosencwaig A., Tabor W. J., Nelson T. J. New domain-wall configuration for magnetic bubbles // Phys. Rev. Lett. - 1972. - V. 29, No. 14. - P. 946.

178. Slonczewski, J. C. Theory of Bloch-line and Bloch-wall motion, J. Appl. Phys. - 1974. - V. 45, No. 6. - P. 2705.

179. Tabor W. J., Bobeck A.H., Vella-Coleiro G. P., Rosencwaig A. A new type of cylindrical magnetic domain (hard bubbles) // AIP Conf. Proc. - 1973. -V. 10. - P. 442.

180. Ilyicheva E.N., Shishkov A.G., Ilyashenko E.I., Fedyunin Yu.N. Determination of the structure of the interdomain boundary using the Faraday effect //FTT. - 1978. - T. 20, No. 8. - P. 2322.

181. Grundy P. J., Hothersall D. C., Jones G. A., Middleton B. K. The formation and structure of cylindrical magnetic domains in thin cobalt crystals // Phys. Stat. Sol. (a). - 1973. - V. 9, No. 1. - P. 79.

182. Grundy P. J., Jones G. A. Lorentz microscopy of bubble domain structure in cobalt at fields approaching saturation // AIP Conf. Proc. - 1973. - V. 10. - P. 364.

183. Grundy P. J., Herd S.R. Lorentz microscopy of bubble domain and charges in domain wall state in hexaferrite // Phys. Stat. Sol. (a). - 1973. - V. 20, No. 1. - P. 295.

184. Suzuki R., Tokahashi M., Kobayashi T., Sugita J. Planar domains and domain wall structures of bubbles in permalloy-coated garnet films // Appl. Phys. Lett. - 1975. - V. 26, No. 6. - P. 342.

185. Dekker P. J., Slonczewski J. C. Switching of magnetic bubbles states // Appl. Phys. Lett. - 1976. - V. 29, No. 11. - P. 753.

186. Mac Neal B. E., Humphrey F. B. Azimuthal angular rotation (ψ) in domain walls during radial motion of bubble domains // J. Appl. phys. - 1979. - V. 50, No. 2. - P. 1020.

187. Hagedorn F.B. Dynamic conversion during magnetic bubble domain wall motion //

J. Appl. phys. - 1974. - V. 45, No. 7. - P. 3129.

188.   Schlomann E. Domain walls in bubble films. I. General theory of static properties // J. Appl. phys. - 1973. - V. 44, No. 4. - P. 1837.

189.   Schlomann E. Domain walls in bubble films. II. Static properties of thick films // J. Appl. phys. - 1973. - V. 44, No. 4. - P. 1850.

190.   190. Schlomann E. Domain walls in bubble films. III. Wall structure of stripe domains // J. Appl. phys. - 1974. - V. 45, No. 1. - P. 369.

191.   Schl"omann E. Domain walls in bubble films. IV. High-speed wall motion in the presence of an in-plane anisotropy // J. Appl. phys. - 1976. - V. 47. No. 3. - P. 1142.

192.   Slonczewski J. C. Theory of domain-wall motion in magnetic films and platelets // J. Appl. phys. - 1973. - V. 44, No. 4. - P. 1759.

193.   Sokolov Yu.F. On the theory of domain grains and stripe domain structure in strongly anisotropic uniaxial ferromagnets // Phys. - 1980. - V. 22, No. 3. - S. 652.

194.   Smetanin B.M. Technical creativity. - Moscow, Young Guard, 1955.

195.   Kostenko I., Mikirtumov E. Flying models. - Moscow, Detgiz, 1951.

196.   VG Bar'yakhtar, Yu.I. Gorobets. Cylindrical magnetic domains and their lattice. - Kiev, Naukova Dumka, 1986.

197.   V. V. Gann, Yu. I. Gorobets. Vibrations of the lattice of cylindrical magnetic domains // FTT. - 1975. - V. 17, No. 5. - S. 1305.

198.   Bar'yakhtar V. G., Gann V. V., Gorobets Yu. I. Waves in a cylindrical lattice domains// FTT. - 1976. - T. 18, No. 7. - S. 1990.

199.   Dikshtein I. E., Lisovskiy F. V., Mansvetova E. G., Chizhik E. S. The development of a reflexive domain structure in the case of unipolar and cyclic magnetization reversal of a uniaxial magnet // ZhETF. - 1991. - V. 100, No. 5. - P. 1606.

200.   Lisovsky F.V., Mansvetova E.G., Nikolaeva E.P., Nikolaev A.V. Dynamic self-organization and symmetry of magnetic distribution moment in thin films // ZhETF. - 1993. - V. 103, No. 2. - P. 213.

201.   201. Logunov M. V., Moiseev N. In. Formation of a cellular domain structure in magnetic tapes // PZhTF. - 1997. - V. 23, No. 9. - S. 46.

202.   202. Logunov MV, Gerasimov MV Formation and evolution of giant dynamic domains in a harmonic magnetic field // Phys. - 2003. - V. 45, No. 6. - S. 1031.

203.   203. Kilshakbaeva Zh.A., Rusinov A.A., Kandaurova G.S. Phase diagrams we have dynamic systems of magnetic domains. Single-domain state // Combined conf. on magnetoelectronics. Abstracts. report 1995. - Moscow, IRE RAS. - S. 39.

204.   Kandaurova G.S. Chaos and order in a dynamic system of magnetic domains. Anger condition // I United. conf. on magnetoelectronics. Abstracts. report 1995. - Moscow, IRE RAS. - S. 41.

205.   Kandaurova G.S. New phenomena in the low-frequency dynamics of the team of magnetic domains // Phys. - 2002. - T. 172, No. 10. - S. 1165.

206.   Kandaurova G.S., Svidersky AE Self-organization processes in suitable magnetic environment and the formation of stable dynamic structures // ZhETF. - 1990. - V. 97, No. 4. - S. 1218.

207.   Gerasimov M.V., Logunov M.V., Nikitov S.A., Nozdrin Yu.N., Spirin A.V., Tokman I.D. Experimental observation of domain wall motion induced by laser pump-pulse // Book of Abstracts of placeCityMoscow International Symposium on Magnetism (MISM). Moscow, 2017. Published by Publishing house Phys. fac. Moscow State University ". Moscow. - P. 36.

208.   Beaurepaire E., Merle J. C., Daunois A., Bigot J. Y. Ultrafast spin dynamics in ferromagnetic nickel // Phys. Rev. Lett. - 1996. - V. 76, No. 22. - P. 4250.

209.   Kirilyuk A., Kimel A.V., Rasing T. Ultrafast optical manipulation of magnetic order

// Rev. Mod. Phys. - 2010. - V. 82, No. 3. - P. 2731.

210. Every A.G. Measurement of the near-surface elastic properties of solids and thin supported films // Meas. Sci. Technol. (Measurement Science and Technology). - 2002. - V. 13. - P. R21.

211. Walowski J., M"unzenberg M. Perspective: Ultrafast magnetism and spintronics // J. Appl. Phys. - 2016. - V. 120, No. 14. - P. 140901 (16).

212. Bigot J.V., Vomir M. Ultrafast magnetization dynamics of nanostructures // Ann. Phys. (StateplaceBerlin). - 2013. - V. 525, No. 1–2. - P. 2.

213. Kashen, Bauer G. E.W. Laser-induced spatiotemporal dynamics of magnetic films // Phys. Rev. Lett. - 2015. - V. 115, No. 19. - P. 197201 (5).

214. Linnik T. I., Scherbakov A.V., Yakovlev D. R., Liu X., Furdina J.K., Bayer M. Theory of magnetization precession induced by picosecond strain pulse in ferromagnetic semiconductor (Ga, Mn) As // Phys. Rev. B. -2011. - V. 84, No. 21. - P. 214432 (11).

215. Jaeger J.V., Scherbakov A.V., Linnik T. I., Yakovlev D. R., Wang M., Wadley P., Holy V., Cavill S.A., Akimov A.V., Rushforth A.W., Bayer M Picosecond inverse magnetostriction in galfenol thin films // Appl. Phys. Lett. - 2013. - V. 103, No. 3. - P. 032409 (5).

216. Jaeger J.V., Scherbakov A.V., Glavin B.A., Salasyuk A. S., Campion R. P., Rushforth A.W., Yakovlev D.R., Akimov A.V., Bayer M. Resonant driving of magnetization precession in a ferromagnetic layer by coherent monochromatic phonons // Phys. Rev. B. - 2015. - V. 92, No. 2. - P. 020404 (5).

217. Kabychenkov A.F. The effect of the light field on the dispersion of the magnetodipole of high waves in ferromagnets // ZhTF. - 1994. - T. 64, No. 8, p. 159.

218. Chernov A.I., Kozhaev M.A., Vetoshko P.M., Dodonov D.V., Prokopov A.R., Shumilov A.G., Shaposhnikov A.N., Berzhansky V.N., Zvez-Dean A.K., Belotelov V.I. Local sounding of magnetic strips with the help of optical excitation of magnetostatic waves // FTT. - 2016. - T. 58, No. 6. - P. 1093.

219. Dreher L., Weiler M., Pernpeintner M., Huebl H., Gross R., Brandt M. S., Goennenwein S. T. B. Surface acoustic wave driven ferromagnetic resonance in nickel thin films: theory and experiment // Phys. Rev. B. - 2012. - V. 86, No. 13. - P. 134415 (13).

220. Thevenard L., Gourdon C., Prieur J. Y., Von Bardeleben H. J., Vincent S., Becerra L., Largeau L., Duquesne J. Y. Surface-acoustic-wave-driven ferromagnetic resonance in (Ga, Mn) (As, P) epilayers // Phys. Rev. B. - 2014. - V. 90, No. 9. - P. 094401 (8).

221. Koopmans B., Malinovski G., Dalla Longa F., Steiauf D., F"ahnle M., Roth T., Cinchetti M., Aeschlimann M. The paradoxical diversity of ultrafast laser-induced demagnetization reconciled // Nature Materials. Supplementary Information. - 2009. - P. 1.

222. Koopmans B., Malinovski G., Dalla Longa F., Steiauf D., F"ahnle M., Roth T., Cinchetti M., Aeschlimann M. Explaining the paradoxical diversity of ultrafast laser-induced demagnetization // Nature Materials. - 2010. - V. 9, No. 3. - P. 259.

223. Janusonis J., Chang C. L., Jansma T., Gatilova A., Vlasov V. S., Lomonosov A.M., Temnov V.V., Tobey R. I. Ultrafast magnetoelastic probing of surface acoustic transients // Phys. Rev. B. - 2016. - V. 94, No. 2. - P. 024415 (7).

224. Janusonis J., Jansma T., Chang C. L., Liu Q., Gatilova A., Lomonosov A.M., Shalagatskyi V., Pezeril T., Temnov V.V., Tobey R. I. Transient grating spectroscopy in magnetic thin films: simultaneous detection of elastic and magnetic dynamics // Scientific reports. - 2016 .-- 6: 29143. DOI: 10.1038 / serp29143. www.nature.com/ scientificreports. - P. 1–10.

225. Chang C. L., Lomonosov A. M., Janusonis J., Vlasov V. S., Temnov V. V., Tobey R.

I. Parametric frequency mixing in a magnetoelastically driven linear ferromagnetic oscillator // Phys. Rev. B. - 2017. - V. 95, No. 6. - P. 060409 (5).

226. Lomonosov A.M., Vlasov V. S., Janusonis J., Chang C. L., Tobey R. I., Pezeril T., Temnov V.V. Magneto-elastic symmetry breaking with surface acoustic wsves // Proceedings of "The 7th International Conference on Metamaterials, Photonic Crystals and Plasmonics "(META-16 Malaga-Spain). ISSN 2429-1390. Metaconferences.org. - P. 1–2.

227. Maznev A.A., Every A.G. Time-domain dynamic surface response of an anisotropic elastic solid to an impulsive line force // Int. J. Engng. Sci. - 1997. - V. 35, No. 4. - P. 321.

228. Vlasov V.S., Makarov P.A., Shavrov V.G., Shcheglov V.I. Orientation characteristics of excitation of magnetoelastic waves by femtosecond pulse of light // Electronic "Journal of Radio Electronics". - 2017. - No. 6. Access mode: http://jre.cplire.ru/jre/jun17/5/text.pdf.

229. Druyvesteyn W. F., Dorleijn J. W. F., Rijnierse R. J. Analysis of a method for measuring the magnetocrystalline anisotropy of bubble materials // J. Appl. Phys. - 1973. - V. 44, No. 5. - P. 2379

230. Tarasenko V. V., Chensky E. V., Dikshtein I. E. Theory of inhomogeneous magnetic states in ferromagnets in the vicinity of phase transitions of the second kind // ZhETF. - 1976. - V. 70, No. 6. - P. 2178.

231. 231. Shimada J., Kojima H., Sakai K. Determination of the anisotropy field of garnet bubble materials from domain observation // J. Appl. Phys. - 1974. - V. 45, No. 10. - P. 4598.

232. Bar'yakhtar V. G., Klepikov V. F. Influence of inhomogeneous states on paramagnet - ferromagnet phase transition // PZhTF. - 1972. - V. 15, No. 7. - P. 411.

233. Bar'yakhtar V. G., Klepikov V. F. Phase transitions in polarized media and the role of inhomogeneous states // Phys. - 1972. - V. 14, No. 5. - P. 1478.

234. Bar'yakhtar V. G., Ivanov B. A., Kvirikadze A. G., Klepikov V. F. About phase transition of a ferromagnet from a single-domain to a multi-domain state// FMM. - 1973. - V. 36, No. 1. - P. 18.

235. Johansen T. R., Norman D. J., Torok E. J. Variation of stripe-domain spacing in a Faraday effect light deflector // J. Appl. Phys. - 1971. - V. 42, No. 4. - P. 1715.

236. V. I. Shcheglov. Some properties of light diffraction by a domain structure tour // FTT. - 1973. - V. 15, No. 4. - P. 1046.

237. Lisovsky F.V., Shcheglov V.I. The influence of the domain structure on the magnetic restriction vibrations in epitaxial films of ferrite garnets // ZhTF. - 1983. - V. 53, No. 3. - P. 596.

238. Lisovsky F.V., Shcheglov V.I. The effect of anisotropy on the magnetostriction vibrations in ferrite-garnet films at orientation phase transitions, Phys. - 1983. - V. 25, No. 12. - P. 3710.

239. A.K. Zvezdin, V.M. Matveev. Features of physical properties of red garnets near the compensation temperature //ZhETF. - 1972. - V. 62, No. 1. - P. 260.

240. 240. Malozemoff A. P., DeLuca J. C. Effect of misorientation on growth anisotropy in [111] -oriented garnet films // J. Appl. Phys. - 1974. - V. 45, No. 10. - P. 4586.

241. 241. Muller M.W. Distribution of the magnetization in a ferromagnet // Phys. Rev. - 1961. - V. 122, No. 5. - P. 1485.

242. Holz A., Kronm"uller H. The nucleation of stripe domains in thin ferromagnetic films // Phys. Stat. Sol. - 1969. - V. 31, No. 2. - P. 787.

243. Goldstein R.M., Muller M.W. Domain nucleation in uniaxial ferromagnets // Phys. Rev. B. - 1970. - V. 2, No. 11. - P. 4585.

244. Shumate P.W., Smith D.H., Hagedorn F. B. The temperature dependence of the anisotropy field of mixed rare-earth iron garnets // J. Appl. Phys. - 1973. - V. 44, No. 1. - P. 449.

245. Muller M.W. Theory of stripe-domain nucleation in garnet films // J. Appl. Phys. - 1974. - V. 45, No. 11. - P. 5050.

246. Dikshtein I. E., Lisovsky F. V., Mansvetova E. G., Tarasenko V. V., Shapovalov V.I., Shcheglov V.I. Domain structure of uniaxial ferrite magnets at phase transitions of the second kind // In the book: All-Union conference on the physics of magnetic phenomena. Abstracts of reports. - Kharkiv, 1979 .-- S. 439.

247. Bar'yakhtar V. G., Borovik A. E., Popov V. A. Intermediate theory of antiferromagnets in a second-order phase transition in an external magnetic field // PZhETF. - 1969. - V. 9, No. 11. - P. 634.

248. Mitsek A.I., Kolmakova N.P., Gaidansky P.F. Metastable states of uniaxial antiferromagnets // Physics and Technology. - 1969. - V. 11, No. 5. - P. 1258.

249. Bar'yakhtar V. G., Galkin A. A., Kovner S. I., Popov V. A. Antiferromagnetic resonance in copper chloride dihydrate at low frequencies at overturning of magnetic moments of sublattices // PZhETF. - 1969. - V. 10, No. 7. - P. 292.

250. Eremenko V.V., Kharchenko N. F., Gnatchenko S. L. Magneto-optical study of the phase diagram and structure near the temperature of magnetic compensation in iron-garnets // Dig. Intermag Conf. - 1974 .-- Toronto. - P. 7.

251. F.V. Lisovskiy, V.I. Shapovalov. The noncollinearity of the sublattices and su- domain structure in high fields of magnetization in $Dy_3Fe_5O_{12}$ near the point of magnetic compensation // PZhETF. - 1974. - V. 20, No. 2. - P. 128.

252. V. G. Bar'yakhtar, D. A. Yablonsky. Domain structure of ferrites in the vicinity of the compensation point // FTT. - 1974. - V. 16, No. 11. - P. 3511.

253. Lisovsky F.V., Mansvetova E.G., Shapovalov V.I. Phase diagram and the structure of domain grains in a uniaxial ferrimagnet near the point of compensation // ZhETF. - 1976. - V. 71, No. 4 (10). - P. 1443.

254. Bogdanov A.N., Yablonsky D.A. Domain structure theory of ferrites in the vicinity of the compensation point // In the book: All-Union conference on physics of magnetic phenomena. Abstracts of reports. - Kharkov, 1979 .-- S. 442.

255. Veselago V. G., Maksimov L. P., Prokhorov A. M. Installation "Solenoid" for obtaining superstrong magnetic fields // Bulletin of the Academy of Sciences of the USSR. - 1968. - No. 12. - P. 58.

256. Kolm G., Freeman A. Strong magnetic fields // Phys. - 1966 .-- V. 88, No. 4. - P. 703.

257. Sampson W., Craig P., Strongin M. Advances in the creation of superconducting magnets // Phys. - 1967. - T. 93, No. 4. - S. 703.

258. S.V. Vonsovsky. Magnetism. - Moscow, Nauka, 1971.

259. Le-Crowe R., Comstock R. Magnetoelastic interactions in ferromag-

260. nitnykhd electrics // In the book: Physical acoustics, ed. Mazo- on U.P. - V. 3B. Lattice dynamics. - Moscow, Mir, 1968 .-- P. 156. TSB. - T. 15. - Moscow, Soviet encyclopedia, 1974. - P. 168. Article Magnetic susceptibility".

261. Korn G., Korn T. Handbook of mathematics for scientific workers and engineers. - Moscow, Nauka, 1973.

262. 262. Sushkevich A.K. Foundations of higher algebra. - M.L .: State. ed. tech n-theor lit, 1941.

263. Vashkovsky A.V., Stalmakhov V.S., Sharaevsky Yu.P. Magneto static waves in microwave electronics. - Saratov: Saratov University Publishing House, 1993.

264. Vlasov V. S., Kotov L. N., Shavrov V. G., Shcheglov V. I. Nonlinear excitation of hypersound in a ferrite plate with ferromagnetic resonance // RE. - 2009. - T. 54,

No. 7. - S. 863.

265.   Vlasov V. S., Kotov L. N., Shavrov V. G., Shcheglov V. I. Excitation coupled agno-elastic oscillations in a nonlinear ferromagnetic reaction resonance // Proceedings of the XVI International Conference "Radiolocation and radio communication ". - Moscow – Firsanovka: MPEI edition, 2008. - P. 197.

266.   Shcheglov V.I., Shavrov V.G., Zubkov V.I., Vlasov V.S., Kotov L.N. Nonlinear precession of the magnetization vector in the normal magnetization of ferrite plate with magnetoelastic properties when orientation phase transition // Proceedings of the XII International conference "Magnetism, long-range and short-range spin-spin interactions consequence ". - Moscow – Firsanovka: Edition of the MEI, 2009. - P. 92.

267.   Suhl H. Ferromagnetic resonance in nickel ferrite between one and two kilomegacycles // Phys. Rev. - 1955. - V. 97, No. 2. - P. 555.

268.   268. Ilyin VA, Poznyak EG Fundamentals of mathematical analysis. - Moscow, Sc ence, 1965.

269.   G. Fikhtengolts Differential and integral calculus course V. 1. - M., L .: State. Ed. Techn.-theory. Lit., 1951.

270.   Dwight G.B. Tables of integrals and other mathematical formulas. - Moscow: Nauka, 1973.

271.   Landau L. D., Lifshits E. M. The theory of elasticity (Theoretical physics. T. VII). - Moscow, Nauka, 1965.

272.   Kittel Ch. Introduction to solid state physics. - Moscow, Nauka, 1978.

273.   Yu.I. Sirotin, M.P. Shaskolskaya. Fundamentals of crystal physics. - Moscow, Nauka, 1979.

274.   Harrison W. Rigid State Theory. - Moscow, Mir, 1972.

275.   Hagedorn F. B., Hewitt B. S. Growth-induced magnetic anisotropy in variously oriented epitaxial films on Sm-YIGG // J. Appl. Phys. - 1974. - V. 45, No. 2. - P. 925.

276.   Hubert A., Malozemoff A. P., DeLuca J. C. Effect of cubic, tilted uniaxial and orthorhombic anisotropy on homogeneous nucleation in a garnet bubble film // J. Appl. Phys. - 1974. - V. 45, No. 8. - P. 3562.

277.   Shumate Jr. P.W. Extension of the analysis for an optical magnetometer to include cubic anisotropy in detail // J. Appl. Phys. - 1973. - V. 44, No. 7. - P. 3323.

278.   278. Ilyashenko E.I., Lisovskiy F.V., Shcheglov V.I., Yurchenko S.E. About the influence of the tilt of the easy magnetization axis on the mobility of domain boundaries in thin films of ferrite-garnets // Phys. - 1977 .-- T. 19, No. 3. - P. 898.

279.   Physical encyclopedia. T. 2. - Moscow, Soviet encyclopedia. 1990. - P. 647. Article "Magnetic atomic structure".

280.   Izyumov Yu.A., Naysh V.E., Ozerov R.P. Neutron diffraction of magnets. - Moscow: Nauka, 1981.

281.   BSE. T. 9. - Moscow, Soviet encyclopedia, 1972. - P. 145. The article "Same climb. "

282.   Corliss L.M., Hastings J.M., Nathans R., Shirane G. Magnetic structure of Cr_2O_3 // J. Appl. Phys. - 1965. - V. 36, No. 3. Pt. 2. - P. 1099.

283.   BSE. T. 29. - Moscow, Soviet encyclopedia, 1978. - P. 470. Article "Spinel".

284.   Lax B., Button K. Ultrahigh-frequency ferrites and ferrimagnettics. - Moscow, Mir, 1965.

285.   BSE. T. 15. - Moscow, Soviet encyclopedia, 1974. - P. 158. Article "Magnetite".

286.   BSE. T. 15. - Moscow, Soviet encyclopedia, 1974. - P. 179. Article "Soft magnetic materials".

287.   BSE. T. 7. - Moscow, Soviet encyclopedia, 1972. - P. 249.

288.   Geller S., Gilleo M.A. The crystal structure and ferrimagnetism of yttrium-iron garnet, Y3Fe2 (FeO4) 3 // Journ. Phys. Chem. Sol. - 1957. - V. 3, No. 1/2. - P. 30–36.

289. Geller S., Guillo M. Crystalline structure and ferrimagnetism of yttrium garnet // In the book: Ferrites in nonlinear high-frequency devices. Ed .: A.G. Gurevich - Moscow: IL, 1961 .-- P. 373

290. Lonsdale K. Crystals and X-rays. - Moscow, IL, 1952.

291. G.E. Shilov, Introduction to the theory of linear n spaces. - M.L .: State. ed. technical-theor. lit., 1952.

292. N. V. Efimov. A short course in analytical geometry. - M.L .: State. ed. technical-theor. lit., 1952.

293. Delone B.N., Raikov D.A. Analytic geometry. Volume 1. —Moscow, OGIZ Gostekhizdat, 1948.

294. Worth Ch., Thomson R. Solid State Physics. - Moscow, Mir, 1969.

295. I.I. Olkhovsky. A course in theoretical mechanics for physicists. - Moscow, Nauka, 1970.

296. Strauss V. Magnetoelastic properties of yttrium iron garnet // In the book: W. Mason (ed.): Physical acoustics. T. 4B. Physical applications of acoustics in quantum physics and solid state physics. - Moscow, Mir, 1970 .-- S. 247.

297. Vlasov V. S., Kotov L. N., Shavrov V. G., Shcheglov V. I. Non-linear dynamics of the establishment of magnetization in a ferrite plate with magnitoelastic properties under the conditions of orientation transition // RE. -2010. - V. 55, No. 6. - P. 689.

298. Vlasov V. S., Kotov L. N., Shavrov V. G., Shcheglov V. I. Installation dynamics magnetization in a normally magnetized ferrite plate with magnetoelastic properties under the conditions of orientation transition // Proceedings of the XXI International Conference "New in Magnetism and magnetic materials (HMMM) ". - Moscow: Publishing house. Moscow State University. - S. 942.

299. Vlasov V. S., Kotov L. N., Shavrov V. G., Shcheglov V. I. The dynamics of the orientation of transition of magnetization in a normally magnetized plane steel from a magnetoelastic dielectric // Proceedings of the IX International the seminar "Magnetic phase transitions". - Makhachkala: Publishing house- Institute of Physics, Dagestan Scientific Center, Russian Academy of Sciences, 2009. - P. 71.

300. Vlasov V. S., Kotov L.N., Shavrov V.G., Shcheglov V. I. The nonlinear reorientation of magnetization vector in magnetoelastic medium during dynamic magnetization // Book of Abstracts International conference "Functional Materials (ICFM-2009) ". Parteni. Crimea. Ucraine. Simferopol: TNU. 2009. - P. 297.

301. V. S. Vlasov, L. N. Kotov, V. G. Shavrov, V. I. Scheglov. Dynamic properties of the orientational phase transition in a ferrite plate with magnetoelastic properties // Abstracts of XXXIII International winter school of theoretical physicists "Kourovka 2010". - EkaterinBurg: IPM RAN, 2010. - P. 95.

302. Vlasov V. S., Kotov L. N., Shavrov V. G., Shcheglov V. I. Nonlinear precession of magnetization under the conditions of orientation transition with asymmetric metric excitation // Abstracts of the XII All-Russian School-Seminar "Wave phenomena in inhomogeneous media (Waves-2010)" (electric collection on CD). - Zvenigorod: Ed. Faculty of Physics. Moscow State University, 2010. Section 8. - P. 12.

303. Vlasov V. S., Kotov L. N., Shavrov V. G., Shcheglov V. I. Non-linear precession of the second order magnetization in the normally magnetized ferrite plate at orientational phase transition // Collection nickname of the proceedings of the XVIII International Conference "Electromagnetic field and materials ". - Moscow-Firsanovka (MPEI): PLANTI, 2010. - P. 122.

304. Vlasov V. S., Kotov L. N., Shavrov V. G., Shcheglov V. I. Multi-mode of the nature of the second-order nonlinear precession of magnetization at orientation transition // Proceedings of the X International seminar "Magnetic phase transitions".

- Makhachkala: Edition of the Institute Physics of the Dagestan Scientific Center of the Russian Academy of Sciences, 2010. - P. 99.

305. Vlasov V. S., Kotov L. N., Shavrov V. G., Shcheglov V. I. The kinetics of restroke between different modes of precession of the magnetization of the second order // Proceedings of the X International Seminar "Magnetic phase transitions ". - Makhachkala: Edition of the Institute of Physics of Dagestan Scientific Center of the Russian Academy of Sciences, 2010. - P. 103.

306. Vlasov V. S., Kotov L. N., Shavrov V. G., Shcheglov V. I. Forced nonlinear precession of the magnetization vector under conditions of orientation transition // RE. - 2011. - V. 56, No. 1. - P. 84.

307. Vlasov V. S., Kotov L. N., Shavrov V. G., Shcheglov V. I. Asymmetrical forced nonlinear precession of magnetization under conditions of orientational transition // RE. - 2011. - V. 56, No. 6. - P. 719.

308. Vlasov V. S., Kotov L. N., Shavrov V. G., Shcheglov V. I. Multi-mode of the nature of the second-order nonlinear precession of magnetization in the condition in the orientation transition // RE. - 2011. - T. 56, No. 9. - P. 1120.

309. Vlasov V. S., Kotov L.N., Shavrov V.G., Shcheglov V. I. Second order magnetization precession by reorientation transition // Book of Abstracts of Moscow International Symposium on Magnetism (MISM). - Moscow, 2011. Published by Fiz. fac. Moscow State University ", Moscow. - P. 782.

310. Vlasov V. S., Kotov L. N., Shavrov V. G., Shcheglov V. I. Asymmetrical excitation of the precession of the second-order magnetization with compensation of the impact of the constant field // Collection of works of the XIX International conference "Electromagnetic field and materials". - Moscow – Firsanovka: NRU MPEI, 2011. - P. 206.

311. Vlasov V. S., Kotov L. N., Shavrov V. G., Shcheglov V. I. Nonlinear preassignment of the equilibrium position of the magnetization vector in anisotropic environment // Proceedings of the XIX International Conference "Electromagnetic field and materials ". - Moscow – Firsanovka: NRU MPEI, 2011. - P. 217.

312. V. Vlasov, L. N. Kotov, V. G. Shavrov, V. I. Scheglov. Asymmetrical excitation of precession of the second-order magnetization under conditions orientational transition // RE. - 2012. - T. 57, No. 5. - P. 501.

313. Vlasov V. S., Kotov L. N., Shavrov V. G., Shcheglov V. I. Asymmetrical excitation of precession of the second-order magnetization in the presence of compensating constant field // Collection of materials XV International People's Winter School-Seminar on Microwave Electronics and radiophysics. - Saratov: Ed. SSU, 2012 .-- P. 38.

314. Vlasov V. S., Kotov L. N., Kirushev M. S., Shavrov V. G., Shcheglov V. I. Nonlinear precession of magnetization in a normally magnetized plane line in a constant field less than the demagnetization field. Treasures of the XXXIV International Winter School of Theoretical Physicists "Colevel 2012 ". - Yekaterinburg: Ed. IPM Ural Branch of the Russian Academy of Sciences, 2012. - P. 22.

315. Grishina M.K., Vlasov V.S., Kotov L.N., Shavrov V.G., Shcheglov V.I. Asymmetric excitation of the second order magnetization precession in a normally magnetized ferrite plate // Abstracts of the XXXIV International winter school of theoretical physicists "Kourovka 2012 ". - Yekaterinburg: Ed. IPM UB RAS, 2012. - P. 106.

316. Kirushev M. S., Vlasov V. S., Kotov L. N., Shavrov V. G., Shcheglov V. I. Second-order precession with forced oscillations of the magnet in terms of orientation transition // Abstracts of the XXXIV International winter school of theoretical physicists "Kourovka 2012 ". - Yekaterinburg: Ed. IPM UB RAS, 2012. - P. 111.

317. Vlasov V. S., Kirushev M. S., Kotov L. N., Shavrov V. G., Shcheglov V. I. Second order precession of magnetization in an anisotropic medium in the context of orientation transition // Proceedings of the XX International home conference "Electromagnetic field and materials". - Moscow: NRU MPEI, 2012 .-- P. 230.

318. Vlasov V. S., Kirushev M. S., Kotov L. N., Shavrov V. G., Shcheglov V. I. Features of the temporal characteristics of the precession of the equilibrium position in an anisotropic ferrite plate // Proceedings of the XXII International People's Conference "New in Magnetism and Magnetic Materials. Astrakhan: Ed. house "Astrakhan University", 2012. - P. 214.

319. Vlasov V. S., Kirushev M. S., Kotov L. N., Shavrov V. G., Shcheglov V. I. Second-order precession in a ferrite layer with magnetoelastic properties states // Proceedings of the XXII International Conference "New in magnetism and magnetic materials (NMMM) ". - Astrakhan: Ed. house "Astrakhan University", 2012. - P. 223.

320. Vlasov V. S., Kirushev M. S., Kotov L. N., Shavrov V. G., Shcheglov V. I. The precession of the second-order magnetization in a medium with a uniaxial magnet anisotropy // Proceedings of the XXII International Conference "New in magnetism and magnetic materials (NMMM)". - Astrakhan: Ed. house "Astrakhan University", 2012. - P. 257.

321. V. Vlasov, M. S. Kirushev, L. N. Kotov, V. G. Shavrov, V. I. Shcheglov. Second-order magnetization precession in a medium with cubic izotropia // Proceedings of the XXII International Conference "New in magnetism and magnetic materials (NMMM) ". - Astrakhan: Ed. house "Astrakhan University", 2012. - P. 260.

322. Vlasov V.S., Pleshev D.A., Kirushev M.S., Kotov L.N., Shavrov V.G., V. I. Sheglov Study of nonlinear precession of magnetization in a perpendicularly magnetized plate under the conditions of orientation transition // Abstracts of the XIII All-Russian School-Seminar on problems of physics of condensed matter (SPFKS-13). - Ekaterinburg: IPM UB RAS, 2012. - P. 16.

323. Vlasov V. S., Kirushev M. S., Kotov L. N., Shavrov V. G., Shcheglov V. I. Second-order precession of magnetization in an anisotropic medium. Part 1. Uniaxial anisotropy // RE. - 2013. - T. 58, No. 8. - P. 806.

324. Vlasov V. S., Kirushev M. S., Kotov L. N., Shavrov V. G., Shcheglov V. I. Second-order precession of magnetization in an anisotropic medium. II. Cubic anisotropy // RE. - 2013. - V. 58, No. 9. - P. 857.

325. Vlasov V. S., Kirushev M. S., Kotov L.N., Shavrov V.G., Shcheglov V. I. Nonlinear second order magnetization precession in anisotropy medium // Book of Abstracts of Moscow International Symposium on Spin Waves 2013 (SW-2013). Saint Petersburg. Russia. Published by "Ed. FTI them. Ioffe ". Moscow. - P. 168.

326. Vlasov V. S., Kirushev M. S., Pleshev D.A., Kotov L.N., Shcheglov V. I., Shavrov V.G., Varser E.V. Investigation of regimes of nonlinear magnetoelastic oscillations in the ferrite layer // Book of Abstracts of Moscow International Symposium on Spin Waves 2013 (SW-2013). Saint Petersburg. Russia. Published by "Ed. FTI them. Ioffe ". Moscow. - P. 170.

327. Vlasov V. S., Kirushev M. S., Kotov L.N., Shcheglov V. I., Shavrov V. G. Investigation of regimes of second order magnetization precession in anisotropy medium // Abstracts of V Euro-Asian Symposium "Trends in MAGnetism. Nanomagnetism "(EASTMAG-2013). Russia. Vladivostok. 2013. - P. 101.

328. Vlasov V. S., Kirushev M. S., Kotov L. N., Shavrov V. G., Shcheglov V. I. Second-order nonlinear precession of magnetization in anisotropic environment // Collection of materials of the X International Winter School-Seminar "Chaotic self-oscillations and the formation of structures" (CHAOS-2013). - Saratov: Ed. Center "Science",

2013. - P. 80.

329. Kirushev M. S., Vlasov V. S., Pleshev D.A., Kotov L.N., Shavrov V.G., Shcheglov
      V. I. Second order precession in the plate with cubic anisotropy and magnetoelastic
      properties // Book of Abstracts of Moscow International Symposium on Magnetism
      (MISM). Moscow, Published by Fiz. Fac. Moscow State University ". - P. 571.

330. Vlasov V. S., Kirushev M. S., Shavrov V. G., Shcheglov V. I. Precession magnetiza-
      tion of the second order in a magnetoelastic medium // Electronic "Journal of Radio
      Electronics". - 2015. - No. 4. Access mode: http://jre.cplire.ru/jre/apr15/16/text.pdf.

331. Kirushev M. S., Vlasov V. S., Pleshev D.A., Asadullin F. F., Kotov L.N., Shavrov
      V.G., Shcheglov V. I. Second order precession in the plate with cubic anisotropy and
      magnetoelastic properties // Solid State Phenomena. - 2015. - Vols. 233-234. P. 73.

332. Vlasov V.S., Kirushev M.S., Shavrov V.G., Shcheglov V.I. Stationary re- second-
      order magnetization precession modes in a medium with magnetoelastic their prop-
      erties // Proceedings of the XXIII All-Russian Conference "Electrotromagnetic field
      and materials ". - Moscow: INFRA-M, 2015 .-- S. 217.

333. Vashkovsky A. V., Locke E. G., Shcheglov V. I. The effect of the induced one-axial
      anisotropy on the domain structure and phase transitions of films yttrium iron garnet
      // FTT. - 1999. - V. 41 No. 11. - P. 2034.

334. Vetoshko P.M., Shavrov V.G., Shcheglov V.I. The role of elastic dissipation in the
      formation of resonance properties of precession of magnetization in nitoelastic me-
      dium // RE. - 2017. - T. 62, No. 4. - P. 364.

335. Vlasov V.S., Ivanov A.P., Shavrov V.G., Shcheglov V.I. Application of separated
      carriers for the analysis of nonlinear excitation of hypersound in a ferrite plate at
      ferromagnetic resonance. Part 1. Basic equations // RE. - 2015. - T. 60, No. 1. - P. 79.

336. Telesnin R. V., Kozlov V. I., Dudorov V. N. Ferromagnetic resonance in epitaxial
      films Y_3Fe_5 - xGa_xO_{12} // Phys. - 1974 .-- V. 16, No. 11. - P. 3532.

337. 337. Gavrilin V.P., Berezin D.G., Miroshnikov Yu.F. Ferromagnetic resonance and
      magnetic crystallographic anisotropy of single crystals of lithium ferrite flakes // Izv.
      Univ. Fiz. - 1973. - No. 9 (136). - P. 86.

338. Hansen P., Schuldt J., Tolksdorf W. Anisotropy and magnetostriction of iridium-
      substituted yttrium iron garnet // Phys. Rev. (B). - 1973. - V. 8, No. 9. - P. 4274.

339. Antonov L.I., Osipov S.G., Khapaev M.M. Domain wall calculation by the estab-
      lishment method // FMM. - 1983. - V. 55, No. 5. - P. 917.

340. Antonov L.I., Ternovsky V.V., Khapaev M.M. About calculating periodic domain
      structures in ferromagnetic materials // FMM. - 1989. - V. 67, No. 1. - P. 57.

341. Antonov L.I., Mukhina E.A., Lukasheva E.V. Two-dimensional magnetic field the
      periodic distribution of magnetization // FMM. - 1994. - V. 78, No. 4. - P. 5.

342. E. V. Lukasheva. Two-dimensional micromagnetic structure of magnetization uni-
      axial agnite films. - Dissertation for a scientific degree Penalty Cand. physical-mat.
      sciences. - Moscow, 1995.

343. Antonov L.I., Lukasheva E.V., Mironova G.A., Skachkov D.G. Dynamic establish-
      ment of an equilibrium period in the structure of magnetization of ferromagnetic
      films // FMM. - 2000. - T. 90, No. 3. - P. 5.

344. Antonov L.I., Lukasheva E.V., Mironova G.A., Skachkov D.G. Dynamic determi-
      nation of the distribution of magnetization in ferromagnetic films // Proceedings of
      the XVII International Conference "New magnetic materials for microelectronics
      (NMMM) ". - Moscow: Publishing house Moscow State University, 2000 .-- P. 507.

345. 345. Antonov L.I., Lukasheva E.V., Mironova G.A., Skachkov D.G. Magnetic field
      of the periodic structure of magnetization in a thin magnetic film // Proceedings of
      the XVII International Conference "New magnetic materials for microelectronics

(NMMM) ". - Moscow: Publishing house Moscow State University, 2000 .-- P. 510.

346. Antonov L.I., Lukasheva E.V., Mironova G.A., Skachkov D.G. Structure inhomogeneities of magnetization in blast furnaces // Sbornik of proceedings of the XVII International conference "New magnetic materials microelectronics (NMMM) ". - Moscow: Moscow State University Publishing House, 2000 .-- P. 514.

347. Antonov L.I., Lukasheva E.V., Popkova M.V. The ideal magnetic uniaxial thin ferromagnetic film // Collection of works of the XVIII International Conference "New magnetic materials for microelectronics (NMMM) ". - Moscow: Publishing house of Moscow State University, 2000 .-- P. 147.

348. Antonov L.I., Mironova G.A., Lukasheva E.V., Skachkov D.G., Parshion Yu.V. Horizontal Bloch lines in a thin domain wall of the ferromagnetic film // Proceedings of the XVIII International Conference lecture "New magnetic materials of microelectronics (NMMM)". - Moscow: Moscow State University Publishing House, 2000. - P. 410.

349. Lukasheva E. V., Popkova M. V., Sinilo P. V. Domain structure evolution of thin agnite films in the process of ideal magnetization // Proceedings of the XVII International Conference "New Magnets for microelectronic materials (NMMM) ". - Moscow: Publishing house of Moscow State University, 2000 .-- P. 510.

350. Antonov L.I., Zhukarev A.S., Polyakov P.A., Skachkov D.G. Magnetization of a uniaxial ferromagnetic film // ZhTF. - 2004. - V. 74, No. 3. - P. 83.

351. Demidovich B.P., Maron I.A. Foundations of Computational Mathematics. - Moscow: Fizmatgiz, 1963.

352. Demidovich BP, Maron IA, Shuvalova EZ Numerical methods of analysis - Moscow, Fizmatgiz, 1963.

353. Chikazumi S. Physics of Magnetism. Wiley. New York, 1964.

354. Rosencwaig A., Tabor W. J. Growth-induced noncubic anisotropy in bubble garnets // AIP Conf. Proc. - 1972. - V. 5. - P. 57.

355. Gyorgy E. M., Sturge M. D., Van Uitert L. G., Heilner E. J., Grodkiewicz W.H. Growth-induced anisotropy of some mixed rare-earth iron garnets // J. Appl. Phys. - 1973. - V. 44, No. 1. - P. 438.

356. Hagedorn F.B., Tabor W. J., Van Uitert L.G. Growth-induced magnetic anisotropy in seven different mixed rare-earth iron garnets // J. Appl. Phys. - 1973. - V. 44, No. 1. - P. 432.

357. Kitamura K., Iyi N., Kimura S., Chevrier F., Devignes J.M., Le Gall H. Growth-induced optical anisotropy of epitaxial garnet films grown on (110) -oriented substrates // J. Appl. Phys. - 1986. - V. 60, No. 4. - P. 1486.

358. Hagedorn F. B., Blank S. L., Barns R. L. Growth-induced magnetic anisotropy in Y2.4Eu0.6Ga1.2Fe3.8O12 // Appl. Phys. Lett. 1973. - V. 22, No. 5. - P. 209.

359. Gyorgy E. M., Rosencwaig A., Blount E. I., Tabor W. J., Lines M. E. General conditions for growth-induced anisotropy in garnets // Appl. Phys. Lett. - 1971. - V. 18, No. 11. - P. 479.

360. Shumate Jr. P.W., Smith D.H., Hagedorn F. B. The temperature dependence of the anisotropy field and coercivity in epitaxial films of mixed rare-earth iron garnets // J. Appl. Phys. - 1973. - V. 44, No. 1. - P. 449.

361. Hagedorn F.B. Instability of an isolated straight magnetic domain wall // J. Appl. Phys. - 1973. - V. 41, No. 3. - P. 1161.

362. Vetoshko P.M., Shavrov V.G., Shcheglov V.I. The role of elastic dissipation in the formation of the decay of the precession of magnetization in the magnetoelastic environment // PZhTF. - 2015. - V. 41, No. 21. - P. 1.

363. 363. Vashkovsky A.V., Locke E.G., Shcheglov V.I. Distribution without exchange

spin waves in ferrite films with a domain structure // Letters to ZhETF. - 1996. - V. 63, No. 7. - S. 544.

364. Vashkovsky A.V., Locke E.G., Shcheglov V.I. The spread of magnesium static waves in unsaturated ferrite films with stripe domain structure // ZhETF. - 1997. - V. 111, No. 3. - P. 1016.

365. Vashkovsky A. V., Locke E. G., Shcheglov V. I. Hysteresis characteristics magnetostatic waves in ferrite films with stripe domains of the magnetization vectors of which are oriented near the plane films // ZhETF. - 1998. - V. 114, No. 4 (10). - P. 1430.

366. Vashkovsky A.V., Locke E.G., Shcheglov V.I. Exchangeless spin waves in films of yttrium iron garnet with stripe domains, magnetic inside which is oriented near the plane of the film // Microelectronics. - 1998. - V. 27, No. 5. - P. 393.

367. Vashkovsky A.V., Locke E.G., Shcheglov V.I. Hysteresis characteristics magnetostatic waves in yttrium iron garnet films with a blast structure // XVI International School-Seminar "New Magnetic materials for microelectronics ". 1998. Moscow, Moscow State University. Abstracts of reports. Part 1. - P. 75.

368. Vlasov V.S., Ivanov A.P., Shavrov V.G., Shcheglov V.I. Application of separated carriers for the analysis of nonlinear excitation of hypersound in a ferrite plate at ferromagnetic resonance. Part 2. Some nonlinear phenomena // RE. - 2015. - T. 60, No. 3. - P. 297.

369. Ivanov A.P., Shavrov V.G., Shcheglov V.I. Self-modulation analysis of oscillations in a magnetoelastic medium based on the model of coupled magnetic and elastic oscillators // Electronic "Journal of Radio Electronics". - 2015. - No. 5. Access mode: http://jre.cplire.ru/jre/may15/4/text.pdf.

370. Ivanov A.P., Shavrov V.G., Shcheglov V.I. Self-modulation analysis of phenomena in a system of coupled agnite and elastic oscillators based on new model of potential // Electronic "Journal of Radioelectronics". - 2015. - No. 6. Access mode: http://jre.cplire.ru/jre/jun15/9/text.pdf.

371. V. I. Shcheglov. Stochastic instability of surface trajectories magnetostatic waves in a "shaft" type field with a spatial mode lation // Electronic "Journal of Radio Electronics". - 2014. - No. 10. Access mode: http://jre.cplire.ru/jre/oct14/1/text.pdf.

372. V. I. Shcheglov. Bistable nonlinear oscillator as a structure model phase transition // Proceedings of the VII International Seminar "Magnetic phase transitions". - Makhachkala: Ed. Institute Physics of the Dagestan Scientific Center of the Russian Academy of Sciences, 2005. - P. 70.

373. V. I. Shcheglov. Dynamic establishment of forced oscillations of bistable nonlinear oscillator as a model of the structural phase transition // Proceedings of the VIII International Seminar "Magnetic phase transitions ". —Makhachkala: Ed. Institute of Physics of Dagestan Scientific Center of the Russian Academy of Sciences, 2007. - P. 71.

374. V. I. Shcheglov. Bistable nonlinear oscillator as a dynamic model of structural phase transition // Collection of materials X Interinternational winter school-seminar "Chaotic self-oscillations and the formation of structures "(CHAOS-2013). - Saratov: Ed. center "Science", 2013. - P. 103.

375. Avaeva I. G., Kopylov Yu. L., Kravchenko V. B., Lisovsky, Mushkarenko Yu.N., Sobolev A.T., Shcheglov V.I. Epitaxial films mixed ferrite-garnets for CMD applications // Proceedings INEUM "Devices of automation and computer technology using the development of the domain structure of magnetic crystals ". - Moscow, 1975.- Issue 46 .-- P. 27.

376. Schlomann E. Ferromagnetic resonance in polycrystalline ferrites with large anisot-

ropy - I: General theory and application to cubic materials with negative anisotropy constant // J. Phys. Chem. Sol. - 1958. - V. 6, No. 2-3. - P. 257.

377. Dikshtein I. Ye., Shcheglov V. I. High frequency susceptibility to compositional medium made on the basis of anisotropic ferriteparticles // In the book: International conference "Physics and technical applications of wave processes ". Abstracts of reports. V. 2. - Samara, 2001 .-- P. 14.

378. 378. V. I. Zubkov, V. I. Scheglov. Compositional environment containing anizotropic ferrite particles in a non-magnetic matrix // Vkn.: 18th International Crimean Conference "Microwave Engineering and Telecommunication (KryMiKo-2008) ". Theses of reports. — Sevastopol. Ukraine, 2008. — P.567.

379. V. I. Zubkov, V. I. Shcheglov. Dynamic sensitivity of the composition of this environment, consisting of arbitrarily oriented anisotropic ferrite particles // Proceedings of the XVI International Conference "Radiolocation and radio communication" .— Moscow-Firsanovka: Edition of MEI 2008. — p.304.

380. V. I. Shcheglov. Magnetic susceptibility of the compositional environment from anisotropic ferrite particles // Proceedings XXI International home of the conference "Novoevmagnetism and magnetic materials Moscow, 2009. Izd-vo MGU. — P.948.

381. Shcheglov V. I., Zubkov V. I. Magnetic susceptibility of composite medium consisted of uniaxial ferrite particles embeddedi n nonmagnetic insulating matrix // International conference "Progress InElectromagnetics Research Symposium (PIERS) ". — Moscow, 2009, Bookof Abstracts. — P.896.

382. V. I. Zubkov, V. I. Shcheglov. Magnetic susceptibility of the compositional environment consisting of arbitrarily oriented anisotropic ferrite particle // Incl.: IX International seminar "Magnetic phase transitions. "- Makhachkala: edition of the Institute of Physics of Dagestan Scientific Center RAS, 2009. — P.63.

383. ShcheglovV. I., Zubkov V. I. The microwave susceptibility of composite medium-consisting of ferrite particles with uniaxial anisotropy // International conference "Functional Materials (ICFM-2009)" .— Partenit.Crimea. Ukraine.BookofAbstracts.TNU.Simferopol. 2009. — P.299.

384. V. I. Zubkov, V. I. Shcheglov. Magnetic susceptibility of the compositional environment with variable parameters, consisting of arbitrary orientation of anisotropic ferrite particles // RE. — 2010. — V.55, No. 4. — P.488.

385. V. I. Zubkov, V. I. Scheglov Formation of the specified properties of the magnetic the susceptibility of a compositional medium consisting of anisotropic ferrite particles // Proceedings of the XVIII International Conference lecture "Electromagnetic field materials". - Moscow-Firsanovka (MEI): PLANTI, 2010. — P.133.

386. V. I. Zubkov, V. I. Shcheglov Magnetic susceptibility of the compositional medium consisting of partially ordered anisotropic ferrite particle under the conditions of an orientation transition // In.: X International seminar "Magnetic phase transitions." - Makhachkala: edition of the Dagestan Scientific Center of RAS, 2010. — P.169.

387. V. I. Zubkov, V. I. Shcheglov. Magnetic susceptibility of the compositional medium consisting of isotropic ferrite particles // In: XIII Russian school-seminar "Wave manifestations in non-homogeneous media (Waves-2011) ". Theses of reports (electronic collection on CD). Faculty of Physics, Moscow State University, 2011, Section 2 (Electrodynamics) .— P. 7-11.

388. ShcheglovV. I., Zubkov V. I. Magnetic susceptibility of composite medium consisted of uniaxial ferrite particles // Book of Abstracts of Moscow International Symposium on Magnetism (MISM) .— Moscow: "Publishing house of Fiz. fac. Moscow State University. "- P.783.

389. V. I. Zubkov, V. I. Shcheglov Reflection of electromagnetic waves location envi-

ronment containing ferrite elements under conditions of orientational transition // Proceedings of the XIX International Conference "Electromagnetic field materials." - Moscow-Firsanovka (MPEI): NRU MEI, 2011. — P.245.

390. V. I. Zubkov, V. I. Scheglov Reflection of electromagnetic waves compositional environment containing anisotropic ferrite elements // In the book: XV International Winter School-Seminar on Electronics high frequencies and radio physics. — Saratov: SSU Publishing House, 2012. — P.56.

391. V. I. Zubkov, V. I. Scheglov Magnetic susceptibility tensor compositional medium of anisotropic ferrite particles // In.: XXXIV International Winter School of Physicists-Theorists - Kourovka, 2012. Abstracts. — Yekaterinburg: IFMURORAN, 2012. — P.139.

392. V. I. Zubkov, V. I. Shcheglov. Reflection and transmission of electromagnetic impact of waves and three-layer compositional structure, containing using gyromagnetic elements // Proceedings of the XX International conference "Electromagnetic field materials." - Moscow: NRU MEI, 2012. — P.260.

393. V. I. Zubkov, V. I. Scheglov Reflection of electromagnetic waves of lattice of anisotropic ferrite spheres // Proceedings of the XXII International Conference "Novoevmagnetism and Magnetic materials (NMMM-2012) ".- Astrakhan (ASU): Publishing house" Astrakhan University of Science ", 2012. — P.277.

394. V. I. Zubkov, V. I. Shcheglov. Magnetic susceptibility of the compositional medium consisting of anisotropic ferrite particles with different by ordering the orientation of the oseianisotropy // RE. — 2013. — V.58, No. 2. — P.143.

395. ShcheglovV. I., Zubkov V. I. Electromagnetic waves propagation in mul-tilayer composite structure consist of anisotropic ferrite elements // Book of Abstracts of-Moscow International Symposiumon Spin Waves 2013 (SW-2013) .— SaintPetersburg.Russia.Publishedby "Publishing House of the Institute of Physics and Technology.. —P.131.

396. V. I. Shcheglov. Interaction of elastic vibrations with precessing magnetic moment // RE. — 1971. — T.16, No. 12. — P.2321.

397. Goldin B.A., Kotov L.N., Zarembo L.K., Karpachev S. H. Spin-phonon-Interactions in Crystals (Ferrites), Leningrad: Nauka, 1991.

398. MigulinV. V., Medvedev V.I., Mustelle. R., Parygin V.N. The basics theory of osci lations. — Moscow: Nauka, 1978.

399. Strelkov S.P. Introduction to the theory of oscillations. — Moscow: Nauka, 1964.

400. Andronov A.A., Vitt A.A., Haykin S. E. Theory of oscillations. — M.: Nauka, 1981.

401. Strelkov S.P. Mechanics. — Moscow: Nauka, 1965.

402. BCE. T.6.— M .: Soviet Encyclopedia, 1971. — P.557. Article "Gyroscope ".

403. Physical encyclopedia. Vol. 1. — M.: Soviet encyclopedia, 1988.— P. 484. Article "Gyroscope".

404. Lifshits E. M., Pitaevsky L. P. Physical kinetics (Theoretical Physics), Moscow: Nauka, 1979.

405. Zilberman P. E., KazakovG. T., Kulikov V. M., Tikhonov V. B. Influence of weak magnetizing fields on the propagation of magnetostatic waves in films of yttrium iron garnet with submicron thickness // RE. —1988. — V.33, No. 2. — P.347.

406. 406. Zilberman P. E., Kulikov V. M., Tikhonov V. V., Shein V. Magnitostatic waves in films of yttrium iron garnet magnetization // RE. — 1990. — T.35, No. 5. — P.986.

407. Zilberman P. E., Kulikov V. M., Tikhonov V. V., Shein V. Nonlinear effects in the spread of surface magnetostatic waves in films of iron-yttrium garnet in weak magnetic fields //ZhETF. — 1991. — V.99, No. 5. — P.1566.

408. Puch Ch. Technique of EPR spectroscopy. — M.: Mir, 1970.
409. Wertz J., Bolton J. Theory and practical applications of the method of EPR.— M .: Mir. 1975.
410. Kurushin E. P., Nefedov E. I. Electrodynamics of anisotropic waves conducting structures — M.: Nauka, 1983.
411. KolotovO. S., Pogozhev V.A., TelesninR. B. Methods and equipment for research of pulse properties of thin magnetic films. — Moscow: Izd. Moscow State University, 1970.

Index

A

For Product Safety Concerns and Information please contact our EU
representative GPSR@taylorandfrancis.com
Taylor & Francis Verlag GmbH, Kaufingerstraße 24, 80331 München, Germany